T0221348

ENVIRONMENTALLY FRIENDLY

AND

BIOBASED

LUBRICANTS

ENVIRONMENTALLY FRIENDLY

AND

BIOBASED LUBRICANTS

edited by
Brajendra K. Sharma
Girma Biresaw

CRC Press
Taylor & Francis Group
Boca Raton London New York

CRC Press is an imprint of the
Taylor & Francis Group, an **informa** business

CRC Press
Taylor & Francis Group
6000 Broken Sound Parkway NW, Suite 300
Boca Raton, FL 33487-2742

First issued in paperback 2019

© 2017 by Taylor & Francis Group, LLC
CRC Press is an imprint of Taylor & Francis Group, an Informa business

No claim to original U.S. Government works

ISBN-13: 978-1-4822-3202-8 (hbk)
ISBN-13: 978-0-367-86958-8 (pbk)

This book contains information obtained from authentic and highly regarded sources. Reasonable efforts have been made to publish reliable data and information, but the author and publisher cannot assume responsibility for the validity of all materials or the consequences of their use. The authors and publishers have attempted to trace the copyright holders of all material reproduced in this publication and apologize to copyright holders if permission to publish in this form has not been obtained. If any copyright material has not been acknowledged please write and let us know so we may rectify in any future reprint.

Except as permitted under U.S. Copyright Law, no part of this book may be reprinted, reproduced, transmitted, or utilized in any form by any electronic, mechanical, or other means, now known or hereafter invented, including photocopying, microfilming, and recording, or in any information storage or retrieval system, without written permission from the publishers.

For permission to photocopy or use material electronically from this work, please access www.copyright.com (http://www.copyright.com/) or contact the Copyright Clearance Center, Inc. (CCC), 222 Rosewood Drive, Danvers, MA 01923, 978-750-8400. CCC is a not-for-profit organization that provides licenses and registration for a variety of users. For organizations that have been granted a photocopy license by the CCC, a separate system of payment has been arranged.

Trademark Notice: Product or corporate names may be trademarks or registered trademarks, and are used only for identification and explanation without intent to infringe.

Library of Congress Cataloging-in-Publication Data

Names: Sharma, Brajendra K., 1969- editor. | Biresaw, Girma, 1948- editor.
Title: Environmentally friendly and biobased lubricants / editors, Brajendra K. Sharma and Girma Biresaw.
Description: Boca Raton : CRC Press, 2016. | Includes bibliographical references and index.
Identifiers: LCCN 2015040347 | ISBN 8791482232028 (hardcover : alk. paper)
Subjects: LCSH: Vegetable oils--Biocompatibility. | Lubricating oils--Biocompatibility. | Synthetic lubricants. | Green chemistry. | Lubrication and lubricants.
Classification: LCC TP680 .E66 2016 | DDC 664/.3--dc23
LC record available at http://lccn.loc.gov/2015040347

Visit the Taylor & Francis Web site at
http://www.taylorandfrancis.com

and the CRC Press Web site at
http://www.crcpress.com

Contents

SECTION I Advanced Environmentally Friendly Base Oils and Feedstocks

SECTION II Biobased Hydraulic Lubricants and Biodegradability

SECTION III Chemically/Enzymatically Modified Environmentally Friendly Base Oils

SECTION IV Vegetable Oil-Based Environmentally Friendly Fluids

SECTION V Additives for Environmentally Friendly Fluids

Preface

Research on environmentally friendly and biobased lubricants is increasing due to environmental concerns, rapidly depleting petroleum reserves, and stringent government regulations. Vegetable oil-based lubricants are finding applications in many environmentally sensitive and total loss applications ranging from hydraulic oils to grease. They possess excellent biodegradability, lower ecotoxicity, higher viscosity index, good boundary tribological properties, lower volatility, and higher flash points compared to petroleum base oils. The global lubricant demand is growing at about 2% per year and it will reach 42.1 million metric tons by 2017. According to a study done by the Freedonia Group, the demand for biobased lubricants is on the rise, from a share of 0.3% in 1997 to 0.6% in 2010, with a projected share of 1.2% by 2017. Clearly, there is room in the market for the expansion of biobased lubricants, where the annual demand is growing at a rate of 16.3% in the European market. This contrasts with the much slower rate of increase of the consumption of lubricants as a whole. Additionally, natural products have an inherent advantage in the food-grade lubricant market, which is projected to expand at about 5% annually, with much of the growth in the United States.

There exists a large amount of scattered information on environmentally friendly and biobased lubricants. However, as in any field of scientific endeavor, there is a need for an occasional pause, assessment, and refocus of research accomplishment and communication in terms of our research paradigms. We hope to at least partially address this need for reflection and consideration through this book. During the Society of Tribologists and Lubrication Engineers (STLE) annual meetings, leading researchers from around the world are engaged in unraveling the importance and the relevance of this area. They present their latest research findings in technical sessions on environmentally friendly fluids and biobased lubricants. These technical sessions were always such a huge success that the need to take the next logical step—a text that synthesizes the burgeoning knowledge in this field—was born during the 2012 STLE Annual Meeting. Allison Shatkin, senior editor of CRC Press/Taylor & Francis invited us to edit a book on this topic, and the editors enthusiastically accepted the challenge and immediately started working on the project.

Selected authors (who presented their work in the environmentally friendly fluids and biobased lubricants technical sessions of STLE Annual Meetings) and other prominent researchers were invited to submit manuscripts on research topics that have been shared at various technical sessions for the past several years. After a thorough peer review process, 21 manuscripts were selected for the current volume. Authors were encouraged to critically analyze respective chapters beyond simple reviews of existing research in their area of expertise. We hope the resulting array of viewpoints provides a valuable tool for those interested in environmentally friendly and biobased lubricants. Varying perspectives of research experiences and concerns have been gathered in this book in order to provide the reader with a sense of both appropriate applications and development needs in the field. Specific case histories demonstrating the uses and values of various commercially available biobased lubricants are included. Laboratory development of newer environmentally friendly fluids, field studies of mature biobased lubricants, biodegradability studies, and future research needs are identified. Most importantly, the chapters provide a critical assessment of gaps and weaknesses in the current state of knowledge on these subjects and also provide direction for future research.

This book offers a review, a synthesis, and a critical assessment of the major issues in the field of environmentally friendly fluids and biobased lubricants by some of the leading scientists in the field. It provides a collective review of developments and approaches being used in this emerging area of lubrication. The 21 chapters are logically grouped into five sections. Section I consists of five chapters examining advanced environmentally friendly base oils and feedstocks. This section covers recent advances in farnesene-derived base oils, estolides, epoxidized soybean oils, isostearic acid, and new plant oil-based monomers and polymers. Section II, which is on biobased hydraulic lubricants and biodegradability, has four chapters. These give the readers a flavor of performance and technical requirements of low-environmental impact lubricants along with their biodegradability and ecotoxicity evaluations; comparison of current standards for environmentally acceptable lubricants to the actual requirements of the environment and reproducibility of biodegradation test methods; and availability of different types of environmentally preferable fluids, their strengths, limitations, and comparable performance; and recommend best maintenance practices for prolonged fluid and equipment life. Section III contains three chapters on chemically/enzymatically modified environmentally friendly base oils. The first two chapters in this section explore the enzymatic route as a biocatalyst for biolubricant production, while the last one presents chemical methodologies for producing ester-based lubricants. Section IV comprises four chapters dealing with vegetable oil-based environmentally friendly fluids. The first two chapters in this section discuss some of the drawbacks of vegetable oil-based lubricants, such as poor thermo-oxidative stability, low-temperature flow properties, biological deterioration, and hydrolytic stability and ways to rectify these issues. The last two chapters give specific examples of applications of vegetable oils for producing lubricant additives by chemical modifications followed by their performance evaluations. Section V contains five chapters on additives for environmentally friendly fluids. This section covers topics such as fatty acid friction in nanoscale contact, additives used in formulations of biodegradable lubricants, nano-TiO_2 as additive for solid and liquid lubricants, biodiesel as lubricant and solvent, and anticorrosion effects of starch–oil dry lubricants.

This book will make an excellent reference for students, chemists, scientists, engineers, tribologists, professionals, and policy makers in the academia, the government, or the corporate world working or entering the field of environmentally friendly fluids and biobased lubricants. With the information contained in this book, it is our hope that future research will lead to more cost-effective commercial biobased products with performance equivalent to or better than that of petroleum-based ones. We are very grateful to a large number of individuals for their assistance with this book. Special thanks to all the contributors, for their enthusiasm, commitment to, and cooperation in the project and for sharing their original research or for their review of the current knowledge. Without their commitment, encouragement, suggestions, and assistance, this project would not have taken shape into this book. We sincerely thank each and every one of the contributing authors, who had to hear our pleas to meet deadlines and worked hard to finally create this book. We are also grateful for the invaluable comments and suggestions made by the reviewers, which significantly improved the quality, the clarity, and the content of the chapters. Last but not least, we extend our sincere thanks to Allison Shatkin, senior editor of CRC Press/Taylor & Francis, who conceptualized this idea, and Jill J. Jurgensen, Laurie Oknowsky, and Hayley Ruggieri of CRC Press for their timely efforts in publishing this book.

Brajendra K. Sharma
Illinois Sustainable Technology Center,
Prairie Research Institute,
University of Illinois–Urbana-Champaign, Illinois

Girma Biresaw
National Center for Agricultural Utilization Research,
Agricultural Research Service,
U.S. Department of Agriculture, Peoria, Illinois

Editors

Brajendra K. Sharma is a senior research engineer at the Illinois Sustainable Technology Center (ISTC), Prairie Research Institute, University of Illinois–Urbana-Champaign, since 2009. At ISTC, he conducts research in support of ISTC's Applied Research on Industrial Environmental Systems Program in a variety of areas, including biobased lubricants and additives, thermochemical conversion technologies, alternative nonfood feedstocks for second- and third-generation biofuels, alternative fuels from pyrolysis of waste plastics, bio-oil upgradation to refinery-ready crude oils; and engineered asphalt binders.

Prior to 2009, Dr. Sharma was a research associate in the Department of Chemical Engineering at Pennsylvania State University and conducted research at the National Center for Agricultural Utilization Research, Agricultural Research Service, U.S. Department of Agriculture, Peoria, Illinois, as a visiting research chemist on joint appointment. In this position, he conducted innovative research on developing biobased lubricants/additives through chemical modifications of vegetable oils for seven years. Before that, he conducted research on petroleum base oils as a postdoctoral research associate at State University of New York–College of Environmental Science and Forestry, Syracuse, New York. Dr. Sharma received a PhD in chemistry from the Indian Institute of Petroleum/Hemwati Nandan Bahuguna Garhwal University (India).

Dr. Sharma is a vice chair of Environmentally Friendly Fluids Technical Committee (EFFTC) of the Society of Tribologists and Lubrication Engineers (STLE) and a member of the American Chemical Society. Earlier, he served as paper solicitation chair for STLE's EFFTC and Lubrication Fundamentals Technical Committees. He is an associate technical editor of American Oil Chemists' Society's *Journal of Surfactants and Detergents*. Dr. Sharma holds five issued patents on biobased lubricants/additives and has authored/coauthored more than 140 peer-reviewed articles, proceedings, book chapters, and technical reports and more than 130 scientific abstracts.

Girma Biresaw received a PhD in physical organic chemistry from the University of California–Davis, and spent 4 years as a postdoctoral research fellow at the University of California–Santa Barbara, investigating reaction kinetics and products in surfactant-based organized assemblies. He then joined the Aluminum Company of America as a scientist and conducted research in tribology, surface/colloid science, and adhesion for 12 years. He joined the Agricultural Research Service (ARS) of the U.S. Department of Agriculture, in Peoria, Illinois, in 1998 as a research chemist, and became a lead scientist in 2002. At ARS, he conducts research in tribology, adhesion, and surface/colloid science in support of programs aimed at developing biobased products from farm-based raw materials. He has received more than 150 national and international invitations, including requests to participate in and/or conduct training, workshops, advisory, and consulting activities. He is the lead organizer of the "Biobased Lubricants" sessions of the Non-Ferrous Technical programs at the STLE Annual Meetings. He is also the senior organizer and chair of the biannual international symposium series "Surfactants in Tribology," held at various locations around the world in conjunction with the "Surfactants in Solution" symposium series. He is a fellow of STLE, and a member of the editorial board of the *Journal of Biobased Materials and Bioenergy*. Dr. Biresaw has authored/coauthored more than 270 invited and contributed scientific publications, including more than 80 peer-reviewed manuscripts, six patents, five edited books, more than 40 proceedings and book chapters, and more than 140 scientific abstracts.

Contributors

Hind Abi-Akar
Caterpillar Fluids and Filters Engineering
  Group
Caterpillar Inc.
Peoria, Illinois

Herman Benecke
Battelle Memorial Institute
Columbus, Ohio

Jakob W. Bredsguard
Biosynthetic Technologies
Irvine, California

Jeff Brown
Novvi LLC
Emeryville, California

Steven C. Cermak
National Center for Agricultural Utilization
  Research (ARS)
U.S. Department of Agriculture
Peoria, Illinois

Archana Charanpahari
Department of Chemical Engineering
Indian Institute of Science, Bangaluru
Karnataka, India

José André Cavalcanti da Silva
Lubricants and Special Products
Petrobras S. A. Research Center (CENPES)
Rio de Janeiro, Brazil

Kenneth M. Doll
National Center for Agricultural Utilization
  Research (ARS)
U.S. Department of Agriculture
Peoria, Illinois

Robert O. Dunn
National Center for Agricultural Utilization
  Research (ARS)
U.S. Department of Agriculture
Peoria, Illinois

George F. Fanta
National Center for Agricultural Utilization
  Research (ARS)
U.S. Department of Agriculture
Peoria, Illinois

Victoria L. Finkenstadt
National Center for Agricultural Utilization
  Research (ARS)
U.S. Department of Agriculture
Peoria, Illinois

Daniel Garbark
Battelle Memorial Institute
Columbus, Ohio

Gerhard Gaule
Hermann Bantleon GmbH
Ulm, Germany

Pranab Ghosh
Department of Chemistry
University of North Bengal
Darjeeling, India

Sayanti Ghosh
Corporate R&D Centre
Bharat Petroleum Corporation Limited
Greater Noida, India

Jasneet Grewal
Department of Chemistry
Indian Institute of Technology Delhi
New Delhi, India

Hyeok Hahn
Novvi LLC
Emeryville, California

Sudeep Ingole
Curt G. Joa Inc.
Sheboygan Falls, Wisconsin

Terry A. Isbell
National Center for Agricultural Utilization
  Research (ARS)
U.S. Department of Agriculture
Peoria, Illinois

Tien-Chien Jen
Department of Mechanical Engineering
University of Wisconsin–Milwaukee
Milwaukee, Wisconsin

Pattathilchira Varghese Joseph
Research and Development Center
Indian Oil Corporation Ltd.
Haryana, India

Satish V. Kailas
Department of Mechanical Engineering
Indian Institute of Science
Karnataka, India

Gobinda Karmakar
Department of Chemistry
University of North Bengal
Darjeeling, India

James A. Kenar
National Center for Agricultural Utilization
  Research (ARS)
U.S. Department of Agriculture
Peoria, Illinois

Sunil K. Khare
Department of Chemistry
Indian Institute of Technology Delhi
New Delhi, India

Gerhard Knothe
National Center for Agricultural Utilization
  Research (ARS)
U.S. Department of Agriculture
Peoria, Illinois

Padmaja V. Korlipara
Centre for Lipid Research
Indian Institute of Chemical Technology
Hyderabad, India

Tyler Kuchta
RSC Bio Solutions, LLC
Indian Trail, North Carolina

Lalit Kumar
Corporate R&D Centre
Bharat Petroleum Corporation Limited
Greater Noida, India

Brian M. Lipowski
Functional Products Inc.
Macedonia, Ohio

Zengshe Liu
National Center for Agricultural Utilization
  Research (ARS)
U.S. Department of Agriculture
Peoria, Illinois

Michael R. Lovell
Department of Industrial & Manufacturing
  Engineering
College of Engineering and Applied Science
University of Wisconsin–Milwaukee
Milwaukee, Wisconsin

Sarah M. Lundgren
Akzo Nobel Surface Chemistry AB
Stenungsund, Sweden

Jagadeesh K. Mannekote
Centre for Emerging Technologies
Jain University
Karnataka, India

Pradeep L. Menezes
Department of Mechanical Engineering
University of Nevada Reno
Reno, Nevada

Bryan R. Moser
National Center for Agricultural Utilization
  Research (ARS)
U.S. Department of Agriculture
Peoria, Illinois

Ben Müller-Zermini
Hermann Bantleon GmbH
Ulm, Germany

Rex E. Murray
National Center for Agricultural Utilization
  Research (ARS)
U.S. Department of Agriculture
Peoria, Illinois

Bharat L. Newalkar
Corporate R&D Centre
Bharat Petroleum Corporation Limited
Greater Noida, India

Helen Ngo
Eastern Regional Research Center (ARS)
U.S. Department of Agriculture
Wyndmoor, Pennsylvania

Anne Otto
RSC Bio Solutions, LLC
Indian Trail, North Carolina

Shivanand M. Pai
Corporate R&D Centre
Bharat Petroleum Corporation Limited
Greater Noida, India

Karin Persson
Chemistry, Materials and Surfaces
SP Technical Research Institute of Sweden
Stockholm, Sweden

Vivek Rathore
Corporate R&D Centre
Bharat Petroleum Corporation Limited
Greater Noida, India

Carlton J. Reeves
Department of Mechanical Engineering
University of Nevada Reno
Reno, Nevada

Bernard C. Roell, Jr., PhD
RSC Bio Solutions, LLC
Indian Trail, North Carolina

Marina Ruths
Department of Chemistry
University of Massachusetts Lowell
Lowell, Massachusetts

Deepak Saxena
Research and Development Center
Indian Oil Corporation Ltd.
Haryana, India

Shailesh N. Shah
Department of Chemistry
The Maharaja Sayajirao University of Baroda
Gujarat, India

Travis D. Thompson
Biosynthetic Technologies
Irvine, California

Suresh Umare
Department of Chemistry
Visveswaraya National Institute of Technology
Nagpur, Maharashtra, India

Daniel M. Vargo
Functional Products Inc.
Macedonia, Ohio

Paula Vettel
Novvi LLC
Emeryville, California

Jason Wells
Novvi LLC
Emeryville, California

Winnie C. Yee
Eastern Regional Research Center (ARS)
U.S. Department of Agriculture
Wyndmoor, Pennsylvania

Section I

Advanced Environmentally Friendly
Base Oils and Feedstocks

1 Farnesene-Derived Base Oils

Jeff Brown, Hyeok Hahn, Paula Vettel, and Jason Wells

CONTENTS

ABSTRACT

This chapter describes the suitability of the renewable hydrocarbon olefin β-farnesene (E-7,11-dimethyl-3-methylene-1,6,10-dodecatriene, Biofene®; hereafter, simply farnesene) as a feedstock for lubricant base oils and the performance of the lubricant products made from farnesene-derived base oils. The farnesene is manufactured from sugars—ubiquitous, cheap, and renewable chemicals that can be found in many forms around the world—by a simple fermentation process using baker's yeast. Farnesene is unique among renewable lubricant feedstocks in that it is directly obtained as a pure hydrocarbon olefin from the biological conversion process. This allows it to directly drop in to the many existing olefin processes and equipment of the petroleum/lubricant industry. The technology thus exists today, with farnesene, to introduce the first environmentally friendly base oils to meet the performance and economic requirements for success.

Farnesene can undergo oligomerization with linear alpha olefins to make isoparaffinic structures. Subsequent hydrogenation and distillation steps produce fully saturated farnesene-derived base oils (FDBOs) that are classified as American Petroleum Institute (API) Group III. FDBOs have significantly higher biodegradability than other hydrocarbon and mineral oil-based oils. Since farnesene is a 100% renewable material obtained from sugar sources, the base oils derived from farnesene are also renewable products. FDBOs have undergone extensive toxicity and ecotoxicity testings for many regulatory registrations, and as a result have been approved for the Ecolabel Lubricant Substance Classification List (LuSC) as 100% Classification A (ultimately aerobically biodegradable, >60%) and 100% Classification D (not toxic). The Ecolabel LuSC is the definitive list of allowed ingredients in Ecolabel-approved lubricants. FDBOs have thus met the stringent requirements to be classified as white oil by the National Sanitation Foundation, and have been granted HX-1 and H-1 approval for use as lubricant ingredients for incidental food contact.

Due to the excellent environmental performance of FDBOs, there are a wide range of industrial lubricant applications in which new previously unattainable levels of performance and applications can be reached. The FDBO formulations allow for finished products with the classical performance of a polyalphaolefin-based fluid, but with the environmental performance of an ester product. Of particular interest are industrial products such as hydraulic fluids, transformer oils, compressor oils, gear oils and greases where there is legislation-driven or environmental need in place for the finished lubricant.

FDBOs can also be used to formulate passenger car motor oils in a range of resource-conserving viscosity grades, meeting API and other industry and original equipment manufacturer requirements. FDBO engine oils deliver performance characteristics comparable to or better than other synthetic engine oils, while redefining performance versus key environmental metrics when compared with classical petroleum-derived base oils.

1.1 INTRODUCTION

The lubricants industry is always looking for new technology to improve product performance. The performance targets shift over time as market demands change and the industry continually reaches new levels. A new subset of performance criteria has now been introduced to the lubricants market, namely, environmental performance. Biodegradable lubricants have been in the market for some time; but it was not until recently, through a combination of government regulation and consumer demands, that an available marketplace for products with high levels of environmental performance has arisen.

The lubricants industry has developed many specifications and regulations to define what an environmentally friendly lubricant product is and the performance specifications it must meet. The industry has aligned on three key characteristics that define an environmentally friendly lubricant [1]:

1. Biodegradability: To be consumed by microorganisms and return to compounds found in nature. The most common lubricant specification for biodegradability is >60% biodegradation in 28 days by Organisation for Economic Cooperation and Development (OECD) 301 [2] or American Society for Testing and Materials (ASTM) methods [3,4].
2. Renewability: An organic natural resource that can replenish in due time compared to the usage. The typical lubricant specification for renewability is carbon content that is <5 years old by ASTM D6866 [5]. Most products must have a minimum value of 25% renewable content to be called renewable or biobased [6].
3. Toxicity: The degree to which a substance can damage an organism. There are many toxicity tests for lubricants, but the most typically used are the OECD 201–203 [7–9] tests for acute toxicity and the OECD 210–211 [10,11] tests for chronic toxicity.

With the definition of environmentally friendly lubricant in hand, it is possible to look at the available technologies to see which can best meet both the environmental and the classical lubricant performance requirements. With the inclusion of renewability, the search must begin at the level of the feedstock, which is used to derive the base oil. When analyzing the feedstock, there are many additional criteria if the product is to meet market requirements. For example,

- Feedstock chemical capabilities and downstream processing requirements
- Cost-effectiveness
- Degree of biodegradability and toxicity
- Food versus fuel
- Sustainability
- Drop-in performance

As performance and technology improve and the market for environmentally friendly lubricants grows, more dimensions will be added, for example, life cycle analysis, drain interval, CO_2 reduction, and recyclability. These all represent additional performance requirements on the feedstock to produce a premium lubricant.

Most next-generation renewable technologies are initially developed for the fuels industry, and as a result, environmentally friendly chemicals often get wrongly grouped with fuels for how they are evaluated for suitability in other fields. This ignores the fact that the lubricants industry has a very specific set of criteria that are different from those of fuels and other industries. Table 1.1 outlines some of these key differences.

Amyris Inc. has commercialized a renewable feedstock material, farnesene, which has many applications in the fuel and chemical world [12]. This chapter will investigate the suitability of farnesene as a feedstock for lubricant base oil and the performance of the lubricant products based on it.

Farnesene is manufactured in a biological fermentation process with any fermentable sugar as the feedstock [12]. Sugar is a ubiquitous chemical that can be obtained from many sources around the world. The farnesene process can be run on any form of sugar, making it highly sustainable for a local region or economy. Thus, Amyris has successfully demonstrated the manufacture of farnesene in the United States from corn syrup [13]; in Spain, from beet sugars [14]; and in Brazil, from sugarcane [12]. In addition, Amyris has demonstrated equivalent farnesene production from cellulosic sugars [15,16]. These characteristics allow farnesene to offer outstanding

---

TABLE 1.1
Criteria for Lubricant Base Oil versus Fuel

	Lubricant Base Oil	Fuel
Molecular weight	Large molecule: 30–80 carbons	Small molecule: C5–C15
Lifetime	Can operate for many years	Once through, burn in use
Performance	Many difficult performance requirements based on application, lubricity, stability, seals, etc.	Burn clean, octane, cetane
Infrastructure	Utilize existing	Utilize existing
Value	Many high-value market opportunities driven by performance	Energy content and quality
Market size	0.4–0.9% barrel of oil	60% of barrel [17]
Environmental	Biodegradability, renewability, toxicity	Renewability, volatile organic compound, NO_x, particulate emission
Political	Biodegradability, renewability, toxicity	Improve security, CO_2 reduction

Source: Brown, J. A., Feedstocks for biolubricants, Paper presented in ICIS London, 2014. With permission.

sustainability in its current production with a road map to the eventual use of cellulose for best cost/performance.

The security and availability of feedstocks is critical and this inevitably leads one to the food versus fuel debate. Farnesene is currently produced in Brazil from sugarcane juice. The lubricants industry is two orders of magnitude smaller than the fuels industry, so, in principle, the fundamental food versus fuel debate does not apply. In fact, when one considers the overall size of the specialty lubricant market, of which environmentally friendly lubricants will be a part, it is another order of magnitude smaller. Nevertheless, it is important to address this issue and understand that it will not be a factor. In fact, the opposite is true; and due to the relatively small size of the environmentally friendly lubricants market, the industry must rely on other industries to provide the scale and the cost reductions to create a viable feedstock technology. The Brazilian sugarcane industry applies here. Sugarcane is the most cost-effective and environmentally friendly source of feedstock for renewable chemicals at scale in 2014. In 2014, the global sugar industry produced 175 Mmt of sugar [18], and the lubricants industry, 39.2 Mmt/year in 2013 [19]. Based on the optimistic projection that the environmentally friendly lubricants will make up 5% of the overall lubricant market, the market size will be 1.9 Mmt/year. This will not have an effect on the pricing or the supply in the global sugar industry.

One of the industry concerns with Brazilian sugarcane is land use and overall environmental sustainability. If one looks at the current production numbers in Brazil, 1% of Brazil's arable land (3.4 million ha) delivers the equivalent of 50% of Brazil's gasoline demand through ethanol and makes Brazil the largest global producer of food sugar (Figure 1.1). There are 25 million ha of degraded, underutilized pastureland available for sugarcane expansion—this is enough to deliver >100% of lubricant demand to the world as farnesene or alternate markets for sugar [20].

Renewability alone is not enough for a feedstock to succeed. The proper land use for feedstock cultivation and the amount of greenhouse gas reduction is also important [21]. Lubricant marketers will require green credentials prior to introduction of green products in the mainstream market. Other items under consideration for feedstock sustainability are the following [22]:

FIGURE 1.1   Sugarcane-producing regions in Brazil. (From Amyris Inc. website, http://www.amyris.com, and http://sugarcane.org/sustainability/preserving-biodiversity-and-precious-resources, accessed May 15, 2015. With permission.)

- Water use, fertilizer, transport, productivity
- Maintaining biodiversity
- Fair land ownership, labor rights, effective agriculture growth
- Further chemical modification of farnesene

To address all these issues, Amyris has sought and obtained in May 2014 the highly coveted Roundtable on Sustainable Biomaterials (RSB) sustainability certification for its production of farnesene at the Brotas plant in São Paulo State, Brazil [23]. The RSB is a global sustainability standard and a voluntary certification system for biomaterials production, which represents a consensus of over 100 organizations around the world including farmers, refiners, governments, United Nations agencies, and nongovernmental organizations [24]. It is designed to ensure the sustainability of biomaterial production while providing user-friendly tools for industry to demonstrate compliance. RSB certification validates sustainable production including greenhouse gas emission reductions, respect for human rights, protection of biodiversity and water, and maintenance of food security.

Downstream processing of the feedstock into a suitable lubricant base oil is important for economical and scale considerations. The lubricants industry is built on the refining of hydrocarbons and a renewable feedstock that can drop in most closely to these processes is desired. Farnesene is unique among renewable feedstocks since it is pure hydrocarbon olefin direct from the biological conversion process. This allows the direct application of the many olefin processes and equipment of the petroleum industry. Farnesene can be treated similar to linear alpha olefin (LAO) as in the production of polyalphaolefin (PAO), using the processes and technologies to produce farnesene-derived base oils (FDBOs). This provides a significant advantage to FDBOs, as it is not necessary to develop new process technology and infrastructure on route to market.

Production of base oil is a scale business—in order to compete on cost, a large plant is necessary to drive down processing and operational costs. For environmentally friendly lubricants to succeed, they must become cost competitive with petroleum-derived products. This is a difficult situation for a new technology. Scale is a requirement to be able to compete, but the introduction of a new technology often has a slow adoption rate. The way to mitigate and allow for success is to have a technology that can drop into existing equipment and processes. Since >98% of lubricant base oil is hydrocarbon chemistry, a drop-in hydrocarbon technology will have the best chance of succeeding. The lubricants industry needs to learn from other industries that have had successful introductions of renewable technology, such as polymers, which showed that drop-in technology is necessary [25].

When calculating the cost to a customer/manufacturer for switching to a new base oil technology, the advantage of a drop-in product becomes clearer. From handling, compatibility, seals, in-use expectations, specifications, disposal, and rerefining perspective, if the base oil is not a drop-in for petroleum products, then each one of these adds cost to the product acceptance and will limit its ability for scale-up. The lubricant market has >1000 esters available for use with many different degrees of environmental and classical technical performance. The main reason for the high cost of the esters is that they are so specialized that the high volume can never be achieved to drive down cost and make them more widely used. The lack of compatibility with petroleum products through the entire supply chain of finished products means that it will be exceedingly difficult to break out of this situation. There are many great ester products for extremely highly demanding applications with performance and best-in-class environmental performance; but economics and scale will limit their ability to have a significant market share.

The lubricants industry can become a leader in the global chemical industry. The nature of the industry pricing structure for high-value/high-performance products, scale, and environmental regulations make it ripe for the success of a new environmentally friendly technology. The technology exists today, with farnesene, to introduce the first environmentally friendly base oil to meet the performance and economic requirements for success.

1.2 SESQUITERPENES AS RENEWABLE BUILDING BLOCKS FOR BASE OILS

Recent developments in synthetic biology and metabolic engineering by several companies, including Amyris, have led to the commercial production of a wide range of chemical building blocks which have been used to replace petroleum-derived molecules including fuels, polymers, lubricant base oils, and additives, among many others [26]. One building block with great possibilities for application in engineered base oils and functional fluids is the $C_{15}H_{24}$ sesquiterpene β-farnesene (*E*-7,11-dimethyl-3-methylene-1,6,10-dodecatriene, Biofene®; Figure 1.2). This substance and other sesquiterpenes are naturally found in trace amounts within plants and insects often functioning as defensive chemical agents or pheromones [26].

β-Farnesene can be synthetically produced from isoprene [27] or by the dehydration of sesquiterpene alcohols [28]; however, these routes tend to produce a low-purity mixture of isomers (β-farnesene and α-farnesene) and other by-products. In contrast, the β-farnesene commercially manufactured by Amyris is obtained as a single isomer at 93–95% purity directly from the fermentation process [26]. A simple flash distillation increases this to 97–98% purity. Industrial scale production of high-purity β-farnesene was made possible in recent years by the extensive genetic modification of the yeast *Saccharomyces cerevisiae* (baker's yeast), an organism used for centuries to make bread, beer, and wine because of its natural adeptness at converting sugar into ethanol and CO_2 [29]. Starting with a wild-type high-ethanol producing yeast strain, Amyris' core technology for high-throughput strain engineering was used to manipulate a metabolic pathway present in all higher eukaryotes, the mevalonate pathway (also known as the HMG-CoA [3-hydroxy-3-methylglutaryl-coenzyme A] reductase pathway or isoprenoid pathway, which is the basis for the biological production of terpenoids and sterols [30]), to convert sugars into β-farnesene via 21 enzymatic steps, 20 of which occur in the yeast's native mevalonate pathway. The addition of the plant enzyme farnesene synthase allows the yeast to produce farnesene from farnesyl diphosphate, a naturally occurring sterol precursor, and excretes it from the cells [31]. By selecting the appropriate nodes in the pathway and other synthases, commercial quantities of a range of isoprenoids [6], from C5 (isoprene) [32], through C10 (monoterpenes) to C15 (sesquiterpenes) or higher, can be produced as illustrated in the diagram in Figure 1.3. This technology received a 2014 Environmental Protection Agency (EPA) Presidential Green Chemistry Challenge Award [33].

1.2.1 CATALYTIC PROCESS FOR FDBOS

Given the similarities between the chemical structures of β-farnesene and isoprene, it might be obvious that many of the routes to higher molecular weight isoprene polymers developed over the last century or more can be applied to farnesene. Cationic polymerizations using strong Lewis acids, free radical polymerizations, and anionic polymerization have all been studied; and a number of farnesene homopolymer and interpolymer compositions and processes have been published and patented [34].

However, when designing a base oil molecule for use as a high-performance lubricant, the fluid property requirements dictate careful control of the chemical structure, linearity, branching, and low molecular weight. A growing demand for improving fuel economy of automobiles and reduced carbon dioxide emissions requires lubricating oils with low viscosity, but excellent wear performance and low volatility.

FIGURE 1.2 Structure of β-farnesene. (From Amyris Inc. website, http://www.amyris.com, and http://sugarcane.org/sustainability/preserving-biodiversity-and-precious-resources, accessed May 15, 2015. With permission.)

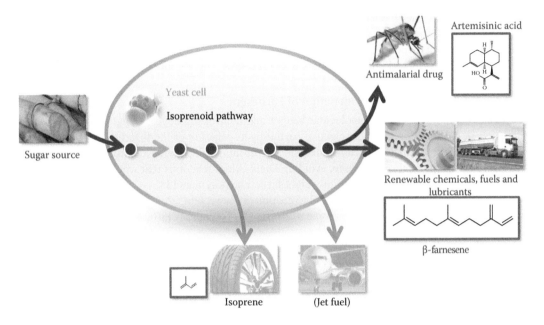

FIGURE 1.3 Mevalonate biosynthetic pathway. (Courtesy of Amyris Inc., Emeryville, California, 2015.)

Figure 1.4 shows the classic structure–property relationships of lubricant base oils. Group III and Group IV isoparaffins such as PAOs or isomerized hydrocarbons contain a well-defined degree and position of branching in the structures. This gives greatly improved low-temperature properties while maintaining excellent viscosity index (VI) and oxidative resistance. As rings are added to the structures in Group II and Group I base oils, VI and oxidative properties decrease. Aromatics, particularly polycondensed aromatics, have poor lubricating properties and are removed in modern

API group	Chemistry	Structure	Characteristics
Group III, Group IV	Branched chain isoparaffins	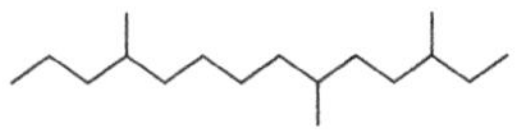	• High VI • Great oxidation resistance • Lowest pour point • Value as base oil: ++++
Group II	Naphthenic rings with short paraffinic side chains	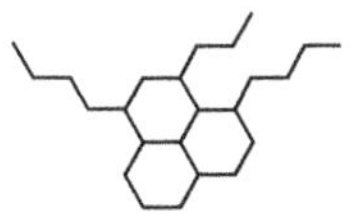	• Moderate VI • Good oxidation resistance • Low pour point • Value as base oil: +++
Group I	Polycondensed aromatic rings	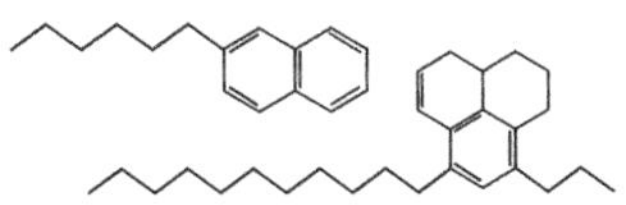	• Moderate VI • Poor oxidation resistance • May have low pour point • Value as base oil: +
Group V	Triglycerides Synthetic esters Polyalkylene glycols Others	Structure varies depending on types of chemistry	• High to low VI • Poor to good oxidation resistance • Low to high pour point • Value as base oil: variable with chemistry • No API base oil interchange

FIGURE 1.4 Base oil structure–property relationships. (Novvi LLC, Emeryville, California, Original work. With permission.)

refining technology. Group V contains natural and synthetic esters with carboxylate groups, which impart excellent lubricity and VI, and possibly good low-temperature properties, but often contain double bonds, which reduce stability. Other base oil chemistries are included in American Petroleum Institute (API) Group V. API does not allow base oil interchange for any Group V base oils.

1.2.1.1 Selective Partial Hydrogenation of Farnesene

In order to produce the desired molecular weight and control the degree, length, and position of branching in the final molecule, a selective partial hydrogenation of the β-farnesene is performed (Figure 1.5). This eliminates the unwanted sites of unsaturation, giving control over the branching, and removes the conjugated diene, avoiding Diels–Alder reactions that would create unwanted cyclic by-products and higher molecular weight farnesene polymers [35].

Selectivity is controlled by optimizing the reaction pressure, temperature, and catalyst and by the careful design of the reactor(s) to overcome mass transfer limitations [36].

1.2.1.2 Boron Trifluoride (BF₃)-Catalyzed Oligomerization

The partially hydrogenated farnesene is reacted with a co-monomer; typically, one or more LAOs are selected to give the desired final chemical structure, branching characteristics, and molecular weight distribution [37]. The oligomerization reaction uses BF_3 as the catalyst and *n*-butanol with butyl acetate as cocatalyst/promotor. The reaction is quenched with caustic and washed, and the lights are stripped. The resulting intermediate or crude oligomer can consist of a mixture of hydrocarbons ranging from C10 to about C80 [37].

1.2.1.3 Final Hydrogenation

The crude oligomer product must be hydrogenated to further eliminate residual unsaturation and increase its thermo-oxidative stability [37]. This is done with a conventional nickel hydrogenation step [36]. Typical levels of final saturation in the crude oligomer are measured by coulometric titration (ASTM D2710 [38]) to be below 100 bromine index.

1.2.1.4 Fractional Distillation

Finally, the fully hydrogenated crude base oil is fractionated into several viscosity grades of base oil having a narrow boiling point distribution [39]. The monomers and dimers are separated using a vacuum distillation column; and the higher oligomers (average expected boiling point: up to 525°C) are recovered using short-path distillation. The process steps and plant for producing FDBOs is schematically shown in Figure 1.6 [39].

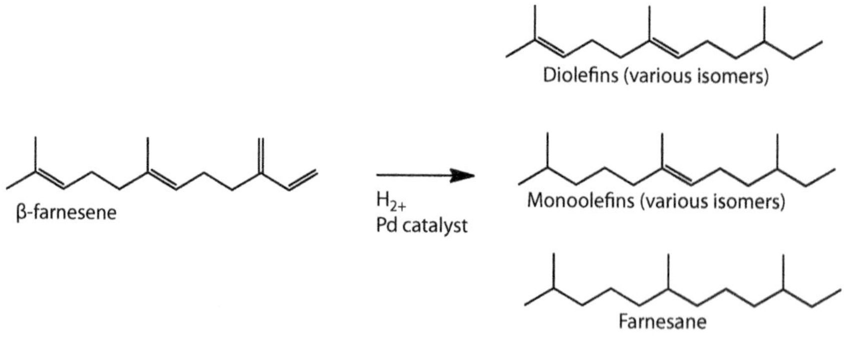

FIGURE 1.5 Reaction scheme: farnesene partial hydrogenation. (Novvi LLC, Emeryville, California, Original work. With permission.)

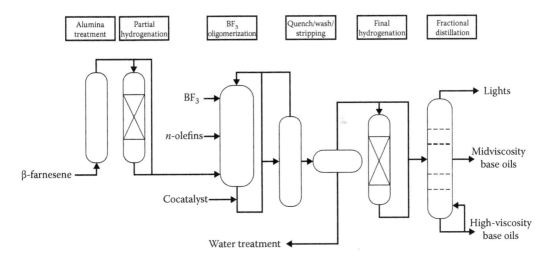

FIGURE 1.6 Schematic for an FDBO plant. (Novvi LLC, Emeryville, California, Original work. With permission.)

1.3 FDBO PROPERTIES

1.3.1 CHEMICAL AND PHYSICAL PROPERTIES OF FDBOS

The API classifies lubricant base oils into groups based on the amount of saturates, sulfur content, and VI (Table 1.2) [40]. Group I base oils have high aromatic content and lower VI. Sophisticated hydrogenation technology is used to produce Group II and Group III base oils. Gas-to-liquid base oil technologies, which produce isoparaffins from natural gas, are also classified as Group III [41]. PAOs make up a very narrow classification for Group IV. Naphthenic, ester, polyalkylene glycol, and most other technology base oils are classified in a catchall category, Group V.

FDBOs have a saturate level of >95% measured by ASTM D2007 [42]. They have a sulfur content of <0.001% by ASTM D2622 [43]; and they have a VI of >120 [39]. These physical characteristics place the FDBOs in API Group III. The base oils resulting from the BF_3-catalyzed oligomerization of farnesene with LAOs have some very similar chemical structure and performance properties as PAO such as high VI and very low pour points and also some characteristic differences caused by the unique chemical structure of the β-farnesene. PAOs contain longer and fewer hydrocarbon branches, while FDBO contains a larger number of methyl branches along with a smaller number of longer branches.

FDBOs contain fully saturated isoparaffinic structures [37]. Figure 1.7 shows the simulated distillation gas chromatograms of one FDBO cut overlaid with a similar viscosity PAO and a

TABLE 1.2
API Base Stock General Categories

Base Oil Category	% Saturates	% Sulfur	Viscosity Index
Group I	<90	>0.03	80–119
Group II	≥90	≤0.03	80–119
Group III	≥90	≤0.03	≥120
Group IV	All PAOs		
Group V	All other base stocks not included in Group I, II, III, or IV		

Source: American Petroleum Institute: Industry Services Department, Appendix E—API base oil interchangeability guidelines for passenger car motor oils and diesel engine oils, 2011. With permission.

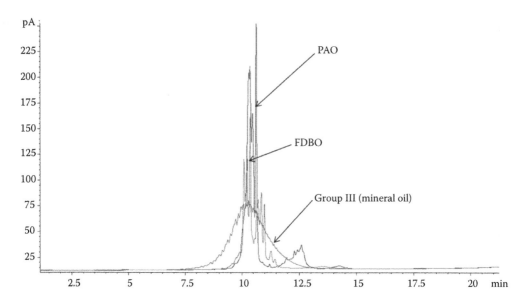

FIGURE 1.7 Gas chromatography (GC) distribution of petroleum group III, PAO, and FDBO base oils. (Novvi LLC, Emeryville, California, Original work. With permission.)

petroleum-derived Group III [39]. Immediately, we see that the PAO and FDBO base oils have much more defined structures containing distinct patterns of oligomers than the petroleum-derived Group III base oil that contains thousands of individual structures along a continuous range of boiling points. Oligomeric base oils do not contain the lower–molecular weight components that result in poorer volatility and higher–molecular weight components that result in low-temperature characteristics of petroleum-derived base oil.

TABLE 1.3
Chemical and Physical Properties of FDBOs

Property	Method	FDBO 3 cSt	FDBO 4 cSt	FDBO 7 cSt	FDBO 12 cSt	FDBO 15 cSt
Appearance	Visual	Bright and clear	Bright and clear	Bright and clear	Bright and clear	Bright and clear
Color	ASTM D1500 [44]	L0.5	L0.5	L0.5	L0.5	L0.5
Density, 15°C (kg/L)	ASTM D4052 [45]	0.82	0.82	0.83	0.84	0.84
Viscosity, 40°C (cSt)	ASTM D445 [46]	13.8	19.2	44.0	104.3	137.6
Viscosity, 100°C (cSt)	ASTM D445 [46]	3.3	4.2	7.2	13.3	15.5
Viscosity index	ASTM D2270 [47]	108	122	125	125	118
Pour point (°C)	ASTM D97 [48]	−57	−42	−48	−27	−42
Flash point (°C)	ASTM D92 [49]	195	220	250	280	290
Bromine index	ASTM D2710 [50]	<200	<200	<200	<200	<200
Water (ppm)	ASTM D6304 [51]	<50	<50	<50	<50	<50
CCS at −25°C (cP)	ASTM D5293 [52]	NA	NA	NA	13,600	NA
CCS at −30°C (cP)	ASTM D5293 [52]	NA	1100	6400	NA	NA
CCS at −35°C (cP)	ASTM D5293 [52]	NA	2000	11,000	NA	NA
Evaporation loss, NOACK (%)	ASTM D5800 [53]	30	13	<3	NA	NA

Source: Novvi LLC, Emeryville, California, Original work.
Note: CCS, cold-cranking simulator.

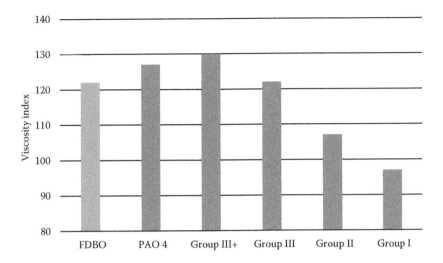

FIGURE 1.8 VI base oil comparison. (Novvi LLC, Emeryville, California, Original work. With permission.)

The narrow molecular weight distribution and the lack of low-boiling components of the FDBOs predict excellent performance characteristics [54]. Novvi LLC has commercialized FDBOs with the trade name NovaSpec. The various viscosity grades of FBDO and their properties are shown in Table 1.3.

Figures 1.8 through 1.11 give a comparison of 4 cSt FDBO with other 4 cSt API Groups I–IV high-volume commercial base oils [39]. Figure 1.8 shows that FDBO has high VI as Groups III–IV base oils. In Figure 1.9, we see that FDBO has low NOACK volatility, as does Group IV. Group III base oil is higher. All these show much better volatility than Group I and Group II base oils.

Figures 1.10 and 1.11 show comparisons of low-temperature pour point and −30°C cold-cranking simulator (CCS) tests for the 4 cSt base oils. FDBO gives values that are intermediate between Group III and Group IV base oils [39].

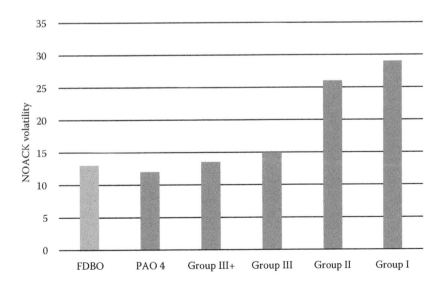

FIGURE 1.9 NOACK volatility base oil comparison. (Novvi LLC, Emeryville, California, Original work. With permission.)

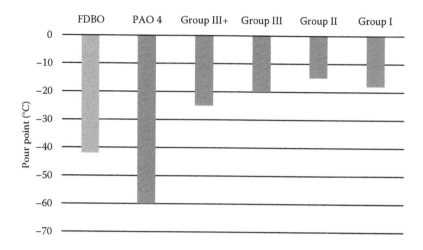

FIGURE 1.10 Pour point base oil comparison. (Novvi LLC, Emeryville, California, Original work. With permission.)

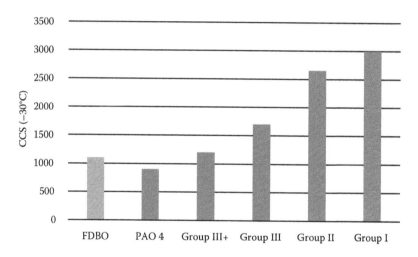

FIGURE 1.11 Cold-cranking simulator case oil comparison. (Novvi LLC, Emeryville, California, Original work. With permission.)

1.3.2 Ecological Properties of FDBOs

The European Union (EU) Ecolabel is the most stringent ecological regulatory system developed for industrial lubricants. It is an expansion of earlier European ecological assessments such as the Nordic Swan [55] and the Blue Angel [56]. The EU Ecolabel is awarded to lubricants that meet the following criteria [57]:

- The product must not contain any excluded or limited substances and mixtures with hazard statements and risk phrases that are designated as *H*- or *R*- according to the Global Harmonized System classification.
- The product must not contain specific substances that are excluded including metallic components with the exception of sodium, potassium, magnesium, and calcium.
- The main components and the mixture must meet aquatic nontoxicity requirements.
- The main components above 0.10% in the mixture must meet biodegradability and bioaccumulation potential requirements.

- The product must contain the required level of renewable raw material.
- The product must meet minimum technical performance requirements.

FDBOs are the only hydrocarbon base stocks that can be used to formulate lubricants that satisfy all the above elements including the technical performance [58].

1.3.3 Biodegradability Properties of FDBOs

FDBOs have significantly higher biodegradability than petroleum-based and PAO hydrocarbon base oils. This is due to the unique nature of the production of the FDBO feedstock farnesene, which is organically made in a cell; thus, the chemical structure is known to nature. Farnesene can be found in nature in the skins of apples and other organic materials. Farnesene is a linear branched olefin that is different from other nonorganic linear olefins due to its unique branching structure. It is this branching structure that allows for the biodegradation of the farnesene.

FDBOs have the formula C_nH_{2n+2} and are chemically similar to a PAO. In addition, both are fully saturated pure hydrocarbons. When a 4 cSt PAO and an FDBO are compared, it is interesting in that, while they are very similar chemically and display many of the same types of technical performance, there is a significant difference in their biodegradability [39]. Each has an average of 30 carbon atoms; but they differ in the number and the type of branching. PAO contains few longer branches, while FDBO contains many methyl branches.

The FDBOs described in this chapter are chemically made by reacting normal petroleum-derived LAOs with farnesene to create a 50% renewable base oil [39]. Squalane is another 4 cSt base oil that is found in nature in shark livers and olive oil and can be made by reacting two farnesene molecules together from end to end [59]. Squalane has 30 carbon atoms just like the 4 cSt PAO and the FDBO. When the biodegradabilities of these three base oils are compared, the effect of the farnesene becomes obvious. All three materials were tested for biodegradation at the same lab at the same time and the same experimental group to minimize any variation in the biodegradability testing using the OECD 301B method (Figure 1.12) [2]. The pure farnesene-structured squalane had the best biodegradability performance at 86%, the hybrid LAO + farnesene-structured FDBO was next at 74%, and the pure LAO-structured PAO was last with 48%. This test demonstrates the biodegradability advantage of the FDBO structure over PAO and other mineral oil materials. This mode of biodegradation, by a recognized branching structure found in nature, is vastly different from how esters achieve biodegradation due to the chemical composition of the oxygen and double bond molecules. This allows the FDBO to provide excellent stability while offering a high degree of biodegradability.

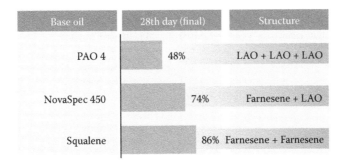

Base oil	28th day (final)	Structure
PAO 4	48%	LAO + LAO + LAO
NovaSpec 450	74%	Farnesene + LAO
Squalene	86%	Farnesene + Farnesene

FIGURE 1.12 OECD 301B biodegradability testing of 4 cSt FDBO and PAO. (Novvi LLC, Emeryville, California, Original work. With permission.)

TABLE 1.4

ASTM D6866 Carbon-14 Testing of 100% Renewable FDBO/PAO Blends

Test Oil	ASTM D6866
Squalane	100% renewable
50:50 4 cSt PAO squalane	51% renewable
4 cSt PAO	0% renewable

Source: ASTM D6866-12, Standard test methods for determining the biobased content of solid, liquid, and gaseous samples using radiocarbon analysis, ASTM International, West Conshohocken, PA, 2012, http://www.astm.org; Novvi LLC, Emeryville, California, Original work.

1.3.4 RENEWABILITY PROPERTIES OF FDBOs

Since farnesene is a 100% renewable material from sugar sources, base oils derived from farnesene are also renewable products. The U.S. Department of Agriculture (USDA) has introduced the BioPreferred program to label products based on their renewable content, which are then given preference in government procurements. The minimum renewal content varies by each type of finished lubricant. Miscellaneous lubricants such as base oils must be 25% minimum renewable to qualify [6]. Renewability is established by ASTM D6866, which determines the renewable content from the ratio of radiocarbon-14 to carbon-12 isotopes [5]. Table 1.4 shows the results of D6866 carbon-14 testing on squalane, 100% renewable FDBO, proving complete renewable content [39]. PAO had no renewable content as expected. A 50:50 blend of squalane and PAO was measured at 51% renewable content, establishing the validity of the test. Finished lubricants based on 50% renewable FDBO can meet USDA BioPreferred labeling requirements depending on the final formulation [6].

1.3.5 LOW TOXICITY PROPERTIES OF FDBOs

FDBOs are synthetic hydrocarbons that are oligomerized from two pure chemical raw materials: farnesene and LAOs, which are >95% isoparaffin hydrocarbon. The process produces base oil that

TABLE 1.5

Regulatory Toxicity Testing of FDBO

Toxicological Property	Test Method	Result	Safety Data Sheet (SDS) Classification
Acute toxicity—oral: rat	OECD 420 [60]	LD_{50} > 2000 mg/kg	Not classified
Acute toxicity—dermal: episkin	OECD 431 [61]	Noncorrosive	Not classified
Acute toxicity—inhalation	–	–	Waived
Skin corrosion/irritation: episkin	OECD 439 [62]	Nonirritating	Not classified
Serious damage/eye irritation: rabbit	OECD 405 [63]	Mild irritant Class 4	Not classified
Respiration sensitization	–	–	Waived
Skin sensitization: mouse lymph node	OECD 429 [64]	Nonsensitizing	Not classified
Chromosome aberration test	OECD 473 [65]	Not clastogenic	Not classified
Germ cell mutagenicity: Ames	OECD 471 [66]	Nonmutagenic	Not classified
Carcinogenicity	–	–	Waived
Genotoxic damage	OECD 474 [67]	Not hazardous	Not classified
Reproductive toxicity	OECD 422 [68]	Not hazardous	Not classified
28-day repeated dose toxicity			

Source: Novvi LLC, Emeryville, California, Original work.

TABLE 1.6
Ecotoxicity Test Results for FDBO

Ecological Property	Test Method	Results—No Observed Effect Concentration (NOEC)	SDS/Ecolabel Classification
Acute toxicity to algae	OECD 201 [7]	NOEC > 100 mg/L	Nontoxic
Acute toxicity to daphnia	OECD 202 [8]	NOEC > 100 mg/L	Nontoxic
Acute toxicity to zebrafish	OECD 203 [9]	NOEC > 100 mg/L	Nontoxic
Respiration of activated sludge	OECD 209 [69]	NOEC > 1000 mg/L	Nontoxic

Source: Novvi LLC, Emeryville, California, Original work.

is free from any aromatic, sulfur, or metal-containing materials. As a result, FDBOs have met the stringent requirements to be classified as a white oil by the National Sanitation Foundation (NSF) [70] and have been granted HX-1 and H-1 approval to be used as lubricant ingredients for incidental food contact (registration numbers 146406 and 149562–149564) [71].

FDBOs also have undergone extensive toxicity testing for many regulatory registrations (Table 1.5). In all tests, they were found to be as safe as white oils, which are known to have a low-toxicity profile [71].

FDBOs have also undergone extensive ecotoxicity testing and were found to be nontoxic in key environmental areas (Table 1.6). As a result, the International Standards Organization (ISO) 34 FDBO has been approved for the Ecolabel Lubricant Substance Classification (LuSC) list as 100% Classification A (ultimately aerobically biodegradable >60%) and 100% Classification D (not toxic) [72]. The Ecolabel LuSC is the definitive list of allowed ingredients in Ecolabel-approved lubricants [57].

1.4 FDBOs IN INDUSTRIAL LUBRICANTS

Due to the unique hydrocarbon and environmental performance characteristics of FDBOs, there is a wide range of industrial lubricant applications where new levels of performance can be reached. The FDBO formulations allow for finished products with the classical performance of a PAO-based product with the environmental performance of an ester product. Of particular interest are the industrial products where there is a legislation or an environmental need in place for the finished lubricant. The U.S. EPA has issued the Vessel General Permit 2013 (VGP 2013) regulation, which requires all ships over 79 ft in length to use environmentally acceptable lubricants in all applications where there is the possibility of leakage into the water [73]. In Europe, the lubricants used in environmentally sensitive areas are regulated by the Ecolabel specification [57].

The hydrocarbon oil characteristics of the FDBO allow for a wider range of formulating opportunities. Most of the off-the-shelf additive technology developed for mineral oils will drop in and provide the expected performance in FDBOs. All of the formulations detailed in this chapter were based on traditional top-tier additive packages used in synthetic and mineral oil base stocks.

1.4.1 HYDRAULIC FLUIDS

FDBOs can be used to formulate many different types of hydraulic fluids, including all industry-standard ISO grades, multigrade, food grade, and Ecolabel products. Hydraulic fluids based on FDBO can be used in the most demanding environments, including those where cold temperatures and environmental concerns push the limits of existing fluids. The mineral oil compatibility that FDBO provides also eliminates complications that can occur with other environmentally friendly hydraulic fluids such as polyalkylene glycols. Operators can top off fluids currently in operation and refill after a drain of a mineral oil product, with no system flushing and no concern about seal compatibility. FDBOs allow these fluids to meet the high-performance requirements of Eaton–Vickers [74], Bosch

TABLE 1.7
Properties of FDBO ISO 32 Hydraulic Fluid

Property	Test Method	FDBO ISO 32 Hydraulic Fluid
Viscosity at 40°C	ASTM D445 [46]	32.82 cSt
Viscosity at 100°C	ASTM D445 [46]	6.07 cSt
Viscosity index	ASTM D2270 [47]	134
Viscosity at 0°C	ASTM D445 [46]	284 cSt
Viscosity at −20°C	ASTM D445 [46]	1433 cSt
Specific gravity	ASTM D4052 [45]	840
Zinc (%)	ASTM D5185 [75]	0
Pour point	ASTM D97 [47]	−42°C
Viscosity at −8°C	ASTM D2983 [76]	381 cP
Flash point	ASTM D92 [48]	243°C
Aniline point	ASTM D611 [77]	119.2°C
Acid number	ASTM D664 [78]	0.63 mg KOH/g
Water content	ASTM D6304 [51]	53 ppm
Rust test (24 h)	ASTM D665A [79]	Pass
Rust test (24 h)	ASTM D665B [79]	Pass
Foam test	ASTM D892 [80]	0/0–10/0–0/0
Emulsion characteristics	ASTM D1401 [81]	40/40/0 (10 min)
Copper corrosion	ASTM D130 [82]	1B
Air release, 50°C	ASTM D3427 [83]	2.6 min
Filterability (dry)	ISO 13357 [84]	
Stage 1		98.2%
Stage 2		94.3%
Filterability (wet) test	ISO 13357 [84]	
Stage 1		90%
Stage 2		79%
Denison filterability test	TP-02100 [85]	
Filter time without water		69.96 s
Filter time with water		101.58 s
Hydrolytic stability	ASTM D2619 [86]	
Total acidity of water		1.6 mg KOH/g
Weight change of copper strip		0.0 mg/cm^2
Copper strip appearance		1B
Oxidation testing for 1000 h	ASTM D943 [87]	
TAN at 0 h	ASTM D664 [78]	0.70 mg KOH/g
TAN at 1004 h	ASTM D664 [78]	1.63 mg KOH/g
ΔTAN		0.97 mg KOH/g
Sludge and corrosion	ASTM D4310 [88]	
TAN 1000 h max	ASTM D664 [78]	1.12 mg KOH/g
Insoluble sludge max		123.7 mg
Total copper max		2.9 mg
Thermal stability	ASTM D2070 [89]	
Sludge max		4.1 mg
Copper wt loss max		−
Copper rod rating		3
Steel rod rating		2
Elastomer compatibility [92]	ISO 6072 [90]	168 h
NBR 1 change in hardness		−4

(Continued)

TABLE 1.7 (CONTINUED)
Properties of FDBO ISO 32 Hydraulic Fluid

Property	Test Method	FDBO ISO 32 Hydraulic Fluid
NBR 1 change in volume		9.2%
NBR 1 change in elongation		−3.9%
NBR 1 change in tensile strength		−11.7%
HNBR change in hardness		−1
HNBR change in volume		2.2%
HNBR change in elongation		−2.3%
HNBR change in tensile strength		2.9%
FKM-2 change in hardness		−1
FKM-2 change in volume		0.9%
FKM-2 change in elongation		−13.3%
FKM-2 change in tensile strength		14.0%
FZG fail load stage	ASTM D5182 [91]	12
Vickers pump test, weight loss	ASTM D7043 [92]	0.8 mg
35VQ25 pump test, weight loss	ASTM D6973 [93]	4, 8, 13 mg
KRL shear test	CEC L-45-99 [94]	0%

Source: Novvi LLC, Emeryville, California, Original work.
Note: FKM: fluorocarbon; HNBR: hydrogenated butyl rubber; NBR: nitrile butyl rubber; TAN: total acid number.

Rexroth [95], ISO 15380 [96], and Denison HF-0 [85], as well as the European Ecolabel environmental specifications of biodegradability, renewability, and nontoxicity [57]. Hydraulic fluids formulated with FDBO are the only biobased products available that can meet all of these specifications. Examples of the excellent technical performance of an ISO 32 hydraulic fluid based on FDBO are shown in Table 1.7. All critical parameters such as corrosion protection, low foaming tendency, demulsibility, copper corrosion, air release, filterability, hydrolytic stability, oxidative stability, thermal stability, and seal compatibility are easily met. Excellent antiwear performance is demonstrated in the D5182 gear test [91], D7043 V104C pump test [92], and D6973 Eaton–Vickers 35VQ25 pump test [93].

Table 1.8 provides comparative studies of ISO 46 FDBO hydraulic fluid relative to other commercial hydraulic fluids [39]. The data show that a renewable hydrocarbon hydraulic fluid meets the most stringent environmental characteristics of renewability and biodegradability while demonstrating equal to or better performance than the industry-leading petroleum products. The PAO and mineral oil hydraulic fluids both provided the classical performance as expected around wet turbine oil stability test (TOST) and oxidative stability but do not meet the environmental specifications. The ester-based hydraulic fluids met one or both of the renewability and biodegradability specifications but fell short in other performance areas such as pour point, oxidation stability, and hydrolytic stability and the wet TOST. The FDBO hydraulic fluid also shows significantly improved oxidation stability compared to any of the commercial products.

Great environmental performance also allows FDBO hydraulic fluids to meet U.S. EPA VGP 2013 specification [73] for marine applications where lubricants may have contact with seawater, providing further environmental protection in the event of leakage. It also offers the best choice in construction, agriculture, and forestry uses by reducing environmental risk without compromising performance.

1.4.2 Transformer Oils

FDBO transformer oil is a new drop-in compatible, environmentally friendly transformer fluid on the market. It meets ASTM D3487 [97] and International Electrotechnical Commission (IEC) 60296 [98] standards for mineral insulating oil used in transformers and switchgears while delivering

TABLE 1.8

Comparison of Hydraulic Fluid Properties of FDBO Hydraulic Fluid and Other Commercial Hydraulic Fluids

Property	Test Method	FDBO	Mineral Oil	PAO 1	PAO 2	Synthetic Ester	Vegetable Oil
Ashless additive		Yes	No	No	No	Yes	Yes
Viscosity at 40°C	ASTM D445 [46]	43 cSt	46 cSt	46 cSt	48 cSt	45.9 cSt	36.8 cSt
VI	ASTM D2270 [47]	130	97	154	192	200	212
Pour point	ASTM D97 [48]	−40°C	−33°C	−54°C	−47°C	−33°C	−34°C
Copper strip corrosion test	ASTM D130 [82]	1B	1A	1B	1A	1A	−
Foam stability	ASTM D892 [80]	50/0	20/0	50/0	0/0/0	0/10/0	−
Oxidative stability by RPVOT	ASTM D2272 [99]	900+ min	152 min	209 min	376 min	258 min	44 min
Oxidative stability by wet TOST	ASTM D943 [87]	Pass	Pass	Pass	Pass	Fail	Fail

Source: Novvi LLC, Emeryville, California, Original work.

Note: min: minutes; RPVOT: rotating pressure vessel oxidation test; TOST: turbine oil stability test.

operational benefits that extend well beyond meeting mineral oil capabilities. FDBO has better heat transfer capabilities than mineral oils [100]; and FDBO transformer fluids inherit the same advantage. Improved heat transfer capabilities enable the more efficient operation by allowing more power throughput at a capacity higher than nameplate power ratings. FDBO transformer oil has excellent dielectric strength (Table 1.9), minimizing power losses. Combining these features leads to improved operating cost. FDBO transformer oil is also biodegradable, renewable, and nontoxic, eliminating the need to compromise between performance and environmental responsibility. While the transformer oil is a fraction of the initial equipment investment, it has a major impact on a transformer's total lifetime through its heat transfer and electrical insulation performances [58].

A fluid's heat transfer capability affects operating cost through power losses and insulating paper degradation, which impacts transformer life expectancy. FDBO transformer oil offers excellent heat transfer capabilities, enabling heat to be removed from the windings and the transformer to transmit more power at a given operating temperature. In order to demonstrate and compare the heat transfer capability of FDBO transformer oil to that of commercially marketed environmentally friendly transformer oil, two 30 kVA class transformers with identical design were filled with each fluid and energized to steady-state operational condition [39]. Figure 1.13 compares the steady-state top-oil temperature of the unit filled with FDBO transformer oil to that of the unit filled with a vegetable oil-based transformer oil. The top-oil temperature of the transformer filled with FDBO transformer rose to 70°C, while that of the transformer filled with vegetable oil-based transformer oil rose to 84°C [39].

FDBO transformer oil has a low viscosity and a pour point of −57°C, comparable to mineral oil transformer oil that is commercially available in the market. This eliminates the gel formation and the need to heat the fluid before a restart. Because FDBO transformer oil is a hydrocarbon, it can also be restored to its original state through standard mineral oil degassing and other reclamation processes so that it can be reused following water, acid, or particulate contamination. A dielectric fluid has a number of properties that affect its ability to function effectively and reliably [98]. These properties include flash and fire points, heat capacity, thermal conductivity, viscosity over a range of temperatures, impulse breakdown voltage, dielectric breakdown voltage, power factor, gassing tendency, and pour point. Power factor is a measure of the dielectric losses in an electrical insulating liquid when used in an alternating electric field and of the energy dissipated as heat. It is typically measured with a standardized method such as ASTM D924 [101]. A low power factor value

TABLE 1.9

Properties of FDBO Transformer Oil Compared to ASTM D3487 Requirements for Mineral Oil-Based Transformer Oil

Property	Method	Requirements	FDBO
Biodegradability	OECD 301B at 28 days [2]	No requirement	80%
Renewability	ASTM D6866 [5]	No requirement	49%
Aniline point	ASTM D611 [77]	>63°C	113.4°C
Color	ASTM D1500 [44]	<0.5	<0.5
Flash point	ASTM D92 [49]	>145°C	178°C
Interfacial tension	ASTM D971 [102] at 25°C	>40 dyn/cm	47 dyn/cm
Pour point	ASTM D97 [48]	<−40	−54°C
Specific gravity	ASTM D4052 [45]	<0.91	0.818
Viscosity at 100°C	ASTM D445 [46]	<3	2.93 cSt
Viscosity at 40°C	ASTM D445 [46]	<12	11.5 cSt
Viscosity at 0°C	ASTM D445 [46]	<76	65 cSt
Dielectric breakdown	ASTM D877 [103]	>30 kV	49 kV
	ASTM D1816 [104] (1 mm gap)	>20 kV	31 kV
Impulse breakdown voltage	ASTM D3300 [105] at 25°C	>145 kV	278 kV
Gassing tendency	ASTM D2300 [106] at 80°C (µL/min)	No requirement	−10 µL/min
Power factor	ASTM D924 [107] at 25°C	<0.05%	0.001%
	ASTM D924 [107] at 100°C	<0.30%	0.009%
Oxidation stability (Acid sludge)	ASTM D2440 [108] at 72 h	%sludge < 0.1%	<0.01%
		TAN <0.3 mg KOH/g	<0.01 mg KOH/g
	ASTM D2440 [108] at 164 h	Sludge <0.2%	<0.01%
		TAN <0.4 mg KOH/g	<0.01 mg KOH/g
Oxidation stability (RPVOT)	ASTM D2112 [109]	>195	700 min
Oxidation inhibitor content	ASTM D2668 [110]	<0.3 wt%	0.27 wt%
Water content	ASTM D1533 [111]	<30 ppm	13 ppm
Neutralization number	ASTM D974 [112]	<0.03 mg KOH/g	<0.01 mg KOH/g

Source: Novvi LLC, Emeryville, California, Original work.

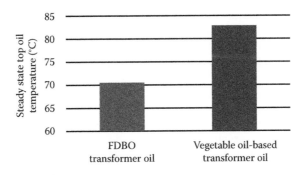

FIGURE 1.13 Steady-state top oil temperature in a 30 kVA distribution transformer. (Novvi LLC, Emeryville, California, Original work. With permission.)

TABLE 1.10

Properties of FDBO Transformer Oil Compared to IEC 60296 Requirements for Mineral Oil-Based Transformer Oil

Property	Method	Requirements	FDBO
Viscosity at −30°C	IEC 61868 [113]	≤1800 cSt	563 cSt
Dielectric breakdown (as received)	IEC 60156 [114]	≥50 kV	73 kV
Dielectric breakdown (after filtration)	IEC 60156 [114]	≥70 kV	96 kV
Dissipation factor at 90°C	IEC 60247 [115]	≤0.005 absolute	0.0001
Oxidation stability	IEC 61125C [116] 500 h	Total acids ≤ 1.2 mg KOH/g	0.99 mg KOH/g
		Total sludge ≤ 0.8%	0.01%
		DDF ≤ 0.5 absolute	0.0443
Polycyclic aromatics content	BS 2000 part 346 [117]	≤3%	0.3%
Polychlorinated biphenyl content	IEC 61619 [118]	<2 mg/kg	<2 mg/kg

Source: Novvi LLC, Emeryville, California, Original work.
Note: DDF: dielectric dissipation factor.

indicates low alternating current dielectric losses. Power factor may be useful as a means of quality control, and as an indication of changes in quality resulting from contamination and deterioration in service or as a result of handling. Gassing tendency is measured using a standardized test method such as ASTM D2300 [119]. It measures the tendency of insulating liquids to absorb or evolve gas under conditions of electrical stress and ionization based on the reaction with hydrogen, the predominant gas in the partial discharge. Insulating liquids shown to have gas-absorbing (H_2) characteristics in the test have been used to advantage in reducing equipment failures, particularly cables and capacitors. However, the advantage of such insulating liquids in transformers is not well defined and the study of its impact on transformer operation is still an evolving science. However, dielectric fluid with gas-absorbing characteristic is gaining popularity in the market due to the popular idea that such liquid will help to prevent gas bubble formation, which can lead to more catastrophic failure of transformers, at least in the short-term operation. Industrial standards use a combination of many parameters (mentioned previously) to provide guidelines on a fluid's efficiency and reliability at any given moment. Such industrial standards include IEC 60296 (mineral oil) [98], ASTM D3487 (mineral oil) [97], CAN/CSA-C50-14 (mineral oil) [120], ASTM D6871 (natural ester) [121], and the Institute of Electrical and Electronics Engineers standard C57.147 (natural ester) [122]. Tables 1.9 and 1.10 show the performance of FDBO transformer oil compared to two of such widely accepted international specifications for mineral insulating oil used in electrical apparatus including transformers (ASTM D3487 and IEC 60296). Mineral oil-based fluid constitutes over 97% of the current transformer oil market; and FDBO transformer oil is in full compliance with the specifications set for mineral oil-based insulating oil. FDBO transformer oil provides premium-grade fluid performances. Tables 1.9 and 1.10 also list the environmental performance of FDBO transformer oil as represented by biodegradability and renewability using OECD 301B [2] and ASTM D6866 [5]. FDBO transformer oil provides a unique combination of dielectric and environmental properties.

1.4.3 COMPRESSOR OILS

Compressor oil performance is very dependent on the characteristics of the base oil. The ideal base oils for compressors have excellent gas/oil separation capabilities and oxidation stability, on top of antiwear and lubricating qualities [123,124]. FDBOs have all of these characteristics, which allows formulators to use FDBO to create differentiated compressor lubricants that increase performance with reduced environmental impact. Compressor oils formulated with FDBO are fit for use in marine applications with U.S. EPA VGP 2013 specifications [73]; NSF H-1 applications for

TABLE 1.11

Properties of FDBO Compressor Oil Compared to Those of Commercial Compressor Oils

Property	Method	FDBO Compressor Oil	Commercial Compressor Oil #1	Commercial Compressor Oil #2	Commercial Compressor Oil #3	Commercial Compressor Oil #4
Base oil chemistry		FDBO	Mineral oil	Diester	PAG	PAO
Oxidative stability at 150°C	ASTM D2272 [99]	4200 min	100 min	1000 min	500 min	2200 min
Air/Oil separation speed	ASTM D3427 [83]	1.1 min	8.8 min	17.9 min	3.7 min	4.5 min
Water/Oil separation speed	ASTM D1401 [81]	10 min	30 min	25 min	30 min	25 min
Four-ball wear scar diameter	ASTM D4172 [125]	0.419 mm	0.526 mm	0.363 mm	0.606 mm	0.457 mm
Rust prevention	ASTM D665B [79]	Pass	Pass	Pass	Pass	Pass
Copper strip corrosion at 100°C	ASTM D130 [82]	Shiny 1B	Shiny 1B	Shiny 1B	Shiny 1B	Shiny 1B
Aquatic toxicity	OECD 201–203 [7–9]	Very low	OK	Low	Low	Very low
Biodegradable	OECD 301 [2]	Yes	No	Yes	Possible	No
Renewable carbon	ASTM D6866 [5]	Yes	No	Possible	No	No

Source: Novvi LLC, Emeryville, California, Original work.
Note: PAG: polyalkylene glycol.

food-grade service with incidental contact [70]; and agricultural, construction, and manufacturing uses with the toughest performance requirements [58].

To demonstrate the performance benefits of FDBO compressor oil, standard ASTM tests were performed on variety of high-performance ISO 46 air compressor oils made from different base oil types and their commercial additive packages (Table 1.11). The FDBO air compressor oil ISO 46 was formulated with the recommended 0.9 wt% treat rate of an off-the-shelf compressor oil additive package. Table 1.11 demonstrates how the FDBO is formulated into a lubricant that outperforms other compressor oils on critical performance parameters and provides key environmental benefits [39].

1.4.4 INDUSTRIAL GEAR OILS

FDBOs can be used to formulate high-performance synthetic gear oils used for industrial and marine applications. FDBOs are compatible with conventional gear oil additive packages, to allow for easy formulation of lubricants that have extreme pressure, antirust, and antioxidation characteristics. Replacing a lower grade (e.g., Group I or II) or synthetic base stock with FDBO leads to gear lubricants with improved operational performance and lower environmental impact, through improved sustainability and biodegradability. Best-in-class oxidation stability of FDBO allows operators to lower the costs of ownership and operation through extending lubricant life and replacement intervals, which helps to reduce maintenance cost. FDBO gear oils have been shown to carry high loads in a simulated FZG test using an SRV technique (Figure 1.14) [126]. Improved performance is obtained compared to mineral oil, PAO, and ester gear oils.

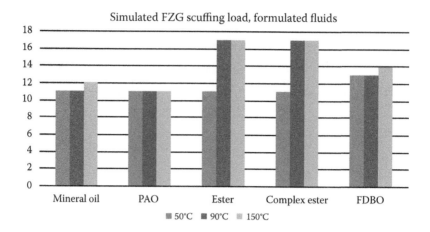

FIGURE 1.14 Simulated FZG base oil comparison in gear oil. (From Galary, J., Study of wear properties of environmentally friendly lubricants for gearing applications as a function of film thickness transition, ASTM STP 1575-EB, 2014. With permission.)

In marine applications where U.S. EPA VGP 2013 [73] requires the use of environmentally acceptable lubricants, FDBO lubricants can be formulated to create gear oils that meet performance expectations where other environmentally friendly replacements fall short.

1.4.5 GREASES

FDBOs have been used to formulate greases with excellent performance, appropriate for use in many applications [39]. They are compatible with a wide variety of high-quality thickening systems,

TABLE 1.12

Properties of FDBO Greases Compared to Those of PAO Greases

Property	Test	FDBO Grease	FDBO Grease	PAO Grease	PAO Grease
Additive		No	Yes	No	Yes
Color	Visual	Light tan	Light tan	Light tan	Light tan
Appearance	Visual	Smooth	Smooth	Smooth	Smooth
Penetration after 0 strokes	ASTM D217 [127]	279 mm	300 mm	263 mm	282 mm
Penetration after 60 strokes	ASTM D217 [127]	276 mm	312 mm	268 mm	280 mm
Penetration after 10,000 strokes	ASTM D217 [127]	354 mm	354 mm	317 mm	329 mm
Drop point	ASTM D2265 [128]	215°C	201°C	216°C	204°C
Oxidative stability by PDSC at 150°C (onset)	ASTM D6186 [129]	25.7 min	>120 min	28.15 min	>120 min
Oil separation	ASTM D6184 [130]	1.56%	0.00%	0.00%	2.00%
Low-temperature apparent viscosity at −41.4°C	Proprietary method [39]	534,000 cP		441,600 cP	
Four-ball antiwear, wear scar diameter	ASTM D2266 [131]	0.584 mm	0.402 mm	0.538 mm	0.432 mm
Four-ball extreme pressure (EP), weld point	ASTM D2596 [132]	<210 kgf	>210 kgf	<210 kgf	>210 kgf
Water washout at 40°C	ASTM D1264 [133]	3.58%	4.78%	4.47%	3.35%

Source: Novvi LLC, Emeryville, California, Original work.
Note: PDSC: pressure differential scanning calorimetry.

TABLE 1.13

Properties of FDBO Grease Compared to Those of Vegetable Oil Grease

	Test	FDBO Grease	Veg. Oil Grease
Drop point	ASTM D2265 [128]	246°C	260°C
Oxidative stability by PDSC at 175°C (onset)	ASTM D6186 [129]	>120 min	5.24 min
Oil separation	ASTM D6184 [130]	1.56%	0.00%
Four-ball antiwear, wear scar diameter	ASTM D2266 [131]	0.389 mm	0.939 mm
Four-ball EP, weld point	ASTM D2596 [132]	>210 kgf	>210 kgf
Water washout at 40°C	ASTM D1264 [133]	4.6%	23.3%
Rust	ASTM D1743 [134]	Pass	Pass
Elastomer compatibility, %volume change/hardness change	ASTM D4289 [135]		
NBR-L		+0.1%/−3	+34%/−25
CR chloroprene		+4.13/−3	+37%/−18

Source: Novvi LLC, Emeryville, California, Original work.

including lithium, lithium complex, and calcium sulfonate, giving formulators a sustainable foundation for high-value products with biodegradable and renewable options.

FDBOs can be formulated and compounded in grease manufacturing in a similar manner to PAO. Testing was performed on FDBO and PAO to demonstrate their performance with a standard lithium stearate thickening system and a standard high-performance grease additive package [39]. Greases were manufactured with and without the additive package to directly compare the performances of thickened base oils. A summary of the data is shown in Table 1.12. The FDBO grease products demonstrated similar performance in manufacturing and characterization tests.

FDBO grease is a good choice for protecting bearings and machinery components against wear and corrosion. It is suitable for high- and low-temperature environments. FDBO grease reduces friction for smooth performance under the toughest conditions and can match PAO grease performance in critical areas [39]. It has good oxidation resistance and hydrolytic stability and will provide long lubricant life and protect the equipment.

FDBO grease is a perfect choice for use in environmentally sensitive areas and can outperform the ester-based grease products, which have historically been used for these applications. Table 1.13 compares FDBO grease with commercial vegetable oil-based grease used in environmentally sensitive applications [136]. The typical vegetable oil problems of stability and elastomer compatibility are shown when compared to the FDBO grease formulation. FDBO grease can be used in Ecolabel applications, food-grade use, and marine settings to comply with U.S. EPA VGP regulations.

Grease manufacturing is a difficult and technical process [137]. FDBOs are the only renewable and biodegradable option that will compound and manufacture the same as petroleum base oils.

1.5 FDBOs IN AUTOMOTIVE LUBRICANTS

1.5.1 ENGINE OILS

FDBOs can be used to formulate passenger car motor oils (PCMOs) in a range of fuel-saving viscosities that meet API SN and other industry and original equipment manufacturer (OEM) requirements. FDBO engine oils deliver performance characteristics comparable or better than those of other synthetic engine oils while redefining performance versus key environmental metrics when compared with petroleum-derived base oils (Table 1.15) [39].

Automakers must meet new regulations without compromising the driving experience. Vehicle designers around the world are hard at work on plans to reduce vehicle weight, improve aerodynamics, and create smaller high-performance engines. These new engines increase performance demands on lubricants, which can be seen in the trends toward lower viscosity fluids that reduce internal friction and circulate better during cold starts [138]. Table 1.14 lists the bench tests required by API on passenger car engine oils for SN and GF-5 specifications by the International Lubricant Standardization and Approval Committee (ILSAC) [40]. A 5W-30 engine oil formulated with FDBO shows excellent performance against listed specifications. The 5W-30 engine oil consists of 82.3 wt% FDBOs, a commercially available engine oil additive package (P5711, Infineum USA L.P.) [139], and VI improver (SV261L, Infineum USA L.P.) [139]. This FDBO-based 5W-30 formulation does not contain any cobase oil nor solvent. Infineum P5711 is designed to meet high-performance

TABLE 1.14

Laboratory/Bench Test Results for FDBO-Based 5W-30 Viscosity Grade Engine Oil and Comparison to Specifications for API SN/ILSAC GF-5 Categories

Test	Method	5W-30 Engine Oil	Limits
FDBO content		82.3 wt%	
P5711, PCMO additive package content		12.5 wt%	
SV261L, VI improver content		5.2 wt%	
Kinematic viscosity at 100°C	ASTM D445 [46]	10.1 cSt	≥9.3 cSt
Cranking viscosity at −30°C	ASTM D5293 [52]	4130 cP	≤6600 cP
High shear rate viscosity at 150°C	ASTM D4683 [140]	3.0 cP	≥2.9 cP
Volatile loss at 250°C	ASTM D5800 [53]	9%	≤15%
Ball rust test, average gray value	ASTM D6557 [141]	116	≥100
Volatile loss at 371°C	ASTM D6417 [142]	1%	≤10%
Engine oil filterability test; % flow change after water/dry-ice/ short thermal aging treatment	ASTM D6795 [143]	−6%	≤50%
Engine oil filterability test after 6 h thermal aging with 0.6% water; % flow change	ASTM D6794 [144]	−4%	≤50%
Engine oil filterability test after 6 h thermal aging with 1.0% water; % flow change	ASTM D6794 [144]	−9%	≤50%
Engine oil filterability test after 6 h thermal aging with 2.0% water; % flow change	ASTM D6794 [144]	−7%	≤50%
Engine oil filterability test after 6 h thermal aging with 3.0% water; % flow change	ASTM D6794 [144]	−9%	≤50%
Phosphorus content	ASTM D4951 [145]	0.075 wt%	0.06–0.08 wt%
Sulfur content	ASTM D2622 [43]	0.212 wt%	≤0.5 wt%
Homogeneity and miscibility	ASTM D6922 [146]	Pass	Pass
Shear stability, kinematic viscosity at 100°C after shear	ASTM D6278 [147]	9.8 cSt	≥9.3 cSt
Moderately high temperature piston deposit test; total deposit weight	ASTM D7097 [148]	23 mg	≤35 mg
Scanning Brookfield; gelation index	ASTM D5133 [149]	<6	≤12
Foaming characteristics	ASTM D892 [80]	0/0–0/0–0/0	≤10/0–50/0–10/0
Sequence IV foaming tendency; static foam	ASTM D6082 [150]	20 mL	≤100 mL
Sequence IV foam stability, 1 min settling	ASTM D6082 [150]	0 mL	0 mL
Emulsion retention with 10% water and 10% E85 after 24 hr at 0°C	ASTM D7563 [151]	No water separation	No water separation
Emulsion retention with 10% water and 10% E85 after 24 h at 25°C	ASTM D7563 [151]	No water separation	No water separation
Oxidative stability by Romaszewski Oil Bench Oxidation	ASTM D7528 [152]	13,200 cP	≤60,000 cP
ILSAC GF-5 seals compatibility test	SAE J2643 [153]	Pass	Pass

Source: Novvi LLC, Emeryville, California, Original work.

engine oil standards, such as API SN/SN-RC, ILSAC GF-5, dexos1, and GM 4718M, when blended with API Group III base oils with approved viscosity modifiers such as SV261L [139]. It is important to note that the formulation shown on Table 1.14 was obtained by interchanging the Group III base oil component of previously tested and approved formulation with FDBO [39]. This interchange resulted in equal or improved performances of the resulting formulation with FDBO. As described in Appendix E of the API Base Oil Interchangeability Guidelines for Passenger Car Motor Oils and Diesel Engine Oils [40], such interchangeability of FDBOs with other Group III base oils in the tested and approved formulations can provide immense benefit to the formulator/producer of engine oil by reducing the number of required physical property and engine tests.

The 5W-30 viscosity grade is currently the most commonly used grade of engine oil, with more than 50% of the market [154]. It is expected that 5W-20 and 0W-XX grades will encroach on this lead as automobile manufacturers increasingly recommend synthetic motor oils to meet technological demands. FDBO offers an ideal base stock to formulate these lower-viscosity higher-performance grades. Table 1.15 shows engine test results of a 0W-30 engine oil formulated using FDBOs, P5711, and SV261L (once again, no cobase oils nor solvent was used). The tested FDBO-based 0W-30 engine oil delivers excellent performance against the listed requirements [39].

FDBO engine oils have passed the latest GF-5 API SN Core Test Program and are being used in an ongoing field trial with Las Vegas taxicabs. FDBO is the only renewable base oil to ever pass a full API SN program as the sole base oil in the formulation (Table 1.15) [39].

Engine oils formulated with FDBO deliver environmental benefits beyond contributing to increased fuel efficiency. FDBO is the only renewably sourced Group III base oil. FDBO engine oil improves sustainability at each step of the life cycle analysis—production, manufacturing, use, and

TABLE 1.15
Engine Test Results of an FDBO-Based 0W-30 Viscosity Grade Engine Oil

Property	Method	0W-30 Engine Oil	Limits
FDBO content		80.0 wt%	NA
P5711, PCMO additive package content		12.5 wt%	NA
SV261L, VI improver content		7.5 wt%	NA
Kinematic viscosity at 100°C	ASTM D445 [46]	10.4 cSt	≥9.3 and <12.5 cSt
Cranking viscosity at −35°C	ASTM D5293 [52]	5200 cP	<6200 cP
High shear rate viscosity at 150°C (mPa s)	ASTM D4683 [140]	3.0 cP	>2.9 cP
Low temperature pumping viscosity at −40°C	ASTM D4684 [155]	27,500 cP	60,000 cP
Sequence IIIG/IIIGA	ASTM D7320 [156]		
Viscosity at 40°C, % increase		45%	≤150%
Weighted piston deposits		4.3	≥4.0
Hot stuck rings		None	None
Average cam wear and lifter wear		41 μm	≤60 μm
Sequence IIIGB	ASTM D7320 [156]		
Phosphorus retention		84%	79% min.
Sequence IVA	ASTM D6891 [157]		
Average cam wear		19 μm	≤90 μm
Sequence VG	ASTM D6593 [158]		
Average engine sludge		9.0	≥8.0
Rocker arm cover sludge		9.5	≥8.3
Average engine varnish		9.4	≥8.9
Average piston skirt varnish		8.7	≥7.5
Oil screen sludge (%)		0	≤15

Source: Novvi LLC, Emeryville, California, Original work.

disposal [22]. It is also compatible with existing recycling and rerefining infrastructure, continuing to reduce petroleum use through responsible collection practices.

1.5.2 TWO-CYCLE AIR AND MARINE ENGINE OILS

Emission controls have reduced two-cycle engine use over the last 30 years, but they are still the high-power, compact engines used in chainsaws, snowmobiles, motorcycles, sprint karts, scooters, and other portable machinery [159]. Low-grade, unburned oil is a big part of the emission problem as the oil mixes with fuel in the combustion chamber, resulting in smoky exhaust filled with unburned hydrocarbons, carbon monoxide, and other particles. FDBO two-cycle engine oils reduce smoke and odor while providing excellent lubrication and improved engine performance [39]. This has been demonstrated in field tests with go-carts at a racing track in California. Figure 1.15 shows

FIGURE 1.15 Comparison of pistons from two-stroke engines—commercial (left) versus FDBO (right). (Novvi LLC, Emeryville, California, Original work. With permission.)

TABLE 1.16
Test Results for FDBO-Based Two-Cycle Engine Oil and Comparison to JASO M345 Specification

Test	ASTM/JASO Test Method	FBDO Two-Cycle Engine Oil	JASO M345 FD Specification
Flash point	ASTM D92 [49]	118°C	≥70°C
Sulfated ash	ASTM D874 [160]	0.14%	≤0.25%
Kinematic viscosity at 100°C	ASTM D445 [46]	10.07 cSt	≥6.5 cSt
Brookfield viscosity at −25°C[a]	ASTM D2983 [76]	7098 cP	<7500 cP[a]
Lubricity	JASO M340 [161]	125	≥95
Initial torque	JASO M340 [161]	99	≥98
Detergency—fundamental part	JASO M341 [161]—60 minutes	100	–
Detergency—piston skirt part	JASO M341 [161]—60 minutes	97	–
Detergency—fundamental part	JASO M341 [161]—180 minutes	128	≥125
Detergency—piston skirt part	JASO M341 [161]—180 minutes	105	≥95
Smoke	JASO M342 [161]	113	≥85
Exhaust system blocking test	JASO M343 [161]	158	≥90

Source: Novvi LLC, Emeryville, California, Original work.
[a] SAE J1536 fluidity—Grade 3.

a comparison of pistons from engines run with a commercial two-stroke engine oil versus FDBO-based two-stroke oil.

FDBO two-cycle engine oils reduce wear, improve performance, and tackle environmental issues. Two-stroke oils formulated from FDBO are also a perfect fit in construction, maintenance, and leisure equipment [159]. FDBO two-cycle engine oil has passed Japanese Automotive Standards Organization (JASO) FD M345 [161] and API TC (ISO EGD) [162] standards and meets Society of Automotive Engineers (SAE) J1536 Grade 3 specification [163] for low-temperature viscosity allowing use in many low-temperature applications (Table 1.16) [39]. As with other FDBO lubricants, FDBO two-cycle engine oils are a drop-in replacement for mineral oil products, with no difference in handling, storage, or equipment compatibility.

1.6 SUMMARY AND CONCLUSIONS

Novel synthetic hydrocarbon base oils are manufactured by oligomerization of farnesene, a C15 branched hydrocarbon that is sourced from renewable plant sugars by novel synthetic biology technology. FDBOs are plant-based synthetic oils designed to deliver high performance while offering major environmental benefits when compared with petroleum-sourced base oils. Ecofriendly advantages include nontoxicity, renewability, and improved biodegradability.

FDBOs have technical performance similar to PAOs and classified as API Group III. This is the only renewable technology available to meet Group III specifications for products marketed in the engine oil segment. FDBOs outperform Group III oils while delivering important environmental benefits. FDBO industrial lubricants meet many demanding industry and OEM specifications.

FDBOs are the only hydrocarbon base oils that can qualify for the Ecolabel program to enable formulators to develop products that meet or exceed both the high-performance and environmental standards that the Ecolabel certification requires. Ecolabel sets a high bar for the lubricants industry by requiring products to meet tough environmental standards without compromising performance. Until now, products could meet performance *or* environmental standards, but not both. Mineral oil products that meet performance requirements fail the environmental test. Formulations from vegetable oils and other natural esters that meet environmental specifications fall short on performance. FDBO formulations are the only lubricants that excel in both sustainability and performance.

REFERENCES

1. Brown, J. A. (2014, November), High performance environmentally friendly lubricants. Paper presented in Independent Chemical Information Service (ICIS) South Africa Presentation.
2. OECD 301. http://www.oecd.org/chemicalsafety/risk-assessment/1948209.pdf (Accessed on May 15, 2015).
3. ASTM D5864-11 (2011), Standard test method for determining aerobic aquatic biodegradation of lubricants or their components. ASTM International, West Conshohocken, PA. http://www.astm.org.
4. ASTM D6731-01(2011), Standard test method for determining the aerobic, aquatic biodegradability of lubricants or lubricant components in a closed respirometer. ASTM International, West Conshohocken, PA. http://www.astm.org.
5. ASTM D6866-12 (2012), Standard test methods for determining the biobased content of solid, liquid, and gaseous samples using radiocarbon analysis. ASTM International, West Conshohocken, PA. http://www.astm.org.
6. USDA website. http://www.biopreferred.gov/BioPreferred/ (Accessed May 15, 2015).
7. OECD 201. http://www.oecd.org/chemicalsafety/testing/1946914.pdf (Accessed May 15, 2015).
8. OECD 202. http://www.oecd.org/chemicalsafety/risk-assessment/1948249.pdf (Accessed May 15, 2015).
9. OECD 203. http://www.oecd.org/chemicalsafety/risk-assessment/1948241.pdf (Accessed May 15, 2015).
10. OECD 210. http://www.oecd.org/chemicalsafety/risk-assessment/1948269.pdf (Accessed May 15, 2015).
11. OECD 211. http://www.oecd.org/chemicalsafety/risk-assessment/1948277.pdf (Accessed May 15, 2015).
12. Amyris Inc. website. http://www.amyris.com; http://sugarcane.org/sustainability/preserving-biodiversity-and-precious-resources (Accessed May 15, 2015).

13. http://www.genengnews.com/gen-news-highlights/tate-lyle-to-manufacture-farnesene-for-amyris/81244196/.

14. http://www.reuters.com/article/2011/03/03/idUS134818+03-Mar-2011+BW20110303.

15. http://www.biodieselmagazine.com/articles/7944/amyris-accelerates-renewable-diesel-biobased-chemicals-efforts.

16. https://amyris.com/amyris-ships-first-truckload-of-biofene-from-its-new-plant-in-brazil/.

17. http://energyalmanac.ca.gov/gasoline/whats_in_barrel_oil.html (Accessed May 1, 2015).

18. USDA Economic Research Service. Sugar and Sweeteners Outlook. SSS-M316, December 16, 2014. Available online: http://www.ers.usda.gov/media/1725277/sss_m_316.pdf (Accessed May 5, 2015).

19. Morvey, G. (2014), Kline & Company "Global Synthetic Lubricants 2013" webinar.

20. Goldemberg, J. (2008), The Brazilian biofuels industry, *Biotechnology for Biofuels* 1 (6), 1–7.

21. Loarie, S., Lobell, D., Asner, G., Mu, Q., and Field, C. (2011), Direct impacts on local climate of sugarcane expansion in Brazil, *Nature Climate Change* 2, 105–109.

22. Brown, J. A. (2014), Feedstocks for biolubricants. Paper presented in ICIS London 2014.

23. http://investors.amyris.com/releasedetail.cfm?releaseid=846031 (Accessed May 5, 2015).

24. Round Table on Sustainable Biomaterials website. http://rsb.org/ (Accessed May 5, 2015).

25. Blad, J. C., and McKinsey & Company (2014), Strategic opportunities in bio-based lubricants and basestocks. Paper presented in 18th ICIS World Base Oils & Lubricants Conference London 2014.

26. McPhee, D. (2013), Sesquiterpenes as chemical building blocks: β-Farnesene. in *Catalytic Process Development for Renewable Materials*, Imhof, P., and van der Waal, J. C. (Eds.). Wiley-VCH: Weinheim, Germany; pp. 56–76.

27. Akutagawa, S., Taketomi, T., Kumobayashi, H., Takayama, K., Someya, T., and Otsuka, S. (1978), Metal-assisted terpenoid synthesis. V. The catalytic trimerization of isoprene to trans-β-farnesene and its synthetic applications for terpenoids, *Bulletin of the Chemical Society of Japan* 51 (4), 1158–1162.

28. Tanaka, S., Yasuda, A., Yamamoto, H., and Nozaki, H. (1975), General method for the synthesis of 1,3-dienes: Simple syntheses of β- and *trans*-α-farnesene from farnesol, *Journal of the American Chemical Society* 97 (11), 3252–3254.

29. http://www.fao.org/docrep/x0560e/x0560e08.htm (Accessed May 5, 2015).

30. Martin, V., Pitera, D., Withers, S., Newman, J., and Keasling, J. (2003), Engineering a mevalonate pathway in *Escherichia coli* for production of terpenoids, *Nature Biotechnology* 21 (7), 796–802.

31. Gardner, T. S., Hawkins, K. M., Meadows, A. L., Tsong, A. E., and Tsehaye, Y. (2014, October 14), Production of acetyl-coenzyme A derived isoprenoids. Assigned to Amyris Inc., U.S. Patent 8859261 B2.

32. McPhee, D. (2013, July 23), Microbial derived isoprene and methods for making the same. U.S. Patent 8492605.

33. http://www2.epa.gov/green-chemistry/2014-small-business-award (Accessed May 5, 2015).

34. (a) Fisher, K., and Woolard, F. (2009, September 22), Farnesene dimers and/or farnesane dimers and compositions thereof. U.S. Patent 7592295; (b) Fisher, K., and Woolard, F. (2010, April 6), Lubricant compositions. U.S. Patent 7691792; (c) McPhee, D. (2012, July 10), Farnesene interpolymer. U.S. Patent 8,217,128; (d) McPhee, D. J., Safir, A., Reeder, C. L., and Doolan, J. G. (2013, November 263), Polyfarnesenes. U.S. Patent 8,592,543.

35. Ohler, N. L. and Vasquez, R. (2012, February 8), Stabilization and hydrogenation methods for microbial-derived olefins. Applicant Amyris Inc., Patent EP 2414311 A2.

36. Ohler, N.; Fisher, K., and Hong, J. K. (2012, February 13), Olefins and methods for making the same. U.S. Patent Application 20140148624 A1.

37. Ohler, N., Fisher, K., and Tirmizi, S. (2012, February 13), Base oils and methods for making the same. U.S. Patent Application 20140221258 A1.

38. ASTM D2710-09 (2013), Standard test method for bromine index of petroleum hydrocarbons by electrometric titration. ASTM International, West Conshohocken, PA. http://www.astm.org.

39. Novvi LLC, Emeryville, CA, Original Work. http://www.novvi.com.

40. American Petroleum Institute: Industry Services Department (2011, September), Appendix E—API base oil interchangeability guidelines for passenger car motor oils and diesel engine oils. Engine Oil Licensing and Certification System, API Publication 1509. (16th edition). http://www.api.org/certification-programs/engine-oil-diesel-exhaust-fluid/publications.aspx (Accessed May 5, 2015).

41. http://gasprocessingnews.com/features/201410/gtl-innovation-produces-clean-base-oils-from-natural-gas.aspx (Accessed May 1, 2015).

42. ASTM D2007-11 (2011), Standard test method for characteristic groups in rubber extender and processing oils and other petroleum-derived oils by the clay-gel absorption chromatographic method. ASTM International, West Conshohocken, PA. http://www.astm.org.

43. ASTM D2622-10 (2010), Standard test method for sulfur in petroleum products by wavelength dispersive X-ray fluorescence spectrometry. ASTM International, West Conshohocken, PA. http://www.astm.org.
44. ASTM D1500-12 (2012), Standard test method for ASTM color of petroleum products (ASTM color scale). ASTM International, West Conshohocken, PA. http://www.astm.org.
45. ASTM D4052-11 (2011), Standard test method for density, relative density, and API gravity of liquids by digital density meter. ASTM International, West Conshohocken, PA. http://www.astm.org.
46. ASTM D445-15 (2015), Standard test method for kinematic viscosity of transparent and opaque liquids (and calculation of dynamic viscosity). ASTM International, West Conshohocken, PA. http://www.astm.org.
47. ASTM D2270-10e1 (2010), Standard practice for calculating viscosity index from kinematic viscosity at 40 and 100°C. ASTM International, West Conshohocken, PA. http://www.astm.org.
48. ASTM D97-12 (2012), Standard test method for pour point of petroleum products. ASTM International, West Conshohocken, PA. http:// www.astm.org.
49. ASTM D92-12 (2012), Standard test method for flash and fire points by Cleveland open cup tester. ASTM International, West Conshohocken, PA. http://www.astm.org.
50. ASTM D2710-09 (2013), Standard test method for bromine index of petroleum hydrocarbons by electrometric titration. ASTM International, West Conshohocken, PA. http://www.astm.org.
51. ASTM D6304-07 (2007), Standard test method for determination of water in petroleum products, lubricating oils, and additives by coulometric Karl Fischer titration. ASTM International, West Conshohocken, PA. http://www.astm.org.
52. ASTM D5293-15 (2015), Standard test method for apparent viscosity of engine oils and base stocks between −10°C and −35°C using cold-cranking simulator. ASTM International, West Conshohocken, PA. http://www.astm.org.
53. ASTM D5800-15 (2015), Standard test method for evaporation loss of lubricating oils by the Noack method. ASTM International, West Conshohocken, PA. http://www.astm.org.
54. Brown, J. A. (2012), Launching a renewable synthetic base oil. Paper presented in ICIS London 2012
55. Nordic Swan website. http://www.nordic-ecolabel.org/ (Accessed May 28, 2015).
56. Blue Angel website. http://www.ecolabelindex.com/ecolabel/blue-angel (Accessed May 28, 2015).
57. EU Ecolabel website. http://ec.europa.eu/environment/ecolabel/ (Accessed May 28, 2015).
58. Novvi website. http://www.novvi.com (Accessed May 15, 2015).
59. http://en.wikipedia.org/wiki/Squalane (Accessed May 15, 2015).
60. OECD 420. http://www.oecd.org/chemicalsafety/risk-assessment/1948362.pdf (Accessed May 29, 2015).
61. OECD 431. http://www.oecd.org/chemicalsafety/testing/43302385.pdf (Accessed May 29, 2015).
62. OECD 439. http://www.oecd.org/chemicalsafety/testing/50305099.pdf.
63. OECD 405. http://www.oecd.org/chemicalsafety/testing/41896718.pdf.
64. OECD 429. http://www.oecd.org/chemicalsafety/testing/43302600.pdf.
65. OECD 473. http://www.oecd.org/chemicalsafety/risk-assessment/1948434.pdf.
66. OECD 471. http://www.oecd.org/chemicalsafety/risk-assessment/1948418.pdf.
67. OECD 474. http://www.oecd.org/chemicalsafety/risk-assessment/1948442.pdf.
68. OECD 422. http://www.oecd.org/chemicalsafety/testing/Draft-updated-TG-422-August-2014.pdf.
69. OECD 209. http://www.oecd.org/chemicalsafety/testing/43735667.pdf.
70. NSF International website. http://www.nsf.org/ (Accessed on May 28, 2015).
71. NSF International website. http://info.nsf.org/USDA/psnclistings.asp (Accessed May 28, 2015).
72. http://ec.europa.eu/environment/ecolabel/documents/lusclist.pdf (Accessed May 28, 2015).
73. EPA website. http://water.epa.gov/polwaste/npdes/vessels/upload/vgp_permit2013.pdf (Accessed May 28, 2015).
74. Eaton–Vickers Lubricant Specification E-FDGN-TB002-E. Eaton Corp.
75. ASTM D5185-13e1 (2013), Standard test method for multielement determination of used and unused lubricating oils and base oils by inductively coupled plasma atomic emission spectrometry (ICP-AES). ASTM International, West Conshohocken, PA. http://www.astm.org.
76. ASTM D2983-09 (2009), Standard test method for low-temperature viscosity of lubricants measured by Brookfield viscometer. ASTM International, West Conshohocken, PA. http://www.astm.org.
77. ASTM D611-12 (2012), Standard test methods for aniline point and mixed aniline point of petroleum products and hydrocarbon solvents. ASTM International, West Conshohocken, PA. http://www.astm.org.
78. ASTM D664-11 (2011), Standard test method for acid number of petroleum products by potentiometric titration. ASTM International, West Conshohocken, PA. http://www.astm.org.
79. ASTM D665-14 (2014), Standard test method for rust-preventing characteristics of inhibited mineral oil in the presence of water. ASTM International, West Conshohocken, PA. http://www.astm.org.

80. ASTM D892-13 (2013), Standard test method for foaming characteristics of lubricating oils. ASTM International, West Conshohocken, PA. http://www.astm.org.
81. ASTM D1401-12 (2012), Standard test method for water separability of petroleum oils and synthetic fluids. ASTM International, West Conshohocken, PA. http://www.astm.org.
82. ASTM D130-12 (2012), Standard test method for corrosiveness to copper from petroleum products by copper strip test. ASTM International, West Conshohocken, PA. http://www.astm.org.
83. ASTM D3427-14 (2014), Standard test method for air release properties of petroleum oils. ASTM International, West Conshohocken, PA. http://www.astm.org.
84. ISO 13357. http://www.iso.org/iso/iso_catalogue/catalogue_tc/catalogue_detail.htm?csnumber=42258.
85. Parker Denison website. http://fluidsinfo.pagesperso-orange.fr/Sommaire.htm (Accessed on May 28, 2015).
86. ASTM D2619-09 (2014), Standard test method for hydrolytic stability of hydraulic fluids (beverage bottle method). ASTM International, West Conshohocken, PA. http://www.astm.org.
87. ASTM D943-04a(2010)e1 (2010), Standard test method for oxidation characteristics of inhibited mineral oils. ASTM International, West Conshohocken, PA. http://www.astm.org.
88. ASTM D4310-10 (2010), Standard test method for determination of sludging and corrosion tendencies of inhibited mineral oils. ASTM International, West Conshohocken, PA. http://www.astm.org.
89. ASTM D2070-91 (2010), Standard test method for thermal stability of hydraulic oils. ASTM International, West Conshohocken, PA. http://www.astm.org.
90. ISO 6072. http://www.iso.org/iso/iso_catalogue/catalogue_tc/catalogue_detail.htm?csnumber=53353.
91. ASTM D5182-97 (2014), Standard test method for evaluating the scuffing load capacity of oils (FZG visual method). ASTM International, West Conshohocken, PA. http://www.astm.org.
92. ASTM D7043-12 (2012), Standard test method for indicating wear characteristics of non-petroleum and petroleum hydraulic fluids in a constant volume vane pump. ASTM International, West Conshohocken, PA. http://www.astm.org.
93. ASTM D6973-14 (2014), Standard test method for indicating wear characteristics of petroleum hydraulic fluids in a high pressure constant volume vane pump. ASTM International, West Conshohocken, PA. http://www.astm.org.
94. CEC L-45-A99. http://www.cectests.org/listdoctypeforsale1.asp?subdoc_type=Lubricants.
95. Bosch Rexroth Fluid Rating (RDE 90235). http://www.boschrexroth.com/fluidrating.
96. ISO website. http://www.iso.org/iso/catalogue_detail.htm?csnumber=54991.
97. ASTM D3487-09 (2009), Standard specification for mineral insulating oil used in electrical apparatus. ASTM International, West Conshohocken, PA. http://www.astm.org.
98. International Electrotechnical Commission website, IEC 60296:2012. Fluids for electrotechnical applications—Unused mineral insulating oils for transformers and switchgear. http://www.iec.ch, 2012 (Accessed April 29, 2015).
99. ASTM D2272-14a, Standard Method for Oxidation Stability of Steam Turbine Oils By Rotating Pressure Vessel.
100. Hahn, H., Vettel, P., Brown, J., and Wells, J. (2014), High performance air compressor oil formulated with biobased hydrocarbon base oil. Session 2J, STLE Annual Meeting, Orlando, FL.
101. ASTM D924-08 (2008), Standard test method for dissipation factor (or power factor) and relative permittivity (dielectric constant) of electrical insulating liquids. ASTM International, West Conshohocken, PA. http://www.astm.org.
102. ASTM D971-12 (2012), Standard test method for interfacial tension of oil against water by the ring method. ASTM International, West Conshohocken, PA. http://www.astm.org.
103. ASTM D877/D877M-13 (2013), Standard test method for dielectric breakdown voltage of insulating liquids using disk electrodes. ASTM International, West Conshohocken, PA. http://www.astm.org.
104. ASTM D1816-12 (2012), Standard test method for dielectric breakdown voltage of insulating liquids using VDE electrodes. ASTM International, West Conshohocken, PA. http://www.astm.org.
105. ASTM D3300-12 (2012), Standard test method for dielectric breakdown voltage of insulating oils of petroleum origin under impulse conditions. ASTM International, West Conshohocken, PA. http://www.astm.org.
106. ASTM D2300-08 (2008), Standard test method for gassing of electrical insulating liquids under electrical stress and ionization (modified Pirelli method). ASTM International, West Conshohocken, PA. http://www.astm.org.
107. ASTM D924-08 (2008), Standard test method for dissipation factor (or power factor) and relative permittivity (dielectric constant) of electrical insulating liquids. ASTM International, West Conshohocken, PA. http://www.astm.org.

108. ASTM D2440-13 (2013), Standard test method for oxidation stability of mineral insulating oil. ASTM International, West Conshohocken, PA. http://www.astm.org.

109. ASTM D2112-01a (2007), Standard test method for oxidation stability of inhibited mineral insulating oil by pressure vessel. ASTM International, West Conshohocken, PA. http://www.astm.org.

110. ASTM D2668-07 (2013), Standard test method for 2,6-*di-tert*-butyl-*p*-cresol and 2,6-*di-tert*-butyl phenol in electrical insulating oil by infrared absorption, ASTM International, West Conshohocken, PA. http://www.astm.org.

111. ASTM D1533-12 (2012), Standard test method for water in insulating liquids by coulometric Karl Fischer titration. ASTM International, West Conshohocken, PA. http://www.astm.org.

112. ASTM D974-14 (2014), Standard test method for acid and base number by color-indicator titration. ASTM International, West Conshohocken, PA. http://www.astm.org.

113. IEC 61868. https://webstore.iec.ch/publication/6046 (Accessed June 5, 2015).

114. IEC 60156. https://webstore.iec.ch/publication/913 (Accessed June 5, 2015).

115. IEC 60247. https://webstore.iec.ch/publication/1150 (Accessed June 5, 2015).

116. IEC 61125. https://webstore.iec.ch/publication/4545 (Accessed June 5, 2015).

117. BS 2000. http://standards.globalspec.com/std/796307/bsi-bs%202000-346 (Accessed June 5, 2015).

118. IEC 61619. https://webstore.iec.ch/publication/5670 (Accessed June 5, 2015).

119. ASTM D2300-08 (2008), Standard test method for gassing of electrical insulating liquids under electrical stress and ionization (modified Pirelli method). ASTM International, West Conshohocken, PA. http://www.astm.org.

120. http://shop.csa.ca/en/canada/electrical-engineering-standards/cancsa-c50-14/invt/27003312014 (Accessed June 2, 2015).

121. ASTM D6871-03 (2008), Standard specification for natural (vegetable oil) ester fluids used in electrical apparatus. ASTM International, West Conshohocken, PA. http://www.astm.org.

122. IEEE website. http://ieeexplore.ieee.org/xpl/articleDetails.jsp?reload=true&tp=&isnumber=4566079&arnumber=4566080&punumber=4566076 (Accessed April 30, 2015).

123. Lilje, K. C. (2006), *Compressors and Pumps in Synthetics, Mineral Oils, and Bio-Based Lubricants, Chemistry and Technology*. Taylor & Francis, New York, NY; pp. 475–492.

124. Mortier, R. M., Fox, M. F., and Orszulik, S. T. (2010), *Chemistry and Technology of Lubricants*. (Third edition). Springer, New York, NY; pp. 261–264.

125. ASTM D4172-94 (2010), Standard test method for wear preventive characteristics of lubricating fluid (four-ball method). ASTM International, West Conshohocken, PA. http://www.astm.org.

126. Galary, J. (2014, October), Study of wear properties of environmentally friendly lubricants for gearing applications as a function of film thickness transition. ASTM STP 1575-EB.

127. ASTM D217-10 (2010), Standard test methods for cone penetration of lubricating grease. ASTM International, West Conshohocken, PA. http://www.astm.org.

128. ASTM D2265-15 (2015), Standard test method for dropping point of lubricating grease over wide temperature range. ASTM International, West Conshohocken, PA. http://www.astm.org.

129. ASTM D6186-08 (2013), Standard test method for oxidation induction time of lubricating oils by pressure differential scanning calorimetry (PDSC). ASTM International, West Conshohocken, PA. http://www.astm.org.

130. ASTM D6184-14 (2014), Standard test method for oil separation from lubricating grease (conical sieve method). ASTM International, West Conshohocken, PA. http://www.astm.org.

131. ASTM D2266-01 (2008), Standard test method for wear preventive characteristics of lubricating grease (four-ball method). ASTM International, West Conshohocken, PA. http://www.astm.org.

132. ASTM D2596-14 (2014), Standard test method for measurement of extreme-pressure properties of lubricating grease (four-ball method). ASTM International, West Conshohocken, PA. http://www.astm.org.

133. ASTM D1264-12 (2012), Standard test method for determining the water washout characteristics of lubricating greases. ASTM International, West Conshohocken, PA. http://www.astm.org.

134. ASTM D1743-13 (2013), Standard test method for determining corrosion preventive properties of lubricating greases. ASTM International, West Conshohocken, PA. http://www.astm.org.

135. ASTM D4289-13(2014)e1 (2014), Standard test method for elastomer compatibility of lubricating greases and fluids. ASTM International, West Conshohocken, PA. http://www.astm.org.

136. Lubricating Specialties Corp. http://www.lsc-online.com/red-i-biodegradable-grease/ (Accessed May 13, 2015).

137. Ishchuk, Y. L. (2008, December 1), *Lubricating Grease Manufacturing Technology* New Age International Pvt Ltd Publishers, New Delhi, India.

138. Canter, N. (2013, September), Fuel economy—The role of friction modifiers and VI improvers in tribology & lubrication technology. 69 (9), 14–27.
139. http://www.infineum.com.
140. ASTM D4683-13 (2013), Standard test method for measuring viscosity of new and used engine oils at high shear rate and high temperature by tapered bearing simulator viscometer at 150°C. ASTM International, West Conshohocken, PA. http://www.astm.org.
141. ASTM D6557-13 (2013), Standard test method for evaluation of rust preventive characteristics of automotive engine oils. ASTM International, West Conshohocken, PA. http://www.astm.org.
142. ASTM D6417-09 (2009), Standard test method for estimation of engine oil volatility by capillary gas chromatography. ASTM International, West Conshohocken, PA. http://www.astm.org.
143. ASTM D6795-13 (2013), Standard test method for measuring the effect on filterability of engine oils after treatment with water and dry ice and a short (30 min) heating time. ASTM International, West Conshohocken, PA. http://www.astm.org.
144. ASTM D6794-14 (2014), Standard test method for measuring the effect on filterability of engine oils after treatment with various amounts of water and a long (6 h) heating time. ASTM International, West Conshohocken, PA. http://www.astm.org.
145. ASTM D4951-14 (2014), Standard test method for determination of additive elements in lubricating oils by inductively coupled plasma atomic emission spectrometry. ASTM International, West Conshohocken, PA. http://www.astm.org.
146. ASTM D6922-13 (2013), Standard test method for determination of homogeneity and miscibility in automotive engine oils. ASTM International, West Conshohocken, PA. http://www.astm.org.
147. ASTM D6278-12e1 (2012), Standard test method for shear stability of polymer containing fluids using a European diesel injector apparatus. ASTM International, West Conshohocken, PA. http://www.astm.org.
148. ASTM D7097-09 (2009), Standard test method for determination of moderately high temperature piston deposits by thermo-oxidation engine oil simulation test—TEOST MHT. ASTM International, West Conshohocken, PA. http://www.astm.org.
149. ASTM D5133-13 (2013), Standard test method for low temperature, low shear rate, viscosity/temperature dependence of lubricating oils using a temperature-scanning technique. ASTM International, West Conshohocken, PA. http://www.astm.org.
150. ASTM D6082-12 (2012), Standard test method for high temperature foaming characteristics of lubricating oils. ASTM International, West Conshohocken, PA. http://www.astm.org.
151. ASTM D7563-10 (2010), Standard test method for evaluation of the ability of engine oil to emulsify water and simulated Ed85 fuel. ASTM International, West Conshohocken, PA. http://www.astm.org.
152. ASTM D7528-13 (2013), Standard test method for bench oxidation of engine oils by ROBO apparatus. ASTM International, West Conshohocken, PA. http://www.astm.org.
153. SAE J2643. http://standards.sae.org/j2643_201312/.
154. 2012 Global Lubricant Trends—Kline & Company. http://www.klinegroup.com/reports/y533.asp.
155. ASTM D4684-14 (2014), Standard test method for determination of yield stress and apparent viscosity of engine oils at low temperature. ASTM International, West Conshohocken, PA. http://www.astm.org.
156. ASTM D7320-15 (2015), Standard test method for evaluation of automotive engine oils in the sequence IIIG, spark-ignition engine. ASTM International, West Conshohocken, PA. http://www.astm.org.
157. ASTM D6891-14 (2014), Standard test method for evaluation of automotive engine oils in the sequence IVA spark-ignition engine. ASTM International, West Conshohocken, PA. http://www.astm.org.
158. ASTM D6593-15 (2015), Standard test method for evaluation of automotive engine oils for inhibition of deposit formation in a spark-ignition internal combustion engine fueled with gasoline and operated under low-temperature, light-duty conditions. ASTM International, West Conshohocken, PA. http://www.astm.org.
159. http://en.wikipedia.org/wiki/Two-stroke_engine (Accessed May 1, 2015).
160. ASTM D874-13a (2013), Standard test method for sulfated ash from lubricating oils and additives. ASTM International, West Conshohocken, PA. http://www.astm.org.
161. JASO M345. http://www.oilspecifications.org/jaso.php (Accessed June 5. 2015).
162. API TC. http://www.oilspecifications.org/api_2t.php (Accessed June 5, 2015).
163. J1536. http://standards.sae.org/j1536_200612/ (Accessed June 5, 2015).

2 Estolides
Bioderived Synthetic Base Oils

*Jakob W. Bredsguard, Travis D. Thompson,
Steven C. Cermak, and Terry A. Isbell*

CONTENTS

ABSTRACT

One novel technology to reach the lubricant market in recent years is the estolide, a class of high-performance, environmentally acceptable lubricant base oils. Estolides have been tested against a set of similar competing base oils from the marketplace, and the results show that they have excellent performance in the areas of oxidative stability, hydrolytic stability, evaporative loss (volatility), viscosity index (VI), and wear protection, in addition to environmental benefits including high renewable content, biodegradability, and nonbioaccumulative nature. These benefits, among others, have lead formulators to begin using estolides in a variety of industrial and automotive lubricant applications. For example, a high-performance estolide-based motor oil formulation has been certified by the API as having met the most current performance specifications for motor oils, API SN-RC (International Lubricants Standardization and Approval Committee [ILSAC] GF-5). In addition, a field trial using estolide-based formulations was conducted in Las Vegas, Nevada, where estolides demonstrated their ability to keep engines looking clean with minimal varnish. Furthermore, an estolide-based motor oil underwent environmental testing to determine the effect on biodegradability, if any, of (1) blending the estolide-based oil with additives and

(2) using the formulation in an engine. The results show that blending the base oil with additives did not have an effect on the biodegradability of the estolide, and using the formulation in an engine appeared to slightly improve the biodegradability of the estolide.

2.1 INTRODUCTION

An exciting technology to reach the lubricant market in recent years are estolides, a class of high-performance, environmentally acceptable lubricant (EAL) base oils. Estolides are currently being used in a variety of industrial and automotive lubricant applications and have garnered recent attention for their use in high-performance motor oil formulations. In addition, they are renewably sourced, biodegradable, and non-bioaccumulative, making them also suitable for environmentally sensitive applications.

Estolides are oligomers of fatty acid monomers derived from the splitting of triglycerides, or vegetable oils. Fatty acid monomers can be synthesized into estolide oligomers using one of two pathways. The first option is to use monounsaturated fatty acids as feed (e.g., oleic acid), whereby a mineral acid is used to catalyze the electrophilic addition of the carboxyl of one fatty acid to the alkene on another (see Figure 2.1) [1]. The second is to use hydroxyl fatty acids as feed (e.g., ricinoleic acid or 12-hydroxystearic acid), whereby a mineral acid is used to catalyze a series of condensation reactions (Fischer esterifications) to generate estolide oligomers [2]. The product from either method will have a free carboxylic acid, which can be esterified with an alcohol to further optimize the compound for use as a lubricant [3].

From the perspective of performance, estolides offer a number of advantages. When compared to a set of similar competing base oils from the industry, estolides showed strengths in the areas of oxidative stability, hydrolytic stability, evaporative loss (volatility), viscosity index (VI), and wear protection, to name a few.

In addition, estolides are a tremendously versatile class of synthetic compounds and can be designed to fit just about any application. Oligomerization can be minimized or maximized, molecules can be functionalized with various moieties, and properties such as oxidative stability, hydrolytic stability, and cold-temperature flow can be tuned to meet specific needs.

Aside from performance, estolides also have superior environmental profiles, making them ideal for environmentally sensitive applications. Increased environmental regulations and legislation in the industry have formulators seeking ingredients which do not fit the traditional "performance versus environment" trade-off. With characteristics of being a high-performance synthetic, along with high biocontent, good biodegradability, and non-bioaccumulative nature, estolides represent one such option for formulators.

The product described throughout this chapter, Biosynthetic SE7B, is a high-performance estolide base oil which is making rapid advances in the industry. It is being tested by numerous companies to formulate the next generation of synthetic lubricant products. The product is currently being used in the development of various formulations, including engine oils, hydraulic fluids, gear oils, greases, metalworking fluids, compressor fluids, and dielectric fluids. Recently, estolides have attracted much attention for their ability to keep engines clean when used in motor oil formulations. These properties, among others, lead to the first estolide motor oil formulations (5W-20 and 5W-30) certified by the American Petroleum Institute (API) which meet the industry's current motor oil standard, API SN Resource Conserving (RC) (International Lubricants Standardization and Approval Committee [ILSAC] GF-5) [4]. In addition, biodegradability tests on an estolide motor oil formulation showed that the estolide base oil in the formulation maintained its biodegradability when blended with additives and tested in an engine for thousands of miles.

The performance properties, the environmental properties, and the chemical versatility of estolides make them attractive not only to developers of environmentally friendly products but also to the oil industry at large. Increasingly stringent performance specifications, along with tighter environmental legislation and regulation in the sector, are all reasons why estolides are poised to take a significant share of the lubricant base oil market in the coming years.

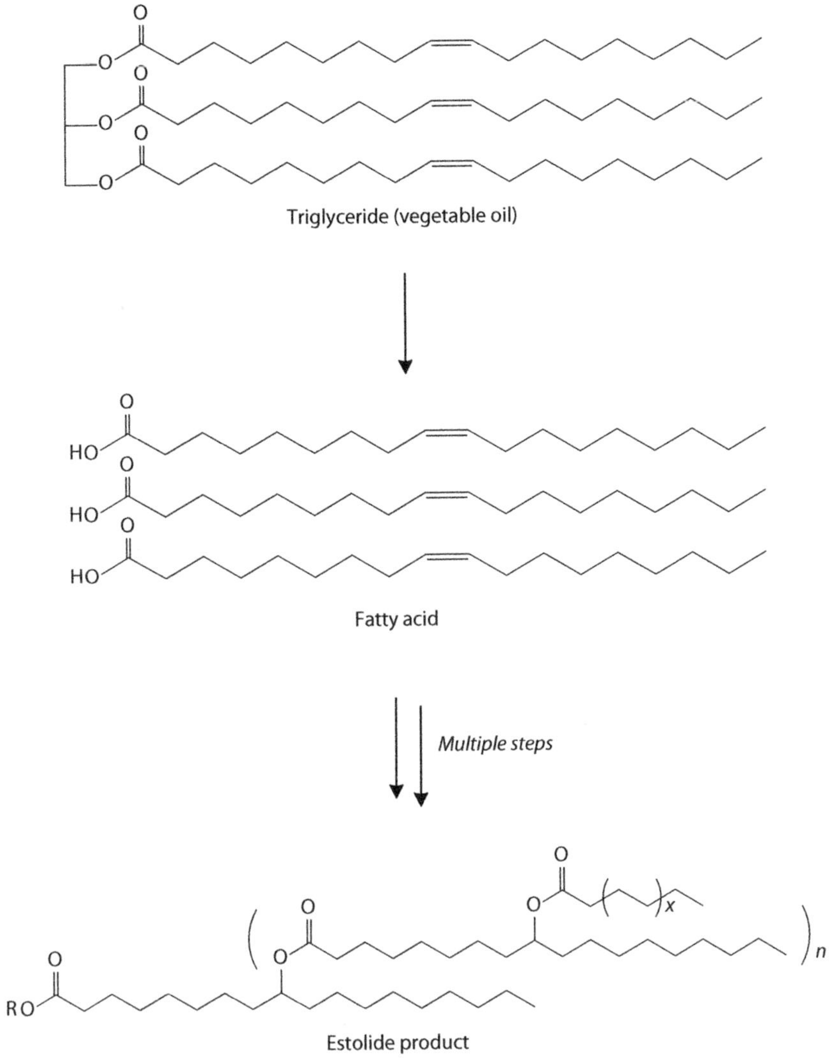

FIGURE 2.1 Reaction schematic of vegetable oil to estolide.

2.2 PERFORMANCE CHARACTERISTICS

Estolides naturally exhibit a number of performance properties that make them ideal as lubricant base oils. This allows them to compete not only with conventional petroleum products, but also with high-end synthetic lubricants. In each of the following subsections, we discuss results of estolides being compared against a set of common lubricant base oils from industry. Table 2.1 provides brief descriptions of the different products used in the comparison studies, and Table 2.2 provides some basic properties of these base oils.

2.2.1 OXIDATIVE STABILITY

Vegetable oils have been used as lubricants for centuries, but they have always come with a marked deficiency in the area of oxidation resistance. The ability of a fluid to resist oxidation, also referred to as oxidative stability, is one of the primary indications used to predict the life span of a lubricant.

TABLE 2.1

Descriptions of the Base Oils Tested and Evaluated in This Work

Base Oil	Description
Group II mineral oil	API Group II—refined, hydrotreated crude
Group III mineral oil	API Group III—refined, hydroisomerized crude
Polyalphaolefin (PAO)	Highly branched isoparaffinic PAO
Polyalkylene glycol (PAG)	Oil-soluble PAG
Diester	Adipate diester containing long-chain branched alcohols
Polyol ester	Dipentaerythritol ester
Biosynthetic SE7B (estolide)	Estolide product

TABLE 2.2

Basic Properties of the Base Oils Tested and Evaluated in This Work

	Unit	Method	Group II Mineral Oil	Group III Mineral Oil	PAO	PAG	Diester	Polyol Ester	Biosynthetic SE7B (Estolide)
Kinematic viscosity, 100°C	cSt	D445	6.6	6.5	7.0	6.5	5.5	8.6	7.2
Kinematic viscosity, 40°C	cSt	D445	44	37	38	32	28	53	35
VI	–	D2270	102	130	146	164	135	135	173
Pour point	°C	D97	−13	−15	−43	−57	−60	−51	−18
Flash point	°C	D92	230	256	264	216	243	282	280

On the molecular level, the instability of vegetable oil originates from the sites of unsaturation, or the olefin content of the oil. Vegetable oils high in unsaturates tend to have good cold-temperature properties (remaining liquid to temperatures < 0°C), but poor oxidative stability [5]. If the olefins are reduced through hydrogenation, many vegetable oils become solid at room temperature, thus rendering them ineffective as a lubricant. For example, while the melting point of soybean oil is −7°C, the melting point of fully hydrogenated soybean oil is 71°C [6]. This trade-off between oxidative stability and cold-temperature performance is one of the main reasons that the use of vegetable oils in lubricants has remained limited to a niche set of applications.

Because estolides have a high level of saturation, the oxidative stability of these fluids is similar to that of other high-end synthetics (see American Society for Testing and Materials [ASTM] D2272 result for the estolide product in Figure 2.2) [7–9]. In addition, because the molecular structure is branched at each of the estolide positions, the oligomers have difficulty crystallizing as temperatures are reduced, resulting in good cold-temperature flow despite low levels of unsaturation.

2.2.2 HYDROLYTIC STABILITY

In the presence of water and a small of amount of catalyst, esters can degrade to form acidic by-products that can cause corrosion to various metals used in bearings, engines, and other equipment. With respect to estolide esters, however, large hydrophobic branches on both sides of each estolide link provide a steric barrier that protects the esters from hydrolytic attack.

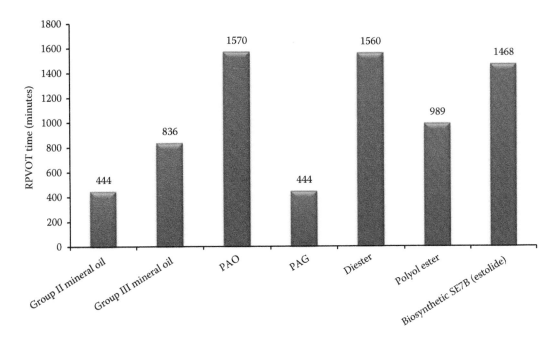

FIGURE 2.2 Oxidative stability of estolides (Biosynthetic SE7B, 7 cSt at 100°C) compared to base oils with similar viscosity (6–8 cSt at 100°C), according to ASTM D2272 rotating pressure vessel oxidation test (minutes). Base oil samples contain 1 wt% aminic/phenolic antioxidant blend (1:1 weight ratio).

To demonstrate this trait, a modified ASTM D2619 hydrolytic stability test (Figure 2.3) was performed, where the test duration was extended from 48 to 144 hours to exaggerate test results—all other test parameters followed method guidelines. As shown in Figure 2.3, these hydrophobic barriers provide estolides with hydrolytic stability better than that of the synthetic esters tested, making their performance more comparable to those of refined mineral oils and synthetic hydrocarbons.

2.2.3 EVAPORATIVE LOSS (VOLATILITY)

Another performance characteristic of estolides is low volatility, resulting in reduced evaporative loss as compared to other high-performance base oils (per the Noack standardized test method, ASTM D5800). Results of this test are often referred to as the Noack of an oil or lubricant formulation. Noack is the weight percent of a fluid evaporated after exposure to 250°C for 60 minutes, and is a critical parameter for lubricants as it indicates the level of evaporative loss a lubricant might experience while in high-temperature service. This characteristic is of particular importance in motor oils, where the heat of an engine can vaporize the lower molecular weight components of the fluid, thereby changing its composition. As the molecular weight distribution of the lubricant changes, the viscosity of the fluid can increase, resulting in poor engine oil circulation, and therefore reduced fuel economy. In addition, another effect of lubricant loss to the atmosphere is that the fluid must then be replaced, or "topped off," between oil changes. In this way, higher evaporative loss can result in increased oil consumption. As shown in Figure 2.4, the estolide product measured has a Noack value lower than those of competing base stocks with similar viscosities.

2.2.4 VISCOSITY INDEX

VI is defined as the ability of a fluid to resist drastic change in viscosity as the temperature of the fluid is either increased or decreased. Because estolides have VIs higher than those of most

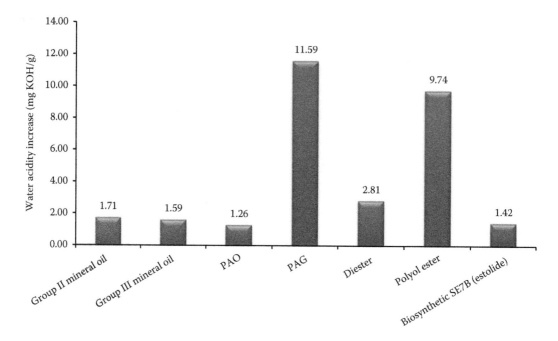

FIGURE 2.3 Hydrolytic stability of estolides (Biosynthetic SE7B, 7 cSt at 100°C) compared to base oils with similar viscosity (6–8 cSt at 100°C) via modified ASTM D2619, measuring water acidity increase after 144 hours (milligrams KOH/gram). Base oil samples were tested without additives.

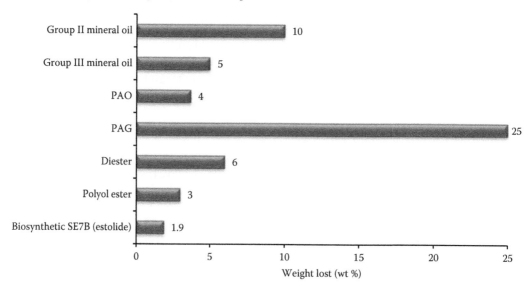

FIGURE 2.4 Evaporative loss (Noack) of estolides (Biosynthetic SE7B, 7 cSt at 100°C) compared to base oils with similar viscosity (6–8 cSt at 100°C), according to ASTM D5800 (weight percent lost during testing). Base oil samples were tested without additives.

products (see Figure 2.5), they have a number of advantages over other base oils. Higher-VI fluids provide increased film thickness at elevated temperatures, resulting in better protection, and in many cases, reduced wear. At lower temperatures, high-VI base fluids display a lower rate of viscosity increase, resulting in reduced viscous drag on moving parts, leading to higher horsepower output and increased energy efficiency [10]. In addition, formulations containing high-VI fluids require less

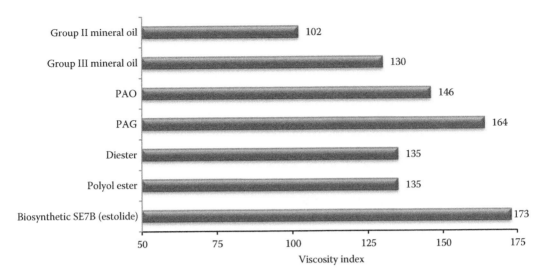

FIGURE 2.5 VIs of estolides (Biosynthetic SE7B, 7 cSt at 100°C) compared to base oils with similar viscosity (6–8 cSt at 100°C), according to ASTM D2270. Base oil samples were tested without additives.

VI improver additives to meet minimum VI requirements—thus, the higher the VI of the lubricant base stock, the less such additives are required.

2.2.5 WEAR PROTECTION

Estolides also exhibit unique properties in the area of wear protection. Because the compounds are polar, they have an increased affinity for metal surfaces, allowing them to form protective barriers between moving parts. This attraction to the metal surface fortifies the surface against wear, as shown in Figure 2.6.

2.3 ENVIRONMENTAL PROPERTIES

Governments from around the globe are paying more and more attention to the environmental impact of lubricants, leading to increased regulation and legislation in the sector. Examples of such programs are the U.S. Environmental Protection Agency's (EPA) Vessel General Permit (VGP), the BioPreferred program, and Europe's Ecolabel, to name a few. These requirements are forcing many lubricant manufacturers to begin investigating alternative ingredients for their formulations. With an environmental profile favorable to many traditional base oils, estolides represent one such option for formulators developing EALs.

2.3.1 RENEWABLE CONTENT

Estolides are one of the few lubricant base oils with high levels of biocontent. Refined mineral oil products, including Groups I–III base stocks, are sourced directly from petroleum, and thus do not contain renewable carbon. In addition, current feed streams for other synthetic hydrocarbons, including PAOs and PAGs, are also petroleum-sourced and do not have renewable content. There are some new technologies being developed, however, which can produce biobased raw materials for products similar to Group III (farnesene-based) and PAO (decene derived from vegetable oil metathesis). Other widely used products with high biocontent include esters that are based on fatty acid or fatty alcohol chemistries, including polyol esters and diesters.

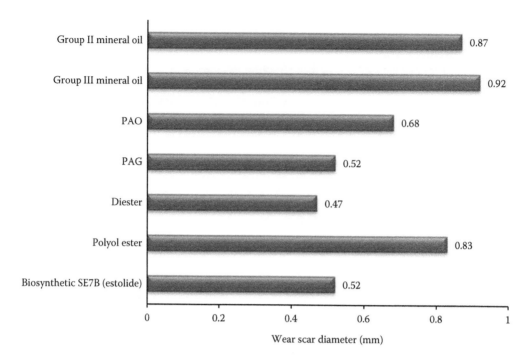

FIGURE 2.6 Four-ball wear properties of estolides (Biosynthetic SE7B, 7 cSt at 100°C) compared to base oils with similar viscosity (6–8 cSt at 100°C), according to ASTM D4172 (wear scar diameter, millimeters). Test parameters were 1 hour, 75°C, 1200 rotations/minute, and 40 kgf. Base oil samples were tested without additives.

Because estolides are oligomers of fatty acid monomers derived from plant, animal, or algal oils, they can have very high levels of biocontent. In fact, estolides can be entirely renewably sourced if the alcohol chosen to terminate oligomerization is also derived from biobased carbon (e.g., biobased ethanol, *n*-butanol, and fatty alcohols).

2.3.2 Biodegradability

Another unique characteristic of estolides is their ability to rapidly biodegrade once released into the environment. As shown in Figure 2.7, Organization for Economic Cooperation and Development (OECD) 301 evaluation shows that neat estolide base oils have a higher potential for biodegradation than many base oils with similar viscosities.

In a follow-up study on the biodegradability of estolide-containing motor oil formulations, it was determined that the individual components of a formulation biodegrade independently from one another. Findings from this study indicated that the estolide fraction of the formulation maintained a proportional amount of biodegradability, even after being used in an engine. See Section 2.4.2 for more detailed information.

2.3.3 Bioaccumulation

Bioaccumulation refers to the accumulation of a contaminant in the tissues of a living organism via any route, including respiration, ingestion, direct contact, sediment, or other means [11]. Thus, a bioaccumulative compound has a rate of tissue uptake which exceeds the rate of elimination. Bioaccumulation is a critical environmental property because compounds with very low environmental concentrations (e.g., water or soil concentration) can build up in the tissues of living organisms, eventually reaching harmful or lethal levels.

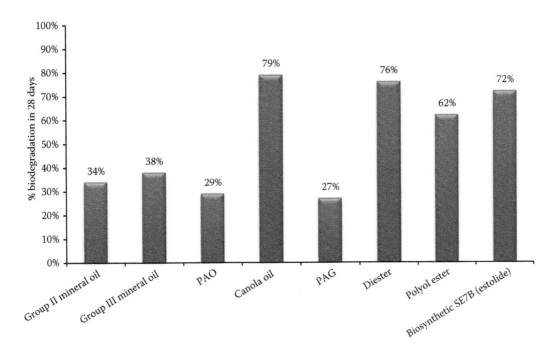

FIGURE 2.7 Biodegradability of estolides (Biosynthetic SE7B, 7 cSt at 100°C) compared of base oils with similar viscosity (6 to 8 cSt at 100°C), OECD 301B (% biodegradation in 28 days). Base oil samples were tested without additives.

The tendency of a substance to bioaccumulate depends on the lipophilicity, the size or molecular weight, and the chemical stability of the compound. First, lipophilic (nonpolar) substances can be more readily absorbed into fatty tissues, potentially promoting the bioaccumulative process. However, highly lipophilic substances can also readily diffuse out of an organism's tissues. Therefore, for a compound to have bioaccumulative potential, its lipophilicity must be sufficient to penetrate fatty tissue, but not so lipophilic that it readily diffuses out of the tissue once absorbed. Second, regarding molecular weight, small molecules can readily enter and exit an organism's tissue, whereas large molecules cannot enter the tissue at all. Thus, similar to lipophilic compounds, midrange molecular weight compounds tend to favor bioaccumulation. Third, a bioaccumulative compound must also be very stable in biological systems. In other words, once it enters an organism, it must be resistant to the biological processes by which the organism would naturally remove contaminants [12].

The primary test methods used to determine bioaccumulative potential are OECD 107 (shake flask method) and OECD 117 (high-performance liquid chromatography). Both methods are ways to estimate the partition coefficient of a substance between water and *n*-octanol phases (P_{OW} = [solute]$_{octanol}$/ [solute]$_{water}$) as a means of determining the degree of lipophilicity of the compound [13,14]. A substance which favors the aqueous phase will yield a small P_{OW} and shows that the substance tested is more polar, or nonlipophilic. Alternatively, favoring the *n*-octanol phase will yield a large P_{OW} and shows that the substance tested is more nonpolar, or lipophilic.

According to the U.S. EPA, a substance is non-bioaccumulative if the log P_{OW} of the sample per OECD 117 is <3 or >7 (not lipophilic enough to bioaccumulate, or too lipophilic to bioaccumulate, respectively). In addition, in some regulations such as the VGP, there are exclusions for compounds above a certain molecular weight or molecular diameter [15]. A sample of low-viscosity estolide product (Biosynthetic SE7B) obtained a result of log P_{OW} > 7 as per these guidelines, indicating that the material is predicted to be too lipophilic to accumulate in the tissues of living organisms.

2.4 ESTOLIDE-BASED LUBRICANT FORMULATIONS

Estolides can be used in a wide range of lubricant formulations, including engine oils, hydraulic fluids, gear oils, greases, metalworking fluids, compressor fluids, and dielectric fluids [16–18]. In addition, because of their favorable environmental profile, they are of particular interest in ecologically sensitive lubricants for the marine, forestry, mining, and petroleum drilling fluid market segments. The focus of this section will be on the use of estolides in passenger car motor oil (PCMO) formulations.

2.4.1 ESTOLIDE-BASED MOTOR OILS

A variety of PCMO field trials have been conducted using lubricants with estolide base oils, in both hot and cold climates. Common to each of these tests has been the observation of enhanced engine cleanliness with estolide-based motor oil formulations compared to conventional petroleum-based motor oil formulations. Even after a field trial of over 100,000 miles, engines using estolide-based motor oils displayed high levels of cleanliness. In fact, in a formulation containing Group II base oil, replacing just 10% of the base stock with an estolide product showed significant improvements on the engine cleanliness measurements of a Sequence IIIG engine test. In Figures 2.8 through 2.10,

(a)

(b)

FIGURE 2.8 Cylinder heads from two Chevy Impala 3.5 L V6 engines used in an 18-month, 150,000 mile field trial in Las Vegas, Nevada. (a) The conventional motor oil formulation had a typical level of varnish at the end of the test, while (b) the estolide formulation showed a high degree of overall cleanliness and minimal varnish.

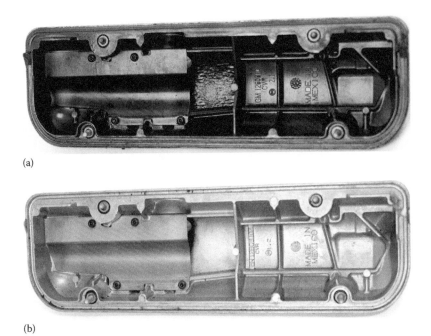

(a)

(b)

FIGURE 2.9   Valve covers from two Chevy Impala 3.5 L V6 engines used in an 18-month, 150,000 mile field trial in Las Vegas, Nevada. (a) The conventional motor oil formulation had a typical level of varnish at the end of the test, while (b) the estolide formulation showed a high degree of overall cleanliness and minimal varnish.

we compare sets of images from two engines (Chevy Impala, 3.5 L V6) after an 18-month and 150,000 mile field trial in Las Vegas, Nevada. The reference engine (Figures 2.8a, 2.9a, and 2.10a) was run using a standard quality GF-5 motor oil formulation, while the test engine (Figures 2.8b, 2.9b, and 2.10b) was run using an estolide formulation. As shown in the figures, the reference engine showed levels of varnish consistent with what is expected from a standard motor oil formulation. The test engine with the estolide formulation, however, showed outstanding overall cleanliness and minimal varnish.

Both 5W-20 and 5W-30 motor oil formulations containing estolide base oils have been certified by the API and met the most current specifications for motor oils, ILSAC GF-5, thus achieving the API SN-RC designation (see Figure 2.11).

2.4.2   Motor Oil Biodegradability

With the use of more biobased materials in motor oils, along with the success of technologies such as estolides, it is becoming important to understand if the environmental properties of the base oils are maintained after (1) being blended with additives and (2) being used in an engine.

To investigate this, motor oil samples were tested for biodegradability according to OECD 301B. The first formulation was 84% estolide with 16% additives, not yet used in an engine, and showed 68.1% biodegradation in 28 days. The same formulation was then used in an engine for 7500 miles, tested on OECD 301B, and showed 74.4% biodegradation in 28 days. This result indicates that the additives do not appear to affect the biodegradability of the estolide base oil, nor does its use in an engine appear to affect the overall biodegradability of the motor oil.

To better understand the theory and logic behind these findings, Dr. Todd Stevens, a renowned expert in environmental microbiology and biogeochemistry [19–21], was consulted. Dr. Stevens has worked in both routine biodegradation testing and research and development (R&D) of novel

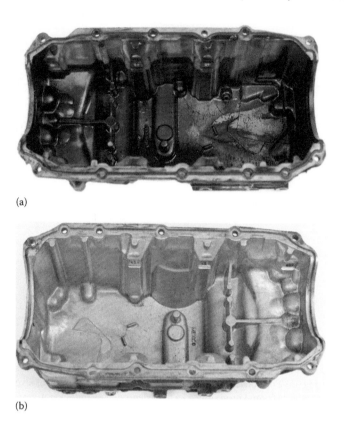

(a)

(b)

FIGURE 2.10 Oil pans from two Chevy Impala 3.5 L V6 engines used in an 18-month, 150,000 mile field trial in Las Vegas, Nevada. (a) The conventional motor oil formulation had a typical level of varnish at the end of the test, while (b) the estolide formulation showed a high degree of overall cleanliness and minimal varnish.

biodegradation and bioremediation processes [22], and is the proprietor of Stevens Ecology, a private laboratory for environmental analysis and R&D consulting, incorporated in Mosier, Oregon.

On the ability of additives to affect the biodegradability of a base oil, Dr. Stevens provided the following comments:

> Hypothetically, there are two ways that such additives could inhibit biodegradability of the base oil. First, they could be toxic or otherwise inhibitory to the microorganisms that carry out the biodegradation process. Second, they could be preferred substrates, such that biodegradation of the base oil is inhibited until biodegradation of the additives is completed. (This is known as "catabolite repression.") However, the mere fact that these additives may or may not be biodegradable has no bearing on whether or not the base oil will be biodegraded.

After testing the biodegradation of this formulation in his laboratory, Dr. Stevens stated, "Based on our experiments, we conclude that the presence of additives had no apparent effect on the biodegradability of the base oil."

Regarding the ability of a formulation to maintain its biodegradability after being used in an engine, Dr. Stevens commented:

> Certainly processes that make chemical alterations to a material can change the biodegradability of that material. The products of such reactions might have increased or decreased biodegradability, or might include some of both. Most processes that involve heat and oxidation tend to increase biodegradability. Reactions that involve polymerization or aromatization can decrease biodegradability.

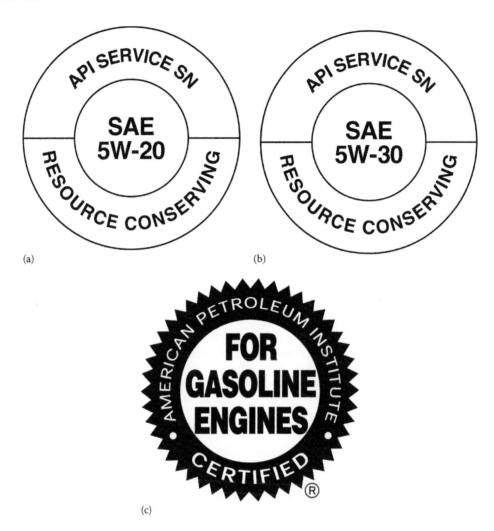

FIGURE 2.11 Two motor oil formulations containing estolide base oils, (a) 5W-20 and (b) 5W-30, have recently been certified by (c) the API as SN-RC quality products (ILSAC GF-5).

During use in an engine, motor oils are known to undergo two of the aforementioned processes: (1) heat/oxidation and (2) subsequent polymerizations of these heat/oxidation products. Therefore, because oxidation can lead to increased biodegradability, and polymerization can lead to decreased biodegradability, the true impact of these opposing processes on biodegradability is difficult to predict. However, based on the OECD 301B data generated by Dr. Stevens on fresh (68.1% in 28 days) and used estolide motor oils (74.4% in 28 days), he concluded, "The use of this oil in an engine did not inhibit, and may have slightly improved the biodegradability of the motor oil" (T. Stevens, personal communication, April 2, 2014).

2.5 CHEMICAL VERSATILITY OF ESTOLIDES

From the perspective of chemical versatility, large variations of estolide products can exist, which make the family of compounds useful for a broad range of applications. For example, estolide products can be customized to enhance properties such as oxidative or hydrolytic stability, or be manipulated to improve characteristics such as cold-temperature flow (with some products achieving pour points as low as −54°C) [2,23]. Estolide oligomerization can be controlled to make products with viscosities ranging from ISO VG 15 to ISO VG 2200 (International Organization for Standardization Viscosity Grades), which allows products to be custom-tailored to fit just about any application [24]. In addition, the product can be functionalized with various moieties to yield compounds with additive-like properties, such as improved antiwear or extreme pressure protection [25,26]. For these reasons, the estolide class of compounds has near limitless potential as a specialty chemical in the lubricants industry for years to come.

2.6 CONCLUSION

Many industry experts consider estolides to be the next generation of high-performance synthetic lubricants. Their potential for swift adoption in the marketplace has stemmed from a combination of strong performance and environmental characteristics, making estolides an essential tool for formulators of the future.

REFERENCES

1. T.A. Isbell, R. Kleiman, and B.A. Plattner, Acid-catalyzed condensation of oleic acid into estolides and polyestolides, *J. Am. Oil Chem. Soc.* 71, no. 2 (February 1994): 169–174.
2. S.C. Cermak, K.B. Brandon, and T.A. Isbell, Synthesis and physical properties of estolides from lesquerella and castor fatty acid esters, *Ind. Crop. Prod.* 23, no. 1 (January 2006): 54–64.
3. S.C. Cermak, J.W. Bredsguard, B.L. John, J.S. McCalvin, T. Thompson, K.N. Isbell, K.A. Feken, T.A. Isbell, and R.E. Murray, Synthesis and physical properties of new estolides esters, *Ind. Crop. Prod.* 46 (April 2013): 386–391.
4. K. Ferrick, Engine oil licensing and certification system, *American Petroleum Institute*. June 17, 2010, http://www.api.org/~/media/files/certification/engine-oil-diesel/forms/whats-new/1509-technical -bulletin-1.pdf?la=en.
5. R. Becker and A. Knorr, An evaluation of antioxidants for vegetable oils at elevated temperatures, *Lubr. Sci.* 8, no. 2 (January 1996): 95–117.
6. M. Kellens and G. Calliauw, Oil modification processes, in *Edible Oil Processing*, eds. W. Hamm, R.J. Hamilton, and G. Calliauw, 153–196 (Chichester: John Wiley & Sons, 2013).
7. S.C. Cermak, J.W. Bredsguard, R.O. Dunn, T. Thompson, K.A. Feken, K.L. Roth, J.A. Kenar, T.A. Isbell, and R.E. Murray, Comparative assay of antioxidant packages for dimer of estolide esters, *J. Am. Oil. Chem. Soc.* 91, no. 12 (December 2014): 2101–2109.
8. S.C. Cermak and T.A. Isbell, Improved oxidative stability of estolide esters, *Ind. Crop. Prod.* 18, no. 3 (November 2003): 223–230.
9. J. Bredsguard, J. Forest, and T. Thompson, Estolide compositions exhibiting high oxidative stability, U.S. Patent 8,372,301, filed May 30, 2012, and issued February 12, 2013.
10. W. Bock, Hydraulic oils, in *Lubricants and Lubrication*, ed. T. Mang and W. Dresel, 274–337 (Weinheim: Wiley-VCH, 2007).
11. U.S. Environmental Protection Agency and U.S. Army Corps of Engineers, *Evaluation of Dredged Material Proposed for Discharge in Waters of the U.S.—Testing Manual: Inland Testing Manual*, by T. Wright, EPA-823-B-98-004 (Washington, D.C.: US EPA, 1998).
12. J.N. Huckins, J.D. Petty, and J. Thomas, Bioaccumulation: How chemicals move from the water into fish and other aquatic organisms, *American Petroleum Institute, Health and Environmental Sciences Department*. 4656 (May 1997).
13. Organisation for Economic Cooperation and Development, Partition coefficient (*n*-octanol/water): Shake flask method, OECD Test No. 107.
14. Organisation for Economic Cooperation and Development, Partition coefficient (*n*-octanol/water): HPLC method, OECD Test No. 117.
15. U.S. Environmental Protection Agency, Vessel general permit for discharges incidental to the normal operation of vessels (VGP), http://water.epa.gov/polwaste/npdes/vessels/upload/vgp_permit2013.pdf (Accessed January 1, 2013).
16. J. Bredsguard, J. Forest, and T. Thompson, Refrigerating fluid compositions comprising estolide compounds, U.S. Patent 8,236,194, filed February 24, 2012, and issued August 7, 2012.
17. J. Forest, J. Bredsguard, and T. Thompson, Electrical devices and dielectric fluids containing estolide base oils, U.S. Patent 8,268,199, filed February 28, 2012, and issued September 18, 2012.

18. J. Bredsguard, J. Forest, and T. Thompson, Food-grade lubricant compositions comprising estolide compounds, U.S. Patent 8,399,389, filed August 16, 2012, and issued March 19, 2013.
19. T.O. Stevens, The deep subsurface biosphere, in *Biodiversity of Microbial Life*, eds. J.T. Staley and A.L. Reysenbach, 439–474 (New York: Wiley-Liss, 2002).
20. J.P. McKinley, T.O. Stevens, J.K. Fredrickson, J.M. Zachara, F.C. Colwell, K.B. Wagnon, S.C. Smith, S.A. Rawson, and B.N. Bjornstad, Biogeochemistry of anaerobic lacustrine and paleosol sediments within an aerobic unconfined aquifer, *Geomicrobiol. J.* 14, no. 1 (1997): 23–39.
21. T.O. Stevens, Optimization of media for enumeration and isolation of aerobic heterotrophic bacteria from the deep terrestrial subsurface, *J. Microbiol. Meth.* 21, no. 3 (March 1995): 293–303.
22. T.O. Stevens, D.L. Crawford, and R.L. Crawford, Biological system for degrading nitroaromatics in water and soils, U.S. Patent 5,387,271, filed July 23, 1993, and issued February 7, 1995.
23. S.C. Cermak and T.A. Isbell, Physical properties of saturated estolides and their 2-ethylhexyl esters, *Ind. Crop. Prod.* 16, no. 2 (September 2002): 119–127.
24. S.C. Cermak, J.W. Bredsguard, B.L. John, K. Kirk, T. Thompson, K.N. Isbell, K.A. Feken, T.A. Isbell, and R. Murray, Physical properties of low viscosity estolide 2-ethylhexyl esters, *J. Am. Oil Chem. Soc.* 90, no. 12 (December 2013): 1895–1902.
25. J. Forest and J. Bredsguard, Sulfurized estolides and methods of making and using the same, U.S. Patent 8,404,867, filed February 8, 2012, and issued March 26, 2013.
26. J. Forest and J. Bredsguard, Expoxidized estolides and methods of making the same, U.S. Patent 8,258,326, filed March 2, 2012, and issued September 4, 2012.

3 Effect of Structure on Viscosity and Pour Points of Epoxidized Soybean Oil Lubricants

Herman Benecke and Daniel Garbark

CONTENTS

ABSTRACT

The low application of triglyceride vegetable oils in lubrication is a reflection of their poor oxidative stability and high pour points. However, their overall biodegradability, lubricity, high VIs, and stable market pricing make them good candidates for environmentally friendly fluids. Some shortcomings of vegetable oils can be improved with the introduction of new varieties or purities of soybean oil such as high oleic and lower saturates that are now increasing in production scale. Another technology is the epoxidation of the triglyceride followed by ring opening of the epoxide with anhydrides to form diesters at the original olefinic site. Furthermore, making changes to the triglyceride backbone can produce improvements in such properties as viscosity and hydrolytic stability. Through such changes, soybean oil-based lubricants having a wide viscosity range of 2200–2700 cSt at 40°C and pour points as low as −39°C have been prepared.

3.1 INTRODUCTION

Lubricants derived from renewable feedstocks such as vegetable oils would be desirable to help reduce the dependence of the United States on foreign oil. Lubricating oils based on renewable sources such as soybean oil have a number of advantages. Soybean oil contains triglycerides with ester carbonyl groups. The polar nature of these ester carbonyl groups leads to strong adsorption on metal surfaces as a very thin film so that the film-forming properties of triglyceride-based lubricants under boundary conditions are particularly advantageous in hydraulic systems. Soybean oil typically has high viscosity indices (VIs) that facilitate its use over wide temperature ranges. Furthermore, triglyceride oils typically include high fume points (e.g., about 200°C) [1] and high flash points (e.g., about 300°C) [1].

The use of vegetable oil-based lubricants will help reduce the depletion of fossil-derived hydrocarbons. Vegetable oil- and fatty acid ester-based lubricants are typically 70% or greater biodegradable owing to their nonaromatic and noncyclic structures [2]. These structures and their biodegradability help mitigate the effect of introducing toxic and nonbiodegradable substances into the environment [2]. This is important since currently about 40% of lubricants used worldwide end up in landfills, rivers, lakes, etc. [3].

There are major performance problems in using vegetable oils as lubricants, including (1) low oxidative stability (requiring large amounts of antioxidants); (2) relatively low viscosity; and (3) tendency to solidify at low operating temperatures as manifested by relatively high pour points (PPs) (temperatures below which they will no longer pour).

Therefore, there is a need for a lubricant based on a renewable feedstock that could be modified to provide the desired properties. The approach taken was to evaluate the effects of change in molecular structure, by altering both the triglyceride backbone structure and the fatty acid distribution, on the viscosity, the VI, and the PP of the lubricant.

3.2   BACKGROUND

The scope of this work was to compare the viscosities and the PP properties of lubricants produced from triglycerides, fatty acid diesters of propylene glycol, and fatty acid esters of 2-butanol, while varying the fatty acid distribution. Some fatty acids were produced using the Colgate–Emery steam hydrolysis process [4], while others were produced by classical acidification of the fatty acid soaps from saponification [5]. The fatty acid distributions were obtained by fatty acid methyl ester gas-chromatography method ASTM D2800. All of the lubricants were produced by first performing an esterification reaction. Next, the fatty acid esters were epoxidized and then ring opened using hexanoic anhydride/potassium carbonate to form dihexanoate esters at the original olefinic site.

3.3   EXPERIMENTAL

3.3.1   MATERIALS

Conventional commodity soybean oil was obtained from Cargill (Minnetonka, Minnesota). Low saturated soybean oil was obtained from Zeeland Farm Services (Zeeland, Michigan). Mid-oleic soybean oil was obtained from the United Soybean Board. Oleic acid (90%), hydrogen peroxide (50%), formic acid, 1,2-propanediol (propylene glycol), potassium carbonate, potassium hydroxide, sulfuric acid (98%), toluene, diethyl ether, ethyl acetate, sodium bicarbonate, magnesium sulfate, Amberlyst™ A26, hexanoic acid, and hexanoic anhydride were all reagent grade and purchased from Sigma-Aldrich Company (St. Louis, Missouri).

3.3.2   SPECTROSCOPY USED

^{1}H nuclear magnetic resonance (NMR) spectra, to determine completion of esterification, epoxidation, epoxide ring opening, and removal of impurities, were obtained using a Bruker Avance 500 spectrometer (Billerica, Massachusetts).

3.3.3   DETERMINATION OF FATTY ACID DISTRIBUTION

The fatty acid distributions of triglycerides and fatty acids were obtained by the ASTM D2800 method by an external laboratory. Soybean oil candidates were chosen to obtain various ranges of oleic and saturated fatty acids. The definitions and fatty acid distributions of the oils can be seen in Table 3.1.

3.3.4   ESTERIFICATION

Dissolved was 390 g oleic acid (90%) with 50 g propylene glycol and 0.90 mL sulfuric acid in 500 mL toluene. The reaction was refluxed for 10 hours using a Barrett tube for azeotropic water removal. Once cool, the mixture was stirred with 160 mL of rinsed Amberlyst A26 for 1 hour. The mixture was then filtered to remove the resin. The resulting filtrate was initially evaporated on a rotary evaporator, followed by a short-path distillation apparatus (a Kugelrohr apparatus) to remove the remaining solvent at 100°C and 0.15 Torr. The final oil product (397.54 g) was shown to be the propylene glycol dioleate as analyzed by ^{1}H NMR.

3.3.5   EPOXIDATION

Reacted was 150.10 g mid-oleic soybean oil with 52 g of hydrogen peroxide (50%) in the presence of 9.69 g formic acid. The epoxidation was performed at 55°C for 4 hours. The product mixture was dissolved in 600 mL diethyl ether and treated with 150 mL of saturated sodium bicarbonate in a separatory funnel followed by two washes with 150 mL deionized water. The organic layer was then dried with magnesium sulfate and filtered. The resulting solution was initially evaporated on a rotary evaporator followed by a short-path distillation (Kugelrohr) apparatus at 30°C and 0.20 Torr to remove any remaining solvent. The final oil product (155.93 g) was identified to be the epoxidized mid-oleic soybean oil using ^{1}H NMR.

3.3.6   EPOXIDE RING OPENING WITH ANHYDRIDE

Reacted was 100.40 g epoxidized propylene glycol dioleate (EPGDO) with 87.84 g hexanoic anhydride in the presence of 2.07 g hexanoic acid and 4.53 g potassium carbonate as catalyst. Hexanoic anhydride, hexanoic acid, and EPGDO were heated to 180°C in a flask with stirring. Potassium carbonate was then added to the mixture and the temperature was maintained for 1.5 hours. The reaction was shown to be complete using ^{1}H NMR. The product mixture was dissolved in 1 L of ethyl acetate and treated consecutively with aqueous sodium hydroxide (10%), aqueous hydrochloric acid (5%), aqueous saturated sodium bicarbonate, and water. The organic layer was dried using

TABLE 3.1
Soybean Oil Fatty Acid Distributions

	Fatty Acid % Weights			
Soybean Oil Description	Oleic	Linoleic	Linolenic	Saturates
Commodity	22.7	52.9	8.0	16.4
Low saturate	22.2	60.1	8.6	9.1
Mid-oleic	53.3	31.1	1.0	14.6
High oleic	75.2	9.5	1.2	14.1
90% oleic	89.4	4.8	2.2	3.6

magnesium sulfate and filtered. The resulting solution was evaporated on a rotary evaporator, followed by distillation on a Kugelrohr apparatus to remove the remaining solvent. The final oil product (144.06 g) was positively identified to be the dihexanoate ester of EPGDO by using ^{1}H NMR.

3.3.7 VISCOSITY

Kinematic viscosities at 40.0°C and 100.0°C were measured following ASTM D445 at Analytical Testing Services, Inc. (Franklin, Pennsylvania). Similar kinematic viscosity values were obtained by calculation from measured density and dynamic viscosity on a rheometer (TA Instruments AR 2000ex).

3.3.8 POUR POINT

Cloud points and PPs were measured according to the ASTM D2500 and ASTM D97 methods, respectively, at Analytical Testing Services, Inc. (Franklin, Pennsylvania).

3.4 RESULTS AND DISCUSSION

We will first focus on the triglyceride work. The soybean oil triglycerides of various fatty acid distributions were epoxidized and then ring opened with hexanoic anhydride following the procedure in U.S. Patent 8357643 [6]. The general schematic can be seen in Figure 3.1.

The viscosities at 40°C and 100°C revealed an expected trend; the higher the percent of polyunsaturated fatty acid esters, the higher the viscosity.

When oleic oil fraction increased, viscosity was reduced. This can be attributed to the difference in the molecular weight of the dihexanoate ester. Commodity soybean oil averages about 4.5 double bonds per molecule, which, after epoxidation and ring opening, will produce nine hexanoic esters. A high-oleic soybean oil, on the other hand, contains about 3 double bonds per molecule and can produce a maximum of six hexanoic esters. From Figure 3.2 it can be seen that there is a large increase in viscosity when linoleic acid is increased from 9.5% to 60.1%. However, there was no appreciable difference in the PPs of the triglycerides with change in fatty acid composition [7].

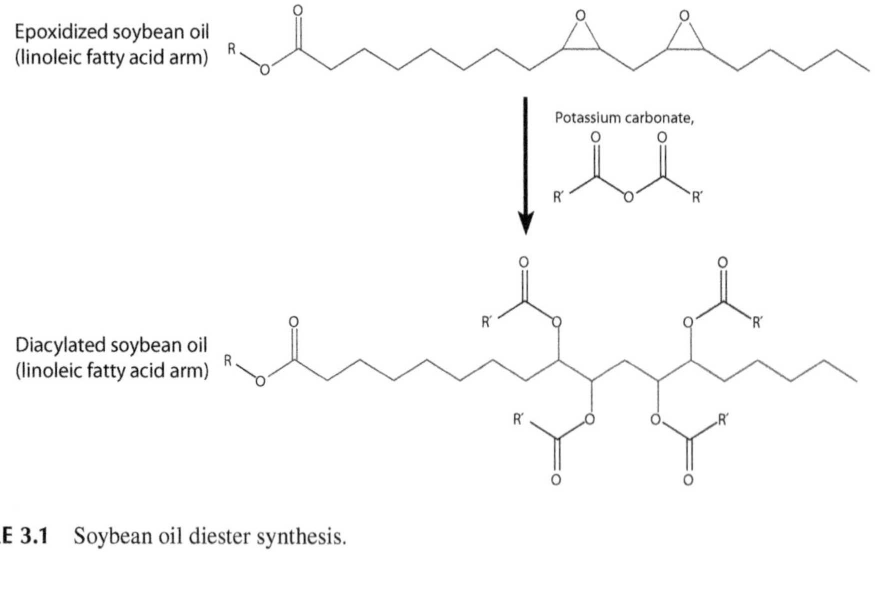

FIGURE 3.1 Soybean oil diester synthesis.

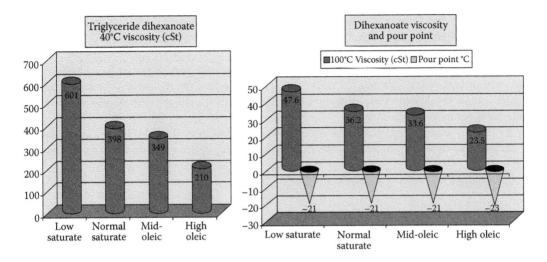

FIGURE 3.2 Viscosities and PPs of dihexanoates from soybean oils of varying fatty acid compositions.

We next considered the effect of using a propylene glycol backbone in place of triglyceride. In Figure 3.3, the same expected reduction in viscosity trend with increasing oleic acid content and decreasing linoleic acid content can be seen.

However, because of the overall lower molecular weight of the dihexanoates, we now see a significant change in PPs. This PP difference can be attributed to saturated fatty acid concentration

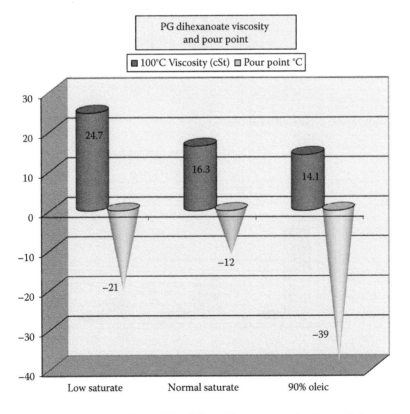

FIGURE 3.3 Viscosities at 100°C (cSt) and PPs (°C) of dihexanoates of soybean oil fatty acids with a propylene glycol backbone. PG: propylene glycol.

(mainly stearic and palmitic acids) in the mixture. Thus, low saturation level, as in the 90% oleic ester (3.6% saturates), gave a low PP of −39°C, whereas normal saturated soybean oil (16.4% saturates) gave a PP of −12°C.

Finally, the triglyceride backbone was converted to an ester of a monoalcohol. The commodity soybean fatty acids were reacted with 2-butanol to produce the monoester. As can be seen in Figure 3.4, we would expect the viscosity to drop because of the change in molecular weight (three fatty acids per molecule to only one).

As can be seen in Figure 3.5, the expected drop in viscosity did take place experimentally. 2-Butanol was chosen to maintain some hydrolytic stability (when compared to a methyl ester of

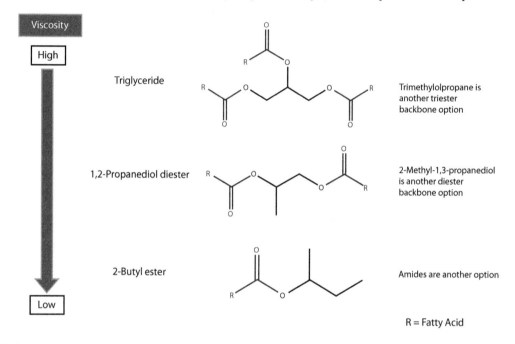

FIGURE 3.4   Variation of soybean oil backbone.

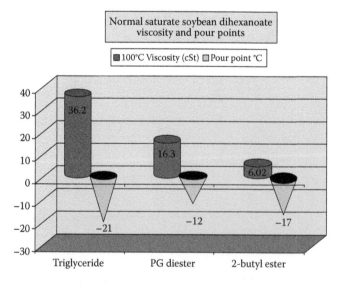

FIGURE 3.5   Normal saturate soybean dihexanoate viscosities and PPs.

biodiesel). We would expect the detrimental effect of saturates on PPs to increase as the molecular weight drops from a triglyceride to a 2-butyl ester. This effect was not seen in regard to the 2-butyl ester. A possible reason is that when high vacuum was applied to the product lubricant, some of the 2-butyl saturated fatty acids were removed owing to the higher volatility of the saturated fatty acid esters versus the ring-opened epoxide product.

3.5   SUMMARY

Fatty acid distribution and backbone modification can effectively control a lubricant's viscosity and PP. The viscosity effect can be tied to molecular weight. Feedstocks containing lower amounts of double bonds per mole of sample (high oleic versus high linoleic) lead to lower amounts of ester content, thereby yielding lower-molecular weight lubricants with lower viscosities. Reduced PPs are mainly a result of the lower quantities of saturated fatty acids, predominately stearic and palmitic acids. However, chain branching and viscosity due to molecular weight must also be considered when evaluating the lubricant PPs. By controlling the fatty acid distribution and the molecular weight through backbone modification, one can dial in the desired lubricant viscosity and pour point. In a complementary work these lubricants were found to be hydrolytically stable, with low PPs, and can be synthesized in a wide viscosity range (2200–2700 cSt at 40°C).

ACKNOWLEDGMENTS

We thank the Ohio Soybean Council and United Soybean Board for their support of this work.

REFERENCES

1. Garmier, W. "Biodegradable lubricant composition from triglycerides and oil soluble antimony," U.S. Patent 5990055, 1999.
2. Morrison, D. "Blend of heavy and light ester oil for engines," U.S. Patent 5378249, 1995.
3. Salas, J.J., Ruiz-Méndez, M.V., and Garcés, R. Prologue: Biodegradable lubricants from vegetable oils, January–March 2011, 7.
4. Bamebey, H.L., and Brown, A.C. Continuous fat splitting plants using the Colgate–Emery process, *J. Am. Oil Chem. Soc.* 1949, 25, 95–99.
5. Crawford, R. "Conversion of fatty acid esters to fatty acids," U.S. Patent 4179457, 1979.
6. Benecke, H. "Lubricants derived from plant and animal oils and fats," U.S. Patent 8357643, 2013.
7. Benecke, H. "Modified vegetable oil lubricants," U.S. Patent Application 20120129746, 2012.

4 Isostearic Acids
Synthesis, Properties, and Process Development*

Helen Ngo, Robert O. Dunn, and Winnie C. Yee

CONTENTS

ABSTRACT

A brief review of the latest developments in isostearic acid technology is summarized, with a particular focus on the rational design of zeolite methods to efficiently and economically produce isostearic acids. Since vegetable oils typically have very high content of unsaturated fatty acids, under extreme conditions (i.e., high temperature and high pressure), the double bonds can easily oxidize to form viscous compounds, oligomers and polymers. These products can be difficult to remove and can impair the performance of the equipment. Chemical modification of the double bonds in the oils to produce isostearic acids has proven to increase the oils' stability. This chapter focuses on the synthesis, characterization, physical properties, purification, and technoeconomic studies of the isostearic acid products from fatty acids.

4.1 INTRODUCTION

There are hundreds of biobased lubricants derived from vegetable oils in the market [1–3], with leading examples of hydraulic fluids from Mobil Corporation (Environmental Awareness Lubricants line) and metalworking additives from Lubrizol Corporation (Lubrizol® 5411) [4].

---

* Mention of trade names or commercial products in this chapter is solely for the purpose of providing specific information and does not imply recommendation or endorsement by the U.S. Department of Agriculture (USDA). USDA is an equal opportunity provider and employer.

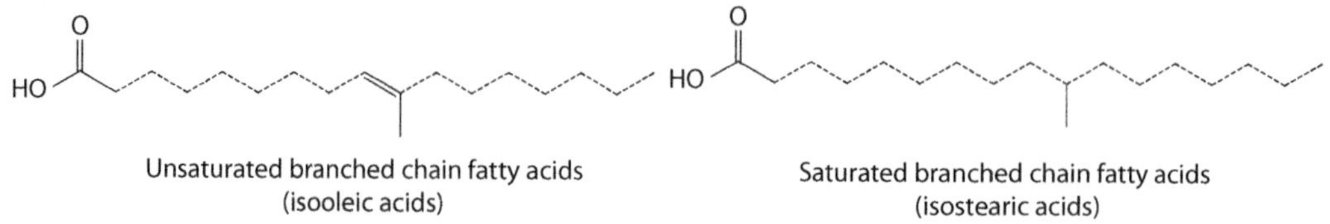

Unsaturated branched chain fatty acids
(isooleic acids)

Saturated branched chain fatty acids
(isostearic acids)

SCHEME 4.1 Proposed structures of isostearic acids and isooleic acids. Dashed lines indicate several possible positions for the branching methyl group.

Although vegetable oils provide a source of environmentally desirable lubricants, they are not widely utilized because of their poor oxidative stability [5–9]. Saturated branched-chain fatty acids, including isostearic acids, have a hydrophilic carboxylic acid headgroup and a lipophilic alkyl chain with a methyl branching group located at various positions along the chain (Scheme 4.1) [10]. Isostearic acids and their simple alkyl esters have the advantages over common fatty acids in that they have much better oxidative stability than unsaturated fatty acids and typically have lower melting points (MPs) and cloud points (CPs) than comparable saturated linear-chain fatty acids [11]. As a result, branched fatty acids are ideal in many applications including body washes [12,13], coatings [14], cosmetics [12,13], lubricant additives [15–18], hydraulic fluids [19–21], and emulsifiers [22].

4.2 VALUE OF ISOSTEARIC ACID PRODUCTS

Isostearic acids are not available from natural sources in appreciable quantities [23] but synthetic sources are common in commercial products. Although isostearic acids have many potential applications because of their superior performance in lubricant properties, they are predominantly used in cosmetics because they are odorless and provide smooth spreading with a nontacky feeling when applied onto skin [24–26]. They have been tested at various concentrations to evaluate their safe use for cosmetic applications. At the tested concentrations, no significant irritation or sensitivity was found in clinical studies [26]. Isostearic acids are liquid at room temperature, which makes them easy to use for blending or mixing purposes. This "oiliness" property, in addition to their excellent thermostability and low flow temperature properties, makes isostearic acids attractive candidates.

4.3 CHEMICAL ROUTES FOR ISOSTEARIC ACID PRODUCTION

4.3.1 Clay Technology

The well-known synthetic route for producing isostearic acids is through a process called clay-catalyzed dimerization of unsaturated linear-chain fatty acids. This technology was discovered in 1970 when den Otter [27,28] carried out the dimerization of unsaturated linear-chain fatty acids (i.e., oleic acid) in the presence of a montmorillonite clay catalyst and water at high temperatures in stainless steel reactors. The reaction predominantly gave oligomers and small amount of monomeric isooleic acids (Scheme 4.1) [27,28]. The proposed structures of oligomeric products are shown in Scheme 4.2, but the structures would vary depending on the catalysts and feedstocks used for their production. In 1983 and 1985, Foglia et al. [29] and Nakano et al. [30] respectively reported a similar process and found that a combination of bentonite clay and 1,2-dichloroethane catalyzed the isomerization of oleic acid to produce the monomers in up to 50% conversion. In 1988, Hayes [31] patented a process for polymerizing the tall oil fatty acids in the presence of catalytic amounts of montmorillonite clay and small amounts of lithium salt and water. This process was also performed under high pressure and high temperatures. The products obtained were also predominantly oligomers (60–80%) and small amounts of monomers (10–40%) [31]. Since isostearic acids are made in

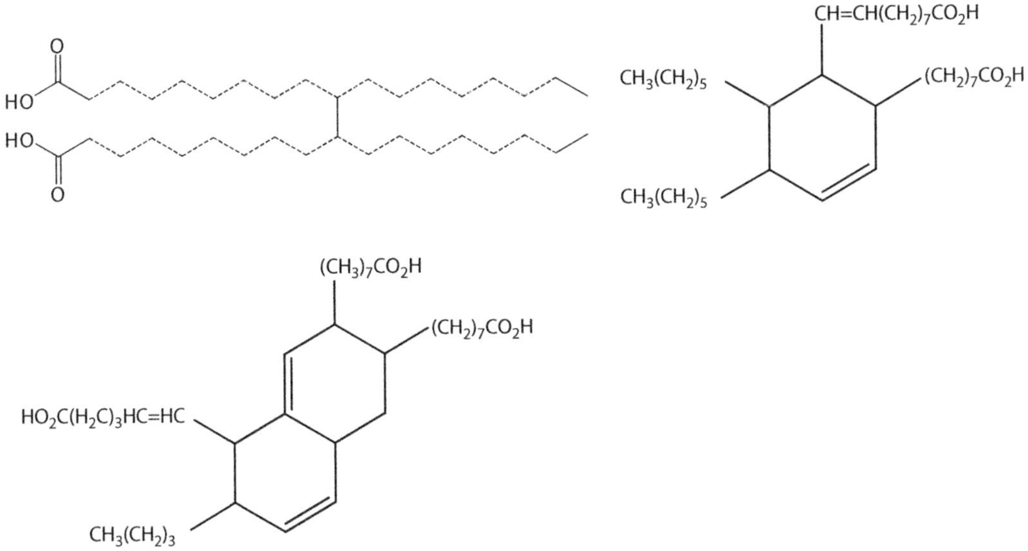

SCHEME 4.2 Proposed structures of dimer and trimer fatty acids.

limited production capacity, they are thereby sold in higher-value markets such as personal care and cosmetics. Two common trade names for these products are Century™ 1105 and 1107 [32]. When the product includes a large proportion of oligomers, its target market is the polyamide sector, which is sold at half of the price of the isostearic acids [33].

It was speculated [34–39] that the low yield of isostearic acids was because of the clay structure. Clays are two-dimensional (2D) layer materials [40] and are ideal catalysts for the production of oligomeric products. This is because when the unsaturated linear-chain fatty acids penetrate the interlayer of the clay to react with the acidic proton and metal cations, the clay forms a fatty acid intermediate with carbon cations. If there is nothing to separate the fatty acid intermediates, then they couple with each other to form dimers and trimers. Thus, the clay catalyst technology is superior in producing predominantly oligomeric products, but the yield of isostearic acids is relatively low.

There are several parameters which should be considered for improving the current clay catalyst technology. First, the initial cost of the catalyst should be lowered; second, the catalyst efficiency should be increased; third, the catalyst should have better recyclability with increased levels of retained activity after regeneration; fourth, the catalyst should be easily recoverable; and finally, the quality of the synthesized product should be better than that of the currently produced.

4.3.2 ZEOLITE DEVELOPMENT

A very promising alternative approach to improving isostearic acid production is through the skeletal isomerization of unsaturated linear-chain fatty acids to isooleic acids followed by hydrogenation to give isostearic acids [34–39,41,42]. Zeolite materials have well-defined pores and channels which can create a size-confinement environment [43]. Because of this well-defined environment, it was found that oleic acid (Scheme 4.3, 1) can be isomerized to isooleic acids in the presence of zeolite catalysts at much higher selectivity and conversions than those obtained with traditional clay, silica, or alumina-type catalysts. Hodgson [34] patented a process for the isomerization of oleic acid in the presence of either H$^+$-mordenite or zeolite L, followed by hydrogenation, which afforded isostearic acids in up to 51% yields. Hodgson found that the ratio of silica to alumina

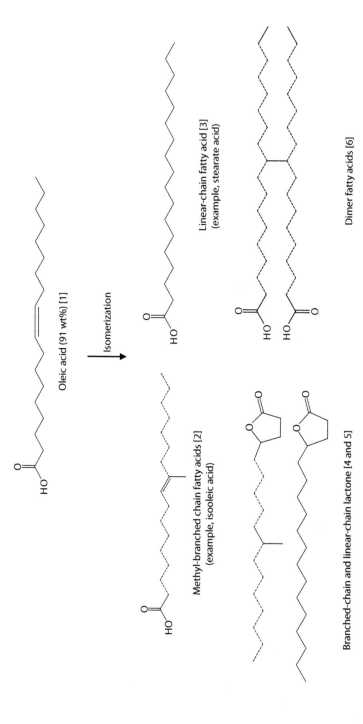

SCHEME 4.3 Isomerization of oleic acid. Dashed lines indicate several possible positions for the branching methyl group.

significantly affected the conversion of oleic acid and the yield of isostearic acids. Lower silica–alumina ratios gave higher activity and selectivity, presumably owing to the presence of more Brønsted acid sites. This isomerization process was further improved by Tomifuji et al. [35] by the addition of small amounts of water or methanol to the reaction mixture, which increased the conversion and selectivity of the skeletal isomerization reaction (up to 70 wt% isostearic acids). More recently, Zhang et al. [36] expanded the scope of the reaction by disclosing further details on the skeletal isomerization reaction of oleic acid using both microporous and mesoporous zeolitic solids. Microporous H^+–beta zeolite was found to isomerize oleic acid to give 45% isostearic acids (after hydrogenation), 5.9% stearic acid, and 14% lactone. About 20% of oleic acid remained after the reaction [36]. In comparison, MAS-5, a highly ordered hexagonal mesoporous aluminosilicate (MAS) with acid strength higher than that of pure silicate MCM-41 (Mobil Composition of Matter No. 41), gave a much lower conversion (~65%) and selectivity for isostearic acids (~30% after hydrogenation). MAS-5 also gave 4% stearic acid and 16% lactone [37–39].

These reports and patents illustrate the potential of solid acid catalysts for the skeletal isomerization of unsaturated linear-chain fatty acids. The Brønsted acids within the channels of these zeolite materials are the likely active sites for the skeletal isomerization reactions. These channels are large enough for the fatty acids to diffuse through to form isostearic acids via isomerization and yet small enough to prevent the formation of oligomeric products. In spite of the great potential of the skeletal isomerization reactions, both the conversion and the yield of the reactions reported in the literature need to be improved before they become useful for commercial production of isostearic acids.

A breakthrough in the isomerization process to significantly boost isostearic acid yields using a modified zeolite ferrierite–Lewis base combination method was reported in 2012 and 2014 [41,42]. A zeolite–base isomerization method was developed that maximizes isostearic acid production and minimizes the bimolecular reactions that produce oligomeric (Scheme 4.3, 6) and other by-products (stearic acid [Scheme 4.3, 3], branched- and linear-chain lactones [Scheme 4.3, 4 and 5]) [41,42]. This process involves application of a combination of protonated ferrierite zeolite (H^+-ferrierite) catalyst and Lewis base (i.e., triphenylphosphine). Ferrierite is commercially available and was previously used in the skeletal isomerization of butene to isobutene [44–46]. It is found to be an excellent catalyst for isomerization because its acidified form possesses strong acidic sites, providing high activity in the skeletal isomerization of fatty acids. Its interconnected open channels help facilitate the transport of the lipid reactants and products. Triphenylphosphine is used to deactivate (i.e., poison) the active acidic sites on the external surfaces of zeolite particles to suppress dimer acid formation, thereby limiting the bimolecular reactions to the interstitial acidic sites of the zeolite catalyst channels.

It was hypothesized [41,42] that in the H^+-ferrierite-catalyzed isomerization reactions, dimer acids formed as a result of the bimolecular reactions catalyzed by the Brønsted acid sites on the external surfaces of micron-sized H^+-ferrierite particles. To produce isostearic acids with desirable fluidity and lubricity, selective poisoning of the active acidic sites on the external surfaces of H^+-ferrierite particles with bulky bases to suppress the formation of dimer products was investigated (Scheme 4.4). The results showed that the addition of small amounts of triphenylphosphine significantly reduced the formation of dimer by-products. For example, in the presence of 10 wt% base additive (relative to the H^+-ferrierite zeolite catalyst), the amount of dimer products was reduced from 14 to 5 wt% (Table 4.1). However, a longer reaction time (22 versus 6 h) was needed to obtain similar conversions. Further increase of the additive concentration to 20 wt% did not lead to further suppression of the dimer formation (Table 4.1). Interestingly, when the reaction was conducted at higher catalyst loading (5.0 wt%), triphenylphosphine (7.5 wt%), and distilled water (1.8 mL) at 260°C for 4 h, it gave 99% conversion and higher selectivity of isostearic acid products (Table 4.1). Furthermore, it was determined that this strategy was general and worked with other types of zeolite catalysts. For example, the H^+-mordenite catalyst used in Tomifuji et al.'s work [35] was the best catalyst for skeletal isomerization reactions prior to Ngo et al.'s [41] and Ngo and Foglia's

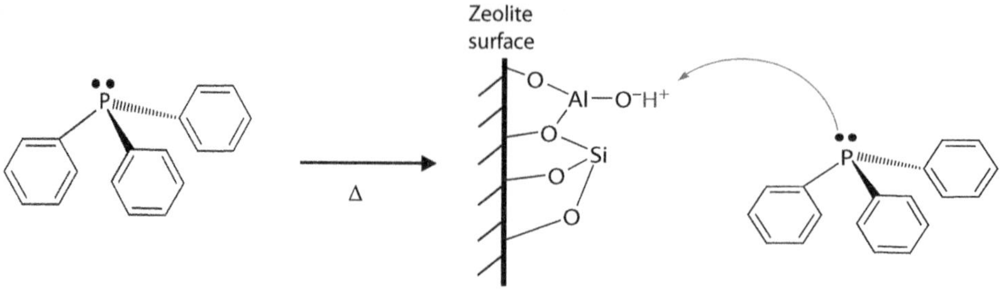

SCHEME 4.4 Deprotonating (poisoning) the external Lewis/Brønsted acid sites of the zeolite with triphenylphosphine.

TABLE 4.1

Weight Percent Compositions of the Isomerized Products Determined by Gas Chromatography (GC)

Catalyst (wt% Loading)	Additive (wt% to Catalyst)	C_{18}-Branched-Chain Fatty Acids	C_{18}-Linear-Chain Fatty Acid	Branched- and Linear-Chain Lactones	C_{36}-Dimer Acid	% Conversion[a]
			GC wt% Composition			
H+-ferrierite [2.5][b]	0	70	6.9	9.1	14	99
H+-ferrierite [2.5][c]	10	70	10	15	5.0	95
H+-ferrierite [2.5][d]	20	75	9.4	12	5.0	96
H+-ferrierite [5.0][d]	7.5	82	7.0	9.9	1.1	99
H+-mordenite [8.0][e]	0	61	6.3	4.7	28	99
H+-mordenite [8.0][e]	10	79	7.7	4.0	9.3	98

Source: Ngo, H.L. et al.: Improved synthesis and characterization of saturated branched-chain fatty acid isomers. *Eur. J. Lipid Sci. Tech.*, 114, 213–221, 2012. Copyright Wiley-VCH Verlag GmbH & Co. KGaA. Reproduced with permission; Ngo, H.L., and T.A. Foglia, Process for preparing saturated branched-chain fatty acids, U.S. Patent 8,748,641 B2, 2014.

Note: Samples were compared against internal standard (C13:0 methyl ester). All products were characterized after isomerization, hydrogenation, and methylation.

[a] Percentage of conversion calculated as 94.3 − (C_{18}-linear-chain fatty acid − 5.74)/94.3 × 100. (94.3% is total unsaturated fatty acids and 5.74% is the total fatty acids which do not contribute to the reaction.)

[b] Reaction was performed at 250°C for 6 h.

[c] Reaction was performed at 250°C for 22 h.

[d] Reaction was performed at 260°C for 4 h.

[e] Reaction was performed at 280°C for 6 h.

works [42]. As shown in Table 4.1, when the isomerization reaction of oleic acid was carried out under the conditions reported by Tomifuji et al., the result obtained was 28 wt% dimer products (Table 4.1). However, with 10 wt% additive, the amount of dimer decreased to <10 wt% (Table 4.1). These results clearly showed that base additives play an important role in deprotonating (poisoning) the external Lewis/Brønsted acid sites of zeolites and lead to higher-purity isostearic acid products compared to all existing methods.

4.4 CATALYST RECOVERY, REGENERATION, AND REUSE

To make the isostearic acid production process economically viable and environmentally sound, the spent zeolite–base catalysts need to be recovered from the reaction mixtures and reused. The use of solid acid catalysts for the skeletal isomerization reactions allows facile catalyst recovery by simple filtration. It was demonstrated that the H[+]-ferrierite–base combination could be reused for up to 10 times by washing the recovered solid with dilute hydrochloric acid at 55°C for 24 h after each use [41,42]. Fresh base additive needs to be added after each acid treatment to maintain the high conversions and isostearic acid selectivity. Recently, it was found that the catalyst regeneration and reuse processes could be further improved by using a new approach where the catalyst was regenerated by heating at 115°C for 20 h after each use. Only after every four or five uses was the recovered solid treated with acid for regeneration [47]. The catalyst was regenerated by acid treatment followed by the addition of fresh triphenylphosphine in the 6th, 12th, and 18th runs in Table 4.2. Thus, by using these improved procedures the catalyst was successfully reused for up to 20 times with no significant decrease in conversions and isostearic acid selectivities [47]. The ability to reuse the solid acid catalysts can reduce the process costs as well as the environmental burden of disposing large quantity of solid catalysts. The improved regeneration protocol is straightforward and scalable.

4.5 PURIFICATION OF ISOSTEARIC ACIDS

The isostearic acid products obtained using zeolite catalysts are a mixture of unsaturated branched-chain isomers and small amounts of dimer and lactone by-products. To ensure accurate assessment of the potential utility of the isostearic acid products as biolubricants, the saturated linear-chain fatty acids and oligomeric by-products should be removed from the crude mixture. The crude isomerized fatty acid products were isolated and purified in two steps: recrystallization followed by distillation. The recrystallization step is designed to remove the saturated acids (stearic and palmitic acids) and the distillation step is used to remove the oligomeric products. First, the crude mixture was dissolved in acetone and placed in a freezer with the temperature set to −15°C, which caused the majority of the saturated acids to precipitate. The recrystallized products (i.e., liquid phase monomer and oligomer mixture) were then transferred to an addition funnel attached to a molecular wiped-film distillation apparatus and slowly added to separate the monomeric products from the oligomeric products. The purified monomeric products are typically very light yellow in color and liquid at room temperature.

4.6 CHARACTERIZATION OF ISOSTEARIC ACIDS

A 2D gas chromatography–time of flight–mass spectrometry (GC×GC-TOF-MS) (LECO, St. Joseph, Michigan) was used to characterize the complicated mixture of isomers in the isostearic acid products. The production of numerous isomeric structures was caused by in part by the migration of the C=C bond at high temperature along the chain. This complexity of isomerized products makes it difficult to characterize the products using the GC-MS instrument. GC×GC-TOF-MS is far more accurate than conventional GC-MS (electron impact). Figure 4.1 displays the GC×GC data for isostearic acid picolinyl ester derivatives eluted in a band across the contour plot. The isostearic acids were converted to picolinyl esters for MS analysis purposes [41]. Each spot represents an isostearic acid isomer. There were at least 28 isomers in the mixture according to the contour plot. The intensity of the peak is represented by a color scale where yellow is low and red is high. Interestingly, the isostearic acid products from the H[+]-mordenite–triphenylphosphine catalyst gave a very different distribution of isomers (Figure 4.2) with substantially more spots in the spectrum. These results show that each zeolite type will probably produce a different ratio of isomers, which subsequently will affect the physical properties of the materials. This highlights the importance of a thorough characterization of the isostearic acid products.

TABLE 4.2
Catalyst Regeneration and Reuse Results

Run	1	2	3	4	5	6[a]	7	8	9	10	11	12[a]	13	14	15	16	17	18[a]	19	20
Isostearic acids (wt%)	82	81	80	77	75	84	77	78	77	75	73	82	81	77	70	78	75	80	80	79
Total % conversion	99	98	98	97	93	99	95	97	96	93	92	98	98	95	89	96	95	98	94	94

Source: Ngo, H.L., and T.A. Foglia, Process for preparing saturated branched-chain fatty acids, U.S. Patent 8,748,641 B2, 2014.

Note: GC data collected on the intact reaction product after isomerization, hydrogenation, and methylation using methyl tridecanoate as internal standard.

[a] Catalyst was pretreated with acid solution.

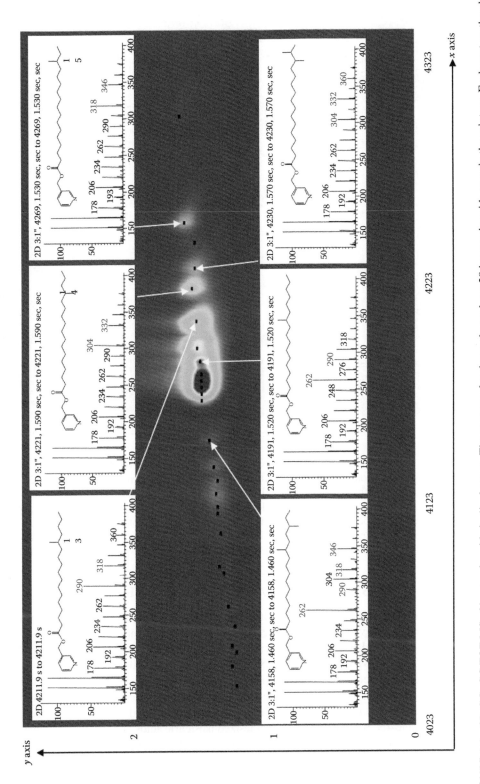

FIGURE 4.1 2D GC×GC-TOF-MS of isostearic acid picolinyl esters. The contour plot shows that at least 28 isostearic acids were in the mixture. Each spot on the plot represents an isostearic acid isomer. Selected MS spectra of the spot are shown. The *x* axis is retention time, on longer timescale, and the *y* axis is retention time, very short, with timescale on the order of seconds. (Ngo, H.L. et al.: Improved synthesis and characterization of saturated branched-chain fatty acid isomers. *Eur. J. Lipid Sci. Tech.*, 114, 213–221, 2012. Copyright Wiley-VCH Verlag GmbH & Co. KGaA. Reproduced with permission.)

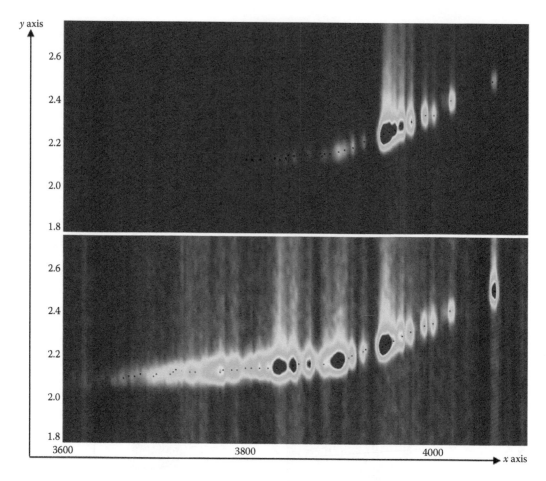

FIGURE 4.2 GC×GC-TOF-MS contour plots of isostearic acid picolinyl esters. *Top*, with 2.5 wt% H⁺-ferrierite at 280°C for 6 h (a total of 28 spots); *bottom*, with 8.0 wt% H⁺-mordenite at 280°C for 6 h (a total of 42 spots). Each spot on the plot represents at least one isostearic acid isomer. The color indicates the concentration of the isomer, with red being the most dominant area. The *x* axis is retention time, on longer timescale, and the *y* axis is retention time, very short, with timescale on the order of seconds. (Ngo, H.L. et al.: Improved synthesis and characterization of saturated branched-chain fatty acid isomers. *Eur. J. Lipid Sci. Tech.*, 114, 213–221, 2012. Copyright Wiley-VCH Verlag GmbH & Co. KGaA. Reproduced with permission.)

4.7 PROPERTIES OF ISOSTEARIC ACIDS AND THEIR ESTER DERIVATIVES

The isostearic acids synthesized following procedures employing a H⁺-ferrierite zeolite catalyst and triphenylphosphine were evaluated for physical properties in lubricant applications. Cloud point (CP), pour point (PP), oxidative induction period, kinematic viscosity, and VI were analyzed according to the appropriate ASTM test method [48–52]. The isothermal (110°C) oxidation induction periods were measured according to test method Cd 12b-92 published by the American Oil Chemists' Society [50]. This test method was applied using an Oxidative Stability Instrument from Omnion Inc. (Rockland, Massachusetts). Results for CP, PP, induction period, kinematic viscosity, and VI are shown in Table 4.3. All results for the synthesized isostearic acids and their ester derivatives were analyzed from purified samples with oligomer concentrations close to zero.

The CP and the PP of the isostearic acid isomers were quite low compared to the MP of pure stearic acid (69.3°C) [53]. This is the result of at least two factors related to chemical structure. First, the C₁₈ "tailgroups" were branched with one methyl group located on the interior of a straight

TABLE 4.3

Physical Properties of Isostearic Acid Isomers and Their Fatty Acid Alkyl Esters

Material	CP (°C)	PP (°C)	IP[a] (110°C) (h)	KV (mm²/s) 40°C	KV (mm²/s) 100°C	VI
Mineral oil	−34.0	−22.0	33.0	16.0	3.5	83
Soybean oil	−5.8	−7.3	3.5	32.0	7.9	236
SMEs	1.4	1.0	3.1	4.2	1.7	—[b]
Isostearic acids	1.5	−2.7	6.7	27.0	5.6	154
Isostearic acid FAMEs	−3.2	−11.0	64.0	6.1	2.1	152
Isostearic acid FAiPREs	−27.0	−34.0	12.0	7.1	2.3	149
Isostearic acid FA2-EHEs	−17.0	−19.0	40.0	10.0	3.1	169
Crude isostearic acids	0.5	0.3	2.1	35.0	6.7	151

Source: Ngo, H.L. et al.: Synthesis and physical properties of isostearic acids and their esters. *Eur. J. Lipid Sci. Tech.*, 113, 180–188, 2011. Copyright Wiley-VCH Verlag GmbH & Co. KGaA. Reproduced with permission.

Note: FAiPrE: fatty acid isopropyl ester; FAME: fatty acid methyl ester; FA2-EHE: fatty acid 2-ethylhexyl ester; IP: oxidation induction period (oil stability index); KV: kinematic viscosity; SME: soybean oil fatty acid methyl ester.

[a] IP was measured by Oxidative Stability Instrument from Omnion Inc. under license from Archer Daniels Midland.

[b] Undefined because KV < 2 at 100°C.

C_{17} chain. Branching in the alkyl chain can decrease the MP relative to that of the normal isomer. Second, the crystallization behavior of isostearic acids is thermodynamically dependent on the concentrations, the MPs, and the enthalpies of fusion of the individual components in the mixture [54]. The isostearic acid mixture contained at least 28 isomers [41,42]. Each component has a unique MP, concentration, and enthalpy of fusion that influence when they begin to crystallize at low temperatures. Thus, the CP of the mixture was a temperature lower than the MP of the isomers in pure form. This phenomenon also applies to the ester derivatives of the isostearic acids. For example, the MP of pure methyl stearate is 37.7°C compared to a CP of −3.2°C for the methyl isostearate isomer mixture in Table 4.3 [11].

Converting the isostearic acids into methyl, isopropyl, and 2-ethylhexyl esters had the effect of further lowering the CP and the PP. Owing to the absence of strong attractive forces (hydrogen bonding), fatty acid alkyl esters generally have better cold flow properties than those of the corresponding fatty acids [53]. It has also been shown that replacing the methyl headgroup with a higher alkyl group improves the cold flow properties of the esters [55–57]. Although the 2-ethylhexyl esters had improved CP and PP, the isopropyl esters demonstrated the best CP and PP among the ester derivatives, suggesting there are limits to how large and bulky the alkyl headgroup should be to improve the cold flow properties.

The induction period data in Table 4.3 were measured isothermally at 110°C with an Oxidative Stability Instrument. This instrument measures the oxidation induction period by applying heat to an oil sample and monitoring the evolution of secondary degradation products (volatile organic acids). The instrument is commonly used in the evaluation of vegetable oils and their derivatives and is mechanically identical to the more widely utilized Rancimat instrument. According to this test, purified isostearic acid isomers demonstrated better oxidative stability than that of soybean oil or soybean oil biodiesel [11]. These results were likely because of the low concentrations of unsaturated components in the isostearic acids. Esterification of the isostearic acids increased the induction period mainly because of the near elimination of the free fatty acids. The methyl isostearate exhibited an induction period comparable to values reported for linear methyl stearate [58].

The kinematic viscosities at 40°C and 100°C of the isostearic acid isomers were lower than that of soybean oil, but higher than that of the isostearic acid esters. It was shown that triacylglycerols with predominantly long-chain fatty acid groups generally have higher viscosities than those of the corresponding free fatty acids [59]. Furthermore, isostearic acid esters do not experience hydrogen bonding, leading to lower viscosities than those of their corresponding linear fatty acids. At 40°C, the kinematic viscosity of methyl isostearates was 6.1 mm²/s, a value comparable to that of linear methyl stearate at the same temperature (5.8 mm²/s) [59]. The VI provides a quantitative measurement of a lubricant's ability to maintain good fluidity over a large temperature range. The VIs of the synthesized isostearic acids and their ester derivatives exceeded 149, values that were significantly greater than the VI of mineral oil [11].

Finally, the isostearic acid isomers and their ester derivatives were investigated for antiwear characteristics using the high-frequency reciprocating rig (HFRR) test method (ASTM D6079 [60]). The HFRR test was performed by immersing a steel disk in oil and rubbing a nonrotating steel ball under 200 g mass load for 75 min at an oil temperature of 60°C. The ball was rubbed with a 1 mm stroke at a frequency of 50 Hz. The ball was removed and cleaned and the dimensions of the major and minor axes of the wear scar were measured under magnification.

In Table 4.4, results for the purified isostearic acids and ester derivatives are compared with those for soybean oil FAMEs (biodiesel), soybean oil triacylglycerols (good antiwear reference oil), mineral oil, and hexadecane (bad antiwear reference oil). The isostearic acids performed better than the reference materials and ester derivatives with respect to reducing wear scar. In addition, the isostearic acids registered the fourth lowest mean coefficient of friction from the HFRR test [11]. Esterification decreases the amphiphilic nature of the fatty acids, which lessens their antiwear properties [61]. This effect is confirmed in the data for the fatty acid alkyl esters in Table 4.4.

Further comparison was made with the antiwear properties of crude isostearic acids (also shown in Tables 4.3 and 4.4). The crude isostearic acids had the same components as the purified isostearic acids plus 14% dimer acids [11]. Although CP and PP were slightly lower, kinematic viscosity was higher and induction period was significantly lower than those of the

---

TABLE 4.4

Lubrication Characteristics of Isostearic Acid Isomers and Their Ester Derivatives by the HFRR Test at 60°C

| Material | Wear Scar (µm) | | | Film[a] (%) | CF[a] |
	Test 1[b]	Test 2[b]	Mean[c]		
Mineral oil	291.5	333.5	312.0	40.0	0.265
Hexadecane	377.5	414.5	396.0	24.5	0.274
Soybean oil	175.0	162.0	168.5	95.5	0.083
SMEs	160.0	209.5	185.0	92.0	0.156
Isostearic acids	51.5	45.0	48.3	99.0	0.130
Isostearic acid FAMEs	197.0	232.0	214.5	90.5	0.121
Isostearic acid FAiPREs	131.0	174.0	152.5	90.5	0.151
Isostearic acid FA2-EHEs	345.5	319.5	332.5	64.0	0.174
Crude isostearic acids	38.5	40.5	39.5	99.0	0.124

Source: Ngo, H.L. et al.: Synthesis and physical properties of isostearic acids and their esters. *Eur. J. Lipid Sci. Tech.*, 113, 180–188, 2011. Copyright Wiley-VCH Verlag GmbH & Co. KGaA. Reproduced with permission.

Note: CF: coefficient of friction; FAME: fatty acid methyl ester; FAiPrE: fatty acid isopropyl ester; FA2-EHE: fatty acid 2-ethylhexyl ester; SME: soybean oil fatty acid methyl ester; Test: ASTM D6079 (two replicate tests).

[a] Averaged values from two tests.

[b] Test n = average value of two scar measurements = $(X_n + Y_n)/2$.

[c] Mean = (test 1 + test 2)/2.

purified isostearic acids. Nevertheless, the lubrication characteristics of the crude isostearic acids were equivalent or better than those of the purified isostearic acids. This suggests that if the oxidative stability could be improved, perhaps by adding antioxidants, crude isostearic acids with up to 14% total dimer acid concentration may be suitable oil for formulating bio-based lubricants.

4.8 CONVERSION EFFICIENCIES AND ECONOMICS

To evaluate the economic feasibility of this zeolite-catalyzed isomerization process, a process simulation model for the production of isostearic acids was developed using the SuperPro Designer process simulation program, version 8.5 (Intelligen Inc., Scotch Plains, New Jersey). Process simulation is a model-based representation of technical processes and unit operations using computer software such as SuperPro Designer and Aspen Plus (Aspen Technology Inc., Burlington, Massachusetts). It is a valuable tool by which scientists and engineers assess the economics, process-development needs, and environmental impacts of a technology before committing to a full-scale implementation and commercialization. This tool has been successfully applied to determine the economic costs and to provide an understanding of the costs associated with a process. It has been used in evaluating the feasibilities of various technology areas including biodiesel production from refined soybean, rapeseed, and algal oils [62–64]; ethanol production by dry-grind process and wet milling process [65–69]; and the production of activated carbon from pecan shells and broiler litter [70,71].

4.8.1 PROCESS MODEL DESCRIPTION

A process model for the production of isostearic acids was developed based on data and information obtained from experimental results and from various technical sources, including the literature, industrial experts, and equipment suppliers. The model was designed to simulate a processing plant in operation for 330 days per year and 24 hours per day with production capacity of 4.5 million kg per year (10 million lb per year) [72,73]. Figure 4.3 shows the schematic flow diagram of the process. Details of the process can be found in a previous publication [47]. The process depicted by the flow diagram in Figure 4.3 is divided into three processing sections, namely, (1) catalyst treatment and regeneration, (2) isomerization, and (3) product recovery.

4.8.2 COST ANALYSIS

A cost analysis of the process was developed from the information generated by the process simulation model. After the process diagram (Figure 4.3) was developed along with the equipment sizing and unit operations, capital and operating costs were then estimated in SuperPro Designer. These methods have generally been used to prepare conceptual cost estimates for industrial processes as recommended by the Association for the Advancement of Cost Engineering [74]. The economic results of the model were linked to the physical flow and unit operations defined in the process simulation model. The capital cost was estimated by summing all the equipment costs to which an installation factor of 3 was applied to cover all costs of the construction material, installation, and engineering and other charges associated with the construction of an industrial facility. Table 4.5 presents the capital costs of the isostearic acid process by section [72,73]. Table 4.6 lists the annual operating costs for the isostearic acid process, which includes raw material, labor, utility, and facility-dependent costs [72,73]. The equipment depreciation cost was calculated based on a 10-year straight-line depreciable life.

Based on the yearly production outputs for the three products, 5,089,626 kg (11,310,280 lb) of isostearic acids, 574,429 kg (1,276,509 lb) of saturated linear-chain fatty acids, and 238,103 kg (529,118 lb) of dimer fatty acids, the estimated unit production cost of the three products is $2.37/kg

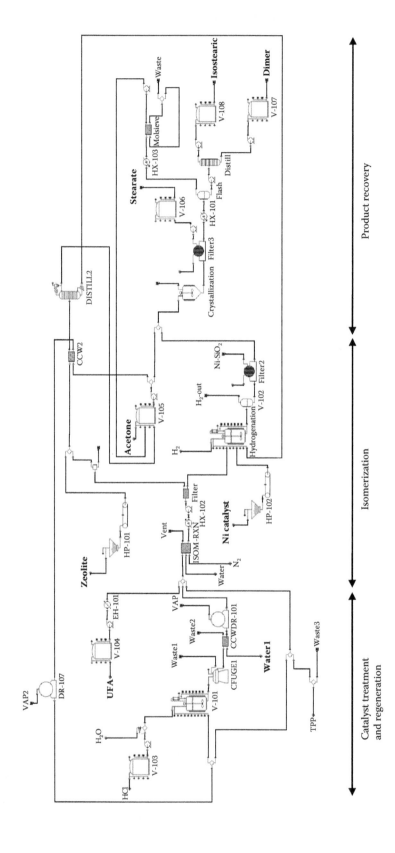

FIGURE 4.3 Simulated process flow diagram for commerical scale manufacture of isostearic acids. (From Ngo, H.L. et al., *J. Agr. Sci.*, 6, 158–168, 2014. With permission.)

TABLE 4.5
Model Estimated Capital Costs for Each of the Major Processing Sections

System	Capital Cost (US$ × 1000)
Catalyst Treatment and Regeneration	
Hydrochloric acid handling and storage	102
Makeup zeolite handling and storage	105
Zeolite activation	108
Zeolite filtering and centrifugation	372
Zeolite drying	276
Unlisted equipment for the zeolite area	168
Catalyst treatment area total	1131
Isomerization and Hydrogenation	
Oleic acid storage and handling	198
Isomerization reactor	1311
Zeolite separation	144
Nickel catalyst storage and handling	105
Hydrogenation reactor	150
Catalyst and solvent removal	180
Unlisted equipment for isomerization area	363
Isomerization and hydrogenation area total	2451
Product Recovery	
Acetone storage and handling	102
Recrystallization reactor	1200
Stearic separation and storage	357
Acetone recovery	999
Isostearic acids/dimer distillation column	168
Isostearic acid storage and handling	102
Dimer storage and handling	102
Unlisted equipment for product recovery area	525
Product recovery area total	3555
Total capital cost	**7137**

Source: Ngo, H.L. et al., *J. Agr. Sci.*, 6, 158–168, 2014.

($1.08/lb) [72,73]. This value of production cost is comparable to the current manufacture cost of commercial isostearic acids and could be improved with further process optimization.

A sensitivity study was also conducted on the model to evaluate the impact of catalyst reuse on production cost. The cost of zeolite constitutes over 58% of the total raw materials cost if the zeolite was discarded after each use, with a predicted production cost of $4.917/kg ($2.231/lb) [72,73]. This value can be significantly reduced by almost 50% ($2.769/kg or 1.256/lb) if the catalyst was reused 5 times (Figure 4.4) [72,73]. The production cost will be reduced as the extent of catalyst reuse increases, leveling off at $2.419/kg ($1.097/lb) after 15 uses [72,73]. At this point the cost of the catalyst becomes insignificant relative to the total operating cost of the process. This demonstrates the importance of catalyst reuse to the cost of production.

TABLE 4.6

Model Estimated Annual Operating Costs of Commercial Isostearic Acid Production

	Operating Cost (US$ × 1000)/Year
Raw Materials	
Oleic acid (91.2 wt%)	9344
Zeolite ferrierite	774
Nickel on silica	862
Acetone	496
Triphenylphosphine	19
Other raw materials	174
Utilities	
Electricity	137
Steam	280
Cooling agents	102
Labor	792
Facility Expenses	
Depreciation	712
Other facility expenses	324
Total operating cost	**14,016**

Source: Ngo, H.L. et al., *J. Agr. Sci.*, 6, 158–168, 2014.

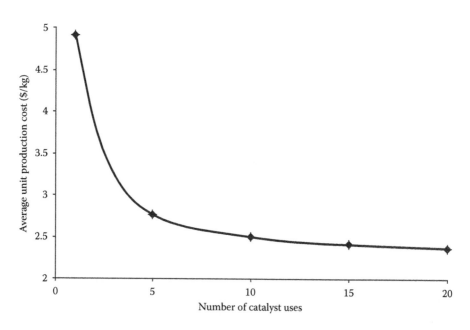

FIGURE 4.4 The impact of catalyst reuse on the cost of production. (From Ngo, H.L. et al., *J. Agr. Sci.*, 6, 158–168, 2014. With permission.)

4.9   CONCLUSION AND PERSPECTIVES

Isostearic acids are unique and important biobased products with superior properties. They are well-established biolubricants, but they are currently not utilized for this application owing to inefficient manufacturing processes and high cost. The zeolite process has the potential technology to replace the existing clay process. At present, the zeolite–base combination method has shown potential in diversifying the outputs of isostearic acid products and boosting the economic viability of this technology. This method to make isostearic acids is a good one as the surface blocking technique is unique. Zeolitic processes represent important new routes for producing isostearic acids, thus increasing consumption and profitability of biobased materials.

ABBREVIATIONS

2D: two dimensional
ASTM: American Society for Testing and Materials
CP: cloud point
FAME: fatty acid methyl ester
GC: gas chromatography
HFRR: high-frequency reciprocating rig
MP: melting point
MS: mass spectrometry
PP: pour point
TOF: time of flight
VI: viscosity index

REFERENCES

1. U. Biermann, W. Friedt, S. Lang, W. Lühs, G. Machmüller, J.O. Metzger, M.R. Klaas, H.J. Schäfer, and M.P. Schneider, New syntheses with oils and fats as renewable raw materials for the chemical industry, *Angew. Chem. Int. Ed.*, 39, 2206–2224 (2000).
2. A. Behr and J.P. Gomes, The refinement of renewable resources: New important derivatives of fatty acids and glycerol, *Eur. J. Lipid Sci. Tech.*, 112, 31–50 (2010).
3. S.Z. Erhan and M.O. Bagby, Vegetable-oil-based printing ink formulation and degradation, *Ind. Crops Prod.*, 3, 237–246 (1995).
4. L.A. Honary, Biodegradable biobased lubricants and greases, *Machinery Lubrication* (2001 September).
5. R.D. O'Brien (Ed.), *Fats and Oils Formulating and Processing for Applications*, Boca Raton, FL, CRC Press (2000).
6. F.D. Gunstone (Ed.), *The Chemistry of Oils and Fats: Sources, Composition, Properties and Uses*, Oxford, Blackwell Publishing (2004).
7. S.J. Randles and M.A. Wright, Environmentally considerate ester lubricants for the automotive and engineering industries, *J. Synth. Lubr.*, 9, 145–161 (1992).
8. N.S. Battersby, S.E. Pack, and R.J. Watkinson, A correlation between the biodegradability of oil products in the CEC L-33-T-82 and modified Sturm tests, *Chemosphere*, 24, 1998–2000 (1998).
9. H.J. Dutton and C.R. Scholfield, Recent developments in the glyceride structure of vegetable oils, *Prog. Chem. Fats Other Lipids*, 6, 313–339 (1963).
10. R.W. Johnson and E. Fritz (Eds.), *Fatty Acids in Industry*, New York, Marcel Dekker (1989).
11. H.L. Ngo, R.O. Dunn, B.K. Sharma, and T.A. Foglia, Synthesis and physical properties of isostearic acids and their esters, *Eur. J. Lipid Sci. Tech.*, 113, 180–188 (2011).
12. G. Weitzel, "Method of Increasing the Water Vapor Porosity of Fat Materials," U.S. Patent 3,035,987 (1962).
13. W. Gunther, "Compositions Comprising Alkyl Branched, Long-Chain, Aliphatic Compounds for Increasing the Water Vapor Porosity of Fat Materials," U.S. Patent 3,335,053 (1967).
14. Shell Chemical Bulletins on Cardura E, Glycidyl Ester of Versatic 911, Shell Chemical Co., Houston.
15. G.J. Benoit, "Lubricant Composition," U.S. Patent 3,110,673 (1963).
16. G.J. Benoit, "Fatty Acid Polyamide," U.S. Patent 3,169,980 (1965).

17. R.T. Schlobohm and H.D. Millay, "Lubricating Oil Compositions," U.S. Patent 3,326,801 (1967).

18. A. Onopchenko and J.G.D. Schulz, "Composition Containing Higher Fatty Acids," U.S. Patent 3,842,106 (1974).

19. R.S. Sayad and A.D. McMahan, "Method and Composition for Reducing Friction on Conveyors," U.S. Patent 3,336,225 (1967).

20. T. Sato, M. Okazaki, Y. Watanabe, and M. Onoda, "Water Soluble Metal Machining Oil," Japan Patent 1978000101854 (1980).

21. H.R. Fife and W.J. Toussaint, "Esters of Polyoxyalkylene Diols," U.S. Patent 2,457,139 (1948).

22. B.W. Landfried, H.J. Bassett, and J.R. Moneymaker, "Frozen Connections Containing Glycerol and Propylene Glycol Monoesters of Isostearic Acid," U.S. Patent 3,515,562 (1970).

23. R.J. Anderson and E. Chargaff, The chemistry of the lipoids of tubercle bacilli: VI. Concerning tuberculostearic acid and phthioic acid from the acetone-soluble fat, *J. Biol. Chem.*, 85, 77 (1929).

24. J.D. Townsend, J. Latus, P.J. McCoy, A.J. Maltby, and D.A. Parker, "Branched Acids," U.S. Patent US 2015/0005211 A1 (2015).

25. G. Weitzel, "Method of Increasing the Water Vapor Porosity of Fat Materials," U.S. Patent 3,035,987 (1962).

26. M.A. Liebert, Final report on the safety assessment of isostearic acid, *Int. J. Toxicol.*, 2, 61–74 (1983).

27. M.J.A.M. den Otter, The dimerization of oleic acid with a montmorillonite catalyst I: Important process parameters; some main reactions, *Fette Seifen Anstrichm.*, 72, 667–673 (1970).

28. M.J.A.M. den Otter, The dimerization of oleic acid with a montmorillonite catalyst II: GLC analysis of the monomer; the structure of the dimer; a reaction model, *Fette Seifen Anstrichm.* 72, 875–883 (1970).

29. T.A. Foglia, T. Perlstein, Y. Nakano, and M. Gerhard, "Process for the Preparation of Branched Chain Fatty Acids and Esters," U.S. Patent 4,371,469 (1983).

30. Y.Y. Nakano, T.A. Foglia, H. Kohashi, T. Peristein, and S. Serota, Thermal alteration of oleic acid in the presence of clay catalysts with co-catalysts, *J. Am Oil Chem. Soc.*, 62, 888–891 (1985).

31. K.S. Hayes, "Polymerization of Fatty Acids," U.S. Patent 4776983 (1988).

32. http://www.arizonachemical.com/en/Products/Chemical-Intermediates/Iso-stearic-Acid/.

33. "What Is the Global Market Outlook for Polymerised Fatty Acid Dimers and Their Monomer—Isostearic Acid," Frost & Sullivan (2006 July).

34. W.R. Hodgson, "Fatty Acid Isomerization," Patent EP 0774451A1 (1996).

35. T. Tomifuji, H. Abe, Y. Matsumura, and Y. Sakuma, "Process for the Preparation of Branched Chain Fatty Acids and Alkyl Esters Thereof," U.S. Patent 5,677,473 (1997).

36. S. Zhang, Z. Zhang, and D. Steichen, "Skeletal Isomerization of Alkyl Esters and Derivatives Prepared Therefrom," U.S. Patent 20030191330 (2003).

37. S. Zhang, Z. Zhang, and D. Steichen, "Skeletal Isomerization of Alkyl Esters and Derivatives Prepared Therefrom," U.S. Patent 6,946,567 (2005).

38. S. Zhang, Z. Zhang, and D. Steichen, "Skeletal Isomerization of Fatty Acids," Patent WO 03/006157 A2 (2002).

39. S. Zhang and Z. Zhang, "Fatty Acid Isomerization with Mesoporous Zeolites," U.S. Patent 6,723,862 (2004).

40. K.Y. Jacobs and R.A. Schoonheydt, Clays: From two to three dimensions, in: *Introduction to Zeolite Science and Practice*, Second Completely Revised and Expanded Edition, H. van Bekkum, E.M. Flanigen, P.A. Jacobs, and J.C. Jansen (Eds.), pp. 299–343, Amsterdam, Elsevier (2001).

41. H.L. Ngo, E. Hoh, and T.A. Foglia, Improved synthesis and characterization of saturated branched-chain fatty acid isomers, *Eur. J. Lipid Sci. Tech.*, 114, 213–221 (2012).

42. H.L. Ngo and T.A. Foglia, "Process for Preparing Saturated Branched-Chain Fatty Acids," U.S. Patent 8,748,641 B2 (2014).

43. L.B. McCusker and C. Baerlocher, Zeolite structures, in: *Introduction to Zeolite Science and Practice*, Second Completely Revised and Expanded Edition, H. van Bekkum, E.M. Flanigen, P.A. Jacobs, and J.C. Jansen (Eds.), pp. 37–67, Amsterdam, Elsevier (2001).

44. H.H. Mooiweer, K.P. De Jong, B. Kraushaar-Czarnetzki, W.H.J. Stork, and B.C.H. Krutzen, Skeletal isomerization of olefins with the zeolite ferrierite as catalyst, *Stud. Surf. Sci. Catal.*, 84, 2327–2334 (1994).

45. K.P. De Jong, Efficient catalytic processes for the manufacturing of high-quality transportation fuels, *Catal. Today*, 29, 171178 (1996).

46. W. Xu, Y. Yin, S.L. Suib, J.C. Edwards, and C. O'Young, Modification of non-template synthesized ferrierite/ZSM-35 for *n*-butene skeletal isomerization to isobutylene, *J. Catal.*, 163, 232–244 (1996).

47. H.L. Ngo, Improved zeolite regeneration processes for preparing saturated branched-chain fatty acids," *Eur. J. Lipid Sci. Tech.*, 116, 645–652 (2014).

48. ASTM D5773: Standard Test Method for Cloud Point of Petroleum Products (Constant Cooling Rate Method), ASTM International, West Conshohocken, PA (2003).

49. ASTM D5949: Standard Test Method for Pour Point of Petroleum Products (Automatic Pressure Pulsing Method), ASTM International, West Conshohocken, PA (2003).

50. American Oil Chemists' Society Cd 12b-92: Oil Stability Index (OSI), *Official Methods and Recommended Practices of the American Oil Chemists' Society*, Fifth Edition, AOCS Press, Champaign, IL (1999).

51. ASTM D445: Standard Test Method for Kinematic Viscosity of Transparent and Opaque Liquids (and the Calculation of Dynamic Viscosity), ASTM International, West Conshohocken, PA (2003).

52. ASTM D2270: Standard Practice for Calculating Viscosity Index from Kinematic Viscosity at 40 and 100°C, ASTM International, West Conshohocken, PA (2003).

53. G. Knothe and R.O. Dunn, A comprehensive evaluation of the melting points of fatty acids and esters determined by differential scanning calorimetry, *J. Am. Oil Chem. Soc.*, 86, 843–856 (2009).

54. H. Imahara, E. Minami, and S. Saka, Thermodynamic study on cloud point of biodiesel with its fatty acid composition, *Fuel*, 85, 1666–1670 (2006).

55. F.D. Gunstone, *An Introduction to the Chemistry and Biochemistry of Fatty Acids and Their Glycerides*, London, Chapman & Hall (1967).

56. L. Hernqvist, Crystal structures of fats and fatty acids, in: *Crystallization and Polymorphism of Fats and Fatty Acids*, pp. 97–137, New York, Marcel Dekker (1988).

57. K. Larsson and P.J. Quinn, Physical properties: Structural characteristics, in: *The Lipid Handbook*, pp. 401–430, London, Chapman & Hall (1994).

58. B.R. Moser, Comparative oxidative stability of fatty acid alkyl esters by accelerated methods, *J. Am. Oil Chem. Soc.*, 86, 699–706 (2009).

59. G. Knothe and K.R. Steidley, Kinematic viscosity of biodiesel fuel components and related compounds: Influence of compound structure and comparison to petrodiesel fuel components, *Fuel*, 84, 1059–1065 (2005).

60. ASTM D6079: Standard Test Method for Evaluating Lubricity of Diesel Fuels by the High-Frequency Reciprocating Rig (HFRR), ASTM International, West Conshohocken, PA (1999).

61. G. Knothe and K.R. Steidley, Lubricity of components of biodiesel and petrodiesel: The origin of biodiesel lubricity, *Energy Fuels*, 19, 1192–1200 (2005).

62. M.J. Haas, A.J. McAloon, W.C. Yee, and T.A. Foglia, A process model to estimate biodiesel production costs, *Bioresour. Technol.*, 97, 671–678 (2006).

63. S.A. Abo El-Enin, N.K. Attia, N.N. El-Ibiari, G.I. El-Diwani, and K.M. El-Khatib, In-situ transesterification of rapeseed and cost indicators for biodiesel production, *Renew. Sust. Energy Rev.*, 18, 471–477 (2013).

64. B. Taylor, N. Xiao, J. Sikorski, M. Yong, T. Harris, T. Helme, A. Smallbone, A. Bhave, and M. Kraft, Techno-economic assessment of carbon-negative algal biodiesel for transport solutions, *Appl. Energy*, 106, 262–274 (2013).

65. J.R. Kwiatkowski, A.J. McAloon, F. Taylor, and D.B. Johnston, Modeling the process and costs of fuel ethanol production by the corn dry-grind process, *Ind. Crops Prod.*, 23, 288–296 (2006).

66. C. Li, L.F. Rodriguez, M. Khanna, A.D. Spaulding, T. Lin, and S.R. Eckhoff, An engineering and economic evaluation of quick germ-quick fiber process for dry-grind ethanol facilities: Model description and documentation, *Bioresour. Technol.*, 101, 5275–5281 (2010).

67. N.P. Nghiem, E.C. Ramirez, A.J. McAloon, W. Yee, D.B. Johnston, and K.B. Hicks, Economic analysis of fuel ethanol production from winter hulled barley by the EDGE (enhanced dry grind enzymatic) process, *Bioresour. Technol.*, 102, 6696–6701 (2011).

68. E.C. Ramirez, D.B. Johnston, A.J. McAloon, W. Yee, and V. Singh, Engineering process and cost model for a conventional corn wet milling facility, *Ind. Crops Prod.*, 27, 91–97 (2008).

69. E.C. Ramirez, D.B. Johnston, A.J. McAloon, and V. Singh, Enzymatic corn wet milling: Engineering process and cost model, *Biotechnol. Biofuels*, 2, 2–10 (2009).

70. C. Ng, W.E. Marshall, R.M. Rao, R.R. Bansode, and J.N. Losso, Activated carbon from pecan shells: Process description and economic analysis, *Ind. Crops Prod.*, 17, 209–217 (2003).

71. I.M. Lima, A.J. McAloon, and A.A. Boateng, Activated carbon from boiler litter: Process description and cost of production, *Biomass Bioenergy*, 32, 568–572 (2008).

72. H.L. Ngo, W.C. Yee, A.J. McAloon, and M. Haas, Process and cost modeling of saturated branched-chain fatty acid isomer production, *Ind. Eng. Chem. Res.*, 51, 12042–12045 (2012).

73. H.L. Ngo, W.C. Yee, A.J. McAloon, and M. Haas, Techno-economic analysis of an improved process for producing saturated branched-chain fatty acids, *J. Agr. Sci.*, 6, 158–168 (2014).
74. Association for the Advancement of Cost Engineering International, Conducting technical and economic evaluations in the process and utility industries, in: AACE Recommended Practices and Standards, Morgantown, West Virginia, AACE International 84 (1990).

5 Producing Monomers and Polymers from Plant Oils*

Kenneth M. Doll, Bryan R. Moser,
Zengshe Liu, and Rex E. Murray

CONTENTS

ABSTRACT

The integration of biobased industrial products into existing markets, where petrochemically derived materials currently dominate, is a worthy objective. This chapter reviews some technologies that have been developed including olefins of various chain lengths, photocurable polymers, vinyl monomers, and biobased oxygen-containing monomers. These products are needed in large markets, such as the lubricants market, where significant economic impact is anticipated. Also, potential uses are possible in products such as absorbent materials, adhesives, coatings, elastomers, latices, paints, plasticizers, plastics, printing inks, resins, surfactants, and textiles. This chapter is divided into several closely related sections on monomers and polymers. The first focuses on hydrocarbon-based monomers from fatty acids, followed by oxygen-containing monomers from lipids, then nonlipids. The final section reviews advanced polymerization of vegetable oils. Overall, the research in this area will eventually lead to the development of direct drop-in biobased replacements for nonrenewable commodity

---

* Mention of trade names or commercial products in this chapter is solely for the purpose of providing specific information and does not imply recommendation or endorsement by the U.S. Department of Agriculture (USDA).

chemicals. This will enhance the agricultural sector while simultaneously expanding our portfolio of renewable materials and reducing dependence on petroleum.

5.1   USE OF NATURAL FEEDSTOCKS TO PRODUCE INDUSTRIAL MATERIALS

The idea that natural materials can be used to replace petroleum products has been around as long as petroleum products themselves. There are several excellent reviews on several of these topics. Some recent examples include biomass processing [1], levulinic acid from carbohydrates [2], and polymer synthesis from caprolactone [3] and carbon dioxide [4,5]. Numerous comprehensive reviews relating to industrial applications of plant oils are also well represented, dating back more than 30 years [6]. Topics such as polymers [7–10] and lubricants [11] are particularly well covered. Other nonfood industrial uses of plant oils include detergents, paints, plasticizers, linoleum, textiles [12], adhesives, cosmetics, and many others [13]. Of course, no discussion of nonfood uses of plant oils would be complete without mentioning biodiesel. This fuel is now a worldwide industry and, along with other renewable fuels, has initiated an economic linkage between the energy and agricultural sectors [14,15]. The emergence of biodiesel has resulted in a dramatic increase in worldwide production of fats and oils, which has more than doubled since 1985 when annual production was only 68.2 million metric tons [16].

Despite these examples of commercial success, the industry has been hesitant to adopt naturally derived products in greater volumes. Much of this is owing to performance and cost differences, real or perceived, that accompany biobased materials. Natural oils and fats are rightly thought to have poor oxidation stability [17,18], a characteristic that carries through into many of the industrial products derived from them. In addition, the performance of many of these renewable materials is lower than those of the conventional petrochemical products they are intended to replace.

The cost competitiveness of natural products is also an issue. Many of these raw materials, although considered commodity products, are more costly than their petroleum counterparts. Additional issues such as the need for new processing technologies, feedstock variability and availability, and the food versus fuel debate leave only a limited pool of willing industrial adopters, even with favorable life cycle assessments [19] and biopreferred programs [20]. All of these potential pitfalls emphasize the importance of targeting specific market areas for incorporation of biobased products. Those are areas where cost competitiveness and performance equality are actually possible.

One approach around this problem is to provide industry with the building blocks that they already use, but made from natural sources as opposed to petroleum. This will overcome unwillingness by industry to use biobased products, because these materials will represent direct drop-in replacements for existing petrochemically derived products. Such an approach has already been applied to liquid transportation fuels [21], and some of the attempts to produce higher-value biobased products, such as olefinic and oxygenated monomers, are detailed in this chapter.

5.2   RENEWABLE HYDROCARBON MONOMERS FROM FATTY ACIDS

The idea of producing hydrocarbons from natural fatty acids has been around for almost 100 years, where early work used alumina catalyst and heat to achieve deoxygenation [22]. Hydrocarbon monomers from renewable feedstocks are of particular interest now, considering the ubiquitous nature of industrial polymers encountered in every facet of modern life.

There are three approaches to preparing hydrocarbons from fatty acids that have received significant attention: hydrogenation, decarboxylation, and decarbonylation (Figure 5.1). All are thermodynamically favored [23,24] to produce the desired hydrocarbon by elimination of carbon dioxide and/or carbon monoxide and/or water.

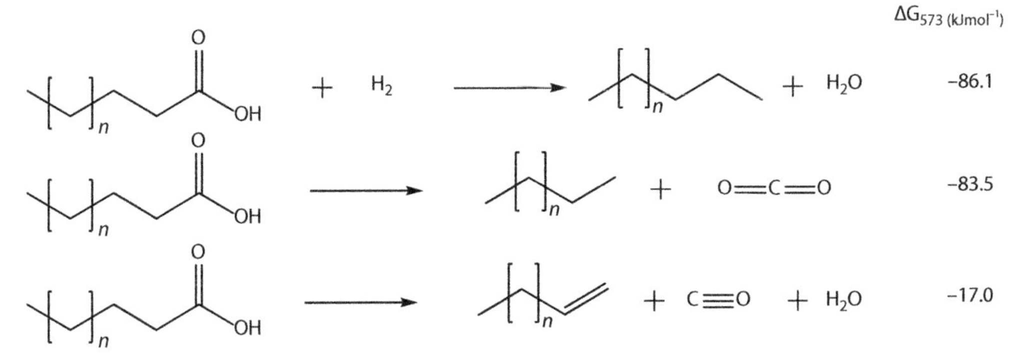

FIGURE 5.1 Thermodynamic data of hydrogenation, decarboxylation, and decarbonylation reactions. (Based on M. Snåre et al., *Ind. Eng. Chem. Res.*, 45, 5708–5715, 2006; M. Czaun et al., *ChemSusChem.*, 4, 1241–1248, 2011.)

The hydrogenation of carboxylic acids and esters is the most straightforward and well studied [21,25]. A major drawback of this method is the requirement of hydrogen, which can be expensive, and its use increases the level of safety required. Additionally, it produces fully saturated hydrocarbons. These are not desirable for either fuel or lubricant applications owing to poor low-temperature properties such as pour point and cloud point.

Decarbonylation produces an unsaturated hydrocarbon as well as carbon monoxide through a mechanism that involves the coordination of a carboxylate moiety to the metal center of a catalyst in an oxidative addition. The first observation of this reaction, catalyzed by ruthenium and osmium, was made approximately 50 years ago [26]. Methodologies using palladium or rhodium catalysts to decarbonylate acids of 8 to 12 carbons were later developed [27,28] and, finally, extensions to biobased fatty acids were made [23,29–37]. Although catalytic, there are some drawbacks to this technology. First, for efficacy, the mechanism requires either the addition of or the in situ production of a hydrogen acceptor such as an anhydride. This disadvantage is best demonstrated by the example of the palladium-catalyzed decarbonylation of decanoic acid. Turnovers of up to 12,370 were observed when equimolar acetic anhydride was added, whereas less than 5 turnovers were noted without the anhydride [28].

Decarboxylation may produce either saturated or unsaturated hydrocarbons along with carbon dioxide as a by-product, which could be considered a drawback, since anthropogenic carbon dioxide emissions are implicated in global climate change. However, it is important to note that this carbon in the substrate that is converted to carbon dioxide is less than 6% of the carbon content of an 18-carbon fatty acid such as oleic acid. In other words, the amount of carbon dioxide produced is trivial compared to the carbon content of the biobased polymers possible using these monomers. Additionally, the carbon dioxide is essentially recycled back into the plant in the following growing season, thus leading to no net change in global atmospheric emissions. In contrast, carbon dioxide produced from petrochemically derived sources contributes to global anthropogenic greenhouse gas emissions because that carbon was sequestered from the atmosphere millions of years ago. Thus, a net increase in atmospheric carbon dioxide is caused by consumption of petroleum-derived products, whereas biobased materials that generate carbon dioxide are neutral or even potentially cable of causing a decrease with respect to carbon dioxide emissions. Also, hidden sources of carbon dioxide, such as those generated by the production of hydrogen, should be considered when comparing competing technologies from a carbon footprint perspective [38]. However, life cycle assessments [39] and the topic of green chemistry [40,41] as they relate to greenhouse gas emissions are beyond the scope of this introduction.

Thermal decarboxylation began with simple oxide materials [22] on various branched [42] and cyclic [43] compounds. A number of zeolites were also catalytically active [44–47] at temperatures ranging from 370°C to 440°C. This technology is effective at producing fuels [14,48–50], but the reaction conditions are not amenable for the production of a narrow range of alkene monomers. More controllable chemistry is required to satisfy such industrial needs.

Addition of a metal catalyst that works in an oxidative mode is an approach that has gained attention. Silver in the +2 oxidation state initially coordinates to a fatty acid, which then undergoes an electron transfer that leads to a reduced silver species, a proton, and a carbonyl radical. Under most conditions, the carbonyl radical is unstable and will spontaneously decarboxylate to yield a hydrocarbon [51]. A copper cocatalyst is needed if an unsaturated hydrocarbon product is desired [52]. Although this reaction works under mild conditions, it leaves the silver and/or copper in an inactive oxidative state. Regeneration of the catalysts, therefore, requires an oxidant such as sodium persulfate.

Another method to reduce the reaction barrier to decarboxylation, without the need for sodium persulfate, is by manipulation of the isomeric structure of the substrate. Specifically, unsaturated carboxylic acids in which the double bond is conjugated or nearly conjugated with the carboxylate moiety exhibit significant enhancement in the rate of decarboxylation compared to isomers where the double bond is not proximate to the carbonyl. This is probably because of the availability of a cyclic transition state for the beta-gamma isomer, which is not possible in other isomers [42]. Therefore, decarboxylation can be conducted at considerably lower temperatures when the carboxylic acid moiety is conjugated with a double bond (Figure 5.2).

Heating plant oils results in indiscriminate isomerization [53], but controlled isomerization is needed to obtain well-defined industrial products. Iodine [13], enzymes [54], and organometallic complexes with complicated phosphine ligands [55–59] isomerize double bonds in plant oil-based materials. Transition metal carbonyls and related complexes are especially effective. In one route, iron carbonyl was used to isomerize fatty acids [60], but this catalyst is difficult to work with because of its volatility. In contrast, triruthenium dodecacarbonyl is easy to handle and will spontaneously form a catalyst precursor when reacted with a carboxylic acid [61]. This precursor facilitates isomerization of alkenes, with [62] or without [63,64] added phosphine ligands. Isomerization of the double bond of an unsaturated fatty acid into a position where subsequent decarboxylation occurs at a lower energy level is possible utilizing this precatalyst.

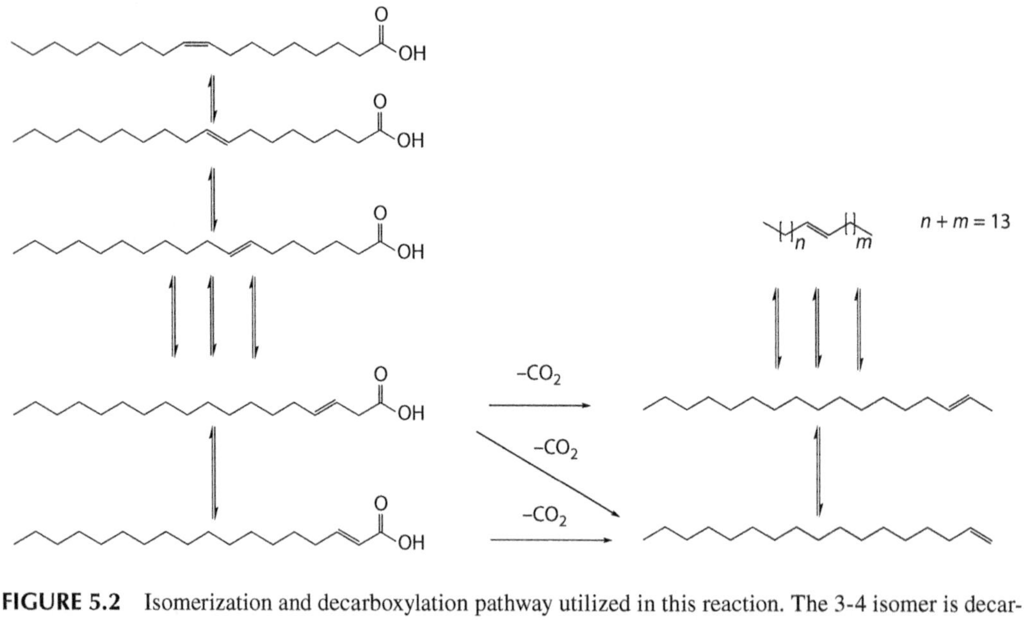

FIGURE 5.2 Isomerization and decarboxylation pathway utilized in this reaction. The 3-4 isomer is decarboxylated preferentially because of its ability to form a cyclic transition state. (Based on R. T. Arnold et al., *J. Am. Chem. Soc.*, 72, 4359–4361, 1950; R. T. Arnold and M. J. Danzig, *J. Am. Chem. Soc.*, 79, 892–893, 1957.)

5.3 RENEWABLE OXYGENATED MONOMERS FROM UNSATURATED LIPIDS

Renewable oxygenated monomers from plant oils have been known for decades. Classic examples include azelaic (nonanedioic), adipic (hexanedioic), and brassylic (tridecanedioic) acids, which are polymer building blocks made from ozonolysis of oleic, petroselinic, and erucic acids, respectively [65–67]. Furthermore, sebacic (decanedioic) acid is obtained via caustic oxidative cleavage of ricinoleic (12-hydroxyoctadec-9-Z-enoic) acid [68]. Linear renewable diacids are used in the manufacture of commercially important polymers as well as fragrances, lubricants, and adhesives [69]. Addition of carbon monoxide to double bonds of fatty acids also yields diacids. A branched 19-carbon diacid is produced from oleic acid. Such diacids have broad applications as monomers for lubricants, plasticizers, polyurethanes, epoxy resins, coatings, and transparent polyamide plastics [70].

5.3.1 MONOMERS BASED ON CARBON MONOXIDE ADDITION

Carbon monoxide reacts with unsaturated lipids in at least three ways: hydroformylation (oxo reaction), the Reppe reaction, and the Koch reaction (Figure 5.3). Hydroformylation yields an aldehyde intermediate that can be readily converted to an acid, alcohol, or amine to provide a 19-carbon diacid, hydroxy acid, or amino acid [70–73]. Hydroformylation of polyunsaturated fatty acids gives polyacids, polyols, or polyamines [73,74]. The Reppe reaction introduces carboxyl groups directly to unsaturated fatty acids, thereby bypassing the aldehyde intermediate [70]. Using ammonia instead of water during the reaction yields a fatty amido acid [75]. The Koch reaction is performed at lower temperature and pressure than those for the Reppe reaction but requires concentrated sulfuric acid as both a coreagent and a solvent [75]. Unlike ozonolysis, which yields linear diacids shorter in length than the starting acid, carbon monoxide addition results in branched diacids with one more carbon than the starting acid.

5.3.2 MONOMERS FROM DIMERIZATION OF FATTY ACIDS

Dimerization affords branched diacids (dimer acids) with twice the number of carbons as the starting acid. Dimer acids are prepared by heating unsaturated fatty acids under pressure in the presence of an acidic clay catalyst. Condensation of dimer acids with diamines yields polyamide resins useful as curing agents and coatings [75]. Additionally, ene reactions of unsaturated fatty acids with maleic

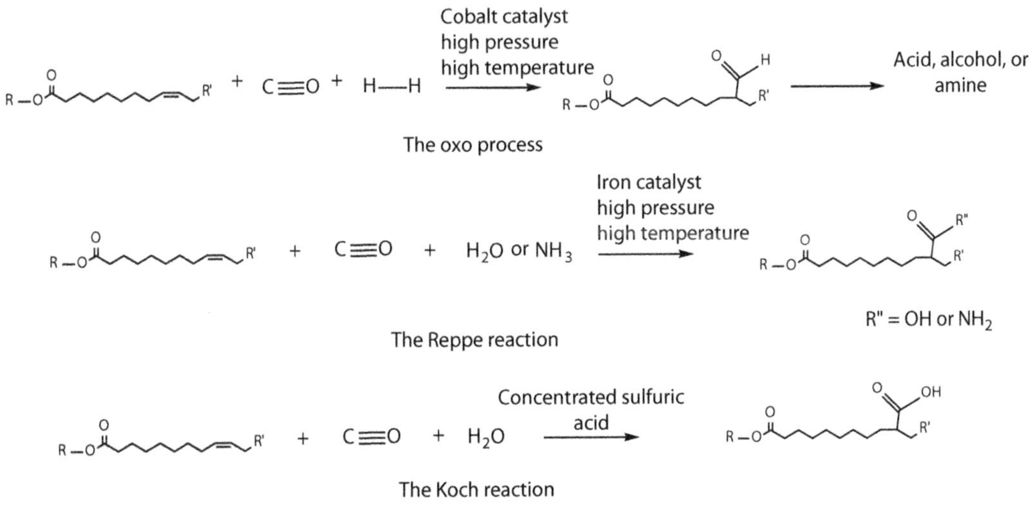

FIGURE 5.3 Reactions between carbon monoxide and lipids which produce lipids with increased functionality.

anhydride yield diacids. For instance, addition of maleic anhydride to oleic acid in the presence of a Lewis acid catalyst results in a 22-carbon branched diacid. Subsequent condensation with polyols affords copolymers useful as adhesives, film formers, and tackifiers [76].

5.3.3 Castor Oil-Based Monomers

The principal fatty acid in castor oil is ricinoleic acid (Figure 5.4), which is both hydroxylated and unsaturated. As a consequence of its unique multifunctional composition, castor oil is used extensively as an industrial oil. In addition to sebacic acid mentioned previously, ricinoleic acid is used to produce a variety of products including estolides via polycondensation; polyols by transesterification with diols or other polyhydroxylated compounds; 10-hydroxydecanoic acid by low-temperature alkali cleavage; and 10-undecenoic (undecylenic) acid from pyrolysis [68,77,78]. Undecylenic acid is used as a platform chemical for the production of numerous monomers [79]. For example, polyamide 11 is obtained from 11-aminoundecanoic acid, which is prepared via an 11-bromo intermediate [68,77,78]. Dienoic monomers suitable for free radical or acyclic diene metathesis polymerization have also been prepared from undecylenic acid. These include allyl, vinyl, acrylic, and methacrylic ethers and esters, all of which have applications in the coatings industry [80]. Dienes suitable for acyclic diene metathesis polymerization include 10-undecenyl undecylenate and undecylenic acid moieties esterified to phosphorus-containing heteroaromatic cores as comonomers for flame-resistant polyesters [81,82]. Finally, a variety of renewable monomers can be prepared via thiol-ene click chemistry in which thiolated compounds are added to undecylenic acid via a free radical mechanism to yield anti-Markovnikov products [83]. Such monomers are useful for production of ultraviolet (UV)–curable resins and coatings. Specific examples of monomers from undecylenic acid include those derived by addition of butanedithiol, thioacetic acid, thioglycerol, 4-hydroxybutanethiol, and methyl 2-mercaptoacetate, among others [83–85]. Methyl 9-decenoate (Figure 5.5 and next subsection), prepared by cross metathesis of methyl oleate with ethene (ethenolysis), represents the 10-carbon analogue of methyl 10-undecenoate [86].

5.3.4 Monomers Synthesized Using Alkene Metathesis

Olefin metathesis is a versatile method for building carbon–carbon bonds and has been exploited by the petrochemical industry for decades using molybdenum- and tungsten-based catalysts.

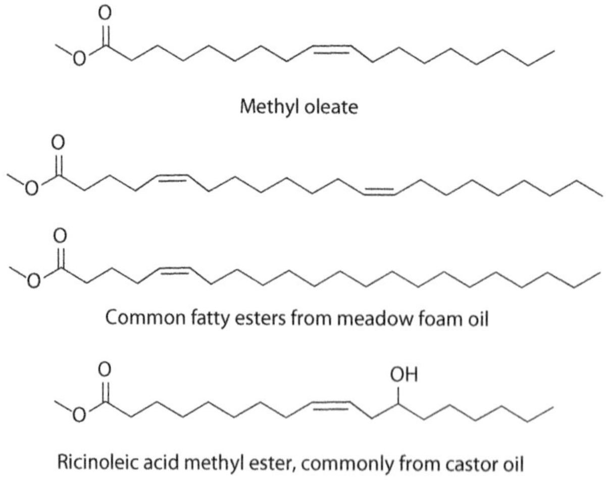

Methyl oleate

Common fatty esters from meadow foam oil

Ricinoleic acid methyl ester, commonly from castor oil

FIGURE 5.4 The structures of methyl oleate and the common fatty materials from castor and meadowfoam oils.

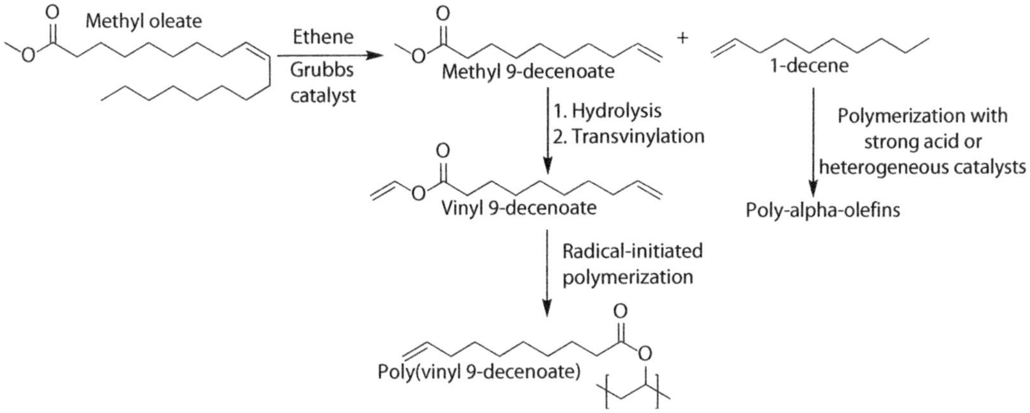

FIGURE 5.5 Renewable oxygenated cross metathesis reaction monomers of methyl oleate and ethene. (Based on S. H. Hong et al., *J. Am. Chem. Soc.*, 127, 17160–17161, 2005; H. Kohashi and T. A. Foglia, *J. Am. Oil Chem. Soc.*, 62, 549–554, 1985.)

The reaction proceeds by the cleavage and reformation of carbon–carbon double bonds and the simultaneous exchange of substituents. The fundamental metathesis transformations include self-metathesis, cross metathesis, ring-closing metathesis, ring-opening metathesis acyclic diene metathesis polymerization, and ring-opening metathesis polymerization. Application of metathesis chemistry to fats and oils was limited until the discovery of active ruthenium catalysts by Grubbs, which, in 2005, earned him the Nobel Prize in Chemistry [87]. Olefin metathesis is an appealing industrial reaction because it is cost effective and catalytic, generates almost no unusable by-products, and is easily scalable [88]. Major petrochemical applications of metathesis include conversion of relatively inexpensive propene into more valuable ethene and butene by the Shell Higher Olefins Process, to produce linear alpha olefins and their derivatives from oligomerized ethene [89].

Self-metathesis of fatty acids by using Grubbs catalysts is an efficient route to unsaturated diacids. For instance, self-metathesis of methyl oleate affords, as one of its products, an 18-carbon α,ω-diester which can be converted to an 18-carbon α,ω-dicarboxylic acid by hydrolysis of the ester groups [90]. Methyl 9-decenoate also yields an 18-carbon α,ω-dicarboxylic acid after self-metathesis and hydrolysis [91]. A mixture of diacids of differing chain lengths is obtained by isomerizing the double bonds in the fatty acids prior to self-metathesis or concurrently with the addition of isomerization catalysts compatible with ruthenium metathesis catalysts [92,93]. Metathesis itself may lead to isomerization, especially if older-generation molybdenum- and tungsten-based catalysts are used. Unwanted isomerization complicates the product mixture and reduces the yield of the intended product [94]. Fortunately, olefin isomerization during metathesis can be prevented with a radical quencher such as 1,4-benzoquinone [95]. Cross metathesis and hydrolysis of unsaturated fatty esters with short-chain unsaturated diesters yield diacids with chain lengths different from those of diacids obtained from self-metathesis. For example, methyl oleate cross metathesized with dimethyl 3-hexenedioate and diisopropyl 4-octenedioate generates 12- and 14-carbon unsaturated diesters, respectively [96].

Cross metathesis of fatty acid alkyl esters with alkenes is a versatile route to fatty esters with double bonds located at specific positions. For instance, metathesis of ethene with methyl oleate yields methyl 9-decenoate and 1-decene (Figure 5.5), which can be used as the starting point for a variety of chemistries [86]. Metathetical cleavage of meadowfoam oil methyl esters (Figure 5.4) with ethene provides methyl 5-hexenoate along with 1-hexadecene. Subsequent self-metathesis of methyl 5-hexenoate affords dimethyl 5-decenedioate as a diester suitable for polymerization [97]. Other alkenes such as 1- and 2-butenes, pentene, 1-hexene, 1-heptene, 2- and 4-octenes, 1-decene, and 1-octadecene have also been cross metathesized with plant oils and fatty esters [75,98,99]. Further transformation of these cross metathesized products at the double bond position yields various α,ω-bifunctional compounds suitable for polymerization, including diacids, amino acids, and cyano acids [100,101]. Additional functional groups can also be introduced by cross metathesizing fatty esters with functionalized alkenes such as methyl acrylate, allyl chloride, acrylonitrile, and fumaronitrile [102–107]. The nitrile group is then readily converted to an acid or amine, thus providing starting materials for the synthesis of polyesters and polyamides [100,101]. Similarly, α,ω-aminoesters are prepared from fatty esters via simultaneous cross metathesis with acrylonitrile and hydrogenation [108]. Lastly, cross metathesis of methyl oleate with 1-allyl-2,3-acetonide-glycerol followed by deprotection (to remove the acetonide-protecting group) and dihydroxylation of the double bond affords a fatty polyol [109].

In the industrial use of metathesis technology, the 18-carbon diacids from self-metathesis of oleic-rich oils serve as precursors to polyamides, polyurethanes, lubricants, and adhesives. Other products from the cross metathesis reaction with ethene of high-oleic plant oils to yield 1-decene and methyl 9-decenoate (Figure 5.5) are also industrially significant [88]. Decene is a valuable α-olefin used in the production of polyalphaolefins (PAOs) and other important chemicals [110]. However, markets for and applications of methyl 9-decenoate are still under development. In addition, practical and scalable methods for production of this emerging biobased chemical intermediate are needed. Existing literature on chemical modification of methyl 9-decenoate includes metathetical dimerization to yield an 18-carbon diacid, identical to that obtained from self-metathesis of methyl oleate [111]. Other chemical modifications include oxidative hydroformylation to yield an 11-carbon diacid for polyesters; derivatization to 10-aminodecanoic acid needed for nylon-10 production; and epoxidation to 9,10-epoxydecanoic acid used as a monomer for epoxy resins [112]. Additionally, transesterification of methyl 9-decenoate with simple diols followed by acyclic diene metathesis polymerization yields unsaturated polyesters which are used in adhesives, coatings, fibers, and resins and which are potentially biodegradable [113]. Nonpolymer applications of methyl 9-decenoate include hydrolysis and hydrogenation to yield decanoic acid or decanol, both of which are used in the synthesis of lubricants and plasticizers [114]. Methyl 9-decenoate is also used to produce fragrances (9-decen-1-ol), pheromones (9-oxo-*trans*-2-decenoic acid), and prostaglandin intermediates (9-oxodecanoic acid) [115,116].

5.4   RENEWABLE OXYGENATED MONOMERS FROM NONLIPIDS

A number of nonlipid, biobased, polymerizable carboxylic acids can be obtained from fermentation of glucose and other simple sugars [25]. This also represents an important avenue in the synthesis of monomers from biobased materials.

5.4.1   ACRYLIC AND METHACRYLIC ACIDS FROM HYDROXYPROPIONIC AND ITACONIC ACIDS

The mose commercially significant acids are succinic, lactic, itaconic, 3-hydrosypropionic and glutamic acids [117–121]. Subjection of these fermentative products to organic synthesis yields monomers such as malonic acid from 3-hydroxypropionic acid, acrylic acid from lactic and 3-hydroxypropionic acids (Figure 5.6), and methacrylic acid from itaconic acid (Figure 5.7) [122–126]. Noncarboxylic acid monomers obtained from simple sugars include 1,2-propanediol from lactic acid, 1,3-propanediol from 3-hydroxypropionic acid, 1,4-butanediol from succinic acid, and acrylonitrile from 3-hydroxypropionic acid [127–129]. Additionally, terephthalic acid can be obtained from simple sugars as well as from lignin [130].

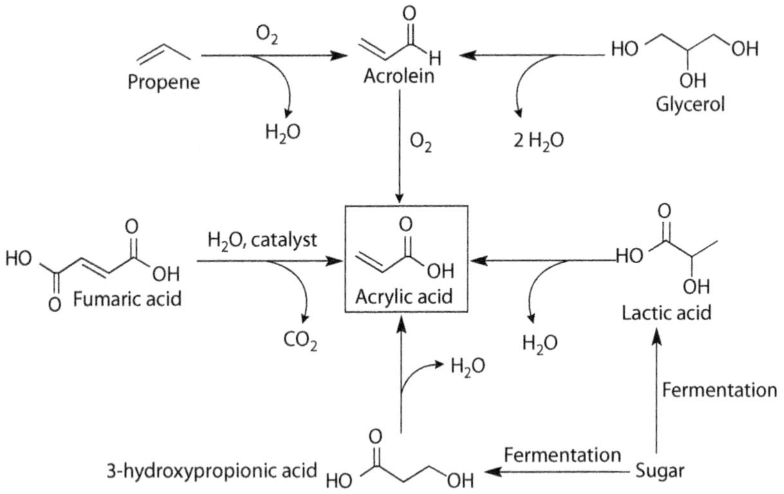

FIGURE 5.6 Biobased routes to acrylic acid from fumaric acid, glycerol, and sugar, as well as the conventional petrochemical route from propene.

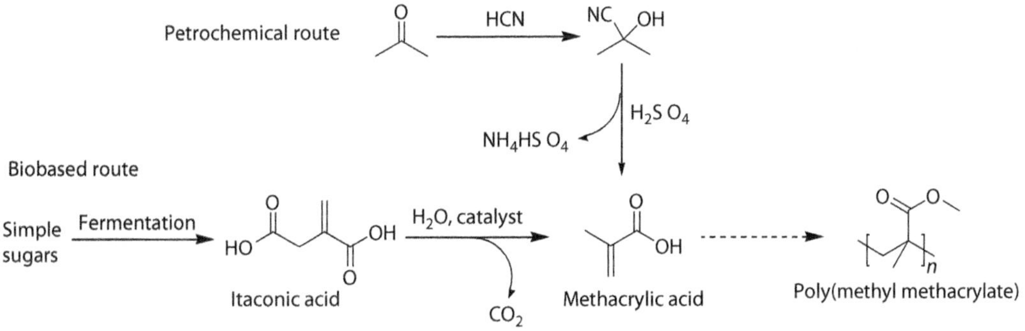

FIGURE 5.7 Methacrylic acid from biobased itaconic acid and from acetone using the conventional petrochemical route.

5.4.2 Applications of Acrylic and Methacrylic Acids

Acrylic and methacrylic acids are of particular interest because their respective polymers have well-established industrial applications. For instance, methyl, ethyl, butyl, and 2-ethylhexyl acrylates are polymerized with several comonomers, including methyl methacrylate, vinyl acetate, styrene, acrylonitrile, and maleic anhydride, to yield commercially important coatings, plastics, adhesives, elastomers, fibers, absorbent materials, polishes, and paints [110,131]. The principal application of methyl methacrylate is homopolymerization to provide poly(methyl methacrylate) (Plexiglas®) as a lightweight and shatter-resistant transparent thermoplastic, which serves as a versatile alternative to glass. It is also an essential component in the polymer dispersion systems of coatings and paints [110,131,132]. Both acrylic and methacrylic acids and their corresponding esters are produced globally on a large scale, each exceeding 3 million metric tons of annual production. In addition, the market for each is projected to increase in coming years, especially in China and India. The rapidly growing construction industry in emerging economies, such as China and India, has propelled the growth of acrylic acid and its derivatives in the Asia-Pacific region. The current market prices for methyl methacrylate and acrylate esters are approximately $1.30/lb and $1.10/lb, respectively. These translate to total market values well in excess of $8 billion annually for each monomer [133,134].

5.4.3 TRADITIONAL ROUTES TO ACRYLIC AND METHACRYLIC ACIDS

The traditional petrochemical route to acrylic acid begins with catalytic oxidation of propene to yield acrolein followed by a second catalytic oxidation to provide acrylic acid (Figure 5.6) [110,131]. The classic route to methacrylic acid begins with condensation of acetone with hydrocyanic acid to give an acetone cyanohydrin intermediate (Figure 5.7). Dehydration and hydrolysis of the nitrile group with a stoichiometric excess of concentrated sulfuric acid at 140°C provide methacrylic acid along with ammonium bisulfate by-product that is produced in a ratio of 1.5:1 relative to methacrylic acid [110,132]. Both of these routes are environmentally unfriendly because they use nonrenewable starting materials, utilize stoichiometric amounts of harmful reagents, generate toxic intermediates, and, in the case of methacrylic acid, produce a large amount of low-value by-product. As a consequence, biobased routes to these industrial chemicals are of interest to overcome the deficiencies in the conventional petrochemical production methods.

Catalytic conversion of lactic acid to acrylic acid through dehydration has long been of interest. Unfortunately, side reactions leading to acetaldehyde, propionic acid, 2,3-pentanedione, and dilactide via decarbonylation/decarboxylation, reduction, condensation, and self-esterification, respectively, result in low yield and low selectivity [135,136]. Several methods of acrylic acid synthesis from lactic acid have been reviewed elsewhere [25,137]. Acrylic acid can also be obtained by catalytic thermal dehydration of 3-hydroxypropionic acid, at temperatures between 140°C and 160°C, in the presence of added copper powder to inhibit polymerization [137]. Methyl acrylate is obtained when methanol is present in this reaction [138]. Use of heterogeneous catalysts, such as NaH_2PO_4 and $CuSO_4$ supported on silica gel and zeolite H-beta, has also been reported [139]. The key problems limiting commercial production of acrylic acid from lactic and 3-hydroxypropionic acids include low yield, high temperatures, high pressures, expensive catalysts, and difficulty in removing prevalent by-products due to poor selectivity [25,135]. Recently, a two-step catalytic process for production of acrylic acid at 300°C has been reported. In the procedure, glycerol was first dehydrated to acrolein using the catalyst $Cs_{2.5}H_{0.5}PW_{12}O_{40}$ supported on Nb_2O_5. This is followed by an oxidation step using vanadium–molybdenum mixed oxides supported on silicon carbide. Unfortunately, this method produces numerous by-products including propionic acid, acetone, allyl alcohol, acetol, acetaldehyde, acetic acid, carbon monoxide, and carbon dioxide [140].

5.4.4 BIOBASED ACRYLIC AND METHACRYLIC ACIDS

Comparatively few biobased syntheses of methacrylic acid have been reported. One such approach is liquid-phase dehydration and decarboxylation of citramalic acid, at temperatures of 250–400°C and pressures of 3.1–27.6 MPa (450–4000 psi), in the presence of basic and/or acidic catalysts. Application of this methodology to maleic acid yields acrylic acid [141]. Another approach relies on biomass-derived ethene, carbon monoxide, and methanol in a two-step procedure referred to as the alpha process to yield methyl methacrylate [142]. The first step entails catalytic carbonylation of ethene in the presence of carbon monoxide to yield methyl propionate. In the second step, methyl propionate is converted to methyl methacrylate by using formaldehyde [143]. However, such a methodology requires considerable processing and purification of the biomass source to obtain the necessary chemical feedstocks, i.e., ethane, carbon monoxide, etc., and the processing steps themselves consume considerable quantities of fossil fuels. Another method involves conversion of citric acid to itaconic acid, followed by decarboxylation of itaconic acid in the presence of bases. Methacrylic acid could be formed under near-critical and supercritical water conditions [125]. However, appreciable accumulation of unwanted by-products such as 2-hydroxybutyric and crotonic acids was noted. In addition to low selectivity, high temperatures (>350°C) were required and yields are relatively low (70% or lower). In an optimized supercritical water procedure, selectivity was improved to over 90% and reaction temperatures were lowered slightly, to 245–270°C, and pressures were lowered to 3.1–20.7 MPa (450–3000 psi). However, stoichiometric amounts of base catalysts were used, and by-products such as crotonic and 2-hydroxybutyric acids were still produced along with propene.

5.5 RENEWABLE POLYMERS FROM PLANT OILS

Polymeric materials prepared from renewable natural resources are becoming increasingly important because of their low cost, ready availability, and possible biodegradability [144]. Plant oils are considered the most important class of agriculturally based raw materials for polymer synthesis.

5.5.1 WORLD PLANT OIL PRODUCTION

Worldwide plant oil production has increased continuously in the past decade, currently reaching more than 129 million metric tons annually [145]. The majority of growth has been in palm oil, representing more than 30% of plant oil production in 2007, with soybean (28%), rapeseed (15%), and sunflower (9%) accounting for most of the remaining production. Soybean oil is reasonably inexpensive and one of the most readily available renewable resource currently available in the United States. In addition to numerous applications in the food industry, soybean oil is utilized extensively to produce industrial products such as coatings, inks, plasticizers, lubricants, fuels, and agrochemicals [146–152]. Within the polymer field, plant oils have been utilized as toughening agents [153,154], polyol components in polyurethanes [155–157], polymeric surfactants [158,159], plastics [160–162], composites [163–165], and pressure-sensitive adhesives [166].

5.5.2 EPOXIDATION OF PLANT OILS

Plant oils can be chemically modified for use in most of these applications. The most common chemical modification is epoxidation, where plant oils are subjected to the Prilezhaev reaction. A percarboxylic acid, such as peracetic or performic acid, is generated in situ by the acid-catalyzed reaction of formic or glacial acetic acid with hydrogen peroxide. The peracid then reacts with olefinic moieties to yield epoxidized plant oils containing three-membered ring epoxy structures, also called oxiranes [75]. Epoxidized triacylglycerols have shown promise for use in industrial applications such as in coatings, inks, and adhesives [167]. Epoxidized plant oils can also be polymerized in the presence of curing agents, such as diamines [168], or by ring-opening catalysts such as Lewis acids [169]. This curing process transforms the relatively low-molecular weight oil into a highly cross-linked network with properties ranging from a high-viscosity liquid to a solid composite panel [170]. Generally, the curing agent attaches to one of the carbons of the epoxide group to yield an ester, ether, or amine linkage. This results in the formation of a hydroxyl group on the other carbon of the epoxide moiety. The most successful use of an epoxidized triacylglycerol is epoxidized soybean oil (ESO), which is commercially available under the brand name Vikoflex 7170 (Arkema). ESO is widely used as a nontoxic stabilizer and plasticizer and potentially can replace phthalates in plastic plasticizers, where it imparts desired properties such as increased flexibility, stability, and improved processing conditions [171].

5.5.3 ACRYLATED EPOXIDIZED SOYBEAN OIL (AESO)

AESO is synthesized by the reaction of acrylic acid with ESO (Figure 5.8) and is commercially available under the brand name Ebecryl 860 (UCB Chemical Company) [172]. AESO has numerous direct applications including surface coatings and films, inks, pressure-sensitive adhesives, and matrix material for fiber composites [173,174]. AESO is also a versatile chemical intermediate owing to the presence of residual reactive epoxide functionalities [172]. However, pure AESO polymers do not exhibit strong mechanical properties. They are amorphous cross-linked rubbers which cannot be processed into useful shapes. An approach to improve the mechanical properties of poly-AESO,

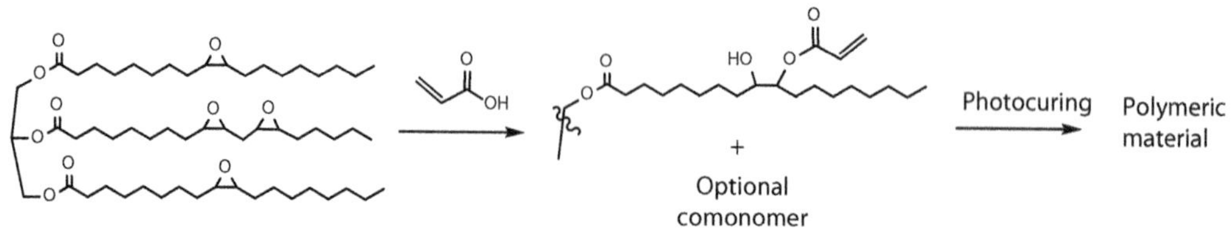

FIGURE 5.8 The synthesis and reaction of AESO.

or other polymeric material for that matter, is to use nanosized fibers and proteins as reinforcing fillers. O'Donnell et al. [175] prepared a composite of AESO resin and natural fiber mats made from flax, cellulose, pulp, and hemp. The composites are cured at room temperature, and a natural fiber reinforcement of 10–50 wt% increases the flexural modulus into a range between 1.5 and 6 GPa. Nanocomposites of AESO and organically modified montmorillonite have also been reported [176,177]. Shibata described the photocuring of AESO with a low-molecular weight organic substance, (R)-12-hydroxystearic acid (HSA), as a gelator [178]. Shibata found that the flexural strength and modulus decreased with increasing HSA concentration because the HSA crystalline phase was distributed heterogeneously in the cross-linked AESO resin matrix. The problem was that the cross-linked AESO has a lower affinity for HSA than that of a cross-linked ESO material. Habib and Bajpai [179] reported the preparation of coating films from AESO and trimethylolpropane trimethacrylate with a benzophenone photoinitiator using a dimethylethanolamine activator. Polymeric films were formed under UV radiation and showed good thermal stability with a decomposition temperature of 200°C. However, no mechanical properties of the films were reported. López and Santiago [180] reported the reaction of AESO with butyl methyl acrylate and acrylamidomethyl cellulose acetate butyrate copolymers. They prepared composites based on these copolymers and carbon black conductive particles. They suggested the use of their composites for electrostatic protection and electromagnetic shields. They predicted that these conductive polymers would be useful for applications such as sensors for vapors, gases, and toxic or inert substances [181–183].

The previous studies indicate that smaller fillers and strong interactions between filler and the polymer matrix tend to improve the composite modulus. For example, soy-based protein is often used as renewable and reliable natural fiber filler. Its efficacy is enhanced by its structure where it forms fractal-like nanoparticle aggregates [184–186] that are rigid ~10 nm protein units with elastic moduli around 2 GPa [187]. It is an effective biomaterial for enhancing the mechanical properties of soft polymers. The conversion of wood and plant fibers into cellulose nanowhiskers/nanofibers has attracted much attention for the application of cellulose in polymer nanocomposites. Three-dimensional nanoporous cellulose gel has been readily prepared by an aqueous alkali hydroxide–urea solvent method developed by Cai et al. [188]. The resulting materials show high mechanical strength, high light transmittance, and high nanometer-scale porosity, owing to strong interchain hydrogen bonding and entanglement of cellulose chains [188–191]. They are useful in producing mechanically tough, foldable, and transparent cellulose bioplastics [192], and can be converted into hydrogels and organogels by solvent exchange processes [188–192].

5.6 CONCLUSION

Currently there is a need to increase the use of biobased materials in industry, especially in the lubricant industry. This review covers some developments relevant to the production of biobased drop-in replacements for petrochemicals. These include alkenes, which are important building blocks for PAOs and poly(internal olefin) lubricants. Also, acrylic acid and methacrylic acid are important in many commercial polymers. Development of biobased routes to existing industrial chemicals avoids the conundrum of new materials in search of a use or application. Also covered in this review is a process of using soybean oil directly through a method of epoxidation followed by reaction with acrylic acid. This process has potential for the manufacture of high-strength materials that can be easily cured into a variety of useful products.

ACKNOWLEDGMENT

This work was a part of the in-house research of the Agricultural Research Service of the USDA.

REFERENCES

1. D. Murzin and I. Simakova, Catalysis in biomass processing, *Catal. Ind.*, 3, 218–249 (2011).
2. J. J. Bozell, L. Moens, D. C. Elliott, Y. Wang, G. G. Neuenscwander, S. W. Fitzpatrick, R. J. Bilski, and J. L. Jarnefeld, Production of levulinic acid and use as a platform chemical for derived products, *Resour. Conserv. Recycl.*, 28, 227–239 (2000).
3. M. Labet and W. Thielemans, Synthesis of polycaprolactone: A review, *Chem. Soc. Rev.*, 38, 3484–3504 (2009).
4. M. North, R. Pasquale, and C. Young, Synthesis of cyclic carbonates from epoxides and CO_2, *Green Chem.*, 12, 1514–1539 (2010).
5. W. B. Kim, U. A. Joshi, and J. S. Lee, Making polycarbonates without employing phosgene: An overview on catalytic chemistry of intermediate and precursor syntheses for polycarbonate, *Ind. Eng. Chem. Res.*, 43, 1897–1914 (2004).
6. E. H. Pryde, Fats and oils as chemical intermediates: Present and future uses, *J. Am. Oil Chem. Soc.*, 56, 849–854 (1979).
7. Y. Xia and R. C. Larock, Vegetable oil-based polymeric materials: Synthesis, properties, and applications, *Green Chem.*, 12, 1893–1909 (2010).
8. M. A. R. Meier, J. O. Metzger, and U. S. Schubert, Plant oil renewable resources as green alternatives in polymer science, *Chem. Soc. Rev.*, 36, 1788–1802 (2007).
9. S. Miao, P. Wang, Z. Su, and S. Zhang, Vegetable-oil-based polymers as future polymeric biomaterials, *Acta Biomater.*, 10, 1692–1704 (2014).
10. U. Biermann, U. Bornscheuer, M. A. R. Meier, J. O. Metzger, and H. J. Schafer, Oils and fats as renewable raw materials in chemistry, *Angew. Chem. Int. Ed.*, 50, 3854–3871 (2011).
11. H. Wagner, R. Luther, and T. Mang, Lubricant base fluids based on renewable raw materials: Their catalytic manufacture and modification, *App. Catal. A*, 221, 429–442 (2001).
12. L. Lazzeri, M. Mazzoncini, A. Rossi, E. Balducci, G. Bartolini, L. Giovannelli, R. Pedriali et al., Biolubricants for the textile and tannery industries as an alternative to conventional mineral oils: An application experience in the Tuscany province, *Ind. Crop. Prod.*, 24, 280–291 (2006).
13. N. Komesvarakul, M. D. Sanders, E. Szekeres, E. J. Acosta, J. F. Faller, T. Mentlik, L. B. Fisher, G. Nicoll, D. A. Sabatini, and J. F. Scamehorn, Microemulsions of triglyceride-based oils: The effect of co-oil and salinity on phase diagrams, *J. Cosmet. Sci.*, 57, 309–325 (2006).
14. J. Dupont, P. A. Z. Suarez, M. R. Meneghetti, and S. M. P. Meneghetti, Catalytic production of biodiesel and diesel-like hydrocarbons from triglycerides, *Energy Environ. Sci.*, 2, 1258–1265 (2009).
15. W. E. Tyner, The integration of energy and agricultural markets, *Agric. Econ.*, 41, 193–201 (2010).
16. H. Baumann, M. Buhler, H. Fochem, F. Hirsinger, H. Zoebelein, and J. Falbe, Natural fats and oils—Renewable raw materials for the chemical industry, *Angew. Chem. Int. Ed.*, 27, 41–62 (1988).
17. M. W. Rigg and H. Gisser, Autoxidation of the saturated aliphatic diesters, *J. Am. Chem. Soc.*, 75, 1415–1420 (1953).
18. J. P. Cosgrove, D. F. Church, and W. A. Pryor, The kinetics of the autoxidation of polyunsaturated fatty acids, *Lipids*, 22, 299–304 (1987).
19. S. A. Miller, A. E. Landis, T. L. Theis, and R. L. Reich, A comparative life cycle assessment of petroleum and soybean-based lubricants, *Environ. Sci. Technol.*, 41, 4143–4149 (2007).
20. K. Collins, Guideline for designating biobased products for federal procurement, *Federal Register*, 70, 1792–1812 (2005).
21. J. Fu, X. Lu, and P. E. Savage, Hydrothermal decarboxylation and hydrogenation of fatty acids over Pt/C, *ChemSusChem*, 4, 481–486 (2011).
22. H. Adkins, The selective activation of alumina for decarboxylation or for dehydration, *J. Am. Chem. Soc.*, 44, 2175–2186 (1922).

23. M. Snåre, I. Kubičková, P. Mäki-Arvela, K. Eränen, and D. Y. Murzin, Heterogeneous catalytic deoxygenation of stearic acid for production of biodiesel, *Ind. Eng. Chem. Res.*, 45, 5708–5715 (2006).
24. M. Czaun, A. Goeppert, R. May, R. Haiges, G. K. S. Prakash, and G. A. Olah, Hydrogen generation from formic acid decomposition by ruthenium carbonyl complexes: Tetraruthenium dodecacarbonyl tetrahydride as an active intermediate, *ChemSusChem*, 4, 1241–1248 (2011).
25. A. Corma, S. Iborra, and A. Velty, Chemical routes for the transformation of biomass into chemicals, *Chem. Rev.*, 107, 2411–2502 (2007).
26. R. H. Prince and K. A. Raspin, Olefin formation from saturated aldehydes and acids by reaction with ruthenium and rhodium complexes, *Chem. Commun.*, 6, 156–157 (1966).
27. D. M. Fenton, "Process for preparation of olefins," U.S. Patent 3530198 (1970).
28. J. A. Miller, J. A. Nelson, and M. P. Byrne, A highly catalytic and selective conversion of carboxylic acids to 1-alkenes of one less carbon atom, *J. Org. Chem.*, 58, 18–20 (1993).
29. T. A. Foglia and P. A. Barr, Decarbonylation dehydration of fatty acids to alkenes in the presence of transition metal complexes, *J. Am. Oil Chem. Soc.*, 53, 737–741 (1976).
30. G. A. Kraus, "Method for producing olefins," U.S. Patent Application 20130102823 (2013).
31. G. A. Kraus and S. Riley, A large-scale synthesis of alpha-olefins and alpha-omega-dienes, *Synthesis*, 44, 3003–3005 (2012).
32. J. G. Immer, M. J. Kelly, and H. H. Lamb, Catalytic reaction pathways in liquid-phase deoxygenation of C18 free fatty acids, *Appl. Catal. A*, 375, 134–139 (2010).
33. Y. Liu, K. E. Kim, M. B. Herbert, A. Fedorov, R. H. Grubbs, and B. M. Stoltz, Palladium-catalyzed decarbonylative dehydration of fatty acids for the production of linear alpha olefins, *Adv. Synth. Catal.*, 356, 130–136 (2014).
34. P. Maki-Arvela, I. Kubickova, M. Snare, K. Eranen, and D. Y. Murzin, Catalytic deoxygenation of fatty acids and their derivatives, *Energy Fuels*, 21, 30–41 (2006).
35. M. O. Miranda, A. Pietrangelo, M. A. Hillmyer, and W. B. Tolman, Catalytic decarbonylation of biomass-derived carboxylic acids as efficient route to commodity monomers, *Green Chem.*, 14, 490–494 (2012).
36. M. Snare, I. Kubickova, P. Maki-Arvela, C. Chichova, K. Eranen, and D. Y. Murzin, Catalytic deoxygenation of unsaturated renewable feedstocks for production of diesel fuel hydrocarbons, *Fuel*, 87, 933–945 (2008).
37. L. J. Goossen and N. Rodriguez, A mild and efficient protocol for the conversion of carboxylic acids to olefins by a catalytic decarbonylative elimination reaction, *Chem. Commun.*, 724–725 (2004).
38. E. Santillan-Jimenez and M. Crocker, Catalytic deoxygenation of fatty acid derivatives to hydrocarbon fuels via decarboxylation/decarbonylation, *J. Chem. Technol. Biotechnol.*, 87, 1041–1050 (2012).
39. A. Azapagic and H. Stichnothe, A life cycle approach to measuring sustainability, *Chim. Oggi*, 27, 44–46 (2009).
40. P. T. Anastas and M. M. Kirchhoff, Origins, current status, and future challenges of green chemistry, *Acc. Chem. Res.*, 35, 686–694 (2002).
41. J. C. Warner, A. S. Cannon, and K. M. Dye, Green chemistry, *Environ. Impact Assess. Rev.*, 24, 775–799 (2004).
42. R. T. Arnold, O. C. Elmer, and R. M. Dodson, Thermal decarboxylation of unsaturated acids, *J. Am. Chem. Soc.*, 72, 4359–4361 (1950).
43. R. T. Arnold and M. J. Danzig, Thermal decarboxylation of unsaturated acids II, *J. Am. Chem. Soc.*, 79, 892–893 (1957).
44. Y. Takemura, A. Nakamura, H. Taguchi, and K. Ouchi, Catalytic decarboxylation of benzoic acid, *Ind. Eng. Chem. Prod. Res. Dev.*, 24, 213–215 (1985).
45. J. J. Scheibel and R. T. Reilman, "Process for decarboxylation of fatty acids and oils to produce paraffins or olefins," U.S. Patent Application 20070281875 (2007).
46. A. Berenblyum, R. Shamsiev, T. Podoplelova, and V. Danyushevsky, The influence of metal and carrier natures on the effectiveness of catalysts of the deoxygenation of fatty acids into hydrocarbons, *Rus. J. Phys. Chem. A*, 86, 1199–1203 (2012).
47. A. S. Berenblyum, T. A. Podoplelova, R. S. Shamsiev, E. A. Katsman, and V. Y. Danyushevsky, On the mechanism of catalytic conversion of fatty acids into hydrocarbons in the presence of palladium catalysts on alumina, *Petroleum Chem.*, 51, 336–341 (2011).
48. S. Bezergianni, A. Kalogianni, and I. A. Vasalos, Hydrocracking of vacuum gas oil–vegetable oil mixtures for biofuels production, *Bioresour. Technol.*, 100, 3036–3042 (2009).
49. W. Charusiri and T. Vitidsant, Kinetic study of used vegetable oil to liquid fuels over sulfated zirconia, *Energy Fuels*, 19, 1783–1789 (2005).

50. S. Bezergianni and A. Kalogianni, Hydrocracking of used cooking oil for biofuels production, *Bioresour. Technol.*, 100, 3927–3932 (2009).

51. W. E. Fristad, M. S. Fry, and J. A. Klang, Persulfate/silver ion decarboxylation of carboxylic acids: Preparation of alkanes, alkenes, and alcohols, *J. Org. Chem.*, 48, 3575–3577 (1983).

52. F. van der Klis, M. H. van den Hoorn, R. Blaauw, J. van Haveren, and D. E. van Es, Oxidative decarboxylation of unsaturated fatty acids, *Eur. J. Lipid Sci. Technol.*, 113, 562–571 (2001).

53. J. C. Cowan, Isomerization reactions of drying oils, *Ind. Eng. Chem.*, 41, 294–304 (1949).

54. L. Alonso, E. P. Cuesta, and S. E. Gilliland, Production of free conjugated linoleic acid by *Lactobacillus acidophilus* and *Lactobacillus casei* of human intestinal origin, *J. Dairy Sci.*, 86, 1941–1946 (2003).

55. R. Quirino and R. Larock, Rh-based biphasic isomerization of carbon carbon double bonds in natural oils, *J. Am. Oil Chem. Soc.*, 89, 1113–1124.

56. D. Grotjahn, "Bifunctional catalysis for extensive isomerization of unsaturated hydrocarbons," U.S. Patent Application 20090143585 (2009).

57. D. Grotjahn, "Catalysts for alkene isomerization and conjugating double bonds in polyunsaturated fats and oils," U.S. Patent Application 20100228031 (2010).

58. D. B. Grotjahn, C. R. Larsen, J. L. Gustafson, R. Nair, and A. Sharma, Extensive isomerization of alkenes using a bifunctional catalyst: An alkene zipper, *J. Am. Chem. Soc.*, 129, 9592–9593 (2007).

59. R. C. Larock, The conjugation of natural oils, *Lipid Technol.*, 15, 58–61 (2003).

60. R. J. Angelici and K-C. Shih, "Process for the synthesis of unsaturated esters," U.S. Patent 5859268 (1999).

61. G. R. Crooks, B. R. G. Johnson, J. Lewis, I. G, Williams, and G. Gamlen, Chemistry of polynuclear compounds: Part XVII—Some carboxylate complexes of ruthenium and osmium carbonyls, *J. Chem. Soc. A*, 2761–2766 (1969).

62. A. Salvini, P. Frediani, and F. Piacenti, Alkene isomerization by non-hydridic phosphine substituted ruthenium carbonyl carboxylates, *J. Mol. Catal. A*, 159, 185–195 (2000).

63. A. Sivaramakrishna, P. Mushonga, I. L. Rogers, F. Zheng, R. J. Haines, E. Nordlander, and J. R. Moss, Selective isomerization of 1-alkenes by binary metal carbonyl compounds, *Polyhedron*, 27, 1911–1916 (2008).

64. Y. Gao, J. Kuncheria, G. P. A. Yap, and R. J. Puddephatt, An efficient binuclear catalyst for decomposition of formic acid, *Chem. Commun.*, 21, 2365–2366 (1998).

65. R. G. Ackman, M. E. Retson, L. R. Gallay, and F. A. Vandenheuvel, Ozonolysis of unsaturated fatty acids: I. Ozonolysis of oleic acid, *Can. J. Chem.*, 39, 1956–1962 (1961).

66. L. I. Placek, A review on petroselinic acid and its derivatives, *J. Am. Oil Chem. Soc.*, 41, 319–329 (1963).

67. H. J. Nieschlag and I. A. Wolff, Industrial uses of high erucic oils, *J. Am. Oil. Chem. Soc.*, 48, 723–727 (1971).

68. F. C. Naughton, Production, chemistry, and commercial applications of various chemicals from castor oil, *J. Am. Oil Chem. Soc.*, 51, 65–71 (1974).

69. Y. Yang, W. Lu, X. Zhang, W. Xie, M. Cai, and R. A. Gross, Two step biocatalytic route to biobased functional polyesters from carboxy fatty acids and diols, *Biomacromolecules*, 11, 259–268 (2010).

70. E. H. Pryde, E. N. Frankel, and J. C. Cowen, Reactions of carbon monoxide with unsaturated fatty acids and derivatives: A review, *J. Am. Oil Chem. Soc.*, 49, 451–456 (1972).

71. E. N. Frankel and F. L. Thomas, Catalytic carboxylation of fats: Carboxy acids and esters from monounsaturates, *J. Am. Oil Chem. Soc.*, 50, 39–43 (1973).

72. E. N. Frankel, Catalytic hydroformylation of unsaturated fatty derivatives with cobalt carbonyl, *J. Am. Oil Chem. Soc.*, 53, 138–141 (1976).

73. E. N. Frankel and E. H. Pryde, Catalytic hydroformylation and hydrocarboxylation of unsaturated fatty compounds, *J. Am. Oil Chem. Soc.*, 54, 873–881 (1977).

74. T. H. Khoe, F. H. Otey, and E. N. Frankel, Rigid urethane foams from hydroxymethylated linseed and polyol esters, *J. Am. Oil Chem. Soc.*, 49, 615–618 (1972).

75. T. W. Abraham and R. Höfer, Lipid-based polymer building blocks and polymers, *Polym. Sci. A*, 10, 15–58 (2012).

76. T. Eren, S. H. Kusefoglu, and R. Wool, Polymerization of maleic anhydride–modified plant oils with polyols, *J. App. Poly. Sci.*, 90, 197–202 (2003).

77. H. Mutlu and M. A. R. Meier, Castor oil as a renewable resource for the chemical industry, *Eur. J. Lipid Sci. Technol.*, 112, 10–30 (2010).

78. J. C. Ronda, G. Lligadas, and V. Cadiz, Vegetable oils as platform chemicals for polymer synthesis, *Eur. J. Lipid Sci. Technol.*, 113, 46–58 (2011).

79. M. Van der Steen and C. V. Stevens, Undecylenic acid: A valuable and physiologically active renewable building block from castor oil, *ChemSusChem*, 2, 692–713 (2009).
80. N. G. Kulkarni, N. Krishnamurti, P. C. Chatterjee, and J. S. Aggarwal, Polymerizable monomers from castor oil: I. Vinyl monomers based on undecenoic acid, *J. Am. Oil Chem. Soc.*, 45, 465–467 (1968).
81. A. Rybak and M. A. R. Meier, Acyclic diene metathesis with a monomer from renewable resources: Control of molecular weight and one-step preparation of block copolymers, *ChemSusChem*, 1, 542–547 (2008).
82. L. Montero de Espinosa, M. A. R. Meier, J. C. Ronda, M. Galia, and V. Cadiz, Phosphorus-containing renewable polyester-polyols via ADMET polymerization: Synthesis, functionalization and radical cross-linking, *J. Polym. Sci. A*, 48, 1649–1660 (2010).
83. O. Turunc and M. A. R. Meier, The thiol-ene (click) reaction for the synthesis of plant oil derived polymers, *Eur. J. Lipid Sci. Technol.*, 115, 41–54 (2013).
84. O. Turunc and M. A. R. Meier, Fatty acid derived monomers and related polymers via thiol-ene (click) additions, *Macromol. Rapid Commun.*, 31, 1822–1826 (2010).
85. N. Koenig and D. Swern, Organic sulfur derivatives: 2. Sulfides, sulfoxides and sulfones from thiols and 10-undecenoic acid, *J. Am. Chem. Soc.*, 79, 362–365 (1957).
86. C. Boelhouwer and J. C. Mol, Metathesis of fatty acid esters, *J. Am. Oil Chem. Soc.*, 61, 425–430 (1984).
87. G. C. Vougioukalakis and R. H. Grubbs, Ruthenium-based heterocyclic carbene-coordinated olefin metathesis catalysts, *Chem. Rev.*, 110, 1746–1787 (2010).
88. J. Van Rensselar, Metathesis powers marketability of biobased lubricants, *Tribol. Lubr. Technol.*, 2–8 (2013).
89. J. C. Mol, Industrial applications of olefin metathesis, *J. Mol. Catal. A Chem.*, 213, 39–45 (2004).
90. H. L. Ngo, K. Jones, and T. A. Foglia, Metathesis of unsaturated fatty acids: Synthesis of long-chain unsaturated dicarboxylic acids, *J. Am. Oil Chem. Soc.*, 83, 629–634 (2006).
91. S. Warwel, C. Demes, and G. J. Steinke, Polyesters by lipase-catalyzed polycondensation of unsaturated and epoxidized long-chain dicarboxylic acid methyl esters with diols, *Polym. Sci. A*, 39, 1601–1609 (2001).
92. T. H. Johnson, "Process for preparing linear alpha, omega difunctional molecules," U.S. Patent 4943396 (1990).
93. D. M. Ohlmann, N. Tschauder, J. P. Stockis, K Goossen, M. Dierker, and L. K. Goossen, Isomerizing olefin metathesis as a strategy to access defined distributions of unsaturated compounds from fatty acids, *J. Am. Chem. Soc.*, 134, 13716–13729 (2012).
94. S. E. Lehman Jr., J. E. Schwendeman, P. M. O'Donnell, and K. B. Wagener, Olefin isomerization promoted by olefin metathesis catalysts, *Inorg. Chem. Acta*, 345, 190–198 (2003).
95. S. H. Hong, D. P. Sanders, C. W. Lee, and R. H. Grubbs, Prevention of undesirable isomerization during olefin metathesis, *J. Am. Chem. Soc.*, 127, 17160–17161 (2005).
96. H. Kohashi and T. A. Foglia, Metathesis of methyl oleate with a homogeneous and a heterogeneous catalyst, *J. Am. Oil Chem. Soc.*, 62, 549–554 (1985).
97. S. Warwel, F. Bruse, C. Demes, and M. Kunz, Polymers and polymer building blocks from meadowfoam oil, *Ind. Crop. Prod.*, 20, 301–309 (2004).
98. R. E. Montenegro and M. A. R. Meier, Lowering the boiling point curve of biodiesel by cross-metathesis, *Eur. J. Lipid Sci. Technol.*, 114, 55–62 (2012).
99. J. Patel, S. Mujcinovic, W. R. Jackson, A. J. Robinson, A. K. Serelis, and C. Such, High conversion and productive catalyst turnovers in cross-metathesis reactions of natural oils, *Green Chem.*, 8, 450–454 (2006).
100. T. W. Abraham, H. Kaido, C. W. Lee, R. L. Pederson, Y. Schrodi, M. J. Tupy, and A. A. Pletnev, "Methods of making organic compounds by metathesis and hydrocyanation," U.S. Patent Application 20090259065 (2009).
101. T. W. Abraham, H. Kaido, C. W. Lee, R. L. Pederson, Y. Schrodi, and M. J. Tupy, "Methods of making organic compounds by metathesis," U.S. Patent Application 20090264672 (2009).
102. T. Jacobs, A. Rybak, and M. A. R. Meier, Cross-metathesis reactions of allyl chloride with fatty acid methyl esters: Efficient synthesis of α,ω-difunctional chemical intermediates from renewable raw materials, *Appl. Catal. A Gen.*, 353, 32–35 (2009).
103. R. Malacea, C. Fischmeister, C. Bruneau, J. L. Dubois, J. L. Couturier, and P. H. Dixneuf, Renewable materials as precursors of linear nitrile-acid derivatives via cross-metathesis of fatty esters and acids with acrylonitrile and fumaronitrile, *Green Chem.*, 11, 152–155 (2009).
104. A. Rybak and M. A. R. Meier, Cross-metathesis of fatty acid derivatives with methyl acrylate: Renewable raw materials for the chemical industry, *Green Chem.*, 9, 1356–1361 (2007).

105. X. Miao, P. H. Dixneuf, C. Fischmeister, and C. Bruneau, A green route to nitrogen-containing groups: The acrylonitrile cross-metathesis and applications to plant oil derivatives, *Green Chem.*, 13, 2258–2271 (2011).
106. C. Bruneau, C. Fischmeister, X. Miao, R. Malacea, and P. H. Dixneuf, Cross-metathesis with acrylonitrile and applications to fatty acid derivatives, *Eur. J. Lipid Sci. Technol.*, 112, 3–9 (2010).
107. U. Biermann, M. A. R. Meier, W. Butte, and J. O. Metzger, Cross-metathesis of unsaturated triglycerides with methyl acrylate: Synthesis of a dimeric metathesis product, *Eur. J. Lipid Sci. Technol.*, 113, 39–45 (2011).
108. X. Miao, C. Fischmeister, C. Bruneau, P. H. Dixneuf, J. L. Dubois, and L. Couturier, Tandem catalytic acrylonitrile cross-metathesis and hydrogenation of nitriles with ruthenium catalysts: Direct access to linear α,ω-aminoesters from renewable, *ChemSusChem*, 5, 1410–1414 (2012).
109. J. A. Zerkowski and D. K. Y. Solaiman, Omega-functionalized fatty acids, alcohols, and ethers via olefin metathesis, *J. Am. Oil Chem. Soc.*, 89, 1325–1332 (2012).
110. H. A. Witcoff, B. G. Reuben, and J. S. Plotkin, *Industrial Organic Chemicals*, Second Ed., John Wiley & Sons, Hoboken, NJ (2004).
111. H. H. Fox, R. R. Schrock, and R. O'Dell, Coupling of terminal olefins by molybdenum (VI) imido alkylidene complexes, *Organometallics*, 13, 635–639 (1994).
112. A. Behr, A. Westfechtel, and J. P. Gomes, Catalytic processes for the technical use of natural fats and oils, *Chem. Eng. Technol.*, 31, 700–714 (2008).
113. S. Warwel, J. Tillack, C. Demes, and M. Kunz, Polyesters of unsaturated fatty acid derivatives, *Macromol. Chem. Phys.*, 202, 1114–1121 (2001).
114. J. C. Mol, Application of olefin metathesis in oleochemistry: An example of green chemistry, *Green Chem.*, 4, 5–13 (2002).
115. J. C. Mol, Metathesis of functionalized acyclic olefins, *J. Mol. Catal.*, 65, 145–162 (1991).
116. C. Boelhouver and J. C. Mol, Metathesis reactions of fatty acid esters, *Prog. Lipid Res.*, 24, 243–267 (1985).
117. K. L. Wasewar, A. A. Yawalkar, J. A. Moulijn, and V. G. Pangarkar, Fermentation of glucose to lactic acid coupled with reactive extraction: A review, *Ind. Eng. Chem. Res.*, 43, 5969–5982 (2004).
118. N. P. Nghiem, B. H. Davison, B. E. Suttle, and G. R. Richardson, Production of succinic acid by *Anaerobiospirillum succiniciproducens*, *Appl. Biochem. Biotechnol.*, 63–65, 565–576 (1997).
119. T. M. Kuo, C. P. Kurtzman, and W. E. Levinson, "Production of itaconic acid by *Pseudozyma antarctica*," U.S. Patent 7479381 (2009).
120. M. A. Batti, "Process for the production of itaconic acid," U.S. Patent 3162582 (1964).
121. Y. C. Tsai, M. C. Huang, S. F. Lin, and Y. C. Su, "Method for the production of itaconic acid using *Aspergillus terreus* solid state fermentation," U.S. Patent 6171831 (2001).
122. K. Edamura and T. Arai, "Malonic and methylmalonic acids from hydroxypropionic acids," Japan Patent 54076518 (1979).
123. R. A. Sawicki, "Catalyst for dehydration of lactic acid to acrylic acid," U.S. Patent 4729978 (1988).
124. P. Tsobanakis, X. Meng, and T. W. Abraham, "Methods of manufacturing derivatives of hydroxycarboxylic acids," WO Patent 2003082795 (2003).
125. M. Carlsson, C. Habenicht, L. C. Kam, and M. J. Antal Jr., Study of sequential conversion of citric to itaconic to methacrylic acid in near-critical and supercritical water, *Ind. Eng. Chem. Res.*, 33, 1989–1996 (1994).
126. D. E. Johnson, G. R. Eastham, M. Poliakoff, and T. A. Huddle, "A process for the production of methacrylic acid and its derivatives and polymers produced therefrom," WO Patent Application 2012069813 (2012).
127. H. S. Broadbent, G. C. Cambell, W. J. Bartley, and J. H. Johnson, Rhenium and its compounds as hydrogenation catalysts: III. Rhenium heptoxide, *J. Org. Chem.*, 24, 1847–1854 (1959).
128. S. P. Crabtree and R. K. Henderson, "Process for the preparation of propane-1,3-diol by vapor phase hydrogenation of 3-hydroxypropanal, beta-propiolactone, oligomers of beta-propiolactone, esters of 3-hydroxypropanoic acid or mixtures thereof," WO Patent 2001070659 (2001).
129. L. Cracium, G. P. Benn, J. Dewing, G. W. Schiver, W. J. Peer, B. Siebenhaar, and U. Siegrist, "Preparation of acrylic acid derivatives from alpha- or beta-hydroxy carboxylic acids," U.S. Patent Application 20050222458 (2005).
130. D. I. Collias, A. M. Harris, V. Nagpal, I. W. Cottrell, and M. W. Schultheis, Biobased terephthalic acid technologies: A literature review, *Ind. Biotechnol.*, 10, 91–105 (2014).
131. J. van Haveren, E. L. Scott, and J. Sanders, Bulk chemicals from biomass, *Biofuels Bioprod. Bioref.*, 2, 41–57 (2008).

132. K. Nagai, New developments in the production of methyl methacrylate, *Appl. Catal. A*, 221, 367–377 (2001).

133. http://www.icis.com/chemicals/acrylic-acid-acrylate-esters/ (Accessed June 19, 2014).

134. http://www.icis.com/chemicals/methyl-methacrylate/ (Accessed June 19, 2014).

135. X. Zhang, M. Tu, and M. G. Paice, Routes to potential bioproducts from lignocellulosic biomass lignin and hemicelluloses, *Bioenergy Res.*, 4, 246–257 (2011).

136. S. Varadarajan and D. J. Miller, Catalytic upgrading of fermentation-derived organic acids, *Biotechnol. Prog.*, 15, 845–854 (1999).

137. B. C. Redmon, "Preparation of acrylic acid," U.S. Patent 246970 (1949).

138. S. R. Charles, B. F. James, B. Thomas, T. K. Heinrich, and W. Less, "Process for the manufacture of methyl acrylate," U.S. Patent 2649475 (1953).

139. P. Tsobanakis, X. Meng, and T. W. Abraham, "Methods of manufacturing derivatives of hydroxycarboxylic acids," WO Patent 2003082795 (2003).

140. R. Liu, T. Wang, D. Cai, and Y. Jin, Highly efficient production of acrylic acid by sequential dehydration and oxidation of glycerol, *Ind. Eng. Chem. Res.*, 53, 8667–8674 (2014).

141. D. W. Johnson, G. R. Eastham, and M. Poliakoff, "Method of producing acrylic and methacrylic acid," WO Patent 2011077140 (2011).

142. R. P. Tooze, G. R. Eastham, K. Whiston, and X. L. Wang, "Process for the carbonylation of ethylene and catalyst systems for use therein," WO Patent 09619434 (1996).

143. J. L. Dubois, "Procede de fabrication d'un methylacrylate de methyl derive de la biomasse," WO Patent 2010058119 (2010).

144. D. L. Kaplan, *Biopolymers from Renewable Resources*, Springer, New York (1998).

145. A. S. Carlsson, Plant oils as feedstock alternatives to petroleum—A short survey of potential oil crop platforms, *Biochimie*, 91, 665–670 (2009).

146. A. Cunningham and A. Yapp, "Liquid polyol compositions," US Patent 3827993 (1974).

147. G. W. Bussell, "Maleinized faty acid esters of 9-oxatetracylco-4.4.1.$^{2.5}$O^{1,6}O$^{8.10}$ undecan-4-ol," U.S. Patent 3855163 (1974).

148. L. E. Hodakowski, C. L. Osborn, and E. B. Harris, "Polymerizable epoxide-modified compositions," U.S. Patent 4119640 (1975).

149. D. J. Trecker, G. W. Borden, and O. W. Smith, "Method for curing acrylated epoxidized soybean oil amine compositions," U.S. Patent 3979270 (1976).

150. D. J. Trecker, G. W. Borden, and O. W. Smith, "Acrylated epoxidized soybean oil amine compositions and method," U.S. Patent 3931075 (1976).

151. D. K. Salunkhe, J. K. Chavan, R. N. Adsule, and S. S. Kadam, *World Oilseeds: Chemistry, Technology, and Utilization*, Van Nostrand Reinhold, New York (1992).

152. C. G. Force and F. S. Starr, "Vegetable oil adducts as emollients in skin and hair care products," U.S. Patent 4740367 (1988).

153. L. W. Barrett, L. H. Sperling, and C. J. Murphy, Naturally functionalized triglyceride oils in interpenetrating polymer networks, *J. Am. Oil Chem. Soc.*, 70, 523–534 (1990).

154. S. Qureshi, J. A. Manson, L. H. Sperling, and C. J. Murphy, Simultaneous interpenetrating networks from epoxidized triglyceride oils: Morphology and mechanical behavior, in: *Polymer Applications of Renewable-Resource Materials*, C. E. Carraher and L. H. Sperling (Eds.), Plenum Press, New York (1983).

155. A. Guo, D. Demydov, W. Zhang, and Z. S. Petrović, Polyols and polyurethanes from hydroformylation of soybean oil, *J. Polym. Environ.*, 10, 49–52 (2002).

156. A. Guo, I. Javini, and Z. Petrović, Rigid polyurethane foams based on soybean oil, *ACS PMSE Preprints*, 80, 503–504 (1999).

157. Z. Petrović, I. Javini, I., A. Guo, and W. Zhang, "Method of making natural oil-based polyols for polyurethanes," U.S. Patent 6443121 (2002).

158. G. Biresaw, G., Z. S. Liu, and S. Z. Erhan, Investigation of the surface properties of polymeric soaps obtained by ring-opening polymerization of epoxidized soybean oil, *J. App. Polym. Sci.*, 108, 1976–1985 (2008).

159. Z. S. Liu and G. Biresaw, Synthesis of soybean oil-based polymeric surfactants in supercritical carbon dioxide and investigation of their surface properties, *J. Agric. Food Chem.*, 59, 1909–1917 (2011).

160. F. Li, M. V. Hanson, and R. C. Larock, Soybean oil-divinylbenzene thermosetting polymers: Synthesis, structure, properties and their relationships, *Polymer*, 42, 1567–1579 (2001).

161. F. Li and R. C. Larock, New soybean oil–styrene–divinylbenzene thermosetting copolymers: II. Dynamic mechanical properties, *J. Polym. Sci. B*, 38, 2721–2738 (2000).

162. F. Li and R. C. Larock, New soybean oil–styrene–divinylbenzene thermosetting copolymers: III. Tensile stress–strain behavior, *J. Polym. Sci. B*, 39, 60–77 (2001).
163. Z. S. Liu, S. Z. Erhan, and J. Xu, Preparation, characterization and mechanical properties of epoxidized soybean oil/clay nanocomposites, *Polymer*, 46, 10119–10127 (2005).
164. Z. S. Liu, S. Z. Erhan, J. Xu, and P. D. Calvert, Development of soybean oil-based composites by solid freeform fabrication method: Epoxidized soybean oil with bis or polyalkyleneamine curing agents system, *J. App Polym. Sci.*, 85, 2100–2107 (2002).
165. S. Z. Erhan, Z. Liu, and P. D. Calvert, "Extrusion freeform fabrication of soybean oil-based composites by direct deposition," U.S. Patent 6528571 (2003).
166. S. P. Bunker, and R. P. Wool, Synthesis and characterization of monomers and polymers for adhesives from methyl oleate, *J. Polym. Sci. A*, 40, 451–458 (2002).
167. M. A. Tehfe, J. Lalevee, D. Gigmes, and J. P. Fouassier, Green chemistry: Sunlight-induced cationic polymerization of renewable epoxy monomers under air, *Macromolecules*, 43, 1364–1370 (2010).
168. L. W. Barrett, L. H. Sperling, and C. J. Murphy, Naturally functionalized triglyceride oils in interpenetrating polymer networks, *J. Am. Oil Chem. Soc.*, 70, 523–534 (1993).
169. Z. S. Liu, K. M. Doll, and R. A. Holser, Boron trifluoride catalyzed ring-opening polymerization of epoxidized soybean oil in liquid carbon dioxide, *Green Chem.*, 11, 1774–1780 (2009).
170. Y. Tanaka and R. S. Bauer, Curing reactions, in: *Epoxy Resins—Chemistry and Technology*, Second Ed., C. A. May (Ed.), Marcel Dekker, New York (1988).
171. A. K. Chakraborti, S. Rudrawar, and A. Kondaskar, Lithium bromide, an inexpensive and efficient catalyst for opening of epoxide rings by amines at room temperature under solvent-free condition, *Eur. J. Org. Chem.*, 2004, 3597–3600 (2004).
172. S. N. Khot, J. L. Lascala, E. Can, S. S. Morye, G. I. Williams, G. R. Palmese, S. H. Kusefoglu, and R. P. Wool, Development and application of triglyceride-based polymers and composites, *J. App. Polym. Sci.*, 82, 703–723 (2001).
173. C. A. Koch, "Pressure sensitive adhesives made from renewable resources and related methods," WO Patent 2008144703 (2008).
174. H. Sulzbach, R. Bemmann, R. Hofer, and M. Skwiercz, "Method for producing radically post-cross-linked polymers," EU Patent 1250372 (2001).
175. A. O'Donnell, M. A. Dweib, and R. P. Wool, Natural fiber composites with plant oil-based resin, *Compos. Sci. Technol.*, 64, 1135–1145 (2004).
176. D. Åkesson, M. L. S. Skrifvars, W. Shi, K. Adekunle, J. Seppälä, and M. Turunen, Preparation of nanocomposites from biobased thermoset resins by UV-curing, *Prog. Org. Coat.*, 67, 281–286 (2010).
177. J. Lu, C. K. Hong, and R. P. Wool, Bio-based nanocomposites from functionalized plant oils and layered silicate, *J. Polym. Sci. B*, 42, 1441–1450 (2004).
178. M. Shibata, Bio-based nanocomposites composed of photo-cured soybean-based resins and supramolecular hydroxystearic acid nanofibers, in: *Soybean—Molecular Aspects of Breeding*, A. Sudaric (Ed.), InTech, Rijeka, Croatia (2011).
179. F. Habib and M. Bajpai, Synthesis and characterization of acrylated epoxidized soybean oil for UV-cured coatings, *Chem. Chem. Technol.*, 5, 317–326 (2011).
180. S. H. López and E. V. Santiago, Acrylated-epoxidized soybean oil-based polymers and their use in the generation of electrically conductive polymer composites, in: *Soybean—Bioactive Compounds*, H. A. El-Shemy (Ed.), InTech, Rijeka, Croatia (2013).
181. Z. S. Liu, S. Z. Erhan, and P. D. Calvert, Solid freeform fabrication of epoxidized soybean oil/epoxy composites with di-, tri-, and polyethylene amine curing agents, *J. App. Polym. Sci.*, 93, 356–363 (2004).
182. Z. S. Liu, S. Z. Erhan, and P. D. Calvert, Solid freeform fabrication of soybean oil-based composites reinforced with clay and fibers, *J. Am. Oil Chem. Soc.*, 81, 605–610 (2004).
183. S. Z. Erhan, Z. S. Liu, and P. D. Calvert, Solid freeform fabrication of soybean oil/epoxy composite with bis or polyalkyleneamine curing agents, *Compos. A App. Sci. Manuf.*, 38, 87–93 (2007).
184. R. A. Badley, C. Atkinson, H. Hauser, D. Oldani, J. P. Green, and J. M. Stubbs, The structure, physical and chemical properties of the soy bean protein glycinin, *Biochim. Biophys. Acta*, 412, 214–228 (1975).
185. G. Qi, K. Venkateshan, X. Mo, L. Zhang, and X. Sun, Physiochemical properties of soy protein: Effects of subunit composition, *J. Agric. Food Chem.*, 59, 9958–9964 (2011).
186. L. Jong, Characterization of defatted soy flour and elastomer composites, *J. App. Polym. Sci.*, 98, 353–361 (2005).
187. L. Jong, Rubber composites reinforced by soy spent flakes, *Polym. Int.*, 54, 1572–1580 (2005).
188. J. Cai, S. Kimura, M. Wada, S. Kuga, and L. Zhang, Cellulose aerogels from aqueous alkali hydroxide-urea solution, *ChemSusChem*, 1, 149–154 (2008).

189. J. Cai, S. Kimura, M. Wada, and S. Kuga, Nanoporous cellulose as metal nanoparticles support, *Biomacromolecules*, 10, 87–94 (2009).

190. J. Cai, S. Liu, J. Feng, S. Kimura, M. Wada, S. Kuga, and L. Zhang, Cellulose–silica nanocomposite aerogels by in-situ formation of silica in cellulose gel, *Angew. Chem. Int. Ed.*, 51, 2076–2079 (2012).

191. J. Cai and L. Zhang, Unique gelation behavior of cellulose in NaOH/urea aqueous solution, *Biomacromolecules*, 7, 183–189 (2006).

192. Q. Y, Wang, J. Cai, L. N. Zhang, M. Xu, H. Cheng, C. C. Han, S. Kuga, J. Xiao, and R. Xiao, A bioplastic with high strength constructed from a cellulose hydrogel by changing the aggregated structure, *J. Mater. Chem. A*, 1, 6678–6686 (2013).

Section II

---

*Biobased Hydraulic Lubricants
and Biodegradability*

6 Performance and Technical Requirements of Low-Environmental Impact Lubricants

Hind Abi-Akar

CONTENTS

ABSTRACT

The volumes of low-environmental impact lubricants are forecast to increase driven by regulations. These regulations include requirements for use of biodegradable hydraulic oils in marine or environmentally sensitive applications. Other drivers are incentives for use of these specialized lubricants, such as the BioPreferred U.S. Department of Agriculture program, similar European programs, or are simply due to the desire of customers to use these products. The cost of these lubricants, however, is still higher than that of their petroleum-based counterparts by an estimated 5% to 10%. The performance requirements for low-environmental impact lubricants are the same as those of the common lubricants. Engine oils have to fulfill all the requirements defined in the API categories, and hydraulic oils have to satisfy the requirements defined by end users. Additionally, low-environmental impact lubricants have to pass the currently defined biodegradability and toxicity tests, the requirements of which are detailed in this chapter. Note that this chapter is focused on the lubricants used in diesel engines.

6.1   BACKGROUND AND DRIVERS

Lubricants are typically petroleum-based products and are considered to be a source of new carbon dioxide through their extraction, production, and postutilization in components and equipment. A cradle-to-gate life cycle analysis of the carbon footprint of hydraulic oil, for example, shows 2.64 kg of CO_2 equivalents for each 1 gal of this oil [1]. Considering that the world demand for lubricants was expected to continue its increase at an estimated annual rate of 2.6% [2] through 2015, the importance of reducing the impact on the environment becomes apparent. (Note that throughout this chapter, *lubricants* and *oils* will be used interchangeably based on the more common usage for the specific application.)

The use of low-environmental impact lubricants is one path to reduce the carbon footprint of these products. Low-environmental impact lubricants are typically derived totally or partially from renewable and biobased resources or developed to be biodegradable, reducing pollution and toxicity. The negative impact can also be reduced when lubricants are rerefined or used as a source of energy in applications such as power generation or industrial kilns, provided that proper emission-scrubbing techniques are employed when burnt for power.

It can also be argued that the negative environmental impacts of lubricants can be significantly reduced when these products are developed to have a long service life, which lessens waste and conserves resources. Avoiding leaks and spills is arguably another pathway to reduce the environmental impact of lubricants.

Low-environmental impact lubricants are still limited in volume due to considerable technical challenges in their development and due to their high cost. In 2010, the world lubricant demand was about 36.7 million tons [2] and was expected to exceed 41.6 million tons in 2015 [2]. Of this volume, biobased lubricants were estimated to be only 1% to 2% [3]. The key driver to the market growth of biobased lubricants is the reduction in their cost, specifically, the reduction in the overall cost to produce them from the raw materials to the final product—also called *finished product*. The cost of these lubricants can be 5% to 10% higher than that of their petroleum-based counterparts [3].

Additionally, legislation mandating the use of biobased lubricants is one of the important growth drivers. An example of such legislation is California Senate Bill 916 [4], which was introduced in January of 2014. This bill mandates that a state agency and anyone contracting with a state agency after January 1, 2016, must purchase only *biosynthetic* lubricant, which is defined as "lubricating oil that contains a biobased product. The amount of biobased content within the lubricating oil is stipulated to be not less than 25 percent and that the biobased content is biodegradable." It is clear that if the bill passes, it can increase the biobased lubricant market by millions of gallons in the state of California.

It has to be kept in mind that the current heightened public awareness of the perils to the environment may play a future role in increasing the use of low-environmental impact lubricants. An examination of various surveys indicates that the outlook of biobased lubricants is very positive [3].

6.2 ECONOMIC DRIVERS

The economic impact of lubricants cannot be emphasized enough. Just as every car owner knows, engine oil has to be changed on a regular basis and throughout the life of the car. Each oil change, and sometimes the disposal of the used oil, is a financial burden to the end user and, hence, can influence the type and the cost of the oil purchased. This economic impact is highly exacerbated in industrial and heavy-duty equipment due to the high frequency and the amount of lubricants used. Some machines and components may require greasing or lubricating every day [5]. Heavy-duty engines may need engine oil change intervals of 250 to 500 hours (oil drain intervals are given in the owner's manual of machine or engines [6]) and the typical oil drain interval for on-highway trucks is 25,000 mi [5]. Furthermore, the hydraulic and power train systems of off-highway machines require large volumes of oil. The recurring maintenance and the large volumes of oils needed for these machines demonstrate the high economic impact of lubricants to the end users. The end users then search for ways to reduce their cost, which can lead them away from typically higher-priced renewable or biodegradable lubricants. To demonstrate the cost to customers, a large excavator [6] may require 8 to 10 engine oil changes per year at 8 gal per oil change. The hydraulic system capacity of this machine is 100 gal and may require an oil change once per year. An increase of 5% to 10% in the cost of an oil volume of 800 to 1000 gal can be significant to the end user.

Generally, low-environmental impact lubricants are used in environmentally sensitive areas, such as in close proximity to aquatic bodies or when required by regulations. In these instances, the higher cost of these lubricants is justified by the potential for fines in case any spillage or leaking occurs and, subsequently, the cost of required remediation. In many cases, spillage or leakage has to be reported to government agencies such as the U.S. National Response Center [7].

6.3 LUBRICANT COMPOSITION

Lubricants are very complex fluids. They are typically composed of base stocks and a myriad of highly specialized chemicals called *additives*. Base stocks contribute specific fundamental performance characteristics to the final lubricant. Additives complement the base stocks and impart additional performance characteristics to the final fluid.

The complexity of lubricants has been increasing steadily. A key driver is the evolving complexity of current engines and machines, which requires advanced lubricants for protection of their components [8]. Another driver is the customer demand for long oil drain intervals [5,9].

This complexity, however, poses a challenge to the production of renewable-based lubricants as well as the biodegradability of lubricants. Some additives used in the lubricants may not be feasibly made from renewable resources, and the biodegradability of the various components in the lubricant varies significantly. Some components of biodegradable lubricants completely degrade into carbon dioxide and water in a relatively short time, while others may not degrade at all.

6.4 PERFORMANCE REQUIREMENTS OF LUBRICANTS

Regardless of its source or biodegradability, a lubricant is developed to offer specific technical performance, it is utilized to protect components, and it allows them to operate as designed. The performance of a lubricant is measured via a series of bench, engine, and/or component tests that are typically followed by field validation tests in the final application. Biodegradable lubricants have to have the same performance as the petroleum-based counterparts; otherwise, customers will not use them regardless of their environmental benefits. In addition to passing all the required performance tests, biobased lubricants have to pass biodegradability and toxicity tests. The latter are not as well known or defined as the former.

Performance requirements are driven and determined by coalitions of organizations that typically include OEMs, petroleum companies, chemical companies, and standards organizations. Examples of these entities include ASTM, ISO, SAE, European Automobile Manufacturers' Association, Engine Manufacturers Association (EMA), API, and American Chemical Council (ACC). In some cases, common specifications that describe customary minimum requirements are developed by a coalition of some of the entities described above. An example is the series of API engine oil performance requirements known as *engine oil categories*. The current diesel engine oil categories are API CH-4, API CI-4, and API CJ-4, where CH-4, CI-4, and CJ-4 are the names of these distinct categories [10,11,12]. Some OEMs may add their own performance requirements or tighten the requirements of certain tests to ensure that their engines or components are adequately protected and/or to differentiate their oils. Well-known OEM engine oil performance specifications include Caterpillar's Engine Crankcase Fluid specifications (ECF) [13], Cummins Engineering Standards specifications (CES) [14], and Detroit Diesel Power Guard Oil specifications (PGO) [15]. These additional OEM specifications requirements are based on the common API industry specifications.

In cases where no industry standard exists, OEMs may develop lubricant standards to ensure that their machines are protected. These requirements become common, drive the design of the lubricants, and in many instances drive the development of industry performance standards for the fluid. An example of an OEM-developed specification is Caterpillar's transmission oil specification Cat TO-4, General Motor's DEXRON transmission oil standard, and Ford Motor Company MERCON transmission oil specifications [16].

The specification development process applies to all lubricants, whether used for machines, engines, or industrial applications. This chapter, however, will focus mainly on hydraulic lubricants and diesel engine oils.

Hydraulic components are more exposed to leaks than other machine components due to high pressures and the rather extensive utilization of hoses in hydraulic systems. Additionally, machine hydraulic systems contain higher volumes of lubricants than other components. Hence, biodegradable and low-environmental impact hydraulic oils are more prevalent and are most often required by regulations.

Diesel engine oil demand is high due to the frequent oil changes for heavy-duty diesel engines. Additionally, the specifications of engine oil are highly defined and established.

6.5 HYDRAULIC LUBRICANT PERFORMANCE REQUIREMENTS

Hydraulic lubricants play a critical role in the operation and the performance of hydraulic systems [17,18,19]. In addition to viscosity and oxidative and thermal stability, hydraulic lubricants must also protect pumps from wear and allow the uninterrupted operation of actuators and valves. Additionally, hydraulic oils have to adequately perform at high temperatures and pressures; perform at close tolerances, in particular, in advanced hydraulic systems; be compatible with a variety of metals and elastomers; protect various types of pumps from wear (piston, gear, vane, etc.); and operate in the presence of moisture that can contaminate the system. In certain operations, hydraulic oils must be fire resistant as well.

At this time, hydraulic lubricants do not have common industry specifications. Rather, the technical requirements of these lubricants are determined by OEMs, pump manufacturers, and lubricant producers. Table 6.1 lists the general performance characteristics and the test methods for hydraulic oils [16,17,20]. It is important to note that the effectiveness of the listed tests may be lower for lubricants based on vegetable oils, or for those based on synthetic base stocks that are from various resources including renewable sources, than the effectiveness for hydraulic lubricants that are petroleum based.

6.6 ENGINE OIL PERFORMANCE REQUIREMENTS

This section will focus on heavy-duty diesel engine oils. Engine oils for the automotive industry follow a very similar process as described below, but the tests are pertinent to the automotive engines [17,18,21].

Performance specifications of heavy-duty diesel engine oils, better known as *API categories*, are well established and defined. The API, EMA, ACC, ASTM, and other organizations have a clear and well-defined process to develop the engine oil categories. The motivations of new category development and/or changes are the following in general:

- Emission regulations that require the development of new engine technologies that in turn require oils of specific chemistries (compatible with emissions reduction strategies)
- General advancement in engine technologies that demand new and more advanced lubricants
- Field issues related to oil performance that requires improved oil quality

TABLE 6.1
Performance Tests of Hydraulic Lubricants

Characteristic	Tests	Typical Limits	Notes and Information
Acid number	ASTM D664/DIN 51558-1/2	1.5 max	Increase in acid number indicates oil degradation
Viscosity	ASTM D2983, ASTM D445	Typical range is 32–68 cSt at 40°C	Varies per viscosity grade (10W, 20W, SAE 30)
VI	ASTM D-2270	100–150 (based on viscosity grade and base stock chemistry)	High-VI oils allow consistent viscosity performance over a wide range of temperatures
Oxidation stability	ASTM D2272, ASTM D943	2000 hours and up (ASTM D943)	May result in poor outcomes for vegetable-based lubricants
Corrosion protection	ASTM D665	Pass/Fail	Ensures the protection of common materials used in hydraulic systems
Copper strip corrosion	ASTM D130	1A–3B	Protection of copper components
Wear protection	ASTM D2882, FZG Gear Test DIN 51534, Vickers M-2952-S, four-ball test ASTM D4172, Denison Hydraulics HF-0	Varies: Vickers: 0.3 mg typical FZG: 12 or lower Four-ball test: 0.3 mm typical	In addition to ASTM test, pump manufacturers developed tests to ensure the proper operation and the durability of their pumps
Pumpability (mPa s)	ASTM D4684/Rexroth A2F pump	Based on temperature (ex.: at −30°C, 15,000 mPa s)	Ensures the appropriate flow through pumps and compatibility with the pump design
Foam	Foam: ASTM D892, Seq. 1, 2, or 3) ISO 6247	Foam: All sequences: 0/0–5/0	Prevent spongy or erratic motion of actuators and poor oil film
Air separation	ASTM D3427/DIN51381	10 minutes max	
Demulsibility	ASTM D2711	Yes/No	When required, water should quickly separate from the oil to reduce rust and ensure wear protection
Material compatibility	ASTM D611/D2240/ D471/6158, DIN 51524/ ISO 11158	Per end user requirements	Oils have to protect various metals (ferrous- and copper-based) and elastomers (seals, gaskets, etc.)
Filterability	ISO 13357-1	Pass/Fail or per end user's requirements	Filter compatibility—Various tests can be also specified
Fire protection (for fire-resistant oils only): flash and fire points	ASTM D92/ISO 7745[a]	Per regulations	Oils are formulated to resist ignition and to resist propagation of a flame from an ignition source

Source: USDA, Minimum biobased content is given under the "Hydraulic Fluids—Mobile Equipment" within the "Product Categories" list, Retrieved from http://www.biopreferred.gov/BioPreferred/faces/pages/ProductCategories.xhtml (Accessed January 2015). With permission.

Note: Tests and requirements other than those given in this table may exist based on OEM, applications, or end-user requirements.

[a] ISO 7745:1989 and BS 7287:1990—Guide for the use of fire-resistant hydraulic fluids.

Note that a category becomes obsolete when tests in this category become unavailable with no replacements.

There are currently three active heavy-duty diesel engine oil categories; they are listed in the following by their names and development dates:

- API CH-4, developed in 1989
- API CI-4 and the related API CI-4 PLUS, developed in 2002 and 2004, respectively
- API CJ-4, developed in 2006

Table 6.2 describes the API category performance criteria and tests. It is worth noting that as this chapter is published, work is ongoing on the development of a new bifurcated category planned for release in the first quarter of 2017 (category names have not been not finalized at the time this chapter is written) [22,23].

The complexity and the performance of the categories increase from one category to the next in order to fulfill the advanced demands of the new engine technologies. These demands include higher temperatures and pressures in the engines, tighter clearances of moving components, new components, higher performance requirements, and tighter emission regulations (which may require emission reduction technologies or strategies). The performance of each category is measured and validated through a series of engine and bench tests [16,17,21]. These tests are codeveloped and fully agreed upon by the various organizations involved in the category development.

The various engine oil categories are designed to be "backwards compatible" [21]. The opposite is not true. The latest category oils, API CJ-4, can be used in all the prior engine technologies. The earlier API CH-4 category oils, on the other hand, may not be used in all the engine technologies that followed it. Engine manufacturers detail the categories that are allowed in their various engine models.

The API has developed certification programs in order to ensure that oils claiming various categories consistently satisfy all the performance requirements of the claimed category.

Low-environmental impact oils, whether biobased, rerefined, biodegradable, or others, that claim an API category have to pass the same test programs and demonstrate performance as defined in the category in order to claim this category.

In addition to the performance criteria listed in Table 6.2, API CJ-4 sets specific chemical concentration limits [24]. These limits are designed to allow compatibility with emission-reducing components (aftertreatment systems), such as diesel particulate filter, diesel oxidative catalyst, and selective catalytic reduction. The chemical limits for API CJ-4 category are the following:

- Sulfated ash: 1.0% maximum mass fraction (ASTM D874)
- Phosphorus: 0.12% maximum mass fraction (ASTM D4951)
- Sulfur: 0.4% maximum mass fraction (ASTM D4951)
- Volatility: 13% maximum (ASTM D5800)

6.7 TESTING FOR RENEWABLE CONTENT

Test methods to determine the percentage of renewable content are very limited. In fact, there is only one widely used standard analytical test procedure to determine the renewable content of fuels and biogenic CO_2 emissions. The test method is ASTM D6866, which now has equivalent European methods, CEN 15591/CEN 15747 [25]. Conceivably, this method can be applied for lubricants as well. ASTM D6866 method uses the same principles of carbon dating that are used for dating archeological artifacts. Carbon-14 analysis of fuels (and of lubricants) reflects the amount of biomass components of the materials. Carbon-14 can be readily distinguished from fossil-based materials since the latter do not contain any carbon-14 isotope. The percentage of the renewable content (biogenic carbon) is easily calculated from the percentage of the total carbon.

TABLE 6.2
Heavy-Duty Diesel Engine Oil Performance Criteria

Performance Criterion	Fuel Sulfur (%wt)	Test Method	Test Specification	API Category and Introduction Date		
				CH-4 1998	CI-4[a] 2002	CJ-4 2006
Oil consumption and aluminum piston deposits	0.4	Caterpillar 1K 1990 Model DI	ASTM D6750	X		
Aluminum piston deposits and oil consumption	0.05	Caterpillar 1N	ASTM D6750			X
Oil consumption and steel piston deposits	0.05	Caterpillar IP	ASTM D6681	X		
Oil consumption and piston deposits	0.05	Caterpillar 1R	ASTM D6923		X	
Aluminum piston deposits and oil consumption	–	Caterpillar 1K or Caterpillar 1N	ASTM D6750		X	
Oil consumption and piston deposits	0.0015	Caterpillar C-13	ASTM D7549			X
Viscosity increase due to 4.8% soot	0.05	Mack T-8E—1991 Model (300 h)	ASTM D5967	X		
Viscosity increase due to 6.0% soot	0.05	Mack T-11	ASTM D7156			X
Used oil viscometrics at low temperature	–	MRV TP-1 Soot[b] Mack T-1QA/T-11A[b]	ASTM D6896		X[b]	X[b]
Ring, liner, and bearing wear and oil consumption	0.0015	Mack T-12	ASTM D7422			X
Ring, liner, and bearing wear and oil consumption	0.05	Mack T-10 (EGR)	ASTM D6987/ D6987M		X	
Ring, liner, and bearing wear	0.05	Mack T-9 1994-12 liter VMAC	ASTM D6483	X		
Valve train slider wear, filterability, and sludge	0.05	Cummins M11 HST 1994	ASTM D6838	X		
Valve train wear, filter deltap, and sludge	0.05	Cummins M11 (EGR)	ASTM D6975		X	
Valve train wear, filter deltap, and sludge	0.05	Cummins ISM	ASTM D7468			X
Valve train wear	0.0015	Cummins ISB	ASTM D7484			X
High temperature/high shear	–	Viscosity at 150°C (mPa s; min)	ASTM D4683/ D4171/D5481		X	X
Shear stability— 90 cycles	–	Bosch injector	ASTM D3945			X
Shear stability— 30 cycles	–	Bosch injector	ASTM D3945	X	X	

(Continued)

TABLE 6.2 (CONTINUED)
Heavy-Duty Diesel Engine Oil Performance Criteria

Performance Criterion	Fuel Sulfur (%wt)	Test Method	Test Specification	API Category and Introduction Date		
				CH-4 1998	CI-4[a] 2002	CJ-4 2006
Roller-follower valve train wear	0.05	GM 6.5 Liter PC—Diesel	ASTM D5966	X	X	X
Oil oxidation	0.1	Sequence III G	ASTM D7320			X
Oil oxidation	–	GM—Liter Gasoline IIIF	ASTM D6984		X	
Oil oxidation	–	GM 3.8 Liter Gasoline IIIE test	ASTM D5533	X		
Corrosion	–	High Temperature Corrosion bench test (135°C)	ASTM D6594	X	X	X
Aeration	0.05	Navistar HEUI 7.3 Liter EOAT	ASTM D6894	X	X	X
Foam	–	Bench test sequences I, II, III	ASTM D892	X	X	X
Volatility	–	NOACK ASTM D5800, Distillation ASTM D2887	ASTM D5800 or D6417 or D6278	X	X	X
Elastomer compatibility[c]	–		ASTM D0471, ref. oil		X	
Total number of engine and bench tests				12	14	15

Source: McGeehan, J. A. et al., API CJ-4: Diesel oil category for both legacy engines and low emission engines using diesel particulate filters, SAE Paper 2006-01-3439, 2006; McGeehan, J. A. et al., The first oil category for diesel engines using cooled exhaust gas recirculation, SAE Paper 2002-01-1673, 2002; McGeehan, J. A. et al., New diesel engine oil category for 1998, SAE Paper 981371, 1998; The Lubrizol Corporation, Ready reference for lubricants and fuels, Version 4, February 2011, p. 139. With permission.

[a] API CI-4 PLUS introduced in 2003 and is the same test as API CI-4 with the addition of Mack T-11 and the 90 cycle Kurt Orbahn injector test for shear.

[b] Low-temperature pumpability tests are incorporated in the Mack T-10/Mack T-11 tests. However, if the MRV TP1 fails, T-10 can be run for CH-4, and T-11A can be run for CJ-4.

[c] Elastomers tested for CI-4 and CJ-4: nitrile, silicone, polyacrylate, and fluoroelastomer. Additionally for CJ-4: Vmac G. Properties tested: volume change, hardness change, tensile strength, and elongation at break change.

6.8 TESTING FOR BIODEGRADABILITY AND TOXICITY OF LUBRICANTS

Lubricants are hydrocarbon based, and, hence, are inherently biodegradable given enough time. However, for low-environmental impact lubricants, it is critical to understand the requirements for biodegradability and impact on the environment of the new and the used lubricant for the intended application of the lubricant. The appropriate tests focus on the time to biodegrade and the toxicological effects on naturally occurring organisms and aquatic life.

Unlike performance testing, which are well defined, testing for the biodegradability and the toxicity of lubricants are not well established. In fact, tests for biodegradability and toxicity of lubricants can still be considered not fully mature, and are still evolving. One of the complexities is the composition of the lubricants themselves: base stocks and a myriad of additives. Each component may have unique biodegradability and toxicity characteristics. As a result, the biodegradability and the toxicity of a finished lubricant are due to the combined impact of all of its components. Tests ideally have to address all of the components of the lubricants, which is a challenge due to the vast differences in chemical properties of these components.

As discussed above, hydraulic fluids potentially have a higher impact on the environment through leaks, spills, or simply the sheer volume of oil used. Two ASTM standards have been developed to guide the biodegradability and environmental impacts of these fluids:

- ASTM D6006, "Standard Guide for Assessing Biodegradability of Hydraulic Fluids"
- ASTM D6046, "Standard Classification of Hydraulic Fluids for Environmental Impact"

6.8.1 BIODEGRADABILITY TESTING

Limited tests exist to assess the biodegradability of lubricants. One of the two most common tests, a shake test, assesses the total biodegradability of the lubricant as measured by the conversion of the lubricant to CO_2. The other test, a Coordinating European Council (CEC) test, assesses the aerobic aquatic biodegradation potential of the lubricant and measures the amount of test material remaining (C–H stretch by IR analysis at 2930 cm^{-1}). Both tests use a growth medium composed of a mineral salt mix and both use the lubricant as the sole test material, which is then compared to a blank that is lubricant free. The tests use unacclimated sewage inoculum obtained in most cases from municipal wastewater treatment plant that has no industrial inputs. A soil inoculum is also used by the shake test. Table 6.3 provides the main tests used for measuring biodegradability. Other tests may exist or may be under development. The biodegradability requirement is typically lower than 100%. In some cases, no more than 50% or 60% biodegradability may be required.

TABLE 6.3
Lubricant Biodegradability Test Methods

Test	Standard	Details
Shake flask test	EPA 560/6-82-003 or OECD301 A and B[a]	Shake test flasks with neoprene stoppers and suspended alkali traps to measure CO_2 evolution under aerobic conditions. Flasks are heavily shaken at 25°C; and CO_2 is periodically measured over 28 days. Note that versions B, F, and D of this test use a respirometer as a measurement method.
Modified Sturm test (similar to shake flask test)	ASTM D5846	Similar to above, and also measures CO_2 evolution as bacteria metabolizes the lubricant samples. Measurement method is respirometer.
Shake test	CEC-L-103-12[b]	Shake test flasks with cotton stoppers. Flasks are heavily shaken at 25°C and periodically extracted with Freon. The C–H stretch is periodically measured over 21 days by infrared spectroscopy (IR) to determine the quantity of test material.
Water quality	DIN EN ISO 14593	Evaluates ultimate aerobic biodegradability of organic compounds in aqueous medium. Analyzes inorganic carbon in sealed vessels.

Source: Beercheck, R., Biodegradability: Which Test is Best? Lube Report, *Lubes'n'Greases*, 2014; Pirro, D. M., and A. A. Wessol, *Lubrication Fundamentals*, Second Edition: Revised and Expanded, ExxonMobil, Marcel Dekker, New York, 2001, pp. 37, 105, and 119; Placek, D., In *Synthetics, Mineral Oils and Bio-based Lubricants*, L. Rudnick (ed.), CRC Press, Boca Raton, 2006, p. 519. With permission.

[a] OECD. Note that OECD 301 has four different versions, A, B, F, and D. Variation among the versions includes measurement techniques.

[b] CEC.

It is important to note that some of the tests do not distinguish between primary and complete degradation. Long and/or branched chains are typically more difficult to break down, while shorter-chain and aromatic components break down more readily. The measured CO_2 does not necessarily distinguish the molecular sources. For example, a product that shows 80% degradation in 21 days may still have most of its paraffinic long-chain molecules intact and the measured CO_2 may be from shorter-chain and aromatic compounds in the sample. Additionally, the reproducibility of most of the tests given in Table 6.3 is poorer than that of the performance tests given in Table 6.2 [3]. These are the considerations to be taken as biodegradability is quantified or new biodegradability tests are developed.

6.8.2 TOXICITY TESTING

The toxicity of environmental lubricants is mainly evaluated by measuring the impact on aquatic life. The main toxicity test is based on the acute aquatic toxicity study defined by EPA 560/7-82-002 test or OECD 203:1-12 (the two tests are very similar). These tests use rainbow trout due to the sensitivity of this fish to environmental changes. Other sensitive aquatic life, such as various species of minnows, mussels, oysters, and shrimp, can also be tested [17]. The main aquatic toxicity test used with lubricants involves the exposure of the aquatic specimens to a dispersion of oil in water under flow conditions that simulate waves and currents. The live specimens are exposed to lubricant dispersions in water at different concentrations reaching a maximum of 5000 parts per million (ppm). The toxicity of environmental lubricants is expressed as the concentration of lubricant (ppm, weight/volume) that is required to kill 50% of the aquatic specimens after 96 hours of exposure. This is typically expressed as median lethal concentration, LC_{50}.

Variations of the above test can be used to test exposure to lower concentrations and to measure various impacts such as development, reproduction, accumulation of toxins, and biomagnification. The tests can also be extended to 168 hours (7 days) of exposure.

Examples of certain acceptable criteria for lubricants that can be used in environmentally sensitive areas that are generally agreed to by the International Maritime Organization and some other global organizations are as follows [17]:

- Ready biodegradability: >60% conversion of test material carbon to CO_2 in 28 days using unacclimated inoculum in the shake flask or the ASTM D5846 test.
- Aquatic toxicity: >1000 ppm (50% minimum survival of rainbow trout).
- Refer to Table 6.4 for example of ecotoxicology data for certain hydraulic fluids.

TABLE 6.4

Example of Ecotoxicology Data for Certain Hydraulic Fluids

Lubricant Base Stock	Trout LC_{50} (ppm)	% Biodegradability	
		Shake Flask[a]	CEC Test[b]
Mineral oil	389 to >5000	42–48	(Not tested)
Vegetable oil	633 to >5000	72–80	>90
Synthetic ester	>5000	55–84	>90
Polyglycol	80 to >5000	6–38	(Not tested)

Source: Pirro, D. M., and A. A. Wessol, *Lubrication Fundamentals*, Second Edition: Revised and Expanded, ExxonMobil, Marcel Dekker, New York, 2001, p. 108. With permission.

[a] The shake test is per EPA method 560/6-82-003.

[b] CEC method is CEC-L-33-T-82.

Other toxicological properties that can be applied to lubricants include the following:

- Mutagenicity index (modified Ames test) per E1687 [25]
- Dimethyl sulfoxide extractables, weight percentage per IP 346 [26]
- Long-term rodent carcinogenicity bioassay, number of tumor-bearing animals/test group (%) [27]

Additional tests may be required by local legislation and regulatory organization.

6.9   ENVIRONMENTAL LABELING

Lubricants that pass certain well-defined environmental criteria can display the symbols associated with these criteria. While there are no global labels and criteria yet, there are some labels and associated criteria that are well known and used. The key criteria are the German Blue Angel label [28] and the European Union (EU) Daisy Ecolabel [29], both of which are described in detail at dedicated websites. The Blue Angel label can be displayed on lubricants that are readily biodegradable and that also have all of their minor components biodegradable per OECD criteria. The EU Ecolabel is based on high standards of environmental performance. The criteria for the Ecolabel consider the whole life cycle of a product, from the extraction of raw materials through manufacture, packaging, distribution, use, and disposal of the product.

In the United States, the USDA has established the BioPreferred program [30] for a myriad of products including hydraulic lubricants. This program includes labels for qualifying products. To qualify for inclusion in this program, a hydraulic fluid for mobile equipment must have 44% minimum biobased content [31].

It is worth mentioning that the ISO has developed the ISO 14020 to 14025 series for environmental labels and declarations [32]. These standards can serve as resources for interested end users.

ACKNOWLEDGMENTS

The author thanks Caterpillar's Fluids Group and technical community for the vast opportunities to learn, grow, and practice in the field of lubricants and fluids in general.

REFERENCES

1. M. C. McManus et al. (2004), LCA of mineral and rapeseed oil in mobile hydraulic systems *Journal of Industrial Ecology*, Volume 7, Numbers 3–4, pp. 163–177.
2. Freedonia Group (2011, July), World lubricants, Industry Study #2771.
3. R. Beercheck (2014, March 19), Biodegradability: Which Test is Best? Lube Report, *Lubes'N'Greases*, p. 18.
4. Bill Number: SB 916 (2014, January 27), Introduced by Senator Correa. Retrieved from http://www.leginfo.ca.gov/pub/13-14/bill/sen/sb_0901-0950/sb_916_bill_20140127_introduced.html (Accessed January 2015).
5. B. Fitch (2012, October), Optimizing oil change intervals in heavy-duty vehicles. Retrieved from http://www.machinerylubrication.com/Read/29117/oil-change-intervals (Accessed January 2015).
6. Caterpillar Products Specification, Excavators examples. Retrieved from http://www.cat.com/en_US/products/new/equipment/excavators.html (Accessed January 2015).
7. U.S. EPA (2014, August 22), Emergency response: National Response Center. Retrieved from http://www2.epa.gov/emergency-response/national-response-center (Accessed January 2015).
8. U.S. EPA (2004), Medium and heavy duty diesel vehicle modeling using a fuel consumption methodology.
9. U.S. Department of Energy (2009, March 27), Motivations for promoting clean diesels.
10. J. A. McGeehan et al. (2002), The first oil category for diesel engines using cooled exhaust gas recirculation, SAE Paper 2002-01-1673.
11. J. A. McGeehan et al. (1998), New diesel engine oil category for 1998, SAE Paper 981371.

12. J. A. McGeehan et al. (1994), The world's first diesel engine oil category for use with low-sulfur fuel: API CG-4, SAE Paper 941939.

13. Caterpillar lubricants specifications. Retrieved from https://parts.cat.com/en/catcorp/machine-fluids (Accessed January 2015).

14. Cummins CES engine oil specifications (2007, May). Retrieved from http://www.kleenoilusa.com/pdf /warranty/Cummins-Oil_ServiceBulletin_May-07.pdf, p. 3 (Accessed January 2015).

15. Detroit diesel PGO specifications (2005). Retrieved from http://www.demanddetroit.com/pdf/vocations /lube-oil-fuel-requirements.pdf. p. 7 (Accessed January 2015).

16. The Lubrizol Corporation (2011, February), Ready reference for lubricants and fuels. p. 139, Version 4.

17. D. M. Pirro and A. A. Wessol (2001), *Lubrication Fundamentals*, Second Edition: Revised and Expanded. ExxonMobil, Marcel Dekker, New York, pp. 37, 105, and 119.

18. W. Givens and P. Michael (2003), *Fuels and Lubricants Handbook*, G. Totten (ed.), ASTM International, West Conshohocken, PA, p. 373.

19. D. Placek (2006), In *Synthetics, Mineral Oils and Bio-based Lubricants*, L. Rudnick (ed.), CRC Press, Boca Raton, FL, p. 519.

20. The Lubrizol Corporation (2011, February), Ready reference for hydraulic and industrial fluids, p. 31, Version 4.

21. ASTM D4485, "Standard Specification for Performance of Engine Oils," http://www.ASTM.org.

22. J. V. Rensselar (2013, January 1), PC-11 and GF-6: New engines drive change in oil specs, Retrieved from http://www.stle.org/resources/articledetails.aspx?did=1671 (Accessed January 2015).

23. J. Rubenstone (2014, June 10), Upcoming diesel engine oil standard promises fuel economy gains, *EngineeringNewsRecord.com*. Retrieved from http://enr.construction.com/products/materials/2014/0616 -upcoming-diesel-engine-oil-standard-promises-gains-in-fuel-economy.asp (Accessed January 2015).

24. J. A. McGeehan et al. (2006), API CJ-4: Diesel oil category for both legacy engines and low emission engines using diesel particulate filters, SAE Paper 2006-01-3439.

25. European Committee for Standardization (2015), Who we are. Retrieved from https://www.cen.eu/about /Pages/default.aspx (Accessed January 2015).

26. Energy Institute (1992, January), "IP 346: Determination of polycyclic aromatics in unused lubricating base oils and asphaltene free petroleum fractions—Dimethyl sulphoxide extraction refractive index method. Retrieved from http://www.energypublishing.org/publication/ip-standard-test-methods/ip-346 -determination-of-polycyclic-aromatics-in-unused-lubricating-base-oils-and-asphaltene-free-petroleum -fractions-dimethyl-sulphoxide-extraction-refractive-index-method (Accessed January 2015).

27. R. H. McKee et al. (2013), Genetic toxicity of high-boiling petroleum substances, *Regulatory Toxicology and Pharmacology*, pp. S75–S85, 67.

28. The Blue Angel (2015), The Blue Angel: Home. Retrieved from https://www.blauer-engel.de/en/home (Accessed January 2015).

29. Europa.eu (2009, February 12), Eco-Label. Retrieved from http://europa.eu/legislation_summaries/other /l28020_en.htm (Accessed January 2015).

30. USDA (2015), Biopreferred. Retrieved from http://www.biopreferred.gov/BioPreferred/ (Accessed January 2015).

31. USDA. Minimum biobased content is given under the "Hydraulic Fluids—Mobile Equipment" within the "Product Categories" list. Retrieved from http://www.biopreferred.gov/BioPreferred/faces/pages /ProductCategories.xhtml (Accessed January 2015).

32. International Organization for Standardization, ISO standards 14020 and 14025. Retrieved from http:// www.iso.org/iso/search.htm?qt=14020&sort=rel&type=simple&published=on&active_tab=standards (Accessed January 2015).

7 Environmental Approach to Hydraulic Fluids

Gerhard Gaule and Ben Müller-Zermini

CONTENTS

ABSTRACT

In this chapter, we compare the position of the current standards for environmentally acceptable lubricants to the actual requirements of the environment. Therefore, we describe the test principle and the different test methods for testing biodegradability of lubricants. Then, there is a comparison of the reproducibility of several lubricant test methods to the reproducibility of biodegradation test methods. Some properties of environmentally acceptable hydraulic fluids, which are not included in today's standards, will be defined. And last, there is a case study about our experiences concerning standard development in Germany.

7.1 INTRODUCTION

Over the past years contradictory information has been spread in the media and in the public about bio-oils and fast biodegrading hydraulic oils. Nevertheless, even today, there is no precise definition for the term *bio-oil*. Discussions on this subject are often politically influenced and are not scientifically sound. However, it is generally agreed that bio-oils must be nontoxic and biodegradable so that they do not pollute the environment. It can also be expected that by using bio-oils, valuable resources are saved and the principle of sustainability is realized. In this chapter, we will point out the particular requirements of the environment, the principles and the reliability of testing biodegradability, and the properties of an environmentally acceptable hydraulic fluid. These shall show the direction for the future development of environmentally acceptable hydraulic oils.

7.2 CURRENT INTERNATIONAL NORMS AND STANDARDS

What is environmentally acceptable? Or what is good for the environment? There are many opinions, norms, and standards which sometimes give contradictory answer to producers and consumers.

In the foreword to the brochure *Environmental Information for Products and Services* [1], published by the German Federal Ministry for the Environment, Nature Conservation, and Nuclear Safety, Sigmar Gabriel, then German Federal Environment minister, described the Integrated Product Policy or IPP:

> The German government regards the development of a comprehensive product-specific environmental policy as being an important step towards the further development of European environmental policy. One of the most important features of IPP is that it takes into consideration the environmental impact during the entire "life cycle" of the product—that is to say from production and application to its disposal. This holistic approach is intended to enable consideration of all environmental impacts that are associated with a product. It should offer a means of reducing pollution and intensifying ecological benefits at every point throughout the product's life cycle.

In the same brochure, there is also a foreword by Jürgen Thumann, the president of the Federal Association of German Industry [1]. According to Thumann, companies have to meet two sets of demands: they have to satisfy the ecological requirements of the public and the legislative body, and they also have to continuously improve the ecological and economic efficiency of the products. Important instruments for product-specific environmental protection are standards, regulations, and self-imposed obligations, as well as environmental management systems and life cycle analyses. These voluntary positive labels offer a means of systematically depicting the environment-friendly aspects of products on the basis of substantiated, convincing information.

Of course, there are ecolabels which show us which products are environmentally acceptable. Unfortunately, there are too many ecolabels. Nearly every country has its own ecolabel. In the case of hydraulic fluids, the key criterion of environmentally acceptable hydraulic fluids is the fast biodegradability of the product. In addition, some ecolabels require a minimal renewable content; others require testing every ingredient separately.

Hydraulic oils can essentially be divided into four groups [2]: mineral oil-based, fire resistant, environmentally acceptable biodegradable, and food grade. This classification is historically developed and it does not mean that a product cannot belong to two different groups. The environmentally acceptable hydraulic fluids are defined according to ISO 15380 [3]. They are divided into more or less water-soluble polyglycols (hydraulic oil environmental polyglycol [HEPG]) and water-insoluble triglycerides (hydraulic oil environmental native triglyceride [HETG]), synthetic esters (hydraulic oil environmental synthetic ester [HEES]), and PAOs (hydraulic oil environmental polyalphaolefin and related products [HEPR]). In these standards, the demand for rapid biodegradability comes first. Toxicity tests on aquatic organisms are also required. But these toxicity tests are only performed with the water accommodated fraction of the oily product; that means that only the water-soluble part of the oil is tested with fish, daphnia, or bacteria. Because the water solubilities of HETG, HEES, and HEPR are very low, there is almost no substance in the aqueous extraction, which comes to the test organisms. However, if hydraulic oil gets into the environment, the whole formulation gets there.

In Europe, the norms' and ecolabels' answer to the question *What is environmentally acceptable?* is that hydraulic fluids are environmentally acceptable if they are:

- Biodegradable—the faster they biodegrade the better,
- Nontoxic for water organisms, and
- Made of renewable materials.

However, there are more questions that need to be answered: Is this rule also applicable if huge amounts of oil get into the environment? What about the toxicity to terrestrial plants and organisms? And is it really true that lubricants made of renewable materials are better for the environment than mineral oil-based products?

Thus, it is very important that a holistic approach be used in environmental regulations. Up to now, life cycle assessments are not required in most ecolabel regulations.

We are living in an era of quick change. Almost every day, we are informed about new discoveries in science, technology, and all other areas of research. Often these new findings revise existing opinions. The same does not apply to existing regulations and standards. They often do not keep in step with the new findings. The development of new regulations and standards takes some time. During this time, old regulations can slow down the introduction of new products and findings in the field of environmental protection.

7.3   TESTING BIODEGRADABILITY

7.3.1   Test Principles

There are many different test methods for testing the biodegradability of organic substances. All of them are set up to simulate the biodegradation process in a laboratory. To do this, a mineral substrate, which is water with dissolved mineral salts, the oil sample, and bacteria are mixed in a bottle or a flask. The bacteria used for biodegradation are mostly taken from a sewage treatment plant. Some methods also allow the testing of surface water from lakes or rivers. The prepared flasks are incubated at room temperature for a certain period. Some tests prescribe 21 days, others 28 days, and still others even longer periods.

The degradation process consists of a chain of very complex biochemical reactions. The net reaction equation can be represented as follows:

$$\text{Organic substance} + \text{oxygen} \rightarrow \text{carbon dioxide} + \text{water.} \tag{7.1}$$

The organic substance is degraded to carbon dioxide and water and for this there is a need for oxygen.

For measuring the degradation rate, the changes in the concentration of reactants or products can be measured. For testing lubricants, it is very common to measure the concentration of the unreacted component and their nonvolatile degradation products (CEC-L-103-12 or CEC-L-33-A-93). Other test methods measure the amount of consumed oxygen, e.g., OECD 301 C or Deutsches Institut für Normung (DIN) European Standards (EN) ISO 9408. Then, there are test methods which measure the carbon dioxide evolution (OECD 301 B or DIN EN ISO 9439).

Every test method specifies special test equipment for measuring organic substance; like in CEC-L-103-12, open flasks on a shaker are used. The laboratory analyzes the concentration of organic substance before and after the 21-day incubation period. If gases are analyzed, the test flask must be airtight. Oxygen consumption can be measured by using a closed bottle or using a respirometer. In the case of a closed bottle, the concentration of oxygen is measured in the water. In the case of the respirometer, a sensor measures the pressure drop because of the microorganism's oxygen consumption and the carbon dioxide absorption by a CO_2 absorber.

Carbon dioxide evolution can be measured by trapping the produced carbon dioxide in a gas wash bottle filled with barium hydroxide solution, which acts like a trap for carbon dioxide. Subsequently, the amount of trapped carbon dioxide is measured by titration.

According to CEC-L-33-A-93, lubricants are considered biodegradable if they reach a degradability of 80% after 21 days. OECD 301 B requires a degradability of 60% within 28 days for the lubricant to be considered degradable.

An interesting approach shows the ASTM D6046. This standard divides hydraulic fluids into four biodegradation categories. This classification allows a better distinction between the individual hydraulic oils. The method goes beyond good (>60%) and bad (<60%). Due to the longer duration of the test run, the method takes into account not only velocity but also completeness of the degradation.

According to the ASTM D6046, hydraulic fluids are classified into four categories [4]:

- Category 1: Biodegradation greater than or equal to 60% after 28 days (4 weeks)
- Category 2: Biodegradation greater than or equal to 60% after 84 days (12 weeks)
- Category 3: Biodegradation greater than or equal to 40% after 84 days (12 weeks)
- Category 4: Biodegradation less than 40% after 84 days (12 weeks)

7.3.2 Viewpoints of Different Biodegradation Test Methods

The process of biodegradation is very complex. There are many reaction steps and, therefore, many intermediates. This makes measuring biodegradation of oils very complex. In the case of a simple mechanism (e.g., radioactive decay), the reaction rate can be determined by measurement of reactant or product concentration. Regarding the biodegradation of hydrocarbons, there is not only one reactant or product but also many intermediates. This fact makes the measurement of degradation rate very complex. Of course, the end product is carbon dioxide and the evolution of carbon dioxide can be measured. However, the formation of carbon dioxide occurs not only at the very end of the reaction chain but also during the intermediate steps. That means the formation of carbon dioxide is not an indicator of the complete degradation of the molecule. It is only an indicator of the complete degradation of a single carbon atom.

The viewpoints of the various test methods are different. This can be visualized in Figure 7.1. In the case of a primary degradation test method, the observer would only regard the reactant molecules as not degraded. Therefore, the test method would need a specific analytical method, which is able to detect only the sample molecules. This could be, e.g., an enzymatic test, like a glucose test strip for the measurement of glucose. In contrast to this, the CEC-L-103-12 and CEC-L-33-A-93 use selective analytical methods which can detect all organic compounds which have C–H bonds and are oil soluble. In the CEC test methods, not only the sample molecules are regarded as not degraded but also the nonvolatile oil-soluble intermediates. Although the CEC test methods are often called *primary degradation tests*, they do not measure primary degradation according to the original definition.

7.3.3 Terms and Definitions

7.3.3.1 Mineralization

There is an International Union of Pure and Applied Chemistry (IUPAC) definition for the term *mineralization*: "In the case of polymer biodegradation, this term is used to reflect conversion to CO_2 and H_2O and other inorganics. CH_4 can be considered as part of the mineralization process because it comes up in parallel to the minerals in anaerobic composting, also called methanization" [5].

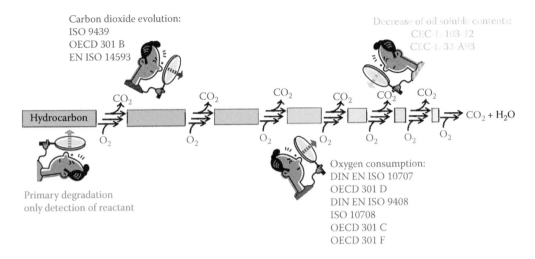

FIGURE 7.1   Viewpoints of different test methods.

That means an organic substance is converted to inorganic degradation products. Carbon dioxide is an inorganic reaction product. By measuring the carbon dioxide evolution, the percentage of mineralization of an organic substance can be measured. Sometimes biodegradation test methods which measure oxygen consumption are also called *mineralization methods*. However, according to the IUPAC definition, mineralization methods do not measure mineralization. They do measure this process only indirectly. A chemical substance can absorb oxygen without any degradation, e.g., rusting of iron or oxidation of alcohol to carboxylic acid.

7.3.3.2 Primary Degradation

There are different definitions of the term *primary degradation*. Mostly, it is defined as the first step of a degradation process. In the detergent industry, detergents are primarily degraded when they have lost their surface activity. Another definition is the following: "measurement of primary degradation is measurement of sample concentration using a specific test method," e.g., using enzymatic glucose strips for detection of glucose.

7.3.3.3 Ultimate Biodegradation

According to the definition in the standard ISO 9439, ultimate biodegradation is the "breakdown of a chemical compound or organic matter by microorganisms in the presence of oxygen to carbon dioxide, water and mineral salts of any other elements present (mineralization) and the production of new biomass" [6].

In the German language, this term is translated to *Vollständiger Abbau*, and also sometimes in English literature the term *complete degradation* is found. However, these terms can be misleading. Actually, there are two possible cases:

- Complete degradation of all the substance; i.e., no test substance is left. The test substance has been completely degraded.
- Complete degradation regarding the degradation steps: One test sample molecule has undergone all the degradation steps and been totally degraded to its end products, in the meaning of ultimate biodegradation.

Figures 7.2 and 7.3 show these two cases in pictures.

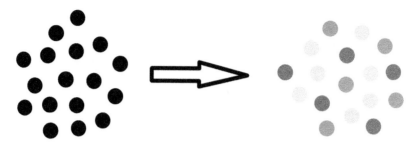

FIGURE 7.2   Complete degradation regarding the amount of a substance.

FIGURE 7.3   Complete degradation regarding degradation steps; ultimate biodegradation.

7.3.4   DOES A TEST METHOD FOR MEASURING ULTIMATE BIODEGRADABILITY ALWAYS SHOW COMPLETE DEGRADATION?

In order to answer this question, we have to distinguish two cases:

- Case 1: Test sample is a blend of many substances.
- Case 2: Test sample is a single pure substance.

7.3.4.1   Case 1: Test Sample Is a Blend

The lubrication industry often uses ultimate biodegradation tests such as OECD 301 B for testing ready-formulated hydraulic fluids. Hydraulic fluids and other lubricants are mixtures of structurally different substances: base oils and additives. In case of a positive test result of 60% or more biodegradation, the product is declared as completely biodegradable. But is this correct?

We will have to look at the following example: A mixture of 70% paraffinic and 30% aromatic hydrocarbons is tested according to OECD 301 B. Paraffinic and aromatic hydrocarbons are structurally different substances. The biodegradability of middle-chain and short-chain paraffinic hydrocarbons is very good. Within 28 days, it is possible that up to 91% of these substances is degraded. However, the biodegradability of aromatic hydrocarbons is very slow. It is possible that bacteria can degrade only up to 3% of these substances within 28 days. However, OECD 301 B measures carbon dioxide evolution. There is no information about the origin of the carbon dioxide.

The OECD 301 B test result of this mixture (70% paraffinic and 30% aromatic hydrocarbons) can achieve, e.g., 65%. This result would be interpreted as the following: The product is biodegradable, although there are components in the mixture which are not biodegradable at all. Figure 7.4 shows the degradation curves of the mixture and the components.

The example shows that the measurements on blends using the OECD 301 B test did not show complete degradation. This phenomenon is known and is called *sequential degradation*. It is taken into consideration in the "Revised Introduction of the OECD Guidelines for Testing of Chemicals Section 3," point 44: "Tests for readily biodegradability are not generally applicable for complex mixtures containing different types of chemicals" [7].

7.3.4.2   Case 2: Test Sample Is a Single Pure Substance

Carbon dioxide evolution tests measure the carbon dioxide from the test substance produced by the microorganisms. The C atoms in carbon dioxide molecules are all equal, even if the organic molecules may be very complex. There are different kinds of C atoms, which can react in a different way. It is possible that some of the C atoms of one molecule are easily broken down by microorganisms

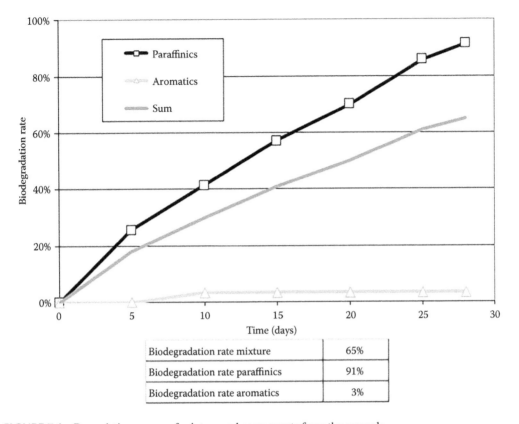

Biodegradation rate mixture	65%
Biodegradation rate paraffinics	91%
Biodegradation rate aromatics	3%

FIGURE 7.4   Degradation curves of mixture and components from the example.

than others. That means sequential biodegradation can happen in one molecule. Figure 7.5 shows an example molecule where sequential degradation can happen.

The molecule in this example consists of an aromatic part and a polyglycol part. Degradation happens step by step. The polyglycol part can be easily and quickly degraded by microorganisms. The end product of polyglycol degradation is carbon dioxide. The aromatic part of the molecule (which, in this example, is dichlorodiphenyl-trichloroethane [DDT], a persistent organic pollutant) is very stable and most microorganisms cannot degrade it at all [8,9].

However, if the molecule in the example is tested using OECD 301 B, it may be classified as biodegradable: The molecule consists of a total of 41 C atoms, of which 27 C atoms are part of the fast biodegradable polyglycol part. Only 14 C atoms are part of the aromatic DDT part. If carbon dioxide evolution is measured, a biodegradation rate of 27/41 = 66% will be found. Even in the case

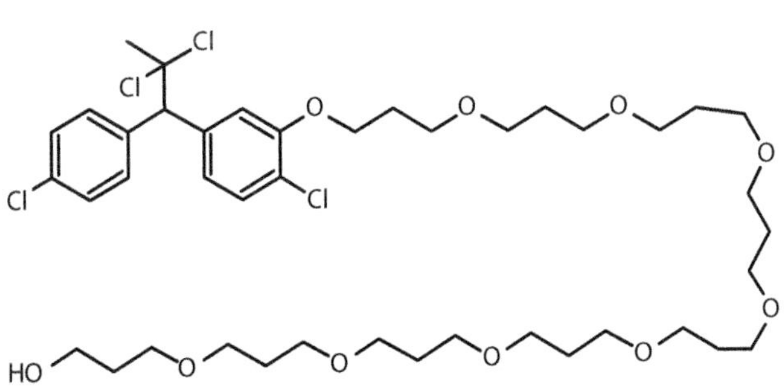

FIGURE 7.5   Molecule which can be sequentially degraded.

of a pure substance, the formation of persistent degradation products is possible, although the biodegradation rate according to OECD 301 B exceeds 60%.

The two abovementioned examples show that it is possible that biodegradation tests, which measure ultimate biodegradability, are not applicable to find persistent intermediates.

7.3.5 Reliability of Biodegradation Test Methods

7.3.5.1 Reproducibility R—Indicator of Reliability of Test Methods

How reliable are these test methods for measuring biodegradability? Compared to usual measurements in engineering science, the precision of biological test methods is very poor.

To determine the precision of a test method, round-robin tests are conducted. This means that several laboratories measure the same sample and compare their results. From round-robin test data, two major statistical values are obtained:

- Repeatability r
- Reproducibility R

The repeatability refers to one laboratory; the reproducibility compares the results of several laboratories. Reproducibility is defined as "the value below which the difference between two test results obtained under reproducibility conditions may be expected to occur with a probability of approximately 0.95 (95%)" [10].

An example can clarify this: In a round-robin test for measuring the open cup flash point, a reproducibility of 17°C had been obtained [11]. Two laboratories measure the flash point of the same sample. Lab A measures 181°C; lab B measures 169°C. The difference between these two results is 12°C. Twelve degrees Celsius is less than 17°C; that means the measurements are in the normal range.

In another case, lab A measures 181°C and lab B measures 163°C. The difference of these two results is 18°C. Eighteen degrees Celsius is higher than 17°C; that means the measurements are out of the normal range. Maybe, this is one case out of the 5% which are not inside the range or an error occurred in one of the laboratories.

7.3.5.2 Reproducibility of Lubricant Test Methods

To compare the reproducibilities of different test methods, the absolute reproducibility values can be related to typical values of the test method to get the relative reproducibility. For example, the reproducibility of density measurement according to ASTM D7042 [12] is 0.0015 g/mL, and

TABLE 7.1
Relative Reproducibilities of Selected Lubricant Tests

Test Method	Standard	R(rel)
Density [12]	ASTM D7042	0.2%
Refractive index [13]	DIN 51423-1	0.4%
Viscosity [12]	ASTM D7042	0.6%
pH [14]	DIN 51369	9%
Saponification number [15]	DIN 51559-1	10%
Flash point (open cup) [11]	DIN EN ISO 2719	11%
Simulated distillation [16]	DIN EN 15199-1	12%
Acid number [17]	DIN 51558-1	15%
X-ray fluorescence spectroscopy (for values < 50 ppm, R = 15 ppm) [18]	DIN 51396-2	30%
Four-ball test (welding load) (for values > 5000 N, R = 1500 N) [19]	DIN 51350-2	30%
Pour point [20]	DIN ISO 3016	33%

a typical value for the density of oil is 0.85 g/mL; the relative reproducibility will then become 100(0.0015/0.85) = 0.2%.

Table 7.1 shows the relative reproducibilities of selected lubricant test methods.

Some lubrication test methods, such as density, refractive index, or viscosity, are very precise. Others, like the four-ball test or the pour point test, are not very precise.

7.3.5.3 Reproducibility of Biodegradation Test Methods

Most current biodegradation test methods do not have any information or data about their reproducibility. Data about reproducibility can be found only in the following test methods: CEC-L-103-12, DIN EN ISO 14593, and ASTM D5864. Data about the reproducibility of the OECD 301 tests can be found in the publication of the round-robin test of 1988 [21]. Table 7.2 shows the relative reproducibilities of some biodegradation test methods.

With the exception of OECD 301 A and CEC-L-103-12, compared to the usual lubrication lab test methods, the reproducibility of the biodegradation test methods is very poor. It is notable that these two methods are not respirometric methods. In OECD 301 A and CEC-L-103-12, the depletion of test sample and that of its nonvolatile intermediates are measured.

It seems that the precision of the respirometric test methods is much poorer. One explanation for this phenomenon could be the complexity of the respirometric methods. This can be shown in an example: Someone is having a meal. The task is to find out how much of the meal has been already eaten. There are two possibilities to solve the task:

1. Measurement of the produced carbon dioxide and calculation of metabolized meal. For this, the person has to be put into a closed system which is purged with carbon dioxide–free air. The out coming air has to be analyzed and the concentration of carbon dioxide has to be measured.
2. Measurement of the meal's weight using a balance. For this, only a suitable balance is necessary.

Figures 7.6 and 7.7 show the test equipment for methods A and B. It is obvious that there can be more mistakes using method A than using method B.

TABLE 7.2

Relative Reproducibilities of Some Biodegradation Test Methods (R(abs) Related to the Limit Value)

Standard	Test Substance	R(rel)	
OECD 301 A [17]	Water soluble	20%	Decrease of sample
R(abs) = 14%, limit 70%			
CEC-L-103-12 [22]	Poor water soluble	24%	
R(abs) = 19%, limit 80%			
DIN EN ISO 14593 [23]	Water soluble	56–75%	Respirometric methods
R(abs) = 34% or 45%, limit 60%			
OECD 301 B [17]	Water soluble	67%	
R(abs) = 40%, limit 60%			
ASTM D 5864 [24]	Poor water soluble	70%	
R(abs) = 42%, limit 60%			
OECD 301 F [17]	Water soluble	117%	
R(abs) = 70%, limit 60%			
OECD 301 D [17]	Water soluble	147%	
R(abs) = 88%, limit 60%			

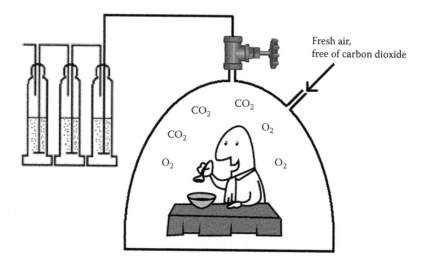

FIGURE 7.6 Measurement of the produced carbon dioxide and calculation of metabolized meal.

FIGURE 7.7 Measurement of the meal's weight using a balance.

Many people think that the CEC-L-33 and CEC-L-103 test methods are out of date, since they do not confirm that mineralization has occurred. But these test methods are the only biodegradation tests which were developed specifically for oil-based lubricants. Therefore, they are not suitable for water-soluble products. The test methods are exactly specified including the sample preparation and introduction. The quality standard of the CEC is very high. There is a constant recording of the test results for the reference oils. CEC places great importance on the repeatability of the test results. There are no longer active working groups for the CEC-L-33. So the CEC gives the L-33 test the status of *unsupported*. But the test method is still available. For CEC-L-103, there is an active working group which meets twice a year and organizes annual round-robin tests.

7.4 ACTUAL REQUIREMENTS OF THE ENVIRONMENT

The norms, standards, and ecolabels of today show that the rate of lubricant degradation velocity matters. As mentioned previously, the ISO norms' and ecolabels' opinion is that the faster the biodegradation of a hydraulic fluid happens, the better. This approach comes from the detergent problem of the 1980s. But can it also be applied to oils? To answer this question, we have to compare hydraulic fluids to detergents: detergents are released in huge amounts to sewage treatment plants. In the case of an oil spill, hydraulic fluids directly enter into the environment. Detergents are water soluble; hydraulic fluids are oil soluble. We can see that the two products, detergents and hydraulic oils, are totally different. Therefore, a transfer of knowledge is not possible, especially if we remember what the German minister of environment Sigmar Gabriel said about product-related environmental information (see Section 7.2).

But what are the actual requirements of the environment? Ecological systems are very complex. Therefore, it is necessary to think about oil spills in a more detailed way.

7.4.1 Environmental Compartments and Scenarios

First, we have to differentiate between the various environmental compartments where the oil spill occurred. The environment is not the same if oil gets into soil, stagnant water, flowing water, mixed compartments, or a sewage treatment plant. Now, a sewage treatment plant is not really an environmental compartment, but it is also simulated in biodegradation tests. We have to consider that every environmental compartment responds differently to the introduction of oil.

Second, we have to distinguish the condition of the oil spill. There are three fundamentally different mining scenarios that need to be distinguished:

- Wastewater treatment
- Loss lubrication outdoors
- Mobile hydraulic accidents

In wastewater treatment plants, waste material is biotechnologically disposed of. These plants have been specifically constructed and optimized for this purpose. Both organic and oxygen inputs are very high. The oxygen saturation is artificially guaranteed by the introduction of air. The faster the biodegradation is effected, the faster the water is purified and can be discharged into nature.

In case of loss lubrication outdoors, as happens, for example, with outboard motors and chainsaws, oil droplets are released into the environment for constructional reasons. A rapid degradation of the oil is tolerated by the environment, since the oxygen depletion can be quickly compensated due to the small amounts of oil.

In case of mobile hydraulics, there is no leakage of oil into the environment during normal operation. However, if an accident occurs, large quantities of oil are suddenly released into the environment. During biodegradation, the oxygen depletion of such large amounts of oil is enormous. Therefore, the spilled oil must be mechanically removed from the contaminated area as quickly and as completely as possible. Here, it is important that the spilled oil is retrievable. This may be a problem if large amounts of a water-soluble operating fluid are spilled into a lake. Oil, which could not be mechanically removed, is then degraded by microorganisms. The more completely the oil has been removed from the environment, the less damage subsequently occurs by oxygen depletion during the degradation of oil residues.

All biodegradability test methods are based on the degradation of very small quantities of oil in a large excess aqueous mineral substrate in the presence of abundant oxygen. However, the reality of most accidents with hydraulic oil is different. Currently, we do not know any standard that takes into account the actual scenario of biodegradation in real spills.

In principle, oil can be completely degraded by bacteria; but in case of an oil spill, this is not possible. Rather, the following criteria must be considered: First, to prevent oil spills at all, the hydraulic fluid must be compatible with the machine. Second, in case of an unavoidable accident, the spilled oil has to be retrievable and removable, so that a mechanical removal of the oil is possible. Third, it is also important that appropriate oil binders for the spilled oil are available. Fourth, nonremovable oil residues are then removed by microorganisms. In this case, however, the velocity rate of the degradation is not a decisive criterion, but rather the completeness of the degradation. Rapid degradation is associated with rapid depletion of oxygen, which is harmful for breathing organisms.

7.4.2 The Problem of Oxygen Consumption

In waters, especially in stagnant water, oxygen is scarce. At 10°C, a maximum of about 11 mg/L oxygen is dissolved in the water [25]. If the temperature rises, the solubility of the oxygen in the water decreases. At 23°C, only about 8 mg/L oxygen is soluble in the water. Bacteria that aerobically degrade organic matter consume oxygen. For the complete degradation of a single droplet of oil, the oxygen in 80 L of water is required [26]. If 1 kg of oil is spilled into the water, the oxygen from 400,000 L of water would be required to completely degrade the oil! The low concentration of 8 mg/L of oxygen in water is very quickly consumed by the aerobic, oil-degrading bacteria. For higher aquatic animals, the concentration of oxygen is no longer sufficient and they will suffocate. With decreasing oxygen content, more anaerobic bacteria get active. They form toxic hydrogen sulfide from sulfate. The result is eutrophication of the water [26].

A sad example of the negative environmental impact of bio-oil can be seen on the pictures of a tanker accident which occurred on July 24, 2006, in Lengenfeld in Saxony (Germany). Twenty thousand liters of bio-oils were spilled into a small river called Göltzsch [27]. The bio-oil, which was biodegradable according to the abovementioned laboratory tests, spread so well in the water that retrieval was very difficult. On the day of the accident, a chemical oxygen demand (COD) of the river water of 35,000 mg/L was measured (normal values are about 7–25 mg/L). The consequences of the accident were terrible. The Göltzsch was contaminated for years. Both fish breeding and crab breeding were completely missing. In over a distance of 10 km, almost all macrosaprobes were suffocated; thus, the whole food chain was interrupted.

The above example shows that the aim must be nature conservation and not very fast biodegradation. We believe that spilled bio-oil must not lead to the death of flora and fauna in case of an accident but it is desirable that flora and fauna continue to exist as unchanged as possible and that nature recovers from the impact as quickly as possible.

The following hypothetical experiment comes to the same conclusion: Imagine a lake with an area of 100 m² and a depth of 2 m. The lake's water has drinking water quality and is clear. It is oligotrophic water. One day, someone puts 20 kg of glucose (grape sugar) into this lake. Glucose is a nontoxic, fast biodegradable, biogenic substance. It is hydrophilic; that means it is water soluble. The lake's volume is 200,000 liters, so the sugar concentration in the lake is 0.1 g/L; you cannot even taste it.

Bacteria will degrade the glucose in a very short time to carbon dioxide and water. But this process needs oxygen and the amount of glucose is high enough to increase the COD of the lake's water by 107 mg/L! The COD of an oligotrophic lake normally lies between 1 and 2 mg/L [28]. Adding 20 kg of sugar would lead to a COD which would be 100 times higher! Stagnant waters which have a COD higher than 65 mg/L are called hypereutrophic. That means the clear lake's water with drinking-water quality becomes a hypereutrophic lake with dead zones only by adding 20 kg of sugar, which is a nontoxic, fast biodegradable substance!

The example shows that any organic substance, whether labeled *bio-oil* or not, is a pollutant for the environment. Even oils from the food sector must be separated with an oil separator and must be separately disposed of. The oxygen depletion during the biodegradation of oils may lead to the eutrophication of water. The faster the biodegradation is effected, the faster the oxygen which is dissolved in the water is consumed.

The biodegradation of small amounts of oil, for example, from outboard motors can be tolerated by nature, since the associated oxygen loss can be compensated. The biodegradation of large quantities of oil, for example, from a hydraulic oil spill requires enormous amounts of oxygen. The following biological degradation is considerably slowed down due to the lack of oxygen; anaerobic bacteria may generate toxic hydrogen sulfide.

Therefore, the scenario *mobile hydraulic accident* cannot be equated with *outdoor lubrication loss*. In case of an outdoor loss, it is desirable and good that the spilled oil can be degraded as quickly as possible by microorganisms. However, in the case of a mobile hydraulic accident, other priorities should be set. Here, if possible, the oil should be completely removed from nature as quickly as possible. The biodegradation of oil residues should be effected as environment friendly as possible and not as quickly as possible.

We think that the previous point of view, where the main criterion for a bio-oil was primarily rapid biodegradability, is no longer up to date. In the total spectrum of environment, biodegradability of a product is only a small segment.

7.5 ENVIRONMENTALLY ACCEPTABLE HYDRAULIC FLUIDS

A holistic approach is necessary to determine which products are really environmentally acceptable. In the case of lubricants, a holistic approach includes technical requirements, component compatibility, fuel consumption, biodegradability, and ecotoxicity to all organisms and not just to water organisms. All these factors can be combined and addressed in a life cycle analysis.

In a life cycle analysis, every input and output from the very beginning to the very end of a product is summarized. This includes the production phase, the utilization phase, and the disposal phase. A life cycle assessment allows a quantification of the effects on factors such as energy consumption, raw material consumption, greenhouse effect, acidification, and waste issues.

For example, the life cycle analysis of vehicles on German roads shows that about 77% of the carbon dioxide emission is caused by the burning of fuels [29]. Another 19% of the carbon dioxide emission is caused by car production. That leaves only 4% for tire and oil consumption. This is graphically shown in Figure 7.8.

A point that has not yet been considered in any ecolabel is the effect of oil degradation products on the growth of terrestrial plants. Already in 1995, the Department of Agriculture and Environmental Science, Newcastle University, conducted a study with wheat plants, where a strong growth inhibition was observed caused by esters of dicarboxylic acids [30]. Within the study, soil was treated with two different esters, mineral oil and vegetable oil. After a year of waiting, wheat grains were planted in the treated beds. At harvest, the number of germinated plants was counted. While the germination of plants in the beds that had been treated with mineral oil, vegetable oil, and trimethylolpropane esters hardly differed from the germination of the untreated beds, in the bed that had been

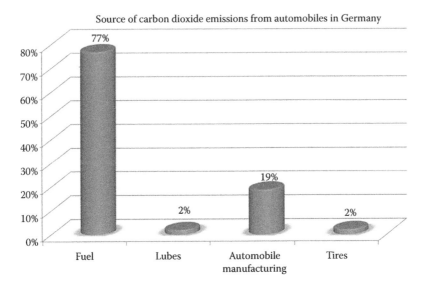

FIGURE 7.8 Example of the result of a life cycle analysis; all values include manufacturing phase and disposal phase.

treated with esters of dicarboxylic acids, not a single plant grew. This effect can also be simulated in a laboratory. Some results of our laboratory test are shown in Figure 7.9.

A real environmentally acceptable hydraulic fluid must have both technical performance and environmental compatibility. Only this combination leads to a positive life cycle analysis.

Some ecolabels require a renewable content in lubricants. But does it make sense to make lubricants from renewable resources? Some politicians want to reduce the carbon dioxide emissions of their countries by adding biodiesel or bioethanol to fuels. Unfortunately, the current agricultural output of biofuels is very small compared to the fuel consumption. Furthermore, most life cycle analysis calculations do not consider the production phase of biofuels. Therefore, it is not clear if we can save carbon dioxide emission by using biofuels.

But how are the actual figures? Germany has a mild climate, much water, and many fields; could we produce our fuel and lubricants on our own fields?

The crude oil consumption in Germany is about 110 million tons per year [31]. The yield of rapeseed is about 1.4 tons of biodiesel per hectare [32]. Nowadays, rapeseed is grown on 1.5 million ha in Germany [33]. That makes 2.1 million tons of biodiesel per year, which is about 2% of the annual crude oil consumption. If all the agricultural lands in Germany were used to grow rapeseed, the yield would cover 22% of the crude oil consumption. And if all the lands in Germany were used for growing rapeseed, only 45% of the crude oil consumption could be produced!

This calculation shows that for using rapeseed as a renewable resource, even the area of Germany will be too small to satisfy the demand of crude oil. For renewable resources to cover more than about 5% of the crude oil consumption, they must be imported from other countries. Transportation requires a high-energy consumption. And deforestation of the rain forests is not sustainable. We must also consider water shortage or the ethical acceptance of biofuels and oils.

Compared to fuels, lubricants are very complex. Native plant oils are not suitable as hydraulic fluids. An extensive chemical transformation process is needed to make lubricants out of plant oil.

There is also the fact that only about 1% of the crude oil production goes into the lubricant production. Compared to fuel, this is a vanishingly small proportion.

Lubricants are an instrument to save as much fuel as possible and to achieve the longest possible machine service life. That is why technical optimizations help the environment much more than the content of renewable resources.

The facts of environmental law can be visualized with the sustainability pyramid; see Figure 7.10. The base of the pyramid is biodegradability and ecotoxicological behaviors. These two points are

Blank solution   Ester-based*

PAO-based*   Mineral oil-based*

FIGURE 7.9   Laboratory test: influence of biodegradation products of hydraulic oils on plants; * = test object.

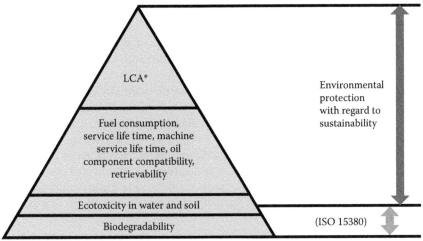

* Life cycle assessment according to ISO 14040ff incl. use phase and end-of-life phase

FIGURE 7.10 The sustainability pyramid for hydraulic fluids.

already included in standard ISO 15380, but regretfully, only the ecotoxicological behavior toward aquatic organisms is considered.

Even if these points are met, it does not mean that a product can be considered as environmentally acceptable and sustainable. There are many other points, which have to be considered, such as fuel consumption, service life of the machine, service life of the oil, component compatibility, and retrievability in case of an accidental spill. Only a comprehensive life cycle assessment can show to which category a product belongs.

7.6 RENEWED PERCEPTION AND UNDERSTANDING OF BIO-OILS IS IMPEDED

Over the last decades, environmental protection sensitivity has progressively increased. Therefore, in Germany and many other countries, there is no longer a need for discussion about the importance of environmental protection. Today, numerous measures of environmental protection are routinely practiced. Examples include the disposal of waste with special, professional disposal procedures; the separation of recyclables; the recycling measures; and the measures for saving energy or reducing waste.

The consideration of the entire life cycle and its balance on the impact of the environment is, however, still mostly neglected or misunderstood. One speaks of zero-emission vehicles and refers to electric vehicles, or programs for the installation of solar panels are offered, suggesting that CO_2 emissions would thus be reduced. For most environmental measures so far, only the individual aspects are considered without regard to the total chain of impacts on the environment, which affects the environment with the implementation of the measure.

Also with respect to the consideration and definition of bio-oils, a maturation process is still necessary in order to actually achieve environmental protection holistically and sustainably. Experience has shown that these necessary changes will be met with resistance, because changes require that people understand.

7.7 CASE STUDY OF ECOLABEL STANDARD IN GERMANY

The Hermann Bantleon GmbH is a middle-sized lubricant manufacturer located in southern Germany. The Hermann Bantleon GmbH has set a target to actively participate in environmental protection. This will be in the field of bio-oils. There is a considerable potential for improvement, in order to optimize bio-oils, particularly in hydraulic applications with regard to their environmental impact according to the latest state of knowledge.

For more than a decade, we have dealt with the subject of what performance characteristics must be provided by a biohydraulic oil and how this performance can be tested in order to make distinctions.

Unfortunately, in doing so, we experienced not just the normally expected resistances when bringing our new findings into the profession and its standards, but much higher than our expectation. We were insulted and discriminated in a university newspaper who tried to ridicule us at expert lectures in Germany. In certain circles, real hate campaigns were carried out against us. Our collaboration in standardization activities was not welcome. Pressure was put on associations so that articles written by us could not be published. A veritable cartel was created against Bantleon and its new findings. We were flooded with lawsuits and attempts were made to mislead our customers and to manipulate them against us. We had the impression that we were living in a dictatorship and even feared personal threats.

Despite all that, we always had the will to continue because we knew that the evidence that guided our actions was properly scientifically proven. Unfortunately, people do not want to see the truth in the evidence, but want to sweep it under the rug and to make it unfactual instead.

Thus, for example, the scientific findings that bio-oils based on esters could be used as herbicides are certainly uncomfortable, but it cannot come to a correction, because things are not allowed to be seen the way they unfortunately actually are.

Tests that are used to determine the biodegradability of lubricants often have no precise specifications, so the results cannot have significance. Consumers are misled because they think that they have received the certificate that the product is biodegradable, having received the results of such tests.

Also for ecolabels, such degradation tests are used without significance concerning the biodegradability of lubricants.

Bantleon never condemned, but took action to correct the error. Bantleon, as a leading laboratory, helped to develop the CEC-L-103-12.

Even the collaboration in the new award guideline RAL UZ 178 of the Blue Angel in 2013 led to discriminatory verbal battles. Our submitted written proposals for improvement were not incorporated and probably have not even been forwarded to the Environmental Label Jury.

All of a sudden, the new award guideline RAL UZ 178 was published, in which none of our suggestions has been taken into account. Therefore, in the spring of 2013, Bantleon filed a petition. This led to a stop of the application of the new award guideline RAL UZ 178.

The petition should have been followed up within 2 months. Later, Bantleon was consoled with the excuse that it was the summer break in the Federal Environment Agency. After the summer break in September/October 2013, they told us that the employee in charge was ill. As of June 2014, we still have not received any further news.

As the new award guidelines were blocked, the Association of Lubricants Industry (VSI) intervened and demanded that the old Blue Angel certificates, according to RAL UZ 79, must remain valid although they had been already limited to 2 years in 2013. The retracted award guidelines RAL UZ 178 caused concern among the owners of the Blue Angel certificates according to RAL UZ 79, because the tests for the new award guidelines require time, and to date, it was not clear whether there would be any changes.

Therefore, the Blue Angel awards currently in use, which are based on the outdated award guidelines RAL UZ 79, continue to be declared as valid. How and when things with respect to the Blue Angel will go on is currently unknown.

At this point, we would like to give special thanks to the colleagues from the mineral oil industry and other areas that could critically support us in our work. We hope that it will proceed faster with respect to standards and ecolabels in the future, so that the latest findings can be considered for better environmental protection.

The labeling of hydraulic oils with the Blue Angel award is currently being supplemented by the attribute "protect the water." This labeling is most questionable because it can be misunderstood by users.

7.8 CONCLUSION

Compared to the requirements of machines, the requirements of the environment are difficult to describe. This fact makes environment protection a political matter. Sometimes, there are even subsidies for a special kind of product. In most cases, ecolabels are used as a marketing instrument. However, we should scrutinize the criteria of ecolabels. It could be possible that standards or ecolabels are misused for economic sake.

REFERENCES

1. Proesler, M. (2008) Environmental information for products and services, requirements, instruments, examples, Federal Ministry for the Environment, Nature Conservation and Nuclear Safety, Federal Association of German Industry and Federal Environment Agency, pp. 3–4. http://www.umweltdaten .de/publikationen/fpdf-l/3701.pdf.
2. Bartz, W. (2010) *Einführung in die Tribologie und Schmierungstechnik*, expert Verlag, Renningen, p. 306.
3. ISO 15380, Lubricants, industrial oils and related products—Specifications for categories HETG, HEPG, HEES and HEPR.
4. ASTM D6046-02 (2006) Standard classification of hydraulic fluids for environmental impact, ASTM International.
5. http://en.wikipedia.org/wiki/Mineralization_(biology) (2013, September 2).
6. ISO 9439 (1999) Water quality—Evaluation of ultimate aerobic biodegradability of organic compounds in aqueous medium—Carbon dioxide evolution test.
7. Revised introduction of the OECD Guidelines for testing of chemicals, Section 3, Point 44 (2006, March 23)
8. http://en.wikipedia.org/wiki/DDT (2013, September 2).
9. http://en.wikipedia.org/wiki/Persistent_organic_pollutant (2013, September 2).
10. http://en.wikipedia.org/wiki/Reproducibility (2013, September 2).
11. International Standard DIN EN ISO 2719 (2003) Bestimmung des Flammpunktes-Verfahren nach Pensky-Martens mit geschlossenem Tigel-ISO 2719.
12. ASTM D 7042-04, Standard test method for dynamic viscosity and density of liquids by stabinger viscosimeter.
13. DIN 51423-1 (2010) Mineralöle—Messung der relativen Brechzahl mit dem Präzisionsrefraktometer.
14. DIN 51369 (2010) Entwurf Kühlschmierstoffe—Bestimmung des pH-Wertes von wassergemischten Kühlschmierstoffen.
15. DIN 51559-1 (2008) Entwurf—Mineralöle—Bestimmung der Verseifungszahl—Verseifungszahl über 2 Farbindikator Titration.
16. DIN EN 15199-1 (2007) Mineralölerzeugnisse—Gaschromatographische Bestimmung des Siedeverlaufs— Teil 1 Mitteldestillate und Grundöle.
17. DIN 51558-1 (1979) Mineralöle—Bestimmung der Neutralisationszahl Farbindikator Titration.
18. DIN 51396-2 (2008) Schmieröle—Bestimmung von Abriebelementen—Wellenlängendispersive Röntgenfluoreszenz Analyse (RFA).
19. DIN 51350-2 (2008) Entwurf—Schmierstoffe—Prüfung im Shell Vierkugel Apparat-Bestimmung der Schweißkraft von flüssigen Schmierstoffen.
20. DIN ISO 3016 (2012) Entwurf—Mineralölerzeugnisse-Bestimmung des Pourpoint.
21. Hashimoto, M. (Ministry of International Trade and Industry) (1988, October 12–14) OECD ring test of methods for determining ready biodegradability, Tokyo, Japan.
22. CEC L-103-12 (2013, April 16) Biological degradability of lubricants in natural environment.
23. DIN EN ISO 14593 (2005) CO_2 headspace test—Biologische Abbaubarkeit mittels Bestimmung des anorganischen Kohlenstoffs in geschlossenen Flaschen.
24. ASTM D5864-05, Standard test method for determining aerobic aquatic biodegradation of lubricants or their components.
25. Solubility of oxygen in water, http://www.Chemiemaster.de, website für den Chemieunterricht, http://www .chemie-master.de/FrameHandler.php?loc=http://www.chemie-master.de/pse/pse.php?modul=tabl3.

26. Widdel, F. (2010) Abbau von Erdöl durch Bakterien, Grundlegendes aus mikrobiologischer Sicht, Max-Planck-Institut für Marine Mikrologie, Bremen, Germany.
27. Glaser, L. (2011, August 29) Ölalarm an der Göltzsch, http://www.igfs-ev.de/Oelalarm.htm, IGFS e.V.
28. Wikipedia (2011, August 29) http://en.wikipedia.org/wiki/Trophic_state_index.
29. Umwelt und Prognose Institut (UPI) (1999) Ökobilanz von Fahrzeugen, UPI-Bericht Nr. 25, 6 Auflage.
30. Haigh, S. (1995) Fate and effect of synthetic lubricants in soil: Biodegradation and effect on crops in field studies, *The Science of the Total Environment* 168, pp.71–83.
31. Wikipedia (2011, August 29) Erdöl/Tabellen und Grafiken, http://de.wikipedia.org/wiki/Erd%C3%B6l /Tabellen_und_Grafiken.
32. Bockey, D. (2006, September) Rohstoffpotentiale für die Produktion von Biodiesel—Eine Bestandsaufnahme, ufop, Union zur Förderung von Öl- und Protein-pflanzen e.V. Berlin, Germany, http://www.ufop.de/downloads/Rohstoffpotenziale_021006.pdf.
33. Wikipedia in Deutschland (2011, August 29) http://de.wikipedia.org/wiki/Landwirtschaft.

8 Environmentally Acceptable Hydraulic Lubricants

Bernard C. Roell, Jr. PhD, Anne Otto, and Tyler Kuchta

CONTENTS

ABSTRACT

The demand for high-performance environmentally acceptable lubricants is increasing due to growth in corporate sustainability initiatives, increasing emphasis on operational risk mitigation, and more restrictive environmental regulations. As fines, cleanup costs, and public awareness increase, many offshore, marine, and land-based fleet and equipment operators are using or considering environmentally safer lubrication products. These types of power fluids can protect the users against regulatory fines, cleanup costs, operational downtime, and reputational risk. However, care must be given in selecting the right product for the specific application to reduce technical risk to equipment and ensuing financial risk that can result from poor fluid choice.

This chapter defines the common terminology used to describe key characteristics of environmental fluids, certifying standards, and labeling programs. It reviews the different types of environmentally preferable fluids available, the strengths, the limitations, and the comparable performance of fluid types and recommends best maintenance practices for prolonged fluid and equipment life.

8.1 INTRODUCTION: ENVIRONMENTALLY ACCEPTABLE LUBRICANTS (EALs)

8.1.1 When to Use EALs and Why

There is growing concern by the public and regulators regarding the environmental impact and associated costs of petroleum-based lubricant operational and accidental discharges. Petroleum components can be persistent and toxic. They can damage living organisms including plants, animals, and marine life and persist in the environment for many years. Federal agencies including the U.S. Coast Guard and the U.S. EPA, as well as state and local governments, are increasingly requiring discharge reporting, imposing fines, and requiring costly cleanup [1]. While fines from regulatory bodies can be meaningful, the costs that arise from the reputational or brand risk that accompanies poor environmental or corporate social responsibility can be far greater and longer lasting.

8.1.2 Oil Releases to the Environment

While small discharges on land or water will not result in a Resource Conservation and Recovery Act (RCRA) cleanup, large spills will. All oil discharges or spills are "reportable events" according to the EPA [1]. These events involve a great deal of cleanup cost, administrative procedures, and punitive fines that can range from tens of thousands to hundreds of thousands of dollars [2–8].

Spilling large quantities of biodegradable hydraulic fluid is still considered to be a reportable event under RCRA [1]; however, agencies are required to evaluate biobased oils differently than petroleum-based oils [9]. As the awareness of biodegradable fluid has increased, state and federal agencies have become more lenient regarding fines and cleanup costs. In fact, there are several case studies of equipment releasing several hundred gallons of readily biodegradable, vegetable-based hydraulic fluid into environmentally sensitive areas with no fines and minimal cleanup expense [2–8]. In most instances, the operator was able to continue working while cleanup efforts were underway, saving significant expense in downtime. Since the fluids were readily biodegradable and minimally toxic, there was no long-term negative effect to the ecosystem.

A 2010 study [10] estimated the global marine lubricant discharges from stern tubes to be 4.6 to 28.6 million L (1.2–7.6 million gal) per annum. In addition, 32.3 million L of oil is introduced to worldwide marine waters from other operational discharges and leaks. In total, operational discharges (including stern tube leakage) input 36.9 to 61 million L (9.7–16.0 million gal) of lubricating oil into marine port waters annually. Leaks of lubricating oil represent 10% of the total oil inputs into marine waters, as estimated in the 2003 NRC Oil in the Sea study (Figure 8.1) [10]. The total annual estimated response and damage costs for these leaks and operational discharges are estimated to be about $322 million worldwide. The total estimated costs for the United States is estimated to be $31 million annually [10].

For many years companies have been proactively trying to minimize risk and exposure from the environmental hazard associated with petroleum-based lubricating oils and fluids. Approaches included reviewing environmental impact from cradle to grave including the exposure from purchase, transport to facility, transport on site, site storage, point of use, transport of used oil to reclaim facility, and, lastly, disposal (recycling) of the product. Critical areas and equipment were identified including reservoir capacities and location of equipment in relation to the containment of any possible leakage. Containment solutions for these critical areas and equipment vary in both manufacture design and approach by geographical region. In some applications, it was determined that with increased filtration and monitoring, oil life could be significantly extended, reducing risk at transit to changeover and disposal. In other areas, the approach was to use environmentally friendlier alternatives. Equipment chosen to use environmentally friendly hydraulic oils often had a design function that forced limitations

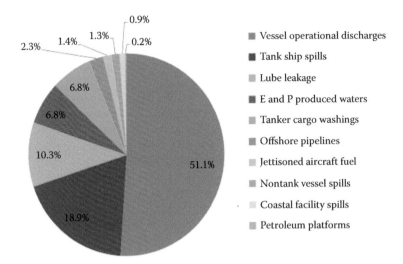

Vessel operational discharges

Tank ship spills

Lube leakage

E and P produced waters

Tanker cargo washings

Offshore pipelines

Jettisoned aircraft fuel

Nontank vessel spills

Coastal facility spills

Petroleum platforms

FIGURE 8.1 Annual oil inputs into the marine environment. (From Etkin, D. S., Worldwide analysis of in-port vessel operational lubricant discharges and leakages, Proceedings of the 33rd Arctic & Marine Oil Spill Program Technical Seminar, Environmental Research Consulting, Cortlandt Manor, New York, 2010, pp. 529–554. With permission.)

on fitting leakage containments and often operated in or near aquatic environments. Equipment most often utilizing environmentally friendly lubricants includes stationary and mobile hydraulics, cranes, lift equipment, work truck hydraulic systems, and vessel propulsion systems.

More and more owners, operators, leaseholders, and regulatory agencies are recognizing the benefits of environmentally friendly fluids. Using these lubricants can potentially save an operator thousands of dollars in terms of fines, cleanup costs, and downtime [2–8].

8.1.3 Environmentally Acceptable Lubricants

There is significant interest in environmentally acceptable lubricants. While regulations and requirements vary throughout the world, there are several specifications of note, particularly from the ISO [11] and ASTM International (formerly known as the ASTM) [12]. There is also a wide variety of performance levels among biodegradable products, which are dependent on the chosen base stocks and additive packages. When an environmentally preferable product is required outside the common temperature or pressure ranges, a biodegradable synthetic is usually required. While offering biodegradation, these products can operate in temperatures in excess of 100°F and still offer extended fluid life. As would be expected, these products are comparable in price to other synthetic oils and significantly more expensive than conventional petroleum fluids [13].

One of the most comprehensive definitions of the key characteristics of environmentally acceptable fluids in use today is the EPA's definition of an EAL contained in Appendix A of the 2013 Vessel General Permit (VGP) revision [14]. Under the revised VGP, all commercial vessels over 79 ft are required to use EALs in oil-to-water interfaces unless technically infeasible or commercially unavailable. Under the revised Small Vessel General Permit, commercial fishing vessels and other nonrecreational vessels less than 79 ft are also required to utilize EALs [15], when this act comes into effect.

According to Appendix A of the VGP revision [14], an EAL is defined as a "lubricant that is biodegradable, exhibits low toxicity to aquatic organisms, and has a low potential for bioaccumulation" [16]. EAL's compliance with VGP can be demonstrated through independent laboratory testing of the above-mentioned three key characteristics of fluid (often referenced as *self-certified*). Alternatively, compliance can be demonstrated through one of the five EU labeling programs that are accepted by the EPA, including Blue Angel [17], EU Ecolabel [18], Oslo Paris Conventions

Regulations (OSPAR) [19], Nordic ECOLABEL [20], and Swedish Standards Institute. These labeling schemes vary in requirements for toxicity, bioaccumulation, and biodegradation, and, in some cases, have the additional requirement for biobased content. While having a biobased content has the benefit of increased sustainability, biobased content levels greater than 25% can result in significant deterioration in fluid performance and durability.

Care must be taken in choosing the appropriate product for the specific application. Responsible suppliers clearly reference their definition of environmentally preferable and offer corroborating support. The Federal Trade Commission (FTC) [21] has been very specific in their recommendations for environmental claims and state to "look for evidence that gives some substance to the claim, the additional information that explains why the product is environmentally friendly." Many suppliers use misleading product environmental claims such as biodegradable, inherently biodegradable, or food grade. Suppliers should be able to support the performance claims with test data. These data can include standard industry tests (ASTM) and field and equipment manufacturer tests. Unless an environmentally friendly supplier specializes in environmentally preferable products, they are probably not an expert in the field.

8.2 STANDARDS FOR ENVIRONMENTAL HYDRAULIC FLUIDS

According to the VGP regulation, a fluid is considered an EAL when it meets the following criteria:

- Biodegradability is more than 60% in 28 days;
- The Ecotoxicity level is considered low (above 100 ppm);
- K_{ow} values are <3 and >7 for bioaccumulation.

The environmental persistence is stated as the rate at which fluids are removed by biological and chemical processes, such as biodegradation, hydrolysis, atmospheric oxidation, and photolysis. The range of biodegradability is measured in the percentage of the test fluid being broken down in 28 days.

Ecotoxicity is based on the survival rate of 50% of the test specimen under exposure of the test concentration. The range of toxicity is stated in the concentration in ppm of the tested fluid.

The bioaccumulation potential of a compound is directly related to its water solubility. Therefore, the measurement of K_{ow} (the logarithm of the partitioning coefficient of a substance in n-octanol and water) is used by the EPA as a measure of bioaccumulation.

8.3 CLASSIFICATIONS OF ENVIRONMENTALLY ACCEPTABLE FLUIDS

As in any emerging industry, companies are offering a variety of products that have a range of environmental benefits and performance attributes. ISO 6743-4 [22] defines the classification of fluids used in hydraulic applications. One of the subcategories is the hydraulic environmental or HE series [22]. The four main ISO 6743-4 classifications of EAL types are as follows:

1. Hydraulic environmental triglyceride—HETG
2. Hydraulic environmental polyalkylene glycol—HEPG
3. Hydraulic environmental synthetic ester—HEES
4. Hydraulic environmental PAO and related hydrocarbon products—HEPR

8.3.1 HETG

Early work in the environmental lubricants field focused on fluids made from vegetable oils (natural esters or HETGs). Vegetable-based fluids are readily biodegradable and have excellent frictional characteristics and VI—especially in cool dry operating environments. Under high-temperature applications, HETG types are more susceptible to oxidation and hydrolytic instability in the presence of >1% water. HETG fluids can typically withstand operating temperatures under 80–100°F. In the past, these types of fluids have been typically recommended for a shorter drain interval; however, with advances in additive technology and high-oleic vegetable oil base stocks and other high-stability oils, the performance of these fluids has significantly increased when systems are maintained with <1% water ingression. The performance of these fluids can significantly vary with specific chemistry, so regular oil sampling and analysis of oil during use are highly recommended.

8.3.2 HEES

The second generation of biodevelopment is focused on HEESs. This class of fluids is one of the most prevalent synthetic biodegradables in the market and is typically more expensive than HETG types, usually 1.5–2 times more expensive. The advantage of synthetic esters is that they can be customized to match the required oxidative stability by varying the degree of saturation of the hydrocarbon backbone and the steric hindrance of the ester moiety. Synthetic esters are also more prone to hydrolytic instability in the presence of water and care must be taken to maintain a dry system (<1% water ingression) to achieve long drain intervals.

Esters are synthesized by the reaction of a fatty acid and an alcohol in the presence of a catalyst. This reaction makes the ester and forms water as by-product, chemically expressed in Figure 8.2.

The double-headed arrows in Figure 8.2 indicate a reversible reaction. When water and heat are present, the ester undergoes a reverse reaction known as hydrolysis. This produces the alcohol and fatty acid base components. In field service, the breakdown of the polyolester leads to a loss of base fluid viscosity and an increase in acid number. The increase in acid number indicates that the fluid in the system may need to be changed. Consequently, ester-based fluids must be maintained in a dry state to obtain maximum longevity.

Both HETG- and HEES-type oils are compatible with most seals and metals as becomes evident from the large number of successful tests resulting in type approvals from seal manufacturers [23].

8.3.3 HEPG

HEPGs have better fire-resistant performance than vegetable oil, synthetic esters, and PAO and related types. However, it is well documented that polyalkylene glycols (PAGs) are not compatible with conventional seals or filters and nor with petroleum-based and vegetable-based oils and esters [23].

When changing from a conventional oil or other EAL to a PAG oil, care must be taken to thoroughly flush the system before introducing the PAG. Most PAGs are miscible with water and, therefore, do not create sheen on water. However, there is a debate on how safe a PAG is as it relates to toxicity.

The water absorption characteristic of a PAG severely limits its use in some industrial applications exposed to water since it draws water into the system and can create rust and wear.

8.3.4 HEPR

HEPRs have excellent frictional and antiwear performances with similar durability to mineral and synthetic oil counterparts. These types of fluids have excellent demulsibility performance, oxidative and hydrolytic stability, and a broad operating temperature range [24].

$$\text{Fatty acid + alcohol} \underset{}{\overset{\text{Catalyst}}{\rightleftharpoons}} \text{Ester + } H_2O$$

FIGURE 8.2 HETG and HEES hydrolysis chemical reaction.

The HEPR group embodies fluids derived from PAOs and related hydrocarbon products. While not all PAOs are readily biodegradable, some are and are often referred to as *biopolyolefin*. Distinct advantages of HEPR-based fluids are they are nonsheening, synthetic, readily biodegradable hydrocarbons [24]. The base fluids do not hydrolyze (break down when mixed with water) and, as such, they are much more stable in hot, wet conditions. HEPRs are typically compatible with nitrile butadiene rubber (NBR), fluoroelastomers (FKMs), and all common elastomers, as well as adhesives used in filters. With stability, seal compatibility, and biodegradability at a high performance level, a drawback is that they are currently derived from a nonrenewable resource. However, biolubricant manufacturers have recently developed PAOs derived from a fermentation process of yeast and sugar that will be commercially available in the future [25].

8.4 HYDRAULIC FLUIDS AND SEAL COMPATIBILITY

There is a misperception in the market that EALs are not compatible with seals and other lubricants. Testing and documentation, however, prove this claim to be unfounded and show that many types of EALs are compatible with many types of sealing materials. Independent third-party testing at RSC Bio Solutions, LLC, using the ASTM D471-06 (Standard Test Method for Rubber Property—Effect of Liquids) [26] and the ASTM D2240 (Standard Test Method for Rubber Property—Durometer Hardness) [27] was done on the four classes of EALs with some of the most common marine elastomers including NBR, hydrogenated nitrile butadiene rubber (HNBR), and FKMs.

Independent laboratory testing of a wide range of EAL fluids, which was commissioned by RSC Bio Solutions, LLC, also found that under some conditions, all of the EAL fluid types can be compatible (Figure 8.3) [23]. Under heat, however, HEPG hydraulic fluids can be stressed, which can lead to less than preferred performance. Water ingress will stress these fluids even more, and laboratory testing has found that the glycols used with FKMs at high temperatures (higher than 100°C) can cause significant problems and should not be utilized [23].

For hydraulic oils, HEPR and petroleum show similar resistance with NBRs; whereas HEPR and HETG are even more comparable, showing similar resistance with NBRs, HNBRs, and FKMs. The changes in elastomer resistance in biodegradable hydraulic oils can be affected by seawater contamination.

8.5 BIODEGRADABILITY

8.5.1 How Is Biodegradation Measured?

Claiming a product is biodegradable means next to nothing in terms of its realistic impact on the environment. Many people often say a product or substance is biodegradable, thinking that such a term indicates that it is less damaging to the environment, but in fact, oil and other potentially harmful things to the environment are biodegradable, too—just after a long period. Biodegradation occurs when a given substance or a fluid's chemical bonds weaken and break apart. In some cases, potentially toxic residue, such as heavy metals, may persist in the environment after the biodegradation of petroleum derived fluids.

There are technical terms, like *inherently biodegradable* and *readily biodegradable*, to describe the rate at which a substance or a fluid degrades and is recognized as a food source by the natural environment. That rate of biodegradation is driven by a number of factors, including the makeup of those original chemical bonds, the temperatures that the substance is being exposed to, the availability of natural enzymes to consume the remains as food, and the presence of water and oxygen. All of these factors need to be present in order for biodegradation to occur.

Readily biodegradable specifically defines a substance, a fluid, or a composition that will degrade 60% or greater within 28 days or less [12]. There are several internationally recognized ASTM tests that confirm this characteristic of a given product and allow companies to back up a readily

	Rubber materials					
	NBR	NBR	HNBR	HNBR	FKM	FKM
	≤60°C	≤100°C	≤60°C	≤100°C	≤60°C	≤100°C
HETG (triglyceride)	E/G	E/G	E/G	E/G	E	E
HEES (synthetic ester)	E/G	E/G	E/G	E/G	E	E
HEPG (polyalkylene glycol)	E	E/G	E	E/G	E/G	P
HEPR (PAO)	E/G	E/G	E/G	E/G	E	E

Excellent   E
Good   G
Poor   P

RSC Bio Solutions developed these data with in-house testing and open-source information

FIGURE 8.3 Hydraulic fluids and seal material compatibility.

biodegradable claim. In fact, the FTC even requires companies who use the term *readily biode-gradable* in describing their products to state the test, for example, ASTM 5864 [28] compliant, in validation of the claim [21]. Biodegradability classifications according to ASTM 5864 [28] are illustrated in Figure 8.4.

EALs are required to be readily biodegradable. Biodegradation is the measure of the chemical breakdown or the transformation of a material caused by organisms or their enzymes [12].

Biodegradation usually takes place by an aerobic biodegradation process. It is a similar mechanism to composting, where food decays and microbes recognize starches as sugars (a source of food) and digest what remains.

Biodegradation follows two routes called *primary* or *ultimate* biodegradation (Figure 8.5) [28]. Primary biodegradation is where the breakdown of the chemical results in the loss of one or more active groups in a chemical compound that renders the compound inactive with regard to a particular function. Primary biodegradation may result in the conversion of a toxic compound into a less toxic or a nontoxic compound. The second route is the ultimate biodegradation, also referred to as *mineralization*, whereby a chemical compound is converted to carbon dioxide, water, and mineral salts [29].

In addition to primary and ultimate biodegradation, biodegradation is also defined by two other operational properties: inherent biodegradability and readily biodegradability. A compound is considered inherently biodegradable so long as it shows evidence of biodegradation in any test for biodegradability. Inherently biodegradable oils are products or base oils that show between 20% and 60% degradation within 28 days (Figure 8.4). Readily biodegradable is an operational definition indicating that some fraction of a compound is ultimately biodegradable within a specific time frame, as specified by a specific test method. Readily biodegradable is defined as degrading 60% or more within 28 days (Figure 8.4). This type of degradation is preferable; because in most cases, the fluid will degrade long before environmental damage has occurred. Because of this, little is required in terms of long-term bioremediation. Vegetable-based lubricants, synthetic ester-based, and some PAO products exhibit ready biodegradation. There are several petroleum-based lubricants that claim inherent biodegradability. These are typically referred to as *environmentally safe*. Inherent biodegradation is defined as having the propensity to biodegrade, with no indication of timing or degree. These types of products can persist in the environment for years, continuing to cause substantial damage. They require long-term remediation due to their environmental persistence. Typically, these products are petroleum based, like conventional lubricants.

Table 8.1 [30] summarizes the different degradation characteristics of different oils. Mineral oil formulations are typically persistent or inherently biodegradable; whereas most of the four main biodegradable hydraulic oil classes—PAGs (HEPGs), PAO and related types (HEPRs), synthetic esters (HEESs), and vegetable oils (HETGs)—are readily biodegradable. HEPR types in a certain viscosity range have been found to be readily biodegradable.

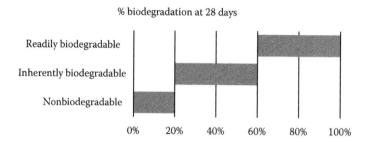

FIGURE 8.4 Biodegradation classifications in accordance with ASTM 5864.

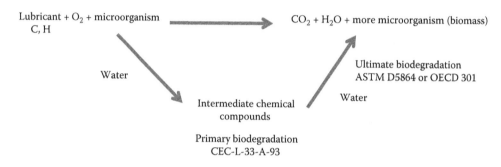

FIGURE 8.5 Primary versus ultimate biodegradation.

TABLE 8.1
Biodegradation Characteristics of Typical EALs

Lubricant Base Oil	Base Oil Source	Biodegradation
Mineral oils	Petroleum	Persistent/Inherently
PAGs	Petroleum-synthesized hydrocarbon	Readily
Synthetic ester	Synthesized from biological sources	Readily
Vegetable oils	Naturally occurring vegetable oils	Readily
PAO-related types	Petroleum-synthesized hydrocarbon	Readily

Source: Mudge, S. M., Comparative environmental fate of marine lubricants, Unpublished manuscript, Exponent, UK, 2010. With permission.

8.5.2 BIODEGRADATION TEST METHODS

Primary degradation measures the reduction of carbon and hydrogen bonds (C–H) in the initial solution; this is the reduction of the amount of the lubricant. The most widely used test that measures this decrease is the CEC-L-33-A-93 [31].

There are a variety of test methods used to measure readily or inherent biodegradability; the characteristics are summarized in Figure 8.6. These methods measure the evolution of CO_2 through biodegradation. The most common test methods for this are the OECD test OECD 301 B [32] and the ASTM D5864 [12].

Illustrated in Figure 8.7 is the percentage of degradation as a function of time of two readily biodegradable products (HETG and HEPR) compared to that of an inherently biodegradable product shown by in-house testing at RSC Bio Solutions [33].

Test type	Test name	Measured parameter*	Pass level**	Method	Ref
Ready biodegradable (a substance is considered to be inherently biodegradable using any of these tests if it shows >20% biodegradability within the test duration)	DDAT	DOC	>70%	OECD 301A	[32]
	Sturm test	CO_2	>60%	OECD 301B	[32]
	MITI test	DOC	>70%	OECD 301C	[32]
	Closed bottle test	BOD/COD	>60%	OECD 301D	[32]
	MOST	DOC	>70%	OECD 301E	[32]
	Sapromat	BOD/COD	>60%	OECD 301F	[32]
	Sturm test	CO_2	>60%	ASTM D-5864	[28]
	Biokinetic model	Chromatography	>60%	ASTM D-7373	[33]
	Shake flask test	CO_2	>60%	EPA 50/6-82-003	[34]
	BODIS	BOD/COD	>60%	ISO10708	[35]
Hydrocarbon degradability	CEC test	Infrared spectrum	>80%	CEC L-33-A-934	[31]
Screening tests (semiofficial)	CO_2 headspace test	CO_2	>60%	ISO 14953	[36]

* DOC: Dissolved organic
CO_2: Carbon dioxide
BOD: Biochemical oxygen demand
COD: Chemical oxygen demand

** The percentage of complete mineralization (or ultimate biodegradation) as indicated by the "measured parameter" that must occur for a product to be classified.

OECD: Organization for Economic Cooperation and Development
EPA: Environmental Protection Agency
ASTM: American Society of Testing and Materials
ISO: International Organization for Standardization
CEC: Coordinating European Council

FIGURE 8.6 Biodegradation test methods.

It is easy to see the difference between a readily biodegradable vegetable-based product (HETG), a PAO-based product (HEPR), and an inherently biodegradable mineral-based product. The EPA and the Coast Guard utilize this differentiation when evaluating an oil release.

Figure 8.7 shows that the percentage degradation after 28 days is greater than 60% for the EAL; whereas the value for the mineral oil is barely over 20%. It is also apparent that the vegetable oil-based EAL is almost 100% degraded in less than 40 days.

8.6 ECOTOXICITY

Another measurement to determine the environmental effect of a lubricant is ecotoxicity. Historically, tests for ecotoxicity have concentrated on the aquatic environment with a number of standard test procedures (Table 8.2). Most typically, the tests are for acute toxicity. This is a measurement of the concentration required to kill half of the various organisms over a short period ranging between 24 and 96 hours. Depending on the test and its end points, the toxicity of a fluid is described by a loading rate that has a 50% effect (EL50) or causes 50% mortality (LL50) after the stated time, that is, the concentration of fluid at which one half of the sample organisms die.

8.7 REGULATORY AGENCIES AND REQUIREMENTS

Predominant marine regulatory agencies and specific requirements related to lubricants usage are summarized in the following.

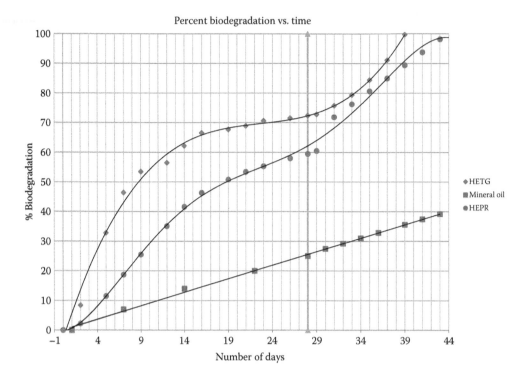

FIGURE 8.7   Ready biodegradation.

Canada's Marine Environmental Protection Agency (CEPA) [34] uses OSPAR (see OSPAR). The Department of Environment Regulation–Government of Western Australia [35] has established recommended environmental quality guidelines (concentration limits) for a number of chemicals to protect marine ecosystems from the effect of toxicants in marine waters. These guidelines are the same as OSPAR and Helsinki Commission (HELCOM). They have also set test limits for acceptable bioaccumulation and biomagnification effects for chemicals. These also follow the OSPAR, HELCOM, and U.S. EPA guidelines.

The Egyptian Environmental Affairs Agency–Ministry of State for Environmental Affairs [36], Egypt, is part of the Middle East Partnership Initiative with the U.S. EPA and employs

TABLE 8.2
OECD Aquatic Toxicity Tests

Test Name	OECD Test Method
Growth inhibition test, algae	OECD 201 [32]
Acute immobilization test, *Daphnia* sp.	OECD 202 [32]
Acute toxicity test, fish	OECD 203 [32]
Prolonged toxicity test: 14-day study, fish	OECD 204 [32]
Respiration inhibition test, bacteria	OECD 209 [32]
Early-life stage toxicity test, fish	OECD 210 [32]
Reproduction test, *Daphnia magna*	OECD 211 [32]
Short-term toxicity test on embryo and sac fry	OECD 212 [32]

Source: OECD, *OECD Guidelines for the Testing of Chemicals*, ISSN: 2074-5761 (online), 1992. With permission.

the U.S. EPA criteria/guidelines for protecting their offshore marine life from potentially toxic chemicals.

The Gulf of Mexico Marine Environmental Protection Agency [37] uses U.S. EPA regulations (see U.S. EPA).

HELCOM–Baltic Marine Environmental Protection Agency regulations [38] have been harmonized with those of the EU and the OSPAR, including an OSPAR list of chemicals for priority action, which requires replacement of said chemicals with others which are less toxic to the marine environment (Texas Institute of Science).

The Brazilian Institute for the Environment and Natural Renewable Resources–Department of Marine [39] uses internationally established guidelines and regulations, such as those from OSPAR, HELCOM, and the U.S. EPA.

Indian Ocean South East Asia [40] utilizes the same requirements as the Partnerships in Environmental Management for the Seas of East Asia (PEMSEA) (see PEMSEA).

OSPAR [41] is a 15-nation environmental cooperative formed to protect the northeast Atlantic (western coasts and catchments of Europe). OSPAR requires companies to complete and submit a Harmonized Offshore Chemical Notification Format questionnaire. This includes sections on data requirements, general information on substances and preparations, fate of the substance/preparation if there is a leak and spill, composition and concentration of a chemical/formulation, general physical properties (materials safety data sheets), ecotoxicological information, partitioning and bioaccumulation potential, biodegradability, aquatic toxicity, and confirmation statement.

Protection of the Arctic Marine Environment (PAME) [42], several members of the 15-nation OSPAR cooperating group, along with a few provinces in northern Canada (CEPA), reside in the Arctic region. Therefore, the PAME group is also guided by the OSPAR and CEPA criteria.

PEMSEA [43] is a legal entity consisting of a partnership of 11 East Asian nations (Cambodia, People's Republic of China, Democratic People's Republic of Korea, Indonesia, Japan, People's Democratic Republic of Lao, Philippines, Republic of Korea, Singapore, Timor-Leste, and Vietnam). This partnership has focused on the integrated management of Southeast Asian coastlines, including the Sustainable Development Strategy for the Seas of East Asia, including developing guidelines and regulations for cleaning up polluted offshore marine sites. PEMSEA uses pollution prevention regulations for chemical spills that have been internationally established, such as OSPAR or U.S. EPA.

U.S. EPA [44] requires completing and submitting a questionnaire similar to that of OSPAR which includes reporting on the quantities, composition, and potential for bioaccumulation or persistence of the chemical/formulation (lubricant) which could potentially spill or leak into the marine environment from offshore platforms. Also, the potential transport of these chemicals and lubricants by biological, physical, or chemical processes is regulated.

This list is not intended to be exhaustive, and it should be noted that within the United States, 23 states have regulations beyond what is federally required.

8.8 COMPARATIVE PERFORMANCE OF EALs

There are many misperceptions regarding EALs versus mineral counterparts. The top two are that EALs are incompatible with seals [45] and they do not perform as well as petroleum-based products [46,47]. High-performance lubricants possess the physical and tribological properties required to exhibit good, reliable, and trouble-free performance under demanding conditions and applications.

Figure 8.8 illustrates the operating temperature ranges of EALs versus those of mineral oils. The data clearly show that the temperature ranges of EALs exceed those of mineral-based counterparts. Key lubricant performance criteria include the following: all-season performance, friction control

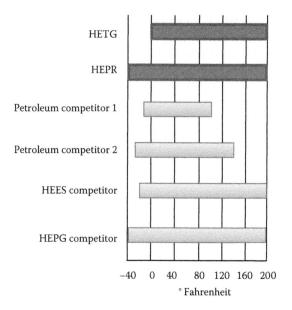

FIGURE 8.8   Operating temperature ranges of EALs versus mineral oils.

and antiwear properties, deposit control properties, excellent hydrolytic and oxidative stability, and component compatibility with seals and multimetals. Data are obtained from various hydraulic fluid suppliers' technical data sheets.

Table 8.3 lists the various performance characteristics of the different classes of environmental fluids and demonstrates the broad differences among the types.

8.8.1   Oxidation Performance of EALs versus Mineral Oil

There is a common misperception that the US Coast Guard approves oils based on the oil not leaving a sheen. The US Coast Guard does not approve, recommend, or specify fluids. Furthermore, the US Coast Guard does not approve or recommend any test procedures, but rather follows US statute

TABLE 8.3
Fluid Performance Characteristics

	HEPR	HEES	HETG	HEPG	Mineral Oil
Readily biodegradable	Yes	Yes	Yes	Yes	No
Ecotoxicity	Low	Low	Low	Low[a]	High
Bioaccumulation potential	No	No	No	No	Yes
Sheen	No	Some	No	No	Yes
Hydrolytic stability	Good	Weak	Weak	Good	Very good
Seal compatibility	Good	Varied	Good	Poor	Good
Wear performance	Very good	Very good	Very good	Very good	Very good
Oxidation performance	Very good	Good	Poor	Very good	Very good
Low-temperature performance	Very good	Good	Poor	Very good	Good
VI	Very good	Very good	Very good	Very good	Poor
Mineral oil compatibility	Good	Good	Good	Poor	Very good

Source: RSC Bio Solutions, in conjunction with Lubrizol, Wickliffe, Ohio, June 2013.
[a]   Solubility may increase the toxicity of some PAGs.

laws. The oil sheen that is frequently referenced is inferred from the Clean Water Act [48] as defining "any substance that leaves a sheen, emulsification, or discoloration, as a pollutant and can be subject to appropriate fines and regulations governing pollutants" [48]. In fact, the US Coast Guard also relies on the guidelines as outlined by equipment manufacturers and highly favors the use of biobased and readily biodegradable fluids.

Additionally, nonsheening means that when the fluid is discharged in water, there is no rainbow sheen or iridescence in the oil expression on the surface. It does not mean there is no expression on the surface, which is often a point of confusion in the EAL market. Table 8.3 summarizes sheen characteristics of EAL types and mineral oil.

8.8.2 PUMP PERFORMANCE

There is a wide variety of performance levels among readily biodegradable products. When an environmentally preferable product is required outside common temperature ranges, a readily biodegradable synthetic is usually required. While offering biodegradation, these products can operate in temperatures in excess of 100°F and offer long fluid life [49]. Figure 8.9 illustrates the oxidative stability of various readily biodegradable hydraulic fluids compared to that of petroleum oil.

Hydraulic fluid's extreme pressure/antiwear performance is measured using industry-standard pump tests. The data in Figure 8.10 show when a vegetable oil-based HETG ISO 46 was tested in a Vickers 35VQ25 vane pump test in accordance with Vickers 35VQ25 pump test procedure [50]. The product showed excellent wear protection. The data showed less than 10 mg weight loss per cartridge against a maximum of 90 mg weight loss. In fact, when the third cartridge was tested for an additional 350 hours, for a total of 500 hours, only less than 16 mg of weight loss was noted.

The same ISO 46 hydraulic fluid was tested in Denison T6CH20 at 5000 psi and 250 hours in accordance with Parker A-TP-30533 [51]. The hydraulic fluid provided excellent results (Figure 8.11).

Additionally, the Vickers 104C vane pump's <2 mg of weight loss was found after 250 hours of testing in accordance with ASTM D7043 [52] (Figure 8.12).

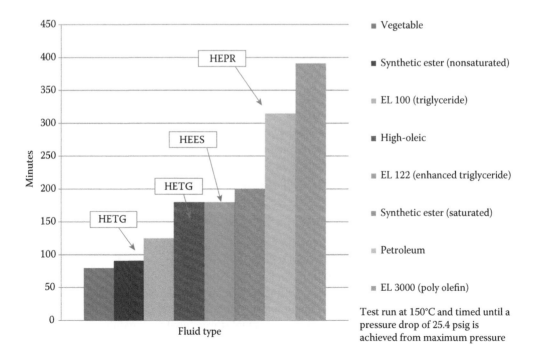

FIGURE 8.9 ASTM D2272 rotary pressure vessel oxidation test on finished hydraulic fluids.

Pump cartridge	Total fluid hours	Ring and vane weight loss	Ring and vane weight loss spec.	Cartridge hours
First	50	8.3 mg	90 mg max	50
Second	100	4.4 mg	90 mg max	50
Third	150	10.0 mg	90 mg max	50
Third*	500	15.8 mg	90 mg max	400

* Ran for an additional 350 hours
RSC Bio Solutions developed these data in accordance to Eaton–Vickers 35VQ25 pump test procedure

FIGURE 8.10 HETG—Vickers 35VQ25 pump test. (From Johnson, H. T., and Lewis, T. I., Vickers 35VQ25 pump test procedure: Tribology of hydraulic pump testing, American Society for Testing and Materials, Conshohocken, PA, 1996, pp. 129–139. With permission.)

Piston weight loss	Piston weight loss spec.	Pin and vane weight loss	Pin and vane weight loss spec.
186.7 mg	300 mg max	1.4 mg	15 mg max

RSC Bio Solutions developed this data in accordance to Parker A-TP-30533

FIGURE 8.11 Denison T6H20C hybrid pump test. (From Parker, D., Parker A-TP-30533 test equipment and instructions for hydraulic fluids performance evaluation on Parker pumps (vane and piston), Verizon, France, 2007. With permission.)

Total hours	Ring weight loss	Ring weight loss spec.	Vane weight loss	Vane weight loss spec.
250	1.9 mg	120 mg max	4.8 mg	30 mg max

RSC Bio Solutions developed this data in accordance to ASTM D7043

FIGURE 8.12 Vickers V104 C pump test. (From ASTM, Standard test method for indicating wear characteristics of non-petroleum and petroleum hydraulic fluids in a constant volume vane pump, ASTM D7043, 2012. With permission.)

HETG-type hydraulics can protect pumps as well as traditional petroleum lubricants can. Extended 35VQ25 testing shows that HETG can continue to protect for extended periods, and HETG-type hydraulic fluid can have a cleaning effect inherent to the base stocks.

From a pump performance perspective, many manufacturers, such as Bosch Rexroth and Danfoss, suggest operating at 10–12% reduction in pump pressure to accommodate the higher specific gravity of these fluids, which makes pump suction conditions worse [53].

8.8.3 COMPARATIVE FIELD PERFORMANCE

While some fluids offer excellent results in the laboratory under controlled laboratory conditions, the real proof of performance is how the fluid withstands the conditions seen during service duty.

The performance levels of an HEPR type and an HEES were put to the test in a long-term field demonstration for a multinational offshore drilling contractor and then analyzed by an independent third-party laboratory. The fluids were monitored in similar applications on similar vessels; both were used in topside rig hydraulics; in-service oil samples were analyzed and the results are summarized in Figure 8.13.

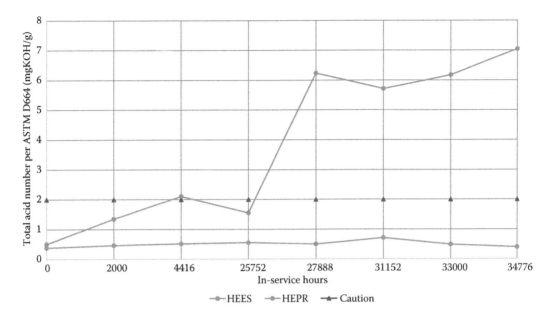

FIGURE 8.13 HEES and HEPR field performance AN data.

One key element of the oil analysis is the measurement of total acid number (TAN), sometimes simply called *acid number* (AN) [54]. This measures the buildup of acids in the system. The industry standard for condemning oil limit is a TAN level of 2.0 units for petroleum products [54]. As previously mentioned, acids can cause multiple problems in a system including accelerated rust and wear and seal degradation. In addition, increased acids shorten the life cycle of the fluid requiring premature oil change outs.

In this field analysis, the synthetic ester reached the break point of 2.0 much sooner than the HEPR did. The HEES in this study completely broke down, jumping up to an AN in excess of 6, almost three times the industry-recognized TAN limit. At an AN of 6 or greater, seal degradation and rust is almost assured. The HEPR maintains a fairly consistent level well below the standard TAN limits [54].

8.9 CONCLUSIONS

As demands on lubricant systems increase, the likelihood of accidental release of fluids increases. Increased operating temperatures, pressures, and working cycles shorten the life of circuit components. The single best approach to protecting the environment, the equipment, and the operation is to prevent leaks and spills through good routine maintenance and use of EALs. A good preventative maintenance program will

- Increase productivity since equipment is utilized more;
- Better utilize in-shop maintenance since there is less emergency work;
- Improve control of spare part inventory and reduce parts usage;
- Reduce equipment downtime;
- Reduce safety hazards;
- Increase equipment life;
- Reduce fines and cleanup costs due to environmental release;
- Reduce downtime related to environmental release.

There are an increasing number of EALs in the market, so it is critical to understand hydraulic system operating requirements and environment for appropriate EAL selection. The key considerations to be evaluated prior to selecting any EAL include the following:

- Operating temperature
- Operating pressure
- Seals and elastomers compatibility
- Water contamination
- Fluid life
- Preventative maintenance cycle
- Spill potential
- Business-driven client preference
- Overall cost of operation (acquisition, maintenance, compliance, fines)

Look to experienced EAL manufacturers who stand behind their products in application and support customers with routine oil monitoring and interpretation of the results. Since EAL products behave differently from conventional petroleum products, one should not exclusively count on the conventional oil analyst's interpretation of lab test results.

There is no substitute for the original equipment manufacturers' stamps of approval for EALs and real-world field experience in the specific application is important.

GLOSSARY AND ACRONYMS

AN: acid number
ASTM: ASTM International, formerly known as the American Society for Testing and Materials
CEC: Coordinating European Council
CEPA: Canada Marine Environmental Protection Agency
EAL: environmentally acceptable lubricant
EPA: United States Environmental Protection Agency
EU: European Union
FKM: fluoroelastomer
FTC: Federal Trade Commission
HEES: as defined by ISO 6743-4, the second phase of biodevelopment focused on synthetic esters; one of the most common synthetic biodegradables in the market
HELCOM: Helsinki Commission
HEPG: as defined by ISO 6743-4, lubricants made from PAGs
HEPR: as defined by ISO 6743-4, embodies fluids derived from PAO-based fluids and related types
HETG: as defined by ISO 6743-4, lubricants made from vegetable oils (natural esters or HETG); readily biodegradable but their performance is most suited to cool and dry operating conditions
HNBR: hydrogenated nitrile butadiene rubber
HOCNF: Harmonized Offshore Chemical Notification Format
ICIS: Independent Chemical Information Service, a business unit of Reed Business Information
ISO: International Standards Organization, a global organization responsible for developing most of the modern international industrial standards
NBR: nitrile butadiene rubber
OECD: Organisation for Economic Cooperation and Development
OSPAR: Oslo Paris Conventions
PAG: polyalkylene glycol

PAME: Protection of the Arctic Marine Environment
PAO: polyalphaolefin
PEMSEA: Partnerships in Environmental Management for the Seas of East Asia
ppm: parts per million
RCRA: Resource Conservation and Recovery Act
RSC: Radiator Specialty Company
TAN: total acid number
VGP: EPA Vessel General Permit

REFERENCES

1. United States Environmental Protection Agency, EPA-550-F-06-006: Oil discharge reporting requirements, Washington, DC: U.S. Office of Emergency Management (5104A) (2006).
2. Etkin, D. S., *A Worldwide Review of Marine Oil Spill Fines and Penalties*, Cortlandt Manor, NY: Environmental Research Consulting (2003).
3. Etkin, D. S., *Estimating Cleanup Costs for Oil Spills*, 1999 International Oil Spill Conference No. 168, Arlington, MA: Oil Spill Intelligence Report, Cutter Information Corp (1999).
4. Montewka, J., Weckström, M., and Kujala, P. A., Probabilistic model estimating spill clean-up cost—A case study for the Gulf of Finland, *Marine Pollution Bulletin*, Volume 76, Issues 1–2, pp. 61–71 (15 November 2013).
5. The International Tanker Owners Pollution Federation Limited, ID No. 83, *Factors That Determine the Cost of Oil Spills*, London: ITOPF (2003).
6. California Coastal Nonpoint Source Program, Water quality fact sheet, non-petroleum hydraulic fluids and biodiesel fuel for construction equipment, http://www.coastal.ca.gov/nps/Non-Petroleum_Hydraulic_Fluids.pdf (2012) (Accessed January 2, 2015).
7. TED Case Studies, Case Number 265: The Russian Arctic oil spill, http://www1.american.edu/ted/KOMI.HTM (1997) (Accessed January 2, 2015).
8. Casey, B., The real cost of fluid power leaks, *Machinery Lubrication 7*, http://www.machinerylubrication.com/Read/902/fluid-power-leaks (2006) (Accessed January 2, 2015).
9. Bliley Jr., T. J., Differentiation among fats, oils, and greases, 104th Washington Congress Report, pp. 104–262 (1995).
10. Etkin, D. S., Worldwide analysis of in-port vessel operational lubricant discharges and leakages, Proceedings of the 33rd Arctic & Marine Oil Spill Program Technical Seminar, pp. 529–554, Cortlandt Manor, NY: Environmental Research Consulting (2010).
11. International Organization for Standardization, Lubricants, industrial oils and related products (class L)—Family H (Hydraulic systems)—Specifications for categories HETG, HEPG, HEES and HEPR, ISO 15380:2011 (2011).
12. ASTM International, Standard classification of hydraulic fluids for environmental impact, ASTM D6046 (2012).
13. Fitch, J., *The Business Case for Lubrication Excellence*, Noria Corporation 2014, http://www.machinerylubrication.com/Read/28752/business-case-for-lubrication-excellence-(2014) (Accessed January 2, 2015).
14. United States Environmental Protection Agency, Vessel general permit for discharges incidental to the normal operation of vessels (VGP), Washington, DC: Headquarters, Office of Water, Office of Wastewater Management (March 2013).
15. United States Environmental Protection Agency, National pollutant discharge elimination system (NPDES) small vessel general permit for discharges incidental to the normal operation of vessel less than 79 feet (sVGP), Washington, DC: Headquarters, Office of Water, Office of Wastewater Management (2013).
16. United States Environmental Protection Agency, EPA 800-R-11-002: Environmentally acceptable lubricants, Washington, DC: Office of Wastewater Management (2011).
17. Blue Angel, Basic criteria for award of the environmental label, biodegradable lubricants and hydraulic fluids, RAL-UZ 178, Sankt Augustin, Germany: RAL gGmbH (2014).
18. European Commission, Commission decision of 24 June 2011 on establishing the ecological criteria for the award of the EU ecolabel to lubricants (notified under document C(2011) 4447), Brussels, Belgium: 2011/381/EU (2011).

19. OSPAR Convention for the Protection of the Marine Environment of the North-East Atlantic, OSPAR list of chemicals for priority action (Revised 2011) (Reference number 2004-12), Paris: OSPAR Convention (2002).
20. Nordic Ecolabel, 2011 Regulations for the Nordic ecolabelling of products, Denmark: Nordic Ecolabelling Board (22 June 2011).
21. Federal Trade Commission, Green guide 77 FR 62124, Washington, DC: FTC (11 October 2012).
22. ISO, Lubricants, industrial oils and related products (class L)—Classification—Part 4: Family H (hydraulic systems), ISO 6743-4 (2014).
23. SKF Group, ASTM D471 (volume change) and ASTM D2240 (durometer hardness), SKF Group, PUB SE/P1 12393/1 EN (April 2013).
24. Saptarshi, R., Rao, P. V., and Choudary, N. V., Poly-α-olefin-based synthetic lubricants: A short review on various synthetic routes, *Lubrication Science*, Volume 24, Issue 1, pp. 23–44 (January 2012).
25. Independent Chemical Information Service (ICIS), Base oils 2014: Green lubricants continue progress, http://www.icis.com/resources/news/2014/02/11/9752026/base-oils-2014-green-lubricants-continue-progress/# (2014) (Accessed January 2, 2015).
26. ASTM, Standard test method for rubber property-effect of liquids, ASTM D471-06 (2006).
27. ASTM, Standard test method for rubber property-durometer hardness, ASTM D2240 (2005).
28. ASTM, Standard test method for determining aerobic aquatic biodegradation of lubricants or their components, ASTM D5864 (2012).
29. Betton, C. I., Mortimer, R., Fox, M., and Orszulik, S. (eds.), Chapter 15: Lubricants and their environmental impact, In *Chemistry and Technology of Lubricants*, Third edition, Dordrecht: Springer, p. 547 (2009).
30. Mudge, S. M., Comparative environmental fate of marine lubricants, Unpublished manuscript. UK: Exponent (2010).
31. CEC, CEC L-33-A-934: Biodegradability of two-stroke cycle outboard engine oils in water, Leicestershire, England: CEC (1997).
32. OECD, *OECD Guidelines for the Testing of Chemicals*, Paris: OECD (1992).
33. RSC Bio Solutions, Internal test of HETG biodegradability, Mentor, OH: RSC Bio Solutions (2012).
34. Canada's Marine Environmental Protection Agency, http://www.ec.gc.ca (Accessed January 2, 2015).
35. Department of Environment Regulation–Government of Western Australia, www.der.wa.gov.au (Accessed January 2, 2015).
36. Egyptian Environmental Affairs Agency–Ministry of State for Environmental Affairs, http://www.eeaa.gov.eg (Accessed January 2, 2015).
37. Gulf of Mexico Marine Environmental Protection Agency.
38. Helsinki Commission, http://www.helcom.fi (Accessed January 2, 2015).
39. Brazilian Institute for the Environment and Natural Renewable Resources, http://www.ibama.gov.br (Accessed January 2, 2015).
40. Indian Ocean South East Asia, http://www.ioseaturtles.org/ (Accessed January 2, 2015).
41. Oslo Paris Conventions, http://www.ospar.org (Accessed January 2, 2015).
42. Protection of the Arctic Marine Environment, http://www.pame.is (Accessed January 2, 2015).
43. Partnerships in Environmental Management for the Seas of East Asia, http://www.pemsea.org (Accessed January 2, 2015).
44. U.S. Environmental Protection Agency, http://www.EPA.gov (Accessed January 2, 2015).
45. Det Norske Veritas, New US requirements as to biodegradable lubricant oils, *Det Norske Veritas Technical eNewsletter* (12 June 2013).
46. Erhan, S. Z., Sharma, B. K., and Perez, J. M., Oxidation and low temperature stability of vegetable oil-based lubricants, *Industrial Crops and Products*, Volume 24, pp. 292–299 (2006).
47. Kabir, M. A., Higgs III, C. F., and Lovell, M. R., A pin-on-disk experimental study on a green particulate-fluid lubricant, *ASME Journal of Tribology*, Volume 13, Issue 4, p. 6 (2008).
48. U.S. Congress, § 1251 et seq., 1972, Federal Water Pollution Control Act as Amended through P. L. 107–303, Washington, DC: U.S. Congress (27 November 2002).
49. RSC Bio Solutions, EnviroLogic 3000 series product description, http://www.rscbio.com (2014).
50. Johnson, H. T., and Lewis, T. I., Vickers 35VQ25 pump test procedure: Tribology of hydraulic pump testing, Totten, G. E., Kling, G. H., Smolenski, D. J. et al., American Society for Testing and Materials, pp. 129–139, West Conshohocken, PA (1996).
51. Danison Vane Technology, Parker A-TP-30533 test equipment and instructions for hydraulic fluids performance evaluation on Parker pumps (vane and piston), Verizon, France (2007).

52. ASTM, Standard test method for indicating wear characteristics of non-petroleum and petroleum hydraulic fluids in a constant volume vane pump, ASTM D7043 (2012).
53. Sauer-Danfoss, Technical information proportional valve group PVG, 32520L0344: Rev HE, Denmark: Danfoss (2014).
54. ASTM, Standard test method for acid number of petroleum products by potentiometric titration, ASTM D664-11a (2011).

9 Biodegradability and Ecotoxicity Evaluation of Lubricants

Jagadeesh K. Mannekote and Satish V. Kailas

CONTENTS

ABSTRACT

The use of vegetable oils for reducing friction dates back to ancient times. The discovery of petroleum and the advent of refinery technology have resulted in versatile products tailored to meet specific performance requirements. Lubricants used for different applications are lost or released into the environment during and after use, causing immense damage. Increased awareness of use and disposal of lubricants, in addition to stringent regulations imposed by various government agencies, has changed the landscape of the lubricant industry. However, developing a completely environmentally friendly lubricant is still a challenging task, requiring a balance between economic possibilities and ecological requirements. Quality standards will be affected if strict restrictions are applied in the use of materials. They also require carrying out risk classification studies to understand the toxic potential and to exclude ecologically questionable materials from the environmentally benign formulations. The algae and daphnia acute tests are used for preliminary evaluation of the candidate ingredients. The new fish acute threshold step-down test can also be used in place of the standard fish toxicity test for further evaluation. In addition, the fish embryo test is also gaining attention as an animal alternative test. Despite this, it has been observed that there are no test procedures available to compare the statistical validity of nonlethal end points with that of lethal end points. Thus, these tests are seldom used for evaluating toxicity of lubricant formulation. It is necessary to fill this gap and modify these tests to the lubricant industry, so toxicity tests are also taken into consideration along with other standard performance evaluation tests. In order to evaluate the persistence of test material in the environment, it is also necessary to determine both primary and ultimate biodegradability characteristics. In any case, for a lubricant formulation to be environmentally friendly, it has to be neutral if not nonlethal to the exposed species. In addition, it must be biodegradable and also fulfill other performance requirements.

9.1   INTRODUCTION

Lubricant is defined as a material used to facilitate relative motion by reducing the friction and the wear between surfaces in contact. The science of using natural products to reduce friction dates back to ancient times. The earliest evidences are seen in the decorations on the inner wall of the Egyptian tomb of Tehuti-Hetep (ca. 1650 BC). Leonardo da Vinci's designs also indicate the use of tallow oil for many applications during his time. It is clear from these evidences that the earliest lubricants were made up of vegetable oils or animal fats which are biodegradable [1,2]. It was only around the 19th century, with the onset of the Industrial Revolution, that the demand for high-performance lubricants increased. The discovery of petroleum provided a new raw material for lubricants in the form of mineral oil. Mineral oils are complex mixtures of C_{20}–C_{50} hydrocarbons containing a range of compounds such as paraffinic, naphthenic, and aromatic species as shown in Figure 9.1. Mineral oils are cheaper, have wider range of viscosities, and are more readily available than natural oils. Further, with the advent of refinery technology, better control in the composition of the mineral base oil was possible. This allows for tailor-made products to meet specific performance requirements.

In a lubricant, the base oil primarily provides a fluid layer to separate the moving surfaces and remove the heat from the interface. Other chemical compounds, mixed in the base oil, called *additives*, fulfill different functional requirements of the lubricant. Examples of additives include antioxidants, antiwear, antifoaming agents, corrosion inhibitors, dispersants, detergents, demulsifying/emulsifying agents, extreme pressure additives, friction modifiers, metal deactivators, and VI improvers [3]. Examples of additives and their mode of action are summarized in Table 9.1.

One of the main constraints with the use of mineral oils is that oils derived from different sources have different characteristics. Volatilization of low-molecular weight components causes the mixture to thicken during use. The presence of low-molecular weight components also reduces their flash point. Other limitations of a mineral oil-based lubricant include the following [4]:

1. It is nonbiodegradable and its disposal is hazardous to the environment.
2. It causes environmental pollution during its production, as oxides of carbon, nitrogen, and other gases are produced and emitted into the atmosphere.
3. It causes skin irritation, oil acne, contact eczema, and irritation in the mucous membranes, upon contact.
4. Even small amounts of mineral oil and their additives can be toxic to aquatic life.

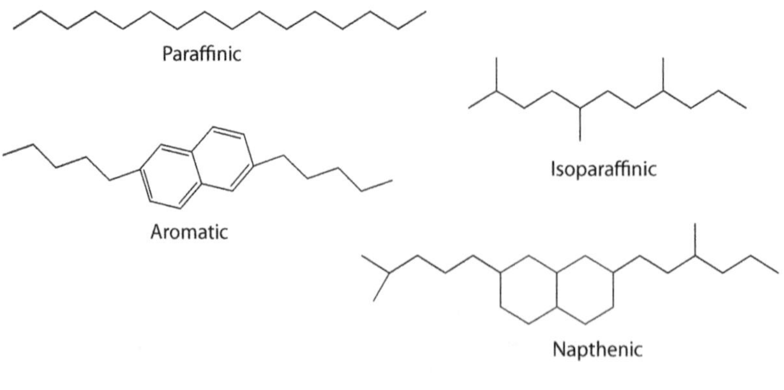

FIGURE 9.1   Schematic representation of different components of mineral oil.

TABLE 9.1

Examples of Additives and Their Modes of Action

Additive	Function and Mode of Action
Antiwear	Additive reacts chemically and forms a film on metal surfaces and minimizes the wear caused by metal-to-metal contact.
Corrosion inhibitor	Corrosion inhibitor forms a film on ferrous metallic parts and protects them from attack by contaminants in the oil.
Detergent	The chemical reaction results in the oxidation products remaining soluble in the oil and not sticking to the metal surfaces.
Dispersant	Oxidation product particles are kept small enough to allow them to float in the oil.
Extreme pressure	When metal-to-metal contact occurs, the generated heat causes the additive to chemically react with the metal, resulting in the formation of a lubricant that reduces friction and prevents welding.
Emulsifier	Emulsifier permits mixing of oil and water by reducing the interfacial tension, resulting in the formation of a stable emulsion.
Foam inhibitor	Foam inhibitor facilitates the combination of small bubbles to dissipate more rapidly and further prevent foaming.
Oxidation inhibitor	Oxidation inhibitor readily reacts with oxygen and retards oxidation of the base oil.
Pour point depressant	Pour point depressant prevents crystallization and facilitates the flow of oil.
Rust inhibitor	Rust inhibitor prevents rusting of ferrous components by forming a film.
VI improver	VI improver reduces rate of decrease of viscosity with increase of temperature; the additive thickens with increasing temperature, preventing oil from thinning out too rapidly.

9.2 ENVIRONMENTALLY FRIENDLY LUBRICANTS

One of the major challenges faced by today's world is damage to the environment as a result of conscious or unconscious actions. Petroleum-based lubricants are one of the products that fall in this category. Their negative effects linked to their use cause contamination of soil, water, and air. In addition to this, technological limitations, such as leak of lubricants, are also becoming a source of pollution [5,6]. This issue can be addressed by the following:

- Improving the machinery systems and their functional integrity
- Decreasing the lubricant consumption and increasing the lubricant life
- Collecting, recycling, and properly disposing used oils
- Replacing the petroleum-based lubricants with environmentally friendly lubricants

However, there are limitations to achieving the first three points mentioned. This led to an increase in the interest in the development of environmentally friendly products and an increase in the awareness on use and proper disposal of lubricants over the last three decades. In addition, various government agencies have also started to impose stringent regulations on use and disposal of the lubricants. These regulations have changed the landscape of the lubrication marketplace, resulting in the appearance of new environmentally friendly lubricants. Different terminologies were proposed for lubricants where the base oil is derived from renewable sources or its derivatives [7]:

- *Environmentally adapted lubricants:* This term is adapted to minimize harm and hazard to nature.
- *Environmentally friendly lubricants:* This term suggests an absolute quality that is unlikely to be attainable in practice.
- *Environmentally compatible lubricants:* This term suggests that the product has no or low negative interaction with the natural surroundings.

One of the important requirements for an environmentally friendly lubricant is biodegradability. Other properties include nontoxicity in its native and degraded form. The following elements were considered while developing environmentally friendly lubricants [8]:

- Environmental cost versus performance
- Compatibility with other lubricating fluids in the process
- Health and safety of operators and users
- Ease of recycling and safe disposal of fluid

Examples of the areas where environmentally friendly lubricants are applied include the following [9]:

- Wire rope and chain saw oils
- Metalworking and metal-forming oils
- Lubricants for food machinery
- Hydraulic oils
- Two-stroke engine oils used in outboard engines of boats

9.2.1 VEGETABLE OILS

Vegetable oils are liquid or semisolid triglycerides of fatty acids and glycerol. The fatty acids have carbon chains with 8 to 24 carbon atoms. The unrefined oil may also contain diglycerides, monoglycerides, free fatty acids, phospholipids, free sterols, sterol esters, and fat-soluble vitamins [10]. Vegetable oils have several advantages over mineral oils. They are renewable, nontoxic, and biodegradable and have high VI, high flash point, low evaporative losses, and good lubricating properties. They have become the primary choice in formulations used in total loss applications and once-through applications, where stability is not an issue but biodegradability is a primary requirement. However, their use in other applications is limited due to their poor thermo-oxidative stability, resulting in reduced operational lifetime [11–15]. This is because the β-hydrogen of the triglyceride can decompose relatively easily through the formation of an intermediate compound with a six-membered ring. The presence of unsaturated double bonds in the fatty acid chain also affects the oxidation stability of these compounds since allylic hydrogen is easily removed by a radical reaction. In addition, the ester group is easily hydrolyzed with water, resulting in formation of free fatty acids. The reactions to which triglyceride molecules are susceptible are shown in Figure 9.2.

Different methods are used to improve the performance of vegetable oils in lubricant formulations. Examples are conventional plant breeding and genetic modification of crops. Chemical modifications of vegetable oils have also been pursued over the last two decades to overcome these

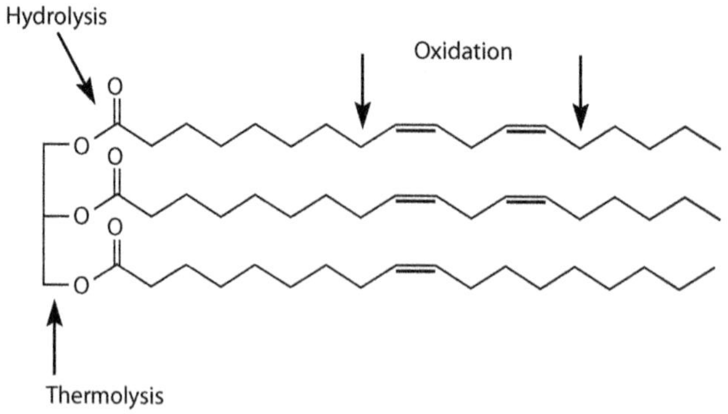

FIGURE 9.2 Schematic representation of triglyceride showing the positions susceptible for specific reactions.

shortcomings. Triglycerides generally have two reactive sites for chemical modification; these are the double bond in the fatty acid chain and the ester group. The question is whether this approach is sustainable. This is where we believe that the thinking has to change from the conventional approach [16–21] in the development of environmentally friendly lubricants.

There are other issues that need to be considered. Since the lifetime of vegetable oil-derived products is relatively short, large volumes would be required. This, in turn, will increase the farmland requirement when there is a demand for such products. Farming, in turn, has to be sustainable and should not create food shortages.

9.3   ADDITIVES FOR ENVIRONMENT-FRIENDLY LUBRICANTS

Additives are important components of modern lubricant formulations, as they impart new characteristics to the formulation and strengthen desirable properties. Thus, the overall performance of a lubricant formulation depends not only on the base oil but also on the additives used in the formulation. The properties of different additives and their modes of action are as shown in Table 9.1. The performance of vegetable oils without additives will be inferior and will also deteriorate the base oil very quickly [22,23]. The additive selection for a given application depends on the several factors including the following:

- Availability of raw materials and cost
- Performance requirements of lubricants
- Solubility in base oil and effective concentration
- Compatibility with other additives and synergic effects
- Compatibility with metals, seals, hoses, and paints
- Stability of the additives

The concentration of additives can vary depending on the application and type of additive, from a few tenths of percent in compressor oil up to 40% in gear oils. Some commercial additives contain phosphorus, sulfur, nitrogen, and zinc as active elements. It is expected that phosphorus will be phased out in the future due to environmental restriction. Further, future lubricants are also expected to be free of chlorine and nitrites.

In order to be environmentally acceptable, additives must be ecotoxicologically harmless. As the present options are limited, it is a challenging to select an additive meeting this requirement of ecological regulations [24]. The availability of ashless additives has given flexibility to solve the problem only to some extent. Further, only for a limited number of additives are complete ecotoxicological data available. In addition, different base oils require different additives to achieve the required performance. It will be difficult to address this issue since most of the additives presently used in lubricant application are not eco-friendly [25–30].

9.4   CONCEPTS OF CLOSED CYCLE AND SUSTAINABLE DEVELOPMENT

We are essentially a part of the universe where matter and energy coexist such that one cannot survive without the other. The oxygen and carbon dioxide conversion, in plants and its link to humans and animals, signifies an important cycle. Humans extract, manufacture, and use different materials from nature and discard them. This is called the *open cycle*. Sometimes, humans reprocess or recondition and use them. This is called the *closed cycle*. For an open cycle, the significance of the timescale of material usage comes in to picture. In an open cycle, it is essential that the extraction and time taken to naturally replenish the disposed materials should be same. An open cycle will not be sustainable if the material being extracted cannot be replenished. Needless to say, the present way of product development, use, and disposal is completely unsustainable, especially in the case of lubricants [31–33].

The earth has limited resources and using these with utmost care is very important. Finding substitutes for limited resources and conserving the environment has led to reliance on biomass sources. Thus, applying a sustainability concept to lubricants makes a lot of sense. Petroleum is a finite resource formed over millions of years, but is being consumed at an alarming rate ever since it has been discovered. The available deposits are fast depleting and access is also highly dependent on complex geopolitical issues, creating price fluctuations. In contrast, oleochemicals are derived from renewable sources by conversion of carbon dioxide via photosynthesis in plants. Further, as mentioned earlier, petroleum derivatives are not biodegradable and will lead to pollution. In addition, the carbon cycle of petrochemical use is not closed but open, leading to an increase in CO_2 level in the atmosphere contributing to global warming. Whereas for vegetable oils, the liberated CO_2 is equal to that of the originally taken up by the plants for photosynthesis and the cycle is closed [34–36].

Quantification of the impact of lubricants on the environment requires a detailed life cycle assessment (LCA). LCA comparisons of different lubricant types based only on their resource requirements during synthesis will give misleading results. This is because lubricants for a given set of applications greatly differ in their performance and a detailed knowledge of product composition is also required for comparison [37,38].

It has been observed that nearly 50% of all lubricants are disposed or lost into the environment through total loss applications, spills, volatility, or other various ways contaminating the air, soil, surface, and groundwater [39]. In such cases, the approach to protect the environment would be to prevent the loss of the lubricant and/or also to use environmentally friendly lubricants.

9.5   STANDARD LUBRICANT TEST METHODS

It is essential to test lubricant formulations and ensure that they meet the required specifications before they are used for the intended application. Most lubricant test methods are developed by one or more of the following bodies [40]:

- ASTM International—http://www.astm.org
- American National Standards Institute—http://www.ansi.org
- API—http://www.api.org
- British Standards—http://www.bsigroup.com/
- Bureau of Indian Standards—http://www.bis.org.in
- Coordinating Research Council Inc.—http://www.crcao.com
- Deutsches Institut für Normung e.V.—http://www.normung.din.de
- Japanese Industrial Standards Committee—http://www.jisc.go.jp/
- Russian Industry Standards—http://www.snip.com/
- Institute of Petroleum—http://www.energyinst.org.uk
- ISO—http://www.iso.org
- National Lubricating Grease Institute—http://www.nlgi.org
- SAE—http://www.sae.org

The present work will focus on the challenges faced while evaluating the biodegradability and the toxicity of lubricant formulations.

9.6   BIODEGRADABILITY AND TOXICITY OF LUBRICANTS

Lubricants find their way into the environment through various means during handling and application including through evaporation, exhaust fumes, faulty operation of equipment, and improper disposal. Environmental authorities have become conscious of the effects on the flora and the fauna, signifying the importance of the biodegradability of lubricant products. Biodegradability measures

TABLE 9.2

Examples of OECD Standard Tests for Biodegradability of Organic Matter

Name of the Test	Parameters Evaluated and Standard Adopted	Comments
Dissolved organic carbon (DOC) die-away test	DOC (OECD 301A)	This method is based on the analysis of DOC. It is suitable for the compounds which are adsorbing in nature.
CO_2 evolution test	CO_2 (OECD 301B)	This method is based on the CO_2 evolution. It is suitable for poorly soluble and adsorbing compounds.
Ministry of International Trade and Industry test	BOD (OECD 301C)	This method is based on evaluating the oxygen consumption and is useful for most of the compounds.
Closed bottle test	BOD/COD (OECD 301D)	This method is based on evaluating the dissolved oxygen and is useful for most of the compounds.
Modified OECD screening test	DOC (OECD 301E)	This method is based on the analysis of DOC. It is suitable for the compounds which are adsorbing in nature.
Manometric respirometry test	BOD/COD (OECD 301F)	This method is based on evaluating the oxygen consumption and is useful for most of the compounds.

Source: OECD Test No. 301: Ready biodegradability, *OECD Guidelines for the Testing of Chemicals*, Section 3, OECD Publishing, 1992. With permission.

the environmental fate of a lubricant component, its decomposition mechanisms, and the possible environmental impact during any stage of its deterioration. Evaluation of the biodegradability of lubricating oils using standard test procedures provides a better knowledge of how the chemical structure influences the biodegradability of lubricants. The OECD is a prominent body that promotes policies to improve the economic and social well-being of the people in member countries. It facilitates common forums in which member governments can work together to share experiences and seek solutions to common problems [41–44]. Examples of standard procedures set by OECD for testing biodegradability are provided in Table 9.2 [45].

The selection of a test procedure for evaluating ultimate biodegradability is difficult since most lubricating base oils do not dissolve in water. The situation is even trickier when biodegradation in soil has to be evaluated. Studies have shown that lubricants based on mineral oils are least biodegradable in soil relative to synthetic oil- and vegetable oil-based formulations. In general, the rate of degradation of all lubricants, including natural vegetable oil, in soil was very slow, taking more than a year to generate extractable residues. In addition, the effects of contamination by different oils on growth rate and yield of spring wheat varied from small reductions to complete inhibition of seed germination [46–48].

All lubricants exhibit common behaviors such as poor solubility and low density. These oils form a layer on the surface of the water reducing oxygen exchange, resulting in depletion of dissolved oxygen. Whenever a spill occurs, the physical and chemical properties of spilled substance determine its fate in the environment. In the case of vegetable oils, the problem is aggravated by their high biological oxygen demand (BOD) during biodegradation. Further, highly viscous oils contaminate feathers of birds and furs of mammals, resulting in loss of buoyancy and alterations in metabolism [49,50].

9.6.1 FRESHWATER BIODEGRADABILITY—AEROBIC AND ANAEROBIC CONDITIONS

Biodegradability gives an account of the environmental fate of a substance and its susceptibility to biochemical breakdown by the action of microorganisms. Since lubricants are insoluble, have less density than water, and are hydrophobic, they spread on the surface of water. The degradation process of lubricant begins with the action of microorganisms and it is first transformed into carboxylated or hydroxylated intermediates that are soluble in water. This leads to the disappearance of visible oil slick on the surface of the water. This primary biodegradability is measured using CEC-L-33-A-93 standard test. In this method, the extent of disappearance of C–H stretching vibrations from the infrared spectrum of the test sample is measured. The method was primarily developed for two-stroke engine oils and relevant only for German environmental label. As the primary degradation only accounts for the disappearance of the parent compound, it does not necessarily imply that the compound has completely degraded. The other limitation of this test is that it does not clearly differentiate between primary and ultimate degradations [51,52].

In order to evaluate the persistence of test material in the environment, it is necessary to determine how readily the biodegradation process occurs. Simple test methods where microbes have the opportunity to acclimatize to the substrate are used for evaluating ultimate biodegradability. The OECD 301B method measures the amount of CO_2 evolved from a flask containing the test substance for over 28 days and compares it to the theoretical value based on the total organic carbon content of the sample. In reality, complete biodegradability will not happen in the OECD 301B test because a portion of the substrate is always converted into biomass. A substance is considered as readily biodegradable if the conversion after 28 days is more than 60%, and if 50% biodegradation occurs within a 10-day period. A few other test procedures, such as ISO 10708 and 14593, are also used to evaluate biodegradability. Vegetable oils are not expected to bioaccumulate, since the ester group is easily susceptible to enzymatic cleavage and the derivatives are metabolized. Among the nonplant derivatives, synthetic esters also have high biodegradability; whereas PAOs are better than mineral oils because of their higher degree of linearity [53].

9.6.2 BIODEGRADABILITY OF LUBRICANTS IN SEAWATER

Biodegradability in seawater is important for the lubricant context as mineral oils, synthetic esters, and PAO are used in offshore drilling equipment and are mainly transported in ships. In the case of offshore drilling equipment, some portion of the lubricant is always left on the seabed after separation from the drilling fluid, leading to contamination. Two standard methods based on the OECD closed bottle test (OECD 301D) and the modified OECD screening tests (OECD 301E) are used for this purpose. These differ from the conventional test procedures in that no inoculum is added and biodegradation occurs by the action of bacteria present in the seawater. The use of a high concentration of test substance in these tests means that the tests do not mimic the conditions normally present in the marine environment [54,55]. Other tests, such as the Oslo and Paris Commission ring test, are equally valid for evaluating the biodegradability of chemicals in seawater.

Studies have shown that readily biodegradable substances in freshwater also biodegrade in seawater, but at a slower rate. With this background, the European Centre for Ecotoxicology and Toxicology of Chemicals recommended a new dosing procedure. In this method, the test duration was increased from 28 to 60 days. It was observed that the extended incubation periods tend to decrease test precision. It also recommended making a hazard assessment of lubricants in seawater using data from freshwater tests [56].

9.7 TOXICITY TESTING AND SAFETY ISSUES

Chemicals used for food can have a wide range of effects on health and there can also be adverse effects from the chemicals used for nonfood applications. Since toxic effects can vary depending on the composition, the best way to compare the toxicities of different chemicals is to measure the lethal concentration or the amount of a material, given all at once, which causes the death of 50% of a group of test species, represented as LC_{50}. Examples of standards toxicity test procedures and relevant information are provided in Table 9.3 [57–61]. Ecotoxicity data give an insight into the toxic effect of chemicals toward various species in the environment. Since it is not feasible to study the

TABLE 9.3
International Standard Ecotoxicological Test Methods

Test Type	Species Used	Duration of Test—Standard Test Method	Reference
Short-term tests (acute tests)	Algae	72 hours—OECD 201	[57]
	Daphnia	48 hours—OECD 202/1	[58]
	Fish	96 hours—OECD 203	[59]
	Bacteria	30 minutes—OECD 209	[60]
Long-term tests (chronic tests)	Daphnia	21 days—OECD 202/2	[61]
	Fish	14 days—OECD 204	[62]

TABLE 9.4
Comparison of OECD Acute Fish Toxicity Test and Acute Threshold Test

Test Parameter	Acute Fish Toxicity Testing	Acute Fish Threshold Test
Fish specifications	Freshwater or marine water species	Freshwater or marine water species
Age of fish species	Adults or juveniles	Adults or juveniles
Number of fish used per test vessel	10 (minimum 7)	5
Number of fish used per test sample	42 (35 + 7 in control)	10 (5 + 5 in control)
Number of test concentrations	5 (+ control sample)	1 (+ control sample)
Dilution series	Applicable	Not applicable
Test duration and end point	96 hours—Mortality within 96 hours	96 hours—Mortality or morbidity within 96 hours
Evaluation of LC_{50}	Probit or binomial method	Binomial method

Source: OECD Test No. 204: Fish, prolonged toxicity test: 14-day study, *OECD Guidelines for the Testing of Chemicals*, Section 2, OECD Publishing, 1984; OECD Guideline No. 126: Short guidance on the threshold approach for acute fish toxicity, ENV/JM/MONO (2010) 17. Series on Testing and Assessment, OECD Publishing, 2010. With permission.

effect on all species, testing is done only on selected species in the food chain such as fish, daphnia, algae, and bacteria [62,63]. Two types of toxicity tests are used; details of the same are provided in Table 9.4 [61,64]:

- The acute tests investigate the effects of high concentrations of test chemical during a relatively short period of exposure to evaluate the LC_{50} or effective concentration (EC_{50}) concentration. Acute test provides preliminary information on the toxic nature of a material for which no other toxicological information is available.
- Subchronic tests investigate long-term effects at low concentrations to find out NOEC. NOEC is the highest tested concentration at which the substance is observed to have statistically no effect on growth when compared with the control, within a given exposure time.

In general for a readily biodegradable substance, if the acute test indicates no toxicity, chronic tests are conducted. For high-volume chemicals such as lubricants, a chronic test is necessary to evaluate a reliable risk assessment [65–67].

9.7.1 Ecotoxicology—Evolution of Fish Toxicity Tests

Use of animals for safety testing represents a dilemma between balancing the safety to humans and the concerns about animals. Tests in the aquatic environment include assessment of acute effects

TABLE 9.5

Specifications of the Fish Species Recommended by OECD for Toxicity Testing

Recommended Fish Species and Total Length of Fish (cm)	Testing Temperature (°C)
Zebrafish (2.0 ± 1.0)	21–25
Fathead minnow (2.0 ± 1.0)	21–25
Common carp (3.0 ± 1.0)	20–24
Rice fish (2.0 ± 1.0)	21–25
Guppy (2.0 ± 1.0)	21–25
Bluegill (2.0 ± 1.0)	21–25
Rainbow trout (5.0 ± 1.0)	13–17

Source: Toxicity Test, *OECD Guidelines for the Testing of Chemicals*, Section 2, OECD Publishing, 1992.

such as acute lethality and chronic effects. They also include tests on development, growth and reproduction, and endocrine modulation of aquatic animals.

Fish represent the oldest and the most diverse class of vertebrates that live in a wide range of aquatic habitats. This diversity makes them important experimental models for toxicity test [68–71]. Specifications of fish species recommended for toxicity tests are provided in Table 9.5 [72]. Fish also possess many characteristics that make them ideal for environmental toxicology tests including the following:

- Fish have intimate contact with the aquatic environment and process specialized functions common with those of higher vertebrates.
- Fish produce eggs in large quantities which can be externally fertilized. Since fish eggs are transparent, embryonic development can be followed, making early life stage assessments easier.
- Many species are suitable for field and laboratory experiments, allowing one to initiate studies in the laboratories and then extend it to in-field tests.

Acute test provides preliminary information on the toxic nature of a material for which no other toxicological information is available. At present, there is no clear consensus on further reducing the number of fish used in a conventional test. There are a number of alternative approaches which have sound scientific evidence to support their use for testing. Some of these alternatives are discussed next in brief.

In the context of effects on the environment, nonlethal end points have become more relevant for hazard evaluation. This is because acute toxicity test only provides information on the percentage of a population that will be dead at a specified concentration and after a certain period. It does not distinguish the percentage of the population, which is technically alive but "ecologically dead." Thus, nonlethal end points, such as NOEC and lowest observed effective concentration, could be used as standard instead of an LC_{50} value. There are arguments against this approach as the NOEC's dependence on test design and variability. In order to overcome this problem, the behavior of the fish, exposed to sublethal concentrations of test sample, can be monitored at multiple and discrete levels. Although all these tests are well established, seldom are they used for evaluating the toxicity of specific lubricant formulation. It is necessary to have required refinements for these tests to suit the lubricant industry so that toxicity tests are also done voluntarily along with the other standard performance evaluation tests [73–75].

9.7.2 Toxic Effects of the Spillage of Vegetable Oils

Contrary to the general belief that vegetable oils are nonpollutant, there exists the possibility that a spillage will have a detrimental effect on the environment [76]. This is because the spilled oil, upon oxidation and polymerization, results in a product that settles on the riverbed. This in turn induces sublethal effects, including the movement of fish away from breeding grounds. In addition, vegetable oil can also adhere to the gas-exchanging organs, reducing their effectiveness and resulting in loss of aquatic life. Spilled vegetable oil is usually consumed by bacteria as long as polymerization does not start. This can be further boosted with the addition of limiting nutrients such as phosphates and nitrates. However, once polymerization starts, the rate of bacterial consumption dramatically drops. Thus, the key to reduce the impact of vegetable oil spillage is to prevent polymerization in the best possible way and allow the natural bacteria to consume the oil as a carbon source [77–79].

9.8 BIOACCUMULATION AND ECOLOGICAL RISK ASSESSMENT

A major drawback of the ecotoxicity and biodegradability tests is that they only determine the inherent properties of a chemical, ignoring the exposure aspects. This has to be addressed by conducting the risk assessment, which compares the predicted environmental concentration (PEC) with the predicted ecotoxicological limit concentration (PNEC) tolerated without adverse effects on most sensitive aquatic organisms. If the ratio PEC/PNEC for a given compound is less than 1, it suggests that the compound is environmentally nontoxic.

This again raises the question as to whether biodegradability and toxicity or biodegradability alone is a better criterion for judging the lubricant formulation as environment friendly. It is indeed very difficult as various parameters play crucial roles in each case. However, it is necessary for the formulation to be environmentally neutral or nonlethal to the exposed species to also be biodegradable.

In the context of the discussion presented in this chapter, we would like to propose few guidelines for developing an environmentally friendly lubricant.

We propose that characterizing the toxicity and the biodegradability should be the first goal in formulating the environmentally friendly lubricants. Unless all the constituent materials pass these two primary requirements, the products should not be used for further evaluation. This means that the toxicity and the biodegradability of both the base oil and the additives should be assured. Currently attention is given only to the base oil, while the additives are often overlooked. Further to this, the renewability and the sustainability of the materials also need to be given priority while developing environmentally friendly lubricants.

9.9 CONCLUSIONS

- Raw materials, including the additives, determined to be toxicologically and ecologically unacceptable should be excluded from lubricant formulations considered as environmentally friendly.
- Different regions of the world have different policies, laws, and regulations to control the production, use, and disposal of lubricants. Unified laws and regulations need to be developed and adopted by all countries.
- Used oils show appreciable deviation with reference to biodegradability and toxicity when compared to freshly formulated oil. This has to be taken into account while evaluating the lubricants.
- There is an urgent need to develop sound predictive models for characterizing the primary biodegradability of lubricant composition based on physicochemical parameters commonly used during the evaluation.
- The best approach to protecting the environment is to prevent leaks and spills in the first place through proper monitoring and scheduled maintenance programs.
- Use of fish acute threshold test in place of standard fish toxicity test will reduce the number of test fish.

- The algae and daphnia acute tests are the important animal alternative tests and can be used in the initial phase of toxicity testing as algae and daphnia are more sensitive to chemicals than fish.
- Fish embryo test is also gaining attention as an alternative to fish toxicity test. This method will provide information on the effect of chemicals on the fish's embryonic stages.
- In addition to being biodegradable, the lubricant must be at least neutral, if not nonlethal, to the species exposed in order to become environmentally friendly.

REFERENCES

1. M. Nosonovsky, Oil as a lubricant in the ancient Middle East, *Tribol Online*, 2 (2), 44–49 (2007).
2. J. S. McCoy, Introduction: Tracing the historical development of metalworking fluids, in *Metal Working Fluids*, J. P. Byers (Ed.), CRC Press, Boca Raton, FL, pp. 1–18 (2006).
3. L. R. Rudnick, *Synthetics, Mineral Oils, and Bio-Based Lubricants Chemistry and Technology*, CRC Press, Boca Raton, FL (2006).
4. J. K. Mannekote, P. L. Menezes, S. V. Kailas, and K. R. S. Chatra, Tribology of green lubricants, in *Tribology for Scientists and Engineers*, P. L. Menezes, S. P. Ingole, M. Nosonovsky, S. V. Kailas, and M. R. Lovell (Eds.), Springer, New York, pp. 495–521 (2013).
5. S. Z. Erhan and S. Asadauskas, Lubricant basestocks from vegetable oils, *Ind Crop Prod*, 11 (2–3), 277–282 (2000).
6. B. Wilson, Lubricants and functional fluids from renewable resources, *Ind Lubr Tribol*, 50 (6), 15–18 (1998).
7. W. J. Bartz, Lubricants and the environment, *Tribol Int*, 31 (1–3), 35–47 (1998).
8. S. Boyde, Green lubricants: Environmental benefits and impacts of lubrication, *Green Chem*, 4, 293–307 (2002).
9. D. R. Kodali, High performance ester lubricants from natural oils, *Ind Lubr Tribol*, 54, 165–170 (2002)
10. J. K. Mannekote and S. V. Kailas, Experimental investigation of coconut and palm oils as lubricants in four stroke engines, *Tribol Online*, 6 (1), 76–82 (2011).
11. A. Willing, Lubricants based on renewable resources—An environmentally compatible alternative to mineral oil products, *Chemosphere*, 43, 89–98 (2001).
12. P. Bondioli, From oil seeds to industrial products: Present and future oleochemistry, *J Synth Lubr*, 21 (4), 331–343 (2005).
13. M. P. Schneider, Plant-oil-based lubricants and hydraulic fluids, *J Sci Food Agr*, 86, 1769–1780 (2006).
14. S. Z. Erhan, B. K. Sharma, and J. M. Perez, Oxidation and low temperature stability of vegetable oil-based lubricants, *Ind Crop Prod*, 24 (3), 292–299 (2006).
15. L. Honary, An investigation of the use of soybean oil in hydraulic system, *Bioresource Technol*, 56, 41–47 (1996).
16. S. Z. Erhan, B. K. Sharma, Z. S. Liu, and A. Adhvaryu, Lubricant base stock potential of chemically modified vegetable oils, *J Agr Food Chem*, 56 (19), 8919–8925 (2008).
17. H. Wagner, R. Luther, and T. Mang, Lubricant base fluids based on renewable raw materials their catalytic manufacture and modification, *Appl Catal A Gen*, 221 (1–2), 429–442 (2001).
18. A. P. Birova and J. Cvngros, Lubricating oils based on chemically modified vegetable oils, *J Synth Lubr*, 18 (4), 291–299 (2002).
19. B. K. Sharma, Z. S. Liu, A. Adhvaryu, and S. Z. Erhan, One pot synthesis of chemically modified vegetable oils, *J Agr Food Chem*, 56 (9), 3049–3056 (2008).
20. M. C. McManus, G. P. Hammond, and C. R. Burrows, Life-cycle assessment of mineral and rapeseed oil in mobile hydraulic systems, *J Ind Ecol*, 7 (3–4), 163–177 (2004).
21. M. Omer, Energy, environment and sustainable development, *Renew Sust Energy Rev*, 12, 2265–2300 (2008).
22. J. K. Mannekote and S. V. Kailas, Studies on boundary lubrication properties of oxidised coconut and soy bean oils, *Lubr Sci*, 21, 355–365 (2009).
23. R. Becker and A. Knorr, An evaluation of antioxidants for vegetable oils at elevated temperatures, *Lubr Sci*, 8 (2), 95–116 (1996).
24. T. Habereder, D. Moore, and M. Lang, Eco requirements for lubricant additives, in *Lubricant Additives Chemistry and Applications*, Second Edition, L. R. Rudnick (Ed.), pp. 647–666 (2009).

25. A. S. Patil, V. S. Pattanshetti, and M. C. Dwivedi, Functional fluids and additives based on vegetable oils and natural products: A review of the potential, *J Synth Lubr*, 15 (3), 193–211 (2002).
26. B. K. Sharma, K. M. Doll, G. L. Heise, M. Myslinska, and S. Z. Erhan, Antiwear additive derived from soybean oil and boron utilized in a gear oil, *Ind Eng Chem Res*, 51 (37), 11941–11945 (2012).
27. B. K. Sharma, J. M. Perez, and S. Z. Erhan, Soybean oil based lubricants: A search for synergic antioxidants, *Energy Fuels*, 21 (4), 2408–2414 (2007).
28. K. M. Doll and B. K. Sharma, Emulsification of chemically modified vegetable oils for lubricant use, *J Surfactants Deterg*, 14 (1), 131–138 (2011).
29. V. L. Finkenstadt, A. A. Mohmed, G. Biresaw, and J. L. Willett, Mechanical properties of green composites with polycaprolactone and wheat gluten, *J Appl Polym Sci*, 110 (4), 2218–2226 (2008).
30. I. Minami, S. Mori, Y. Isogai, S. Hiyoshi, T. Inayama, and S. Nakayama, Molecular design of environmentally adapted lubricants: Antiwear additives derived from natural amino acids, *Tribol Trans*, 53, 713–721 (2010).
31. J. R. Ehrenfeld, Industrial ecology: A framework for product and process design, *J. Cleaner Prod*, 5 (1–2), 8795 (1997).
32. K. U. Oldenburg and K. Geiser, Pollution prevention and … or industrial ecology? *J. Cleaner Prod*, 5 (1–2), 103–108 (1997).
33. N. Nes and J. Cramer, Product lifetime optimization: A challenging strategy towards more sustainable consumption patterns, *J Cleaner Prod*, 14, 1307–1318 (2006).
34. P. T. Anastas and R. L. Lankey, Life cycle assessment and green chemistry: The yin and yang of industrial ecology, *Green Chem*, 2, 289 (2000).
35. D. Tilman, K. G. Cassman, P. A. Matson R. Naylor, and S. Polasky, Agricultural sustainability and intensive production practices, *Nature*, 418, 671–677 (2002).
36. A. D. Kurzons, D. J. C. Constable, D. N. Mortimer, and V. Cunningham, So you think your process is green, how do you know?—Using principles of sustainability to determine what is green–A corporate perspective, *Green Chem*, 3, 1–6 (2001).
37. B. Cunningham, N. Battersby, W. Wehrmeyer, and C. Fothergill, A sustainability assessment of a biolubricant, *J Ind Ecol*, 7 (3–4), 179–192 (2004).
38. P. Nagendramma and S. Kaul, Development of ecofriendly/biodegradable lubricants: An overview, *Renew Sust Energ Rev*, 16, 764–774 (2012).
39. M. P. Schneider, Plant-oil-based lubricants and hydraulic fluids, *J Sci Food Agr*, 13, 1769–1780 (2006).
40. J. K. Mannekote, P. L. Menezes, S. V. Kailas, and K. R. S. Chatra, Tribology of green lubricants, in *Tribology for Scientists and Engineers*, P. L. Menezes, S. P. Ingole, M. Nosonovsky, S. V. Kailas, and M. R. Lovell (Eds.), Springer, New York, pp. 495–521 (2013), pp. 503–504.
41. M. Voltz, N. C. Yates, and E. Gegner, Biodegradability of lubricant base stocks and fully formulated products, *J Synth Lubr*, 12 (3), 215–230 (1999).
42. M. J. Scott and M. N. Jones, The biodegradation of surfactants in the environment, *Biochim Biophys Acta*, 1508, 235–251 (2000).
43. N. S. Battersby, The ISO headspace CO_2 biodegradation test, *Chemosphere*, 34 (8), 1813–1822 (1997).
44. N. S. Battersby and P. Morgan, A note on the use of the CEC L-33-A-93 test to predict the potential biodegradation of mineral oil based lubricants in soil, *Chemosphere*, 35 (8), 1773–1779 (1997).
45. OECD Test No. 301: Ready biodegradability, *OECD Guidelines for the Testing of Chemicals*, Section 3, OECD Publishing (1992).
46. K. Richterich, H. Berger, and J. Steber, The two phase closed bottle test—A suitable method for the determination of ready biodegradability of poorly soluble compounds, *Chemosphere*, 37 (2), 319–326 (1998).
47. N. S. Battersby, D. Ciccognani, M. R. Evans, D. King, H. A. Painter, D. R. Peterson, and M. Starkey, An inherent biodegradability test for oil products: Description and results of an international ring test, *Chemosphere*, 38 (14), 3219–3235 (1999).
48. N. J. Novick, P. G. Mehta, and P. B. McGoldrick, Assessment of the biodegradability of mineral oil and synthetic ester base stocks using CO_2 ultimate biodegradability tests and CEC-L-33-T-82, *J Synth Lubr*, 13 (1) (2000).
49. ASTM D6139: Standard test method for determining the aerobic aquatic biodegradation of lubricants or their components using the Gledhill shake flask, ASTM International, West Conshohocken, PA (2013).
50. ASTM D7373: Standard test method for predicting biodegradability of lubricants using a bio-kinetic model, ASTM International, West Conshohocken, PA (2013).

51. http://www.oecd-ilibrary.org/environment/oecd-guidelines-for-the-testing-of-chemicals-section -2-effects-on-biotic-systems_20745761 (as on 22 December 2014).

52. ASTM D6081: Standard test method for aquatic toxicity testing of lubricants: Sample preparation and results interpretation, ASTM International, West Conshohocken, PA (2013).

53. http://www.oecd-ilibrary.org/environment/oecd-guidelines-for-the-testing-of-chemicals-section -3-degradation-and-accumulation_2074577x (as on 22 December 2014).

54. F. Haus, J. German, and G. Junter, Primary biodegradability of mineral oil base oils in relation to their chemical and physical characteristics, *Chemosphere*, 45, 983–990 (2001).

55. F. Haus, O. Boissel, and G. Junter, Multiple regression modelling of mineral base oil biodegradability based on their physical properties and overall chemical composition, *Chemosphere*, 50, 939–948 (2003).

56. J. Steber, C. P. Herold, and J. M. Limia, Comparative evaluation of anaerobic biodegradability of hydrocarbons and fatty acid derivatives currently used as drilling fluids, *Chemosphere*, 31 (4), 3105–3118 (1995).

57. OECD Test No. 201: Freshwater alga and cyanobacteria, growth inhibition test, *OECD Guidelines for the Testing of Chemicals*, Section 2, OECD Publishing (2011).

58. OECD Test No. 202: Daphnia sp. acute immobilisation test, *OECD Guidelines for the Testing of Chemicals*, Section 2, OECD Publishing (2004).

59. OECD Test No. 203: Fish, acute toxicity test, *OECD Guidelines for the Testing of Chemicals*, Section 2, OECD Publishing (1992).

60. OECD Test No. 209: Activated sludge, respiration inhibition test (carbon and ammonium oxidation), *OECD Guidelines for the Testing of Chemicals*, Section 2, OECD Publishing (2010).

61. OECD Test No. 204: Fish, prolonged toxicity test: 14-day study, *OECD Guidelines for the Testing of Chemicals*, Section 2, OECD Publishing (1984).

62. N. S. Battersby, The biodegradability and microbial toxicity testing of lubricants—Some recommendations, *Chemosphere*, 41, 1011–1027 (2000).

63. M. Balulescu and J. M. Herdan, Ecological and health aspects of metalworking fluids manufacture and use, *J Synth Lubr*, 14, 35–45 (2001).

64. OECD Guideline No. 126: Short guidance on the threshold approach for acute fish toxicity, ENV/JM/MONO (2010) 17. Series on Testing and Assessment, OECD Publishing (2010).

65. E. Lammer, G. J. Carr, K. Wendler, J. M. Rawlings, S. E. Belanger, and Th. Braunbeck, Is the fish embryo toxicity test (FET) with the zebrafish (*Danio rerio*) a potential alternative for the fish acute toxicity test? *Comp Biochem Physiol*, Part C 149, 196–209 (2009).

66. E. U. Ramos, C. Vermeer, H. J. Vaes, and L. M. Hermens, Acute toxicity of polar narcotics to three aquatic species and its relation to hydrophobicity, *Chemosphere*, 37 (4), 633–650 (1998).

67. S. Jeram, J. M. Riego Sintes, M. Halder, J. B. Fentanes, B. S. Klüttgen, and T. H. Hutchinson, A strategy to reduce the use of fish in acute ecotoxicity testing of new chemical substances notified in the European Union Regulatory, *Toxicol Pharmacol*, 42, 218–224 (2005).

68. R. P. H. Schmitz, A. Eisentrager, T. Lindvogt, M. Moller, and W. Dot, Increase in the toxic potential of synthetic ester lubricant oils by usage: Application of the aquatic bioassays and chemical analysis, *Chemosphere*, 36 (7), 1513–1522 (1998).

69. S. E. Belanger, E. K. Balon, and J. M. Rawlings, Saltatory ontogeny of fishes and sensitive early life stages for ecotoxicology tests, *Aquat Toxicol*, 97, 88–95 (2010).

70. L. U. Sneddon, The evidence for pain in fish: The use of morphine as an analgesic, *Appl Anim Behav Sci*, 83, 153–162 (2003).

71. E. Lammer, H. G. Kamp, V. Hisgen, M. Koch, D. Reinhard, E. R. Salinas, K. Wendler, S. Zok, and T. Braunbeck, Development of a flow-through system for the fish embryo toxicity test (FET) with the zebrafish, *Toxicol In Vitro*, 23 (7) 1436–1442.

72. OECD Test No. 202: Daphnia sp. acute immobilisation test, *OECD Guidelines for the Testing of Chemicals*, Section 2, OECD Publishing, p. 6 (2004).

73. M. R. Embry, S. E. Belanger, T. A. Braunbeck, M. G. Burgos, M. Halder, D. E. Hinton, M. A. Léonard, A. Lillicrap, T. N. King, and G. Whale, The fish embryo toxicity test as an animal alternative method in hazard and risk assessment and scientific research, *Aquat Toxicol*, 97, 79–87 (2010).

74. Alternative testing approaches in environment safety assessment, Report No. 97, European Centre for Ecotoxicology and Toxicology of Chemicals, Brussels, Belgium (2005).

75. D. M. Woltering, The growth response in fish chronic and early life stage toxicity tests: A critical review, *Aquat Toxicol*, 5, 1–21 (1984).

76. S. M. Mudge, Deleterious effects from accidental spillages of vegetable oils, *Spill Sci Technol Bull*, 213 (2), 187–191 (1995).
77. P. Campo, Y. Zhao, M. T. Suidan, A. D. Venosa, and G. A. Sorial, Biodegradation kinetics and toxicity of vegetable oil triacylglycerols under aerobic conditions, *Chemosphere*, 68, 2054–2062 (2007).
78. F. D. Gunstone, Environment-danger from edible oil spills debated again, *Lipid Tech*, 6, 107–108 (1994).
79. S. M. Mudge, M. A. Salgado, and J. East, Preliminary investigations into sunflower oil contamination following the wreck of the M. V. Kimva, *Mar Pollut Bull*, 26, 40–44 (1993).

Section III

Chemically/Enzymatically Modified
Environmentally Friendly Base Oils

10 Biolubricant Production Catalyzed by Enzymes

José André Cavalcanti da Silva

CONTENTS

ABSTRACT

Worldwide, 12 million tons/year of lubricants are disposed of in the environment through leakages, exhausted gases, inadequate disposal, water–oil emulsions, etc. Some of them are resistant to biodegradation, thus representing an environmental threat. One solution to modify this situation is the replacement of the mineral oils by biodegradable synthetic lubricants (esters). While the market of finished mineral-based lubricants is stagnant, biolubricants have shown an average growth of 10% per year in the last 10 years. Oleochemistry, regarding the development of vegetable oil-based lubricants, represents the main element of the challenge of bioproducts chemistry. There are not many works in the literature about biolubricant production through the usage of enzymes as catalysts. However, most of them use lipases from different origins, at several reaction conditions. This chapter includes discussion of several technologies related to biolubricant synthesis, including the route developed by the author, which showed excellent results regarding conversion and physicochemical properties of the final products (biolubricants). Moreover, the main scientific and technological trends related to this enzymatic catalysis production are presented.

10.1 INTRODUCTION

Biolubricants are biodegradable lubricants generally obtained from vegetable oils. They are used in applications where there is an environmental risk of leaks. The world's finished lubricants market is about 35,000,000 tons per year [1] and biolubricants represent approximately 1% of this total [2]. Unlike the stagnant mineral-based finished lubricants market, the biolubricants market has been growing on average at 10% per year within the last 10 years [3].

Regarding the development of vegetable oil-based lubricants, oleochemistry represents one of the main elements, which constitute a great challenge in the bioproducts chemistry. All the required technologies from seed crushing to oil refining, fractionation, and chemical transformation are in place. In general, the industrial system is organized through the independent operation of the four process units (crushing, refining, fractionation, and chemical transformation). This kind of organization allows higher flexibility and scale economy. However, the most critical challenge to producing biolubricants is in the chemical transformation step [4].

Lubricants must present certain physicochemical properties within determined specifications, such as viscosity, acid number, corrosion, and pour point. For biolubricants, because they are generally composed of organic esters, oxidative, thermal, and hydrolytic stabilities are also very important. However, as a high oxidation-resistant molecule is synthesized, with the double bonds being removed or protected by steric hindrance, one must take into account that the biodegradation with microorganism action will also be damaged. Thus, the biggest challenge to overcome in this area is to find a balance, or in other words, to synthesize new biolubricants with higher oxidation resistance and good biodegradability [5].

10.2 HISTORICAL APPLICATION OF BIOLUBRICANTS

The first lubricants used by human beings were oils from animals and vegetables. In the 19th century, petroleum-based lubricants, called *mineral oils*, gradually replaced natural triglycerides. In turn, such mineral oils have been replaced by synthetic oils (produced by chemical reactions) due to stricter quality standards and requirements, which are imposed by performance specifications and environmental legislation [6].

According to Lämsa's work [7], the development of synthetic esters as lubricants has begun in the 1930s in the United States and Germany, mainly due to the Second World War and the petroleum production deficit. Diesters and polyolesters were developed in Germany for use as jet turbine lubricants. Their benefits were good low-temperature properties and thermal and oxidation resistances. Since the 1960s, the aviation industry has used neopentyl polyolesters as lubricants for jet turbines. Their most important characteristic is the flexibility in terms of physicochemical properties, obtained through changes in the chemical structures of the raw materials, as chain length, branching, carbon atom number, and alcohol type used in the synthesis.

In the 1960s, synthetic esters were developed for arctic conditions, for use as engine and gear oil lubricants, hydraulic fluids, and greases. In the 1970s, automotive engine oils developed further, starting with the semisynthetic oils, including diesters. The main benefits were good low-temperature properties and low volatility. The first engine oil with esters was launched in 1977, composed mainly by PAOs. The first ester-based nautical two-stroke engine oil was launched in 1982 in Switzerland and in Germany. The main base fluid was a neopentyl polyolester of branched fatty acids [7]. The molecule structures of some of these synthetic oils are in Figure 10.1.

The first laws regarding biodegradable lubricants were published in Portugal in 1991 by the Ministry of Environment. The law required the usage of such lubricants in two-stroke nautical engine oils, whose biodegradability should be at least 66%, according to the test CEC-L-33-T-82. This test, developed by CEC, was the first to measure the biodegradation of lubricants. It is based on the determination of the lubricant's primary biodegradation using infrared equipment to detect physicochemical property changes of such oils by microorganism action. OECD tests have also been used since the beginning of the 1990s [7]. Nowadays, OECD 301B and CEC-L-103-12 are the most widely adopted methods for measuring the relative biodegradability of lubricants.

From the two-stroke engine oils, the following step was the development of biodegradable chain saw oils. These were launched in the mid-1980s. The first products were based on natural esters, rapeseed triglycerides. During this decade, other lubricant applications for these products have emerged, such as hydraulic fluids, metalworking fluids, bearing oils, jet turbine oils, industrial gear oils, and compressor oils. In addition, they were also used for personal care, textiles and fiber industries, thermal exchange liquids, and plastic industry.

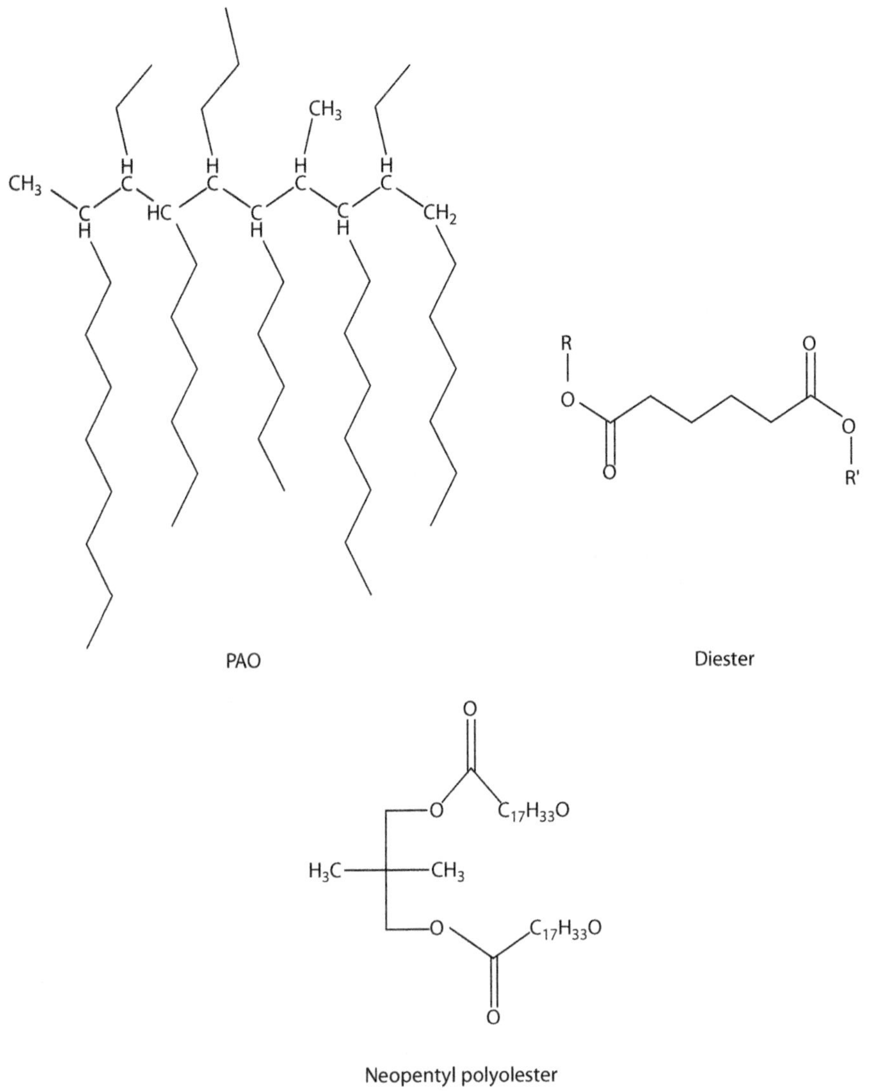

FIGURE 10.1 Molecule structures of some synthetic oils.

Fatty acids used to produce synthetic esters were obtained from vegetable oils or animal fats. Pure fatty acids or partially purified ones were obtained from fractionation, distillation, cracking, dewaxing, etc. The alcohols may vary from short-chain alkanols to polyols. In 1989, the first jet turbine oil, based on trimethylolpropane (TMP) and monoeritrol, was developed [7].

In the 1990s, several American companies followed the European trend of being in accordance with environmental regulations [8]. Among them, Lubrizol Co., from Ohio, invested in research and development to create a series of high-efficiency additives, from sunflower oil with high content of oleic acid. At this same time, the National Ag-Based Lubricants Center was founded at the University of Northern Iowa, resulting in the development and the patent of the first soybean oil-based tractor hydraulic fluid.

10.3 LUBRICANTS AND BIOLUBRICANTS

10.3.1 Lubricants

The word *oil* refers to a wide variety of liquid materials with different properties, which are used, in general, in the modern technology of equipment. Among these liquids, one can highlight the lubricants. Base oils are the higher-proportion constituents in lubricant formulations and are obtained, in general, from petroleum processing. They can be primarily classified as minerals or synthetics, depending on the source or their production process [9]. In several lubricant applications, the required increasingly stricter quality standards (viscosity grades, oxidation resistance, pour point, etc.) may not always be reached by conventional mineral oils. Thus, alternative routes to produce mineral oils have been developed for obtaining products with higher durability and lower environmental impact.

Mineral oils are produced by petroleum distillation and refining and are classified as paraffinics or naphthenics, depending on the predominant hydrocarbon type in their composition. These hydrocarbons have in average 20 to 50 carbon atoms per molecule, and such molecules can be of paraffinic chains (linear or branched alkanes), naphthenics (cycloalkanes with side chains), or aromatics (mainly alkylbenzenes) [10]. Paraffinic base oils show high VI, being useful for engine oil formulations. Yet naphthenic oils present low VI, becoming harder when used in engine oils. They are more suited for metalworking, shock absorbers, and electrical transformer fluids. Finally, aromatic fluids find applications as extender oils in the rubber industry [11].

Mineral base oils can also be classified according to their production process. Solvent extraction is the most traditional or conventional process, where undesirable compounds such as polyaromatics and compounds with heteroatoms (nitrogen and sulfur) are removed with a suitable solvent, in order to increase the VI and also improve the product stability. To remove the aromatics, solvents such as furfural, phenol, or *n*-methylpyrrolidone are the most used. This step occurs in extraction equipment that are called *rotary disk contacting* extraction towers. The low-density flow (from the oil distillation) is fed in the bottom of the tower, while the solvent (higher density) is fed on the top. Thus, by gravity action, the solvent is in continuous contact with the ascendant oil flow, solubilizing the aromatic and polar fractions. This conventional process also includes dewaxing (using methyl isobutyl ketone or methyl ketone/toluene as solvents), to improve the pour point, and hydrotreating. The nonconventional process includes more severe hydrocracking steps, in order to improve VI and stability [11].

Synthetic base oils are produced by chemical reactions in order to achieve the required lubricant properties. About 80% of the world's synthetic lubricants market is composed of three synthetic oils: PAOs (45%), organic esters (25%), and polyglycols (10%) [12]. The main advantages of synthetic products are their high thermal and oxidative stability, better low-temperature properties, and lower volatility. However, synthetic base oils are more expensive than mineral base oils.

The API classifies base oils in five groups, based on source, percentage of saturates, percentage of sulfur, VI, and other properties, as shown in Table 10.1.

Lubricant performance is evaluated by the ability of the lubricant to reduce friction, to avoid formation of deposits, and to resist oxidation, wear, and corrosion. Most problems associated with lubricants are related to oil degradation and its contamination with the by-products of fuel combustion inside the engine. To reduce such problems, finished lubricants are obtained by formulating base oils with additives that provide antioxidant, antiwear, detergent/dispersant, and other functions. Therefore, designing a lubricant for these functions is a complex task and involves careful evaluation of the properties of potential base oils and additives.

10.3.2 Biolubricants

A biolubricant is a biodegradable lubricant capable of being decomposed by microbial action within a determined time. In general, biodegradability means that a lubricant will be metabolized by

TABLE 10.1
API Classification of Base Oils

Category	Saturates (%)[a]		Sulfur (%)[b]	VI[c]
Group I	<90	and/or	>0.03	80–120
Group II	≥90	and	≤0.03	80–120
Group III	≥90	and	≤0.03	>120
Group IV		PAOs		
Group V		All others not included in Groups I, II, III, and IV		

Source: Lubrizol Co., Wickliffe, Ohio, 2013.
[a] ASTM D2007.
[b] ASTM D2622 or ASTM D4294 or ASTM D4927 or ASTM D3120.
[c] ASTM D2270.

microorganisms within 1 year. When the biodegradation is complete, it means that the lubricant has essentially returned to nature, and when the biodegradation is partial, one or more components of the lubricant are not degradable [2].

Through microorganisms, biodegradability results in nontoxic autodecomposing products (CO_2 and water). The activity of such microorganisms widely depends on their structures. Most vegetable oil-based hydraulic fluids are readily biodegradable. Vegetable oils are typically 99% biodegradable by themselves and 90–98% after mixing with additives. On the other hand, the biodegradability of mineral oils is only 20% [13].

Toxicity and biodegradability standards are different for each country. For example, the specifications for biodegradable lubricants are much stricter in Germany than in Switzerland. For several years, the ASTM has been involved in the definition of the word *biodegradability*, since this was the main focus of such lubricants. The environmental compatibility emphasis on these products has meant a smaller focus on their economic aspects and, in some cases, on their performance. As a result, the first products showed high cost and/or low performance. Even then, they still found a market niche. However, by the end of the 1990s, many American companies reduced or eliminated their green projects and the research has continued only in academic and governmental laboratories.

The initial technology for biolubricants was based on vegetable oils, which underwent few chemical treatments and blended with additives to improve their performance. Perhaps, the most important development for the U.S. biolubricants market was the introduction of high-oleic acid soybean by Dupont Co., at the beginning of the 1990s [8]. This genetically modified soybean oil had a fatty acids profile richer in oleic acid than conventional soybean oils [8]. Initially designed for frying, this oil showed, in the OSI method, an oxidative stability approximately 27 times higher than that of the conventional one. This characteristic helped to develop a large number of lubricants and greases [8].

In most cases, the high-oleic acid oils, with improved oxidative stability, were mixed with other oils (synthetics in general) to improve their pour points. In addition, in lubricants for outside applications, it was required that they be blended with several esters, which could negatively affect the final product's compatibility with elastomers (present in seals and gaskets of the lubricated equipment). Recently, chemically modified vegetable oils were developed to allow a better flexibility in their use as lubricant base oil, with improved performance of oxidation stability or pour point. Estolides (Figure 10.2) are a good example of such new chemical transformation technologies. These compounds, synthesized from unsaturated fatty acids, have shown excellent pour points properties (−40°C) and high oxidative stability in the rotary pressure vessel oxidation test (RPVOT) (200 to 400 min) [8].

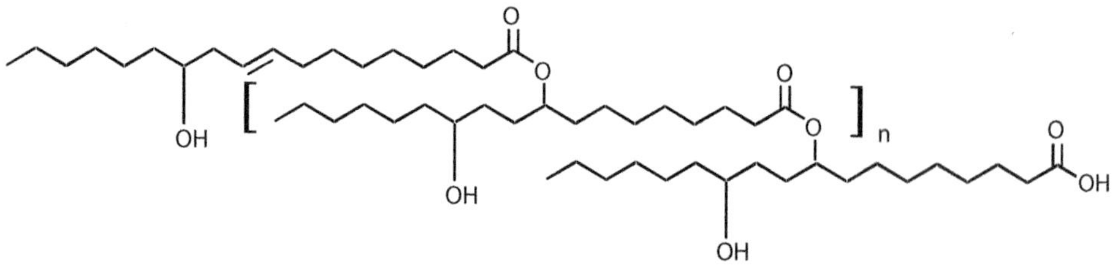

FIGURE 10.2 Estolide general molecule.

In the last few years, the popularity of biofuels has produced increased investment of public and private capital in oilseed development [14]. Although different in terms of final usage, many biofuel production processes are useful for biolubricants.

Biodegradable lubricant demand is related to a growing awareness about the impact of technology on the environment. This awareness results both from a combination of local and national regulations and from consumer influence. European countries, mainly Germany, Austria, and Scandinavia, have devoted great effort in this area.

There is a great effort in Europe to develop materials from agriculture. The Interactive European Network for Industrial Crops and their Applications is composed of 14 countries with the objective of creating a synergy between industrial cultures of the European Union, developing scientific, industrial, and market opportunities for applications in these cultures [8]. One of the main areas of focus for these efforts is biolubricants. The environmentally friendly nature of vegetable oil-based lubricants and their market benefits have generated the involvement of a large number of companies in this market. For example, Mobil Chemical has implemented a clean line of lubricant production as part of the program Agriculture for Chemistry and Energy. Shell and British Petroleum have signed an agreement with the French National Railway Company, a French train company, to develop biodegradable lubricants. Additional efforts are underway to develop vegetable oils as concrete demolding lubricants. In 2002, Western Europe's total lubricants market was 5,020,000 tons/year, of which 50,000 tons/year was based on vegetable oils. On the other hand, the North American lubricants market was 8,250,000 tons/year, of which only 25,000 tons/year are vegetable oil based [8].

Due to their better lubricity, good anticorrosion properties, better viscosity–temperature characteristics, and lower volatility, vegetable oils are already being used as lubricants for industrial applications such as rolling, cutting, drawing, and quenching operations. In these applications, the vegetable oils are used pure or mixed with mineral oils. Vegetable oils are readily biodegradable and environmentally safe, when compared to mineral oils, because of the fatty acids in their composition. Some other advantages of vegetable oils include the facts that they are produced from renewable sources and they do not present dependency on imported oils from foreign countries. From the environmental viewpoint, their importance becomes evident, mainly in total loss lubrication, in military applications, and in outside activities, such as in forestry, mining, railroads, fishing, digging, and agricultural hydraulic systems. However, the extensive use of vegetable oils does not occur due to their inefficient performance at low temperatures and low hydrolytic and thermo-oxidative stabilities. These problems can be mitigated through chemical modifications of the oil structure [15].

Synthetic base oils offer better oxidative stability and performance characteristics than mineral oils. Most biodegradable synthetic oils are esters that show superior thermal and oxidative stabilities. The prices of these products are higher than that of vegetable oils and much higher than that of mineral-based oils. In spite of the higher cost of vegetable oil-based lubricants compared to mineral-based ones, the environmental advantages during short, medium, and long periods can minimize the cost difference. Moreover, using biolubricants can contribute to improve the public image of companies (ISO 14000) and to expand to new markets [16,17].

10.3.3 BIOLUBRICANT SYNTHESIS VIA CHEMICAL CATALYSIS

In the production of simple esters, by esterification (Figure 10.3) and transesterification (Figure 10.4) reactions, from short-chain alcohols such as methanol, ethanol or butanol, and natural triglycerides, the frequently used catalysts are sodium or potassium hydroxides, metal hydroxides and alkoxides, carbonates, acetates, and several acids. An economically feasible monoester production in industrial scale must meet the following minimum criteria: good yield, few by-products, positive energy balance, low environmental impact, and simplicity. However, very few published processes are able to meet these criteria [7].

The main types of esters used as biolubricants are diesters, phthalates, trimelitates, C36 dimerates, and polyolesters (Figure 10.5).

Diesters are produced through the chemical reaction between two monohydric alcohols and one dicarboxylic acid or between one diol and one monocarboxylic acid. For example, azelaic, sebacic, and malic acids are frequently used diacids. The most frequently used alcohols are 2-ethylhexanol, isotridecyl alcohol, and a mixture of C8 to C10 alcohols. Diester lubricants are used in gas turbines, compressors, hydraulic systems, and two-stroke engines. The first lubricants used in gas turbines were dioctyl sebacates. They were later replaced by azelates and adipates and, more recently, by polyolesters. Diesters show excellent low-temperature properties and high VI [7].

Polyolesters are produced from the reaction of polyhydric alcohols with monocarboxylic or dicarboxylic acids. They are also called *hindered esters* and are used in jet turbines and automotive engine oil formulations. This class of products offers an extraordinary thermal stability due to the absence of a secondary hydrogen at the β position and the presence of a quaternary central carbon atom [18]. The frequently used polyalcohols in polyolester synthesis are TMP, neopentyl glycol (NPG), and pentaerythritol (PEN). The acids used in polyolester synthesis can be long or short chain, saturated or unsaturated, and with linear or branched chains. The properties of the final polyolester depend on molar mass, acyl group size, polyol functionality, and method of ester preparation or of esters mixing.

According to Rudnick's work [14], fatty acids are essentially long-chain aliphatic acids, not branched, with carbon atoms bonded to hydrogen and other groups, and with a carboxylic acid at the end of their chain. The most common fatty acids in nature have an even number of carbon atoms in their linear chain, ranging from 14 to 22, the most abundant being those with 16 or 18 carbon atoms. Short-chain fatty acids are water soluble. As the chain increases, the fatty acid becomes progressively less miscible with water, and assumes more lipophilic characteristics. In positions where adjacent carbon atoms lose hydrogen atoms, there is a double bond. If these double bonds occur in multiple carbons (up to a maximum of 6), the fatty acid is said to be polyunsaturated. Unsaturated fatty acids have a lower melting point than saturated fatty acids, and are more abundant in living organisms.

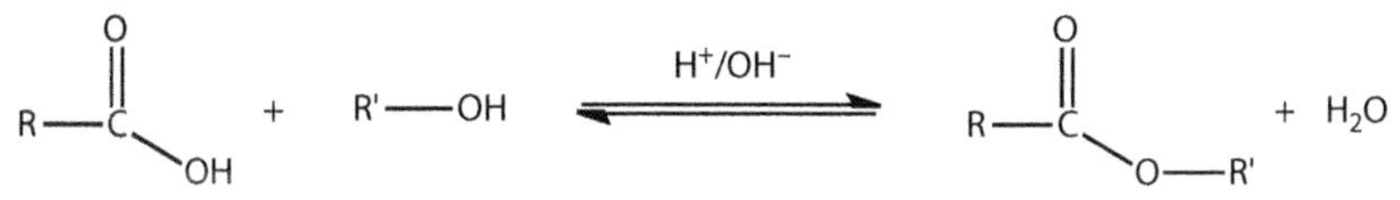

FIGURE 10.3 Esterification reaction of a carboxylic acid with an alcohol.

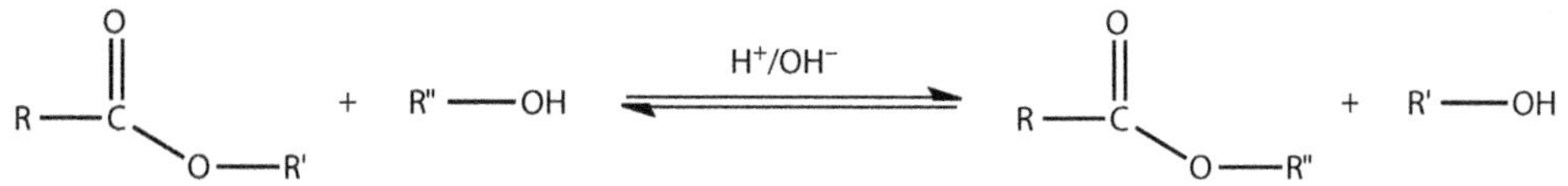

FIGURE 10.4 Transesterification reaction of an ester with an alcohol.

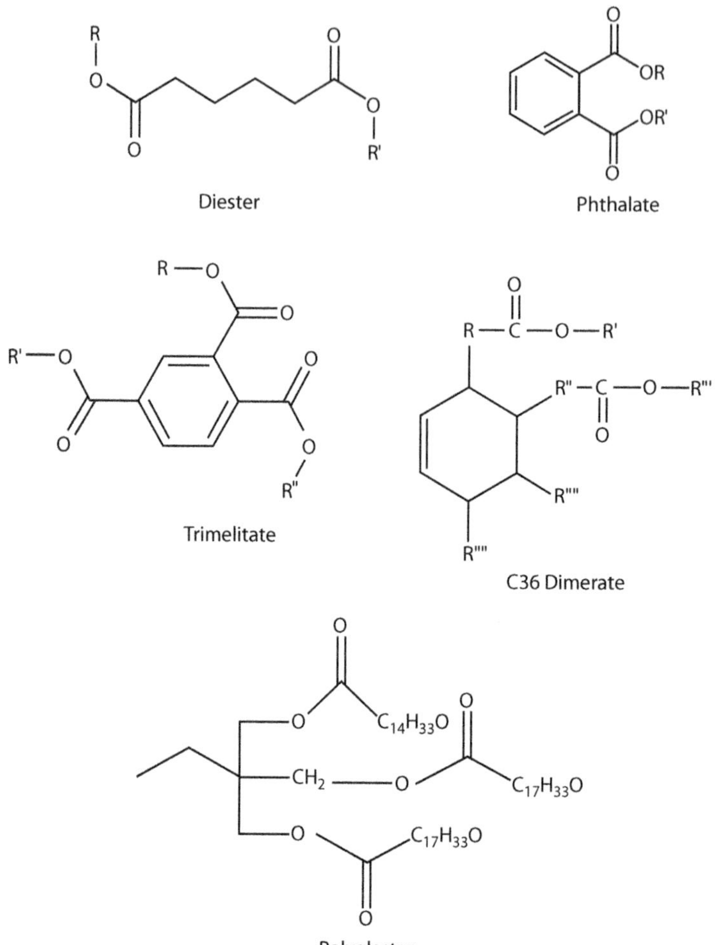

FIGURE 10.5   Structures of the most common esters used as biolubricants.

There is a relationship between the degree of unsaturation of the fatty acid structure and its properties and performance. It is a direct consequence of the curved structure of a fatty acid due to *cis* unsaturation (like vegetable oils), and is related to the liquid nature of the oil. A vegetable oil that contains only linear saturated fatty acids or a high percentage of them is, normally, solid at room temperature (Figure 10.6). As a consequence, these materials do not present advantages as lubricants when subjected to a wide temperature range. For example, if the predominant fatty acids of a vegetable oil are saturated, its performance at low temperatures will be worse than that of vegetable oils with monounsaturated or polyunsaturated fatty acids.

Soybean oil is the largest volume (29%) of vegetable oil produced in the world and represents an important renewable source. In the United States, its production is 2.5 billion gal per year and about 6 billion gal per year worldwide. Crude soybean oil has a viscosity close to that of mineral oil (29 cSt at 40°C), a high flash point (325°C), and a high VI (246). All crude vegetable oils contain some natural impurities, such as unsaponified matter, gum, and fatty matter. These impurities can affect the stability, the hydrocarbon solubility, the chemical reactions, and the pour point properties of vegetable oils. Thus, a purification step is necessary to produce refined vegetable oils. However, to modify the fatty acid chain of the vegetable oil, it is necessary to know its exact composition, in addition to its thermal and oxidative stability. The vegetable oil comprises triacylglycerol structure

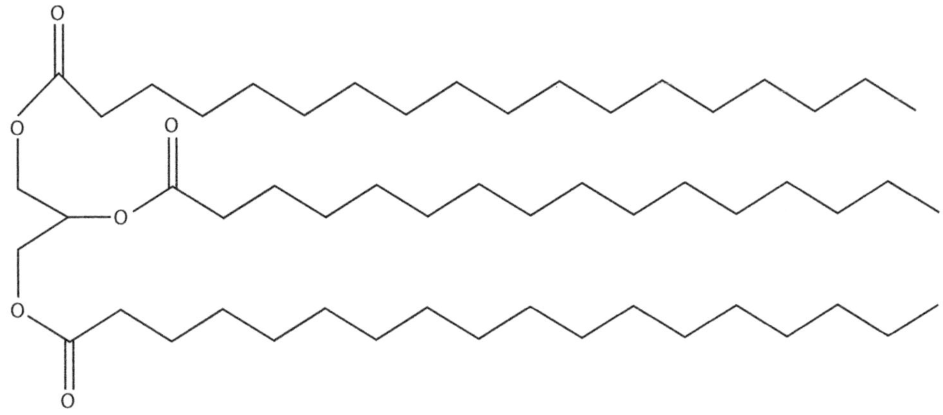

FIGURE 10.6 Structure of a fully saturated vegetable oil.

connected to different fatty acid chains. The presence of unsaturations in the triacylglycerol molecule (oleic, linoleic, and linolenic acids) works as active sites for several oxidation reactions. Saturated fatty acids have an oxidative stability that is relatively higher [14].

More than 90% of the chemical modifications occur in the carboxylic groups of the fatty acids, while less than 10% involve reactions in the fatty acid chain [19]. Without sacrificing the favorable viscosity–temperature and lubricity characteristics, unsaturated vegetable oils can be converted to thermo-oxidative stable products, by means of saturation of the carbon–carbon double bonds using arylation, alkylation, cyclization, hydrogenation, epoxidation, and other reactions. On the other hand, chemical modifications in the carboxyl group of the vegetable oils include esterification, transesterification, and hydrolysis reactions [19].

Transesterification is a process that uses an alcohol (e.g., methanol or ethanol) in the presence of a catalyst (e.g., sodium or potassium hydroxide) to chemically convert the vegetable oil molecule to methyl or ethyl esters, and glycerin as by-product. Transesterified vegetable oil is called *biodiesel*, due to its usage as diesel engine fuel. A great number of transesterification reactions have been reported using alcohols such as methanol, ethanol, and isopropanol, to obtain esters for commercial uses as biodiesel, cosmetics, and lubricants [20].

Few reported transesterification reactions using longer-chain alcohols with 8 to 14 carbons are used for lubricant production. Glycerin in the triacylglycerol structure is not desired, due to the presence of the hydrogen atom in the carbon located at β position relative to the ester bond. This makes the ester more susceptible to elimination reaction, leading to subsequent molecular degradation. The low stability of the β carbon of the glycerol can be suppressed by transesterification reaction with more resistant polyhydric alcohols. Such alcohols have a neopentyl structure without hydrogen on β carbon, as, for example, isosorbitol or neopentyl polyols, like PE, NPG, or TMP (Figure 10.7). Sodium alkoxides are the best transesterification catalysts and produce monoesters yields ranging from 80% to 90%. Other catalysts that can also be used include calcium and lead oxides [21].

Most parts of the longer-chain alcohol esters are obtained through a two-step process. In the first step, the vegetable oil is hydrolyzed to its correspondent fatty acid. In the second step, the fatty acids are esterified with the corresponding alcohol, using in situ produced sodium alkoxide and *p*-toluenesulfonic acid or cation exchange resins as catalysts [22].

10.3.4 Biolubricant Synthesis via Enzymatic Catalysis

The industrial interest for enzymatic technology has been gradually growing, mainly in the areas of protein engineering and nonconventional media enzymology, which have considerably increased the potential application of enzymes as industrial process catalysts. Enzyme-catalyzed processes

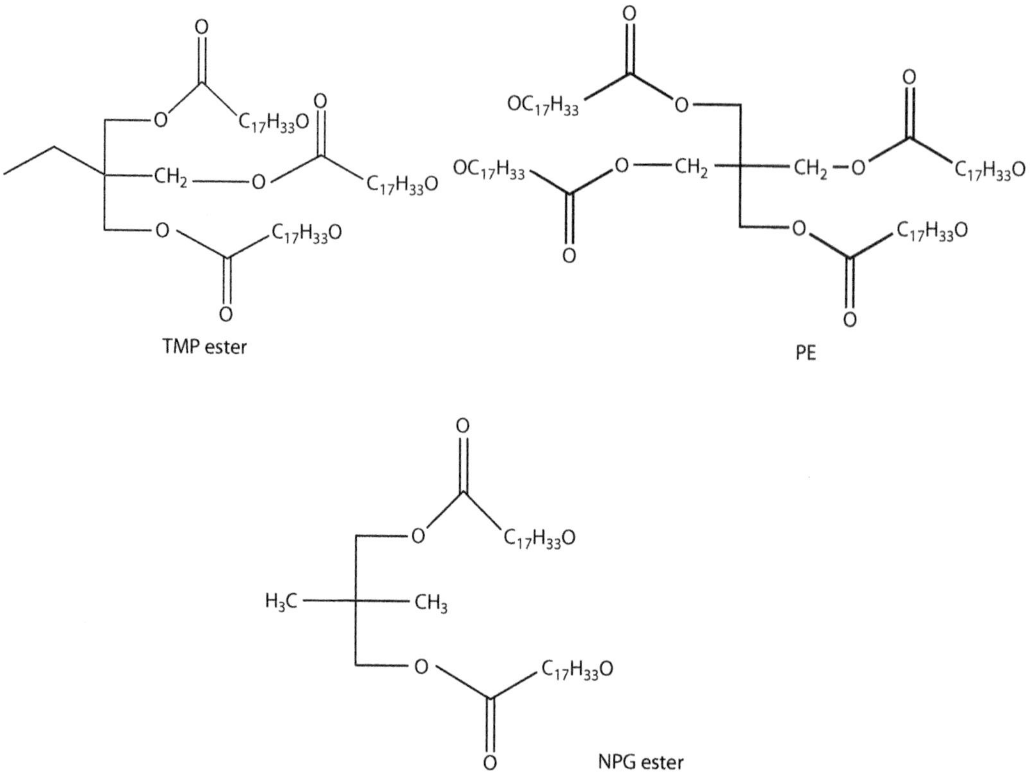

FIGURE 10.7 Molecule structures of polyolesters derived from ricinoleic acid.

of high interest include lipid hydrolysis, synthesis, and interesterification reactions. The reasons for the huge biotechnological potential of enzymatic catalysis (lipase) include the following: (i) have high stability in organic solvents; (ii) do not require presence of cofactors; (iii) have wide substrate specificity; and (iv) exhibit a high enantioselectivity [23]. Recognition of these advantages has provided a considerable increase in the production and commercialization of lipases, resulting in the development of alternative technologies for using it in industrial scale.

Today, although industrial lipase applications are concentrated in the detergent industry, new applications have also been developed in several industries, including pharmaceutical, fine chemicals, cosmetics, oleochemistry, leather, cellulose, paper, foods, and waste treatment. Thus, lipases have gained an increasing share of the industrial enzymes market [24,25].

Lipases are enzymes classified as hydrolases (glycerol ester hydrolases, Enzyme Commission [E.C.] no. 3.1.1.3) and act on the ester bond of several compounds, including acylglycerols, which are their best substrates [26]. Lipases are commonly found in nature and can be obtained from animal, vegetable, and microbial sources. Microorganisms are their preferred source, since this allows production in higher yields and quantity, and this allows broader versatility and simplicity of the environment and genetic manipulation of their productive capacity. Lipases from distinct sources are currently available in the market and, since 1979, a large-scale use of lipases as process catalysts has already occurred. Initially, lipases were produced from animal pancreas and used as digestion auxiliary for human consumption. Due to the low yield of the fermentation process, the cost of microbial lipases was much higher than that of other sources, such as proteases and carboxylases. However, subsequent advances in molecular biology have allowed enzyme producers to launch into the market microbial lipases of higher activity and lower cost [27]. Currently, microbial lipases are produced by several companies worldwide, including Novozymes, Amano, and Gist Brocades [23].

In the lubricants industry, lipases have been gradually applied, mainly for synthesis reactions, such as esterification and transesterification. Several factors influence the conversions of an enzymatic transesterification, including used substrate (vegetable oil or alcohol), molar ratio between the substrates, water content in reaction medium, presence of solvents, temperature, form of enzyme (powder or immobilized), lipase concentration, and so on. Although some literature papers describe biolubricant synthesis using different lipases, it is hard to state any generalization about the optimum reaction conditions, since lipases from different sources tend to respond differently to changes in the reaction medium [28].

Application of lipases in ester synthesis from TMP and rapeseed oil was studied by Linko et al. [29]. The transesterification for the synthesis of triesters of TMP from rapeseed oil fatty acids was performed at atmospheric pressure in closed or opened 13 mL test tubes and also at reduced pressure (2.0–13.3 KPa). After the reaction, a sample was extracted with acetone and the precipitated enzyme was removed by centrifugation. Conversions of greater than 95% in 24 hours were observed.

The importance of controlling water concentration (water activity) in the lipase-catalyzed ester synthesis has been greatly emphasized [7,30]. Water is essential to enzymatic reactions. Normally, a minimum amount of water is necessary to initiate the enzymatic catalysis, and the ester synthesis is allowed to progress under restricted water availability (low water activity). Ester synthesis and hydrolysis are reversible processes, and their equilibrium can be shifted toward the synthesis by an excess of a substrate or by a control of the water concentration in the reaction mixture. Water is essential to the lipase, which becomes inactive due to structural changes in absence of water. It is proposed that in an environment with too low water activity, the conformational space of the enzyme is restricted. In a completely anhydrous environment, there is no space for the conformational changes that are necessary for substrate binding. Lämsa [7] used 3% water to get a 90% conversion in the reaction of rapeseed oil and 2-ethyl-1-hexanol, catalyzed by *Candida rugosa* lipase.

In the transesterification reaction between TMP and rapeseed methyl ester (RMe), Lämsa [7] initially used the immobilized commercial lipase from *Rhizomucor miehei* Lipozyme IM 20, getting TMP ester conversion of about 90%, without water addition. Then, the author used the powder lipase from *Candida rugosa* without any additional organic solvent. Absence of solvent allows higher substrate and product concentrations, simplifies the postreaction processes, and increases operational safety. At 47°C and 15% added water, a conversion in TMP triester of 75% in 24 hours and of 98% in 68 hours was observed. At these conditions, by-product formation did not occur and also no residual RMe was observed.

The chemical and enzymatic transesterifications of TMP with RMe were also studied by Uosukainen et al. [31]. In chemical catalysis, sodium methylate (0.7% in weight) was used and the reaction was carried out at 120°C for 10 hours, producing a conversion of 98% TMP triesters. In enzymatic catalysis, *Candida rugosa* lipase in powder was tested with several immobilization supports. The reaction conditions were milder, under reduced pressure, and at 47°C. After 68 hours, a conversion of 95% TMP triesters was achieved.

Based on previous studies that clearly showed that RMe conversion in TMP esters can be increased by using an immobilized lipase, several support materials for *Candida rugosa* immobilization were investigated. The highest conversion (about 95%) was achieved with lipase immobilized on Celite R-630. Other supports, such as Duolites ES-561 and ES-762, GDC 200, GCC, and HPA 25, gave conversions of approximately 70%. Commercial immobilized lipase *Rhizomucor miehei* Lipozyme IM 20 gave a conversion of 90%.

Lämsa [7] performed the chemical synthesis of rapeseed oil 2-ethyl-hexylester, using acid and alkaline catalysts. The progress of the reaction was followed by thin-layer chromatography (TLC) and high-performance liquid chromatography (HPLC). The final product was obtained by neutralizing the rapeseed oil 2-ethyl-hexylester with acid or alkaline water, depending on the used catalyst, and by washing it three times with water at 50°C. This same material was also synthesized by enzymatic catalysis for 72 hours, and the lipase catalyst (*Candida rugosa*) was separated by centrifugation. Using 3.3% of lipase, a conversion of 80% was achieved in 3 hours. With 14.6% of lipase catalyst, the conversion increased to 100% in 1 hour. Besides *Candida rugosa*, lipases from *Rhizomucor miehei*, *Pseudomonas fluorescens*, and *Chromobacterium viscosum*, all of them immobilized in Amberlyte XAD-7, were also tested.

Lämsa [7] also performed the chemical synthesis of rapeseed oil TMP ester as follows: RMe was reacted with TMP at 120°C for 8 hours, using sodium methylate as catalyst. After cooling, the product was neutralized with aqueous acid, washed with warm water (50°C), and dried with anhydrous sodium sulfate. This reaction was followed by TLC and HPLC. This same synthesis was also carried out at 42°C with lipase (40% w/w). However, in this case, besides methyl ester and TMP, water (10% w/w) was added. All reactions were conducted at reduced pressure. The best conversion (80%) was achieved after 24 hours of reaction. An experimental design was conducted during the synthesis, to determine the optimum conditions (catalyst, catalyst concentration, methyl ester/ TMP molar ratio, temperature, reaction time, and pressure) to maximize conversion (Table 10.2). In the investigation of enzymatic catalysis, in addition to lipase type (*Candida rugosa* showed best results), added water concentration was also investigated (Table 10.2).

In summary, there is plenty of literature evidence about the feasibility of using lipases as biocatalysts for ester synthesis from vegetable oils for biodiesel and for biolubricants applications. Works at Petrobras in cooperation with the Federal University of Rio de Janeiro had resulted in the novel use of the *Candida rugosa* lipase in biolubricant synthesis reactions derived from castor oil [25]. Biolubricants were produced in high yields and showed excellent properties, leading to a patent application [32]. In this invention, the enzyme-catalyzed production of synthetic base lubricants, from castor (methyl ricinoleate), soybean, and jatropha methyl biodiesels as raw materials was investigated. The influence of process variables (enzyme type and concentration, water concentration, temperature, biodiesel concentration, reaction time) was studied in 50 mL glass reactors

TABLE 10.2
Optimum Conditions and Conversions in Rapeseed Oil TMP Ester Synthesis

Variable	Optimum Value/Type	Conversion (%)
Chemical Process		
Chemical catalyst	Sodium methoxide	90.5
Catalyst concentration	0.7% (w/w)	
Molar ratio ester/TMP	3.2:1.0	
Temperature	80°C–110°C	
Reaction time	4 h	
Pressure	3.3 MPa	
Enzymatic Process		
Lipase	*Candida rugosa*	97.5
Catalyst concentration	40.0% (w/w)	
Molar ratio ester/TMP	3.5:1.0	
Temperature	42°C	
Reaction time	24 h	
Pressure	2.0 MPa	
Water concentration	15.0%	

Source: M. Lämsa, "Environmentally Friendly Products Based on Vegetable Oils," DSc thesis, Helsinki University of Technology, Helsinki, Finland, 1995.

under reduced pressure and controlled temperature. Product and substrate conversions of the trans-esterification reactions were determined by HPLC. Temperature variation had the most influence on the percentage of conversion. The best conversion was greater than 95%, after 24 hours of reaction, using *Candida rugosa* lipase as biocatalyst. The products had the following properties: viscosity, 290.2 cSt at 40°C and 28.46 cSt at 100°C; VI, 132; pour point, −39°C; and RPVOT (oxidative stability), 42 min. Such properties qualify the product for potential biolubricant applications, in industrial or marine formulations. Among the investigated raw materials, castor biodiesel had the most promising physicochemical properties. The results of this work are very important from the socioeconomical viewpoint and can lead to more investment in the poorest Brazilian regions to stimulate castor cultivation by small farmers, in order to produce castor oil as an important biodiesel source. The usage of castor biodiesel to produce biolubricants is very noble and profitable [25].

10.3.5 Biolubricant Applications and Uses

The main functions of lubricants are to control friction and reduce wear, corrosion, and deposit formation during the operation of mechanical systems. Thus, lubricant quality depends on several physicochemical properties such as viscosity, pour point, lubricity, thermo-oxidative stability, hydrolytic stability, and additive solvency. Base oils constitute about 80% of the finished lubricant composition, responsible for most of the physical properties of the oil. Vegetable oils outperform mineral oils in lubricity, volatility, and VI. On the other hand, their antiwear protection, load capacity, corrosion prevention, foaming, and demulsibility are similar to those of mineral oils, and can be improved with additives. Other characteristics that are considered critical include detergency, compatibility with other base oils, stability during long-term storage, hydrolytic stability, acidity, and elastomer compatibility. It is known that the oxidative stability of vegetable oils depends on the unsaturation degree of fatty acids. The higher the degree of chain unsaturation, the lower the oxidative stability. The presence of branching and unsaturation and the length of the chain of fatty acids also affect its properties. The higher the degree of branching, the lower its pour point, the higher its hydrolytic stability, and the lower its VI. The opposite is observed with nonbranched chains. In the case of double bonds, the higher the degree of saturation, the better the oxidative stability and the higher its pour point [18]. Synthetic esters based on longer-chain alcohols (TMP, PE, and NPG) and vegetable oils, other than castor oil, have excellent properties and can be easily found in the market.

The worldwide annual vegetable oil production is not enough to meet the demand for biodiesel and lubricants. Thus, it is useful to consider the use of vegetable oils in lubricant applications where properties and performance are more equilibrated. Vegetable oils are suitable for applications whose operation temperatures are lower than 120°C. On the other hand, vegetable oils have poor low-temperature properties compared to synthetic lubricants, mineral oils, and chemically modified mineral oils. Many of these lubricants show excellent low-temperature properties and can be used in arctic conditions for long period. As a result, vegetable oils must be used in applications where the process temperature is higher than −40°C [14].

Lubricants may be classified into two main categories: automotive and industrial. Automotive oils comprise more than 70% of the total lubricants market volume, and the rest are grouped as industrial lubricants. Due to their higher cost, the best applications for vegetable oil-based lubricants are those where their environmental advantages may be maximized [14].

Low-temperature viscosity is one of the most important properties of modern engine lubricants. Cold cranking causes engine wear and can be overcome with the use of products that produce an immediate effective lubrication. In order to meet these specifications related to energy efficiency, low-viscosity and low-evaporation oils have been introduced into the market. Transesterified vegetable oils are good candidates to be used as engine oil due to their superior thermal stability compared to vegetable oils. They also have low viscosity, low deposit formation (longer oil changing intervals and cleaner systems), and better low-temperature properties. Vegetable oils show some disadvantages for engine oil application, such as increasing viscosity during usage, because of oxidation; shorter oil change intervals; incompatibility with mineral oils, which requires purging the engine before oil changing; and limitations at low-temperature properties for particular formulations. Engine manufacturers are not inclined to accept an engine oil with a short oil changing interval [14].

A hydraulic fluid must have the following characteristics: power transmission with minimum power losses; lubrication of sliding surfaces; and corrosion protection. The trend of biolubricants' usage in this area is growing due to their biodegradability, recyclability, fire resistance, thermal stability, good wear performance, and low- and high-temperature performances. Vegetable oils present many of the required properties of hydraulic fluids, except their poor performance at low temperatures and low oxidative and hydrolytic stability, which can be overcome through convenient chemical modifications [33].

A significant amount of heat is generated during metalworking. Lubricants, called *rolling fluids*, usually remove part of the heat. These fluids reduce not only heat in the worked metal but also friction between the cylinders (or rollers), which are used in the metalworking process, and the workpiece of metal. This is important to protect the equipment and decrease the energy required for rolling. Vegetable oils are excellent lubricants and have been successfully used in rolling fluids formulations [14].

The variety of chemicals used to improve the properties and the performance of lubricants is huge. The development of lubricants formulated with biodegradable base oils requires the use of additives that are biodegradable, nontoxic, and compatible with the base oils. Very often, the fate of additives in the environment is unknown and they are usually selected based on previous experience with mineral oil or synthetic formulations [34].

Oils that show low oxidative stability can be used in applications such as dust suppressor systems and for wood treatment. In the first case, the oil must remain liquid and stable until it is applied in the area that needs dust control. Its oxidation or polymerization property becomes an advantage after application, since it forms a thin layer over the dust [35].

The USDA has proposed the implementation of several regulations to require the preferential use by federal agencies of bioproducts, including biolubricants. This effort will help the United States to become less dependent on foreign oil importation and, at the same time, will provide incentive for the usage of environmentally friendly products. This regulation will also help to increase the volume of biolubricants and make them more economically feasible. This program will help federal agencies to significantly increase their use of bioproducts. Federal agencies must purchase these products, wherever they are available, unless their cost is significantly higher than that of a similar nonbiodegradable product. These products must also show similar performance as conventional petroleum-based products. USDA must certify that the bioproduct contains the minimum amount of biodegradable component, before being listed in the required category. This proposed regulation requires that governmental agencies buy lubricants with the required amount of renewable components, including engine oils, metalworking fluids, and hydraulic fluids [14]. It is also important to mention the 2013 VGP and Small VGP regulations from the U.S. EPA. These regulations are aimed at reducing the environmental impact of ships, tugs, barges, and other commercial vessels on American watercourses. The regulation requires that all vessel owners and operators who trade in the U.S. waters must use environmentally friendly lubricants on their equipment [36].

10.4 CONCLUSION AND PERSPECTIVES

Factors responsible for the 10% per year growth of the biolubricants market [3] include increasing environmental awareness about the application of environmentally friendly products and governmental incentives and regulations. Still, compared to the petroleum lubricants market, biolubricant usage is very low and concentrated in few European countries and the United States. The biggest challenge for improving this situation is reducing production costs of such products and making their prices more attractive. Chemical processes have low cost, but also have low conversions. On the other hand, enzymatic processes have higher conversions, and also higher costs. Recent technology development of lipase immobilization may contribute to reduce these costs, making these processes more cost competitive.

Several recent patents and publications have reported the technical feasibility of the enzymatic process for biolubricant (esters) synthesis, using lipases as biocatalyst. In some cases, the enzymatic process has resulted in higher conversions (approximately 98%) compared to the chemical process (approximately 60%), for the same reactions [25,37].

Another important question related to biolubricant use refers to their physicochemical properties. Biolubricants present superior properties such as viscosity, VI, and pour point, compared to petroleum-based lubricants. Efforts have been underway to develop products that show oxidative stability similar to that of mineral oils. This can be achieved by chemical modification of biolubricant molecules or blending specific additives to the formulations. One problem is that such additives must be biodegradable too, in order to not change the biodegradability of the whole final product formulation. The lubricant and additive industries have been working together in order to develop environmentally friendly products.

The use of diverse raw materials, from different countries (such as castor oil in Brazil), is another interesting point, for both economic and social reasons. In the case of Brazil, small farmers from the poorest regions are stimulated to cultivate castor plant, which easily grows in the Brazilian weather. These farmers can sell their castor seeds to the oil and biodiesel producers, who are able to produce biolubricants. This provides a way of promoting social inclusion in developing countries. Another very important characteristic of castor seed is that it is not edible, and its use in lubrication does not cause food shortage.

Finally, biolubricants will play a very important role in future industries, due to their social, economic, and environmental potentials. Among the possible future scenarios for the full implementation of biolubricant production technologies are the following:

1. Higher awareness from consumption market about the benefits of biolubricants usage, and a consequent increase of the market share due to increasing awareness and governmental incentives and regulations. A recent example is a regulation, in the United States, by the EPA, which makes it mandatory to use biolubricants in all commercial vessels in American rivers, lakes, and coasts, effective in December 2013.
2. Biolubricant production with similar or higher quality than that of mineral-based lubricants.
3. Production of biolubricants from renewable raw material, typical of every region of the world, preferably not edible, with minimized cost of production, using advanced synthetic routes, equipments, and catalysts.
4. Development of biodegradable additives to minimize the main disadvantages in terms of biolubricant properties.

REFERENCES

1. L. Lindemann and A. Gosalia, The impact of the global raw material landscape on the worldwide lubricants market or vice-versa? In *Proceedings of the 18th TAE International Colloquium Tribology*, Ostfildern, Germany, 1: 2 (2012).
2. R. D. Whitby, Understanding the global lubricants business—Regional markets, economic issues and profitability, Course Notes, The Oxford Princeton Programme, Oxford, England (2005).
3. S. Z. Erhan, B. K. Sharma, Z. Liu, and A. Adhvaryu, Lubricant base stock potential of chemically modified vegetable oils, *J. Agric. Food Chem.*, 56, 8919–8925 (2008).
4. P. Bondioli, From oil seeds to industrial products: Present and future oleochemistry, *J. Synth. Lubr.*, 21 (4), 331–343 (2005).
5. B. Kolwzan and S. Gryglewicz, Synthesis and biodegradability of some adipic and sebacic esters, *J. Synth. Lubr.*, 20 (2), 100–107 (2003).

6. P. Bondioli, L. Della Bella, and A. Manglaviti, Synthesis of biolubricants with high viscosity and high oxidation stability, *OCL*, 10, 150–154 (2003).

7. M. Lämsa, "Environmentally Friendly Products Based on Vegetable Oils," DSc thesis, Helsinki University of Technology, Helsinki, Finland (1995).

8. L. A. T. Honary, Biolubricants: A global overview, *AOCS Inform.*, 20 (4), 256–259 (2009).

9. L. F. M. Lastres, Lubrificantes e Lubrificação em Motores de Combustão Interna, Course Notes, Petrobras/CENPES, Rio de Janeiro, Brazil (2003).

10. A. Caines and R. Haycock, Chapters 1 and 2, in *Automotive Lubricants Reference Book*, First Ed., Warrendale, PA, SAE (1996).

11. N. I. Do Brasil, M. A. S. Araújo, and E. C. M. De Sousa, Chapter 17, *Processamento de Petróleo e Gás*, First Ed., Rio de Janeiro, Brazil, LTC (2011).

12. W. R. Murphy, D. A. Blain, and A. S. Galiano-Roth, Benefits of synthetic lubricants in industrial applications, *J. Synth. Lubr.*, 18 (18–4), 301–325 (2002).

13. I. Makkonen, *Environmentally Compatible Oils*, FERI, Canada Pointe-Claire, Quebec, Canada (1994).

14. L. R. Rudnick, Chapters 21, 22, and 24, in *Synthetics, Mineral Oils, and Bio-Based Lubricants— Chemistry and Technology*, Boca Raton, FL, CRC Press (2006).

15. R. L. Goyan, R. E. Melley, P. A. Wissner, and W. C. Ong, Biodegradable lubricants, *Lubr. Eng.*, 54, 10–17 (1998).

16. R. C. Gunderson and A. W. Hart, *Synthetic Lubricants*, pp. 103–150, New York, Reinhold (1962).

17. L. R. Rudnick and R. L. Shubkin, Chapters 1, 3–5, in *Synthetic Lubricants and High-Performance Functional Fluids*, New York, Marcel Dekker (1999).

18. H. Wagner, R. Luther, and T. Mang, Lubricant base fluids based on renewable raw materials: Their catalytic manufacture and modification, *Appl. Catal. A Gen.*, 221, 429–442 (2001).

19. H. J. Richtler and J. Knaut, Challenges to a mature industry: Marketing and economics of oleochemicals in Western Europe, *J. Am. Oil Chem. Soc.*, 61 (2), 160–175 (1984).

20. O. N. Anand, J. Mehta, and T. S. R. R. Prasada, Lubricants components from vegetable oils of Indian origin, *J. Synth. Lubr.*, 15, 97–106 (1998).

21. H. Zimmermann and E. Schaaf, On the chemistry of metal-ion catalyzed transesterification reactions, *J. Prakt. Chem.*, 312, 660–668 (1970).

22. M. C. Dwidevi and S. Sapre, Total vegetable-oil based greases prepared from castor oil, *J. Synth. Lubr.*, 19 (3), 229–241 (2002).

23. H. F. Castro, A. A. Mendes, J. C. Santos, and C. L. Aguiar, Modification of oils and fats by biotransformation, *Quím. Nova*, 27 (1), 146–156 (2004).

24. D. M. G. Freire and L. R. Castilho, Lipases em biocatálise, in *Enzimas em Biotecnologia. Produção, Aplicações e Mercado*, E. P. Da Silva, M. A. Ferrara, and M. L. Corvo (Eds.), 1, pp. 367–383, First Ed., Rio de Janeiro, Brazil, Interciência (2008).

25. J. A. C. Da Silva, "Obtenção de um Lubrificante Biodegradável a partir de Ésteres do Biodiesel da Mamona via Catálise Enzimática e Estudos de Estabilidades Oxidativa e Térmica," D Sc thesis, COPPE/ UFRJ, Rio de Janeiro, Brazil (2012).

26. K. E. Jaeger and M. T. Reetz, Microbial lipases form versatile tools for biotechnology, *TIBTECH.*, 16, 396–403 (1998).

27. E. N. Vulfson, *Lipases: Their Structure, Biochemistry and Application*, P. Woolley and S. B. Petersen (Eds.), p. 271, Cambridge, Cambridge University Press (1994).

28. D. M. G. Freire, J. S. Sousa, and E. A. C. Oliveira, Biotechnological methods to produce biodiesel, in *Biofuels*, A. Pandey, C. Larroche, S. C. Ricke, C. G. Dussap, and E. Gnansounou (Eds.), pp. 315–337, Burlington, MA, Academic Press (2011).

29. Y. Y. Linko, M. Lämsa, W. Xiaoyan, E. Uosukainen, J. Seppälä, and P. Linko, Biodegradable products by lipase biocatalysis, *J. Biotechnol.*, 66, 41–50 (1998).

30. X. Y. Wu, S. Jäaskeläinen, and Y. Y. Linko, An investigation of crude lipases for hydrolysis, esterification and transesterification, *Enzym Microb. Technol.*, 19, 226–231 (1996).

31. E. Uosukainen, Y. Y. Linko, M. Lämsa, T. Tervakangas, and P. Linko, Transesterification of trimethylolpropane and rapeseed oil methyl ester to environmentally acceptable lubricants, *J. Am. Oil Chem. Soc.*, 75 (11), 1557–1563 (1998).

32. J. A. C. Da Silva, A. C. Habert, D. M. G. Freire, and V. F. Soares, "Produção de Biolubrificantes a partir de Biodiesel Metílico de Mamona e de Biodiesel Metílico de Pinhão-manso via Catálise Enzimática," Brazilian Patent INPI N° BR 10 2012 009727-3 (2012).

33. W. J. Bartz, Synthetic hydraulic fluids for high performance applications, in STLE 55th Annual Meeting, Nashville, TN (2000).

34. L. R. Rudnick, *Additives: Chemistry and Applications*, pp. 1–27, New York, Marcel Dekker (2003).
35. L. A. T. Honary and E. Richter, *Biobased Lubricants and Greases*, pp. 92–101, First Ed., Hoboken, NJ, Wiley (2011).
36. M. Miller, EPA to require green lubes for vessels, *Lubes'n'Greases*, 19 (2), 22–26 (2013).
37. J. A. C. Da Silva, "Desenvolvimento de um Lubrificante Biodegradável a partir de Ésteres do Biodiesel da Mamona," MSc dissertation, COPPE/UFRJ, Rio de Janeiro, Brazil (2006).

11 Lipases as Biocatalyst for Production of Biolubricants

Jasneet Grewal and Sunil K. Khare

CONTENTS

ABSTRACT

Dwindling petroleum reserves, their increasing prices, and significant health and environmental hazards posed by their poor biodegradability have stimulated interest in the development of ecofriendly lubricants derived from natural sources. Vegetable oil-based esters and synthetic esters are promising candidates for manufacturing biolubricants due to their nontoxicity and biodegradability. However, to improve their physicochemical and tribological properties for effective application and consumer acceptance, multiple approaches of chemical and biotechnological modifications and addition of additives are required. Enzymatic esterification using lipases as biocatalyst is now being extensively studied as an alternative to chemically catalyzed esterification. Nevertheless, the cost, the technical performance, and the scope of application of biobased lubricants manufactured by enzymatic routes face tough competition from conventional mineral oil-based lubricants, and thus requires more research efforts.

    This chapter highlights the advantages and the disadvantages of natural plant oils and synthetic esters as potential feedstocks for biolubricants. Their modifications through chemical and enzymatic methods are also addressed. This chapter also elaborates enzymatic hydrolysis, esterification/transesterification, and epoxidation by lipase biocatalysis in both free and immobilized forms to produce biolubricants. The challenges and the recent developments in formulation of biolubricants obtained by enzymatic route are discussed.

11.1 INTRODUCTION

Lubricants reduce friction between moving parts, ensuring smooth functioning and reducing wear. They are an integral part of the automotive, industrial, and manufacturing sectors. Lubricants can be used in three different physical forms: solid, liquid, and gaseous. Mineral oils derived from petroleum sources form the base stock for nearly all the lubricants currently used in different sectors. However, due to depletion of the world's crude oil resource, there has been an increasing interest in developing renewable biobased lubricants as a viable alternative. Thus, for sustainable development, biobased lubricants, which are nontoxic and also able to meet economic and performance challenges laid down by depleting nonrenewable sources, need to be developed.

Biobased lubricants derived from natural sources such as vegetable oils are eco-friendly, renewable, and biodegradable. Besides vegetable oil-based natural esters, synthetic esters are also considered potential candidates for formulating eco-friendly biolubricants. This is attained by enzymatic and chemical methods for esterification/transesterification of base oils to yield biodegradable ester biolubricants. Lipases have been commonly used by researchers for enzymatic synthesis of biolubricants.

11.2 LUBRICANTS

Lubricants are formulated from 70–90% base oils combined with functional additives, to improve properties that are necessary for desired performance in their varied applications [1]. The base oil can be naturally derived from animal fat or vegetable oil, refined/mineral oils, or synthetic oils.

Currently mineral oils derived from petroleum constitute the major share (95%) of lubricant formulation [2]. Mineral oils are complex mixtures of C_{16}–C_{50} hydrocarbons. These include a range of linear alkanes (waxes), branched alkanes (paraffins), alicyclic (naphthenics), olefinic, and aromatic species, containing significant amounts of sulfur and other heteroatoms. The lower flash point of mineral oils as compared to that of natural oils of the same viscosity causes fire risks [3]. However, the limiting factors forcing the shift from this conventional nonrenewable base stock to biolubricants are the former's finite and depleting reserves and toxicity to the environment due to very low (20–40%) biodegradability [4,5].

11.3 BIOLUBRICANTS

Biobased lubricants use base stock of natural renewable biological content like agricultural materials, e.g., vegetable oils. All lubricants that rapidly biodegrade and are nontoxic to humans and the environment can be classified as biolubricants. They may be biobased like plant oils or synthetic lube oils, satisfying the criteria of rapid biodegradability and nontoxicity. Some synthetic esters derived from petroleum may be biodegradable but cannot be classified as biobased. The use of natural and biodegradable lubricants (e.g., vegetable oil-based) and sustainable energy applications are among the 12 principles of green tribology suggested by Nosonovsky and Bhushan [6].

Other advantages of biolubricants excluding biodegradability and nonuse of petroleum resources are high viscosity index (VI), low evaporative losses, and high flash point. The disadvantages on the other hand include poor oxidative stability, lower cold-flow performance, and higher cost as compared to conventional lubricants.

11.4 BASE STOCKS FOR BIOLUBRICANTS

11.4.1 VEGETABLE OILS

Agricultural products, like vegetable oils, are renewable and nontoxic and thus, form an ideal base stock for both biobased and biodegradable lubricants. These triglycerides, i.e., triesters of long-chain fatty acids combined with glycerol, are amphiphilic in nature due to the presence of both non-polar hydrocarbon and polar ester groups. They have great adhesion to metal surfaces, compatibility with additives, complete biodegradability, and good lubrication properties [7–9].

Both edible and nonedible vegetable oils have been in use as base stock for biobased lubricants. However, use of edible vegetable oils like coconut, soybean, sunflower, cottonseed, olive, sesame, palm, peanut, safflower, corn oils, etc., as base stocks for biolubricants and biofuels triggers food versus fuel debate. Thus, the untapped potential of nonedible vegetable oils seems quite attractive to produce biolubricants. *Ricinus communis* (castor), *Pongamia pinnata* (karanja), *Madhuca indica* (mahua), *Jatropha curcas* (jatropha), *Azadirachta indica* (neem), *Simmondsia chinensis* (jojoba), *Hevea brasiliensis* (rubber), and *Nicotiana tabacum* (tobacco) oils are most common nonedible oils which have been explored for producing biolubricants and biofuels.

Properties of biolubricants derived from the base stock of vegetable oils are affected by their constituent fatty acids, chain length, position, and the number of unsaturated bonds. Apparently, vegetable oils with high oleic acid content compete well with mineral oil-based lubricating oils and synthetic esters [7,10]. The physicochemical properties of selected edible and nonedible vegetable oils are summarized in Table 11.1 [11,12].

11.4.2 MODIFICATIONS OF NATURAL VEGETABLE OILS

Despite the mentioned advantages, the direct effective applications of natural vegetable oils as lubricants are limited due to poor oxidative and thermal stability, lower filterability as compared to mineral oils, and poor low-temperature fluidity [13–17]. Biobased lubricants prepared from vegetable oils show poor corrosion protection and high susceptibility to hydrolytic breakdown. These kinds of biolubricants are mainly suited for total-loss applications, viz., hydraulic fluids with very low thermal stress, chain saw lubricants, concrete mold release oils, drilling mud, and lightly loaded gear drives [18,19].

These limitations can be overcome by modifying the vegetable oil through chemical modifications, genetic engineering, and incorporating appropriate antioxidants and additives. Sharma et al. [20] investigated different additive combinations for antiwear/antioxidant synergism to improve oxidative stability of vegetable oil-based lubricants at elevated temperatures. Oxidative stability can be enhanced by epoxidation, estolide formation, and transesterification of plant oils with polyols [21]. Functional groups such as ester group, hydroxyl groups, allylic carbons, and carbon–carbon double bonds are quite amenable to chemical modifications. Epoxidized vegetable oil like soybean oil finds use as high-temperature lubricants [22]. The readily functionalized epoxy or oxirane group of vegetable oils has been extensively modified by many chemical routes to yield base stocks or additives for biobased lubricants [23,24]. Canola oil is a promising base stock for biobased lubricants because it has 60% of oleic acid, which is advantageous for epoxidation reaction. Madankar et al. [25] synthesized biolubricants from canola oil by its epoxidation with Amberlite IR120 followed by ring opening reaction with different alcohols using Amberlyst 15 (dry) as catalyst. The synthesized epoxy derivatives formulated good biolubricants as they exhibited better friction reduction, enhanced thermal stability, and improved low-temperature properties. Thus, many combinations of Friedel–Crafts alkylation, acylation, hydroformylation, acyloxylation, hydrogenation, dimerization/oligomerization, and metathesis have been used to manufacture better biolubricants [11,21].

TABLE 11.1
Different Vegetable Oils (Both Nonedible and Edible) with Their Physicochemical Properties

Vegetable Oil	Viscosity (cSt) (40°C)	Viscosity (cSt) (100°C)	VI	Pour Point (°C)	Cloud Point (°C)	Flash Point (°C)	Oil Stability Index (h)	Specific Gravity at 25°C (g/cm³)	FFA (%)	Percent of Saturation (%)	Percent of Unsaturation (%)	Saponification Value (mg KOH/g)
Babassu oil	28.65	6.13	170	–	–	112	57.8	0.879	0.06	79–87	13–21	225.6
Castor oil	203.92	34.92	130–140	–31.7	–	260	105.13	0.961	0.90	3.5	96.5	180–182
Coconut oil	36.2	6.76	140–150	20	27	240	75.38	0.915	2–2.05	90–94	6–10	248–265
Corn oil	32.4	8.06	220	–26	–11	280	3.73	0.916	0.05–2.5	10–16	84–90	189–195
Cottonseed oil	34.23	7.91	215	–6	–3.7	320	4.35	0.920	0.05	25–27	73–75	190–198
Jatropha oil	24.5–30.68	–	–	–8	–	117–133	–	0.919	1.76	22–26	74–78	198
Mustard oil	44.1	9.4	205	–	–	316	–	0.918	1.07–1.11	5–6	94–95	168–177
Palm oil	42.66	8.65	189	–	32.55	304	21.52	0.910	6.99	53	47	225–227
Rapeseed oil	32.6	7.90	145	–31.7	–3.9	278	5	0.910	0.5–1.8	6.8	93.2	168–181
Rice bran oil	36.49	8.177	208	–9	–3.9	327	20.82	0.921	0.05	15–20	80–85	180–195
Rubber seed oil	66.2–76.4	–	–	–	–	198	–	0.910	2.6	20–21	79–80	210
Safflower oil	37.9	8.32	206	–22	0.4	322	17.98	0.920	0.03	8–12	88–92	191
Sesame oil	34.1	7.92	216	–10	1.0	328	5.8	0.916	2	11–18	82–89	195
Soybean oil	32.93	8.08	170	–12.2	–3.9	254	17.67	0.916	0.05	16	84	180–200
Sunflower oil	37.1	7.68	172–176	–15	7.20	274	10.23	0.918	0.10	9–11	89–91	188–194

11.4.3 Synthetic Oils

Synthetic oils such as Polyalkylene glycols (PAGs) and synthetic esters are commonly used as feedstocks for manufacturing biodegradable lubricants and are discussed below.

11.4.3.1 PAGs

PAGs are manufactured by the polymerization of ethylene or propylene oxide in the presence of monohydric or polyhydric initiator and sodium or potassium hydroxide catalyst. They are soluble in either oil (propylene oxide) or water (ethylene oxide) depending on the precursor [26]. In general, they have high VIs, low pour points, high flash points, high thermal and oxidative stabilities, and excellent friction control leading to high performance in lubrication. Their limitations include incompatibility with mineral oils and additives and very high costs for production. Studies have proven them to be well suited for applications such as biodegradable fire-resistant fluids, hydraulic fluids, compressor lubricants, gear oils, and metalworking fluids [26,27].

11.4.3.2 Synthetic Esters

Esters can be used as the primary base stock or as an additive to formulate high-performance biodegradable lubricants. Esters were initially developed for use as jet engine lubricants. However, due to their vast possibilities, 90% of all lubricants can be produced from esters. Monoesters, diesters, polyolesters can be synthesized by esterification of appropriate carboxylic acids and alcohols and, thus, synthesis can be tailored according to desired applications. Synthetic ester base stocks have high VIs, thermal and oxidative stability, biodegradability, fire resistance, and good antiwear properties which make them suitable for eco-friendly formulations.

Table 11.2 [28–30] compares the physicochemical properties of various base stocks used for the production of lubricants.

11.5 ESTER-BASED LUBRICANT FORMULATIONS

Esterification of vegetable oils with polyol alcohols increases the stability of the biolubricant base stock by eliminating the hydrogen atom in β-position of the triglyceride. Thus, many polyolester-based biolubricants have been prepared by esterification/transesterification of fatty acids or of fatty acid esters with polyols like trimethylolpropane (TMP), pentaerythritol (PE), and

TABLE 11.2
Physicochemical Properties of Commonly Used Base Stocks in Lubricant Production

Property	Mineral Oil	Vegetable Oils	Glycols	Synthetic Esters	PAOs
Biodegradability (CEC) (%)	Poor (10–30)	Excellent (70–100)	Very good (10–99)	Very good/Excellent (10–100)	Poor
Oxidation stability	Good	Moderate	Good	Very good	Good
Hydrolytic stability	Excellent	Poor/Moderate	Good	Moderate	Excellent
Pour point (°C)	0 to −60	−9 to −21	−30 to −50	−20 to −80	−20 to −60
Solubility in water	Immiscible	Immiscible	Good	Immiscible	Poor
Cold-flow behavior	Good	Poor	Very good	Very good	Very good
Sludge-forming tendency	Good	Poor	–	Moderate	–
Miscibility with mineral oil	–	Good	Immiscible	Good	Very good
VI	100	100–200	100–200	120–220	120–170
Seal swelling tendency	Slight	Slight	Shrinking	Moderate	–
Shear stability	Good	Good	Good	Good	Good
Cost relative to mineral oil	1	2–3	6–10	4–20	3–5

neopentylglycol (NPG) [31–35]. Kamalakar et al. [36] prepared biolubricant from rubber seed oil. They hydrolyzed rubber oil and the obtained fatty acids were esterified with branched-chain alcohol, 2-ethyl-1-hexanol (2-EtH) and polyols, i.e., TMP, PE, and NPG. *p*-Toluenesulfonic acid was used as a catalyst in esterification carried out at 135–140°C in xylene medium. The polyolesters had better VIs (205–222) compared to branched esters (187). PE (170 kg) and TMP (160 kg) esters exhibited good weld load behavior compared to NPG (130 kg) and 2-EtH (120 kg) esters. Thus, all the esters were suitable for hydraulic fluid formulations.

NPG esters are used as lubricants in jet engines. Among the most widely used materials for hydraulic fluids are the TMP esters of oleic acid. Biobased esters generally have high viscosity, good shear stability, moderate oxidative stability, and rapid biodegradability, which make them environment friendly. By varying the type of alcohol, free fatty acid (FFA), molar ratio of alcohol to fatty acid, water content, reaction temperature, duration, and catalyst type, improved yield of ester products can be obtained [37].

11.5.1 Synthesis of Ester-Based Biolubricants by Chemical and Enzymatic Processes

Chemical esterification is carried out at high temperatures (>150°C) using homogeneous or heterogeneous catalysts [38]. Although chemical synthesis has been the method of choice for synthetic esters, the need for sustainable green process has shed light on enzymatic transesterification. The major disadvantages of chemical esterification include high energy consumption, low catalyst selectivity, undesirable side reactions, corrosion (with sulfuric acid and sodium hydroxide catalysts), low conversion rate (average ~40%), foaming (basic catalysts), and harsh operating conditions [39]. Enzymatic methods offer milder operating conditions, high selectivity, purity of products, lower energy consumption, and environment-friendly nature [40–42].

11.6   LIPASE-CATALYZED BIOLUBRICANT SYNTHESIS

Lipases (triacylglycerol acyl hydrolases, E.C. 3.1.1.3) are serine hydrolases which catalyze the hydrolysis of triacylglycerols to release FFAs, diglycerides, monoglycerides, and glycerol [43,44]. They work very well at organic–aqueous interface and this interfacial activation phenomenon allows them to catalyze synthesis in organic media [45]. Thus, apart from their natural function of catalyzing hydrolysis, they also catalyze esterification, interesterification, transesterification, alcoholysis, and epoxidation in nonaqueous (low-water) media [44,46].

The ability of these robust enzymes to catalyze various reactions in heterogeneous media allows them to compete with chemical catalysts in the production of lubricants in different nonaqueous media [47]. Thus, lipases have been used as biocatalyst in the production of lubricants in organic solvent systems, ionic liquid media, and supercritical fluid systems [48–50]. Their excellent promiscuity provides them enormous potential to be exploited as biocatalyst for sustainable, selective, and clean biolubricant production.

Although lipases are found in plants and animals, their microbial sources have been majorly used as biocatalyst in various forms, viz., whole-cell (intracellular lipases) or purified enzyme as free or immobilized formulations [47,51]. Lipases from various microorganisms like *Candida rugosa*, *Chromobacterium viscosum*, *Rhizomucor miehei*, *Pseudomonas cepacia*, *Bacillus subtilis*, *Pseudomonas fluorescens*, and *Candida antarctica* have been effectively used to catalyze industrial esterification and transesterification processes with high regioselectivity and stereospecificity [37,52]. They can use oils from different origins, including waste oils with high acidity and water content, as a substrate for esterification [47,53]. *Candida antarctica* lipase B (CALB) (Novozym 435) has been majorly used for biolubricant synthesis. Nonetheless, the cost of the enzyme and its stability and reusability are the critical factors for determining the economic feasibility of its use as biocatalyst in the production of biolubricants.

11.6.1 Immobilization of Lipases

To withstand the tough synthesis conditions of biolubricants and sustain optimum enzymatic activity and reusability, lipases have been mostly used in immobilized form, although there is a recent report about using free (liquid) lipase for biolubricant synthesis [54].

Immobilization offers several advantages including increased stability, reusability, ease of product separation, better control over catalysis, and cost-effectiveness of operation [46]. Adsorption, covalent bonding, entrapment, encapsulation, and cross-linking are the different methods employed for enzyme immobilization. Biocatalyst cost and efficiency can be effectively modulated by selection of the appropriate method, the choice of suitable matrix, and the amount of enzyme immobilized on the matrix [55].

Novozym 435 is possibly the most widely used commercial lipase [56,57]. This lipase from *Candida antarctica* (CALB) is immobilized via interfacial activation on a moderately hydrophobic resin, Lewatit VP OC 1600 [58]. Lipozyme TL IM® is lipase from *Thermomyces lanuginosus* immobilized on a cationic silicate via anion exchange [59]. Lipase from *Rhizomucor miehei*, immobilized on Duolite ES 562, a weak anion exchange resin based on phenol–formaldehyde copolymers is available as commercial Lipozyme RM-IM [60].

11.6.2 Biolubricant Synthesis by Lipase-Catalyzed Hydrolysis of Vegetable Oils

Biolubricant synthesis can be carried out by following two steps, i.e., (i) direct transesterification of vegetable oils and (ii) hydrolysis of plant oils by lipases followed by esterification of resultant FFAs [61]. Lipase-catalyzed hydrolysis is more advantageous as compared to physical/chemical processes such as alkaline hydrolysis and high-pressure steam splitting. It is more selective, operates under mild conditions (35°C and atmospheric pressure), and leads to high-purity products [62].

Syaima et al. [63] synthesized biolubricant from palm oil mill effluent (POME) by lipase (*Candida* sp.)-catalyzed enzymatic hydrolysis followed by noncatalytic esterification. Fifty percent (v/v) of POME, 20 U/mL of enzyme loading at 40°C, and pH 7.0 with agitation speed of 650 rpm gave optimum hydrolysis rate (0.1639 mg/s L). Noncatalytic esterification at the rate of 0.0018 mg/s L was attained with alcohol to fatty acid ratio of 3:1, at 75°C and 950 rpm. The 100% waste-derived feedstock and the green process of lipase-catalyzed hydrolysis, coupled with good tribological properties, make this lubricant a lucrative candidate for creating its share in market. Apart from its use in the automotive sector, this biolubricant could also be used for biomedical applications like lubrication of joints.

Biolubricant has even been synthesized from waste cooking oil (WCO) by Chowdhury et al. [64], using two different lipases. In the first step, WCO was hydrolyzed into FFAs by *Candida rugosa* lipase, followed by transesterification by Novozym 435 to generate octyl esters as biolubricant. Taguchi L_9 orthogonal design was used to optimize process conditions to achieve 95.19% conversion in 3 h at 60°C with enzyme amount 5 wt% of FFA and molar ratio of 2.5:1 for octanol/FFA. The physicochemical characterization of synthesized octyl esters showed improved VI, flash point, and oxidation stability.

Zamratul et al. [65] also carried out synthesis of biolubricant from *Jatropha curcas* oils by lipase (Polipazyme™)-catalyzed enzymatic hydrolysis and noncatalytic esterification. Enzymatic hydrolysis was carried out at 16% v/v ratio of jatropha oil to buffer at 40°C, pH 7, and agitation speed of 650 rpm.

The extraction of fatty acids obtained by the hydrolysis of triglycerides of plant oils is difficult, thus another approach of transesterification of methyl/ethyl esters or direct transesterification is further discussed.

11.6.3 BIOLUBRICANT SYNTHESIS THROUGH LIPASE-CATALYZED TRANSESTERIFICATION REACTIONS

Both soluble and immobilized lipases have been used with a variety of vegetable oils as feedstocks for producing biolubricants by esterification/transesterification reactions which are discussed in the following and summarized in Table 11.3.

Recently, Trivedi et al. [54] enzymatically synthesized biolubricant (C_{12}–C_{36} esters) in a single-step reaction in a solvent-free medium. Free soluble lipase Callera Trans L (*Thermomyces lanuginosus*, EP 258068) was used for esterification using different combinations of acids and alcohols. Esterification reactions were effectively carried out at room temperature with no by-product formation. Optimization of chain type and length of alcohols, pH, temperature, and initial water concentration led to ~99% conversion. The process can be called more cost effective as cost of immobilization or chemical catalysts is avoided.

Novozym 435, packed in microchannel reactors HCube™ and X-Cube™, was used for continuous production of biolubricant (isoamyl oleate) in solvent-free media by Madarasz et al. [66]. Novozym 435 showed great stability as synthesis of isoamyl oleate continued up to 144 h in HCube reactor. Ninety-eight percent conversion was achieved in X-Cube reactor.

Novozym 435, due to its stability in ionic liquids, was explored as biocatalyst by Bányai et al. [50] in the production of biolubricant in ionic liquid media. Use of ionic liquids is advantageous over conventional organic solvents due to their stability, nonvolatility, nonflammability, and catalytic activity. Fusel oil and isoamyl alcohol were used as substrates for the production of oleate esters in different ionic liquids. Thus, this green process using ionic liquids and Novozym 435 helped to achieve complete conversions in shorter time with no side reactions.

A process for producing biolubricant from castor oil has been described by Malhotra et al. [67]. They used a mixture of immobilized *Mucor miehei* (RMIM) and Novozym 435 to catalyze direct transesterification of castor oil with 1-dodecanol, 1-octanol, and 1-hexanol to get 80–95% conversions.

Kleinaté et al. [68] showed that Lipozyme TL IM effectively carried out transesterification between biodiesel and 2-EtH. The biolubricant (2-EtH oleate) was synthesized by this green process in a solvent-free medium. Lipozyme retained its activity until 10 cycles and 100% conversion was achieved even in 50 L scale-up in 10 h at 60°C.

Happe et al. [72] screened eight different lipases for catalyzing biolubricant synthesis in a microwave barrel reactor. Immobilized *Candida antarctica* lipase was chosen for scale-up and 95% conversion into TMP esters was attained from oleic acid and TMP. Purified biolubricants were biodegradable and their structure analysis was done by Fourier transform infrared spectroscopy and nuclear magnetic resonance.

In an interesting comparison, Akerman et al. [1] synthesized esters from TMP and carboxylic acids (C5 to C18) by (i) heterogeneous catalysts (silica sulfuric acid and Amberlyst-15) and (ii) biocatalyst Novozym 435. Silica sulfuric acid was found to be a better catalyst as compared to Amberlyst-15 for short-chain carboxylic acids. Novozym 435 exhibited increased activity with increase in length of the fatty acid chain. Thus, in the synthesis of C18-ester, enzyme was as good as silica sulfuric acid and gave better quality product. The biolubricant had good cold-flow properties for low-temperature applications. Pour point (−75°C to −42°C) and VI (80–208) increased with increasing alkyl chain length.

Candida sp. 99–125 lipase immobilized on a textile membrane was also used for synthesizing TMP esters by direct esterification between caprylic acid and TMP [71]. Uosukainen et al. [32] synthesized TMP esters in both bench and pilot scales by using *Candida rugosa* lipase as well as alkaline catalysts (sodium methoxide and sodium ethoxide). They achieved complete conversion in both cases, although high temperature was required in chemical synthesis. Hydraulic fluid prepared from rapeseed oil-derived TMP esters showed excellent friction and wear characteristics, good viscosity, and oxidation stability at elevated temperatures.

Castor oil, due to its higher hydroxylated fatty acid content, has been widely used as feedstock for biolubricant production [25,76]. The methanolysis of castor oil for biolubricant synthesis by Novozym 435 has been studied by Hajar and Vahabzadeh [69]. Modeling in this case was performed using an artificial neural network. Experimentation based on the model led to 98%

TABLE 11.3

Lipase-Mediated Biolubricant Synthesis

Lipase	Oil/Fatty Acid	Alcohol	Performance	Biolubricant: Properties and Application	Reaction Conditions	Reference
Novozym 435	Oleic acid	Isoamyl alcohol	98% conversion in X-Cube microfluidic reactor	Isoamyl oleate, continuously obtained in the HCube microfluidic reactor for 144 h without any loss of enzyme catalytic activity	Temperature: 45°C; Isoamyl alcohol/oleic acid molar ratio: 6:5; Enzyme amount: 105 mg; Residence time: 2.43 min.	[66]
Callera Trans L (*Thermomyces lanuginosus*, EP 258068)	Oleic acid	Oleyl alcohol, octanol, hexanol, 3,7-dimethyl-1-octanol, 2-EtH	99% conversion with oleyl alcohol	Long-chain biodegradable esters with good lubrication properties	Solvent-free system; Temperature: 30°C; Acid to alcohol molar ratio: 1:1; Enzyme amount: 1% w/w; Reaction time: 18 h; External water added: 0.9% w/w; pH 6.	[54]
Novozym 435 and RMIM (immobilized *Mucor miehei*)	Castor oil	1-Hexanol, 1-octanol, and 1-dodecanol	95%, 90%, 80% conversions with 1-hexanol, 1-octanol, and 1-dodecanol, respectively	Fatty acid esters produced by direct transesterification of castor oil, bypassing additional conversion step of oil into methyl/ethyl ester	Temperature: 50°C; Oil-to-alcohol molar ratio: 1:3 (solvent-free condition); Enzyme amount: 5–7.5% w/w of oil; Reaction time: 48 h; Stirring rate: 200 rpm.	[67]
Novozym 435	Castor oil	Methanol	Castor oil methyl ester yield, 98%	Castor oil methyl ester usage as lubricity enhancer for diesel fuel and viscosity modifier in preparing drilling fluids in petro-industries; kinematic viscosity at 40°C (mm^2/s): 14; at 100°C (mm^2/s): 3.7; VI: 165; density at 15°C (g/cm^3): 0.925; acid value (mg KOH/g): 0.05; pour point (°C): −27; cloud point (°C): −18; flash point (°C): 186; water content (ppm): 5; refractive index at 20°C: 1.4510	Temperature: 42°C; Methanol-to-oil molar ratio: 6:1; Reaction time: 39 h.	[69]

(Continued)

TABLE 11.3 (CONTINUED)
Lipase-Mediated Biolubricant Synthesis

Lipase	Oil/Fatty Acid	Alcohol	Performance	Biolubricant: Properties and Application	Reaction Conditions	Reference
Lipozyme TL IM	Biodiesel from rapeseed oil	2-EtH	100% conversion in 10 h of reaction time	2-ethyl-1-hexyl oleate	Solvent-free and water-free system; Temperature: 60°C; Enzyme amount: 10% w/w; Reaction time: 10 h.	[68]
Novozym 435	Rapeseed oil (fatty acid methyl ester)	TMP NPG 2-EtH	98% conversion	TMP, NPG, and 2-EtH esters have high thermo-oxidative stability, low pour point, and good viscosity properties, rendering their application as components of environmentally friendly lubricants. Kinematic viscosity (cSt) at 40°C and 100°C, respectively: 7.8, 2.7 (2-EtH ester); 17.4, 4.7 (NPG ester); 38.2, 8.4 (TMP ester) VI: 224 (2-EH ester); 209 (NPG ester); 205 (TMP ester) Pour point (°C): −31.3 (2-EtH ester); −19.5 (NPG ester); −18.0 (TMP ester)	Temperature: 35°C; Enzyme amount: 2%; Fatty acid methyl esters/alcohol molar ratio: 1. 3:1 (FAME/TMP); 2. 2:1 (FAME/NPG); 3. 1:1 (FAME/2-EtH); Stoichiometric; Stirring rate: 250 rpm.	[70]
Candida sp. 99–125 (immobilized on a textile membrane)	Caprylic acid	TMP	96% conversion of fatty acids to total TMP esters and 93% formation of trisubstituted TMP esters	TMP esters (direct esterification without release of toxic methanol)	Temperature: 40°C; Enzyme amount: 0.4 g (for 2 g caprylic acid); TMP to acid molar ratio: 1:10; Water content controlled under 0.8% (w/w).	[71]
Immobilized *Candida antarctica* lipase	Oleic acid	TMP	96% conversion using microwave induction	TMP oleates 1–3	Temperature: 70°C; TMP/oleic acid molar ratio: 1:3; Biphasic nonsolvent process in the microwave barrel reactor.	[72]

(Continued)

TABLE 11.3 (CONTINUED)
Lipase-Mediated Biolubricant Synthesis

Lipase	Oil/Fatty Acid	Alcohol	Performance	Biolubricant: Properties and Application	Reaction Conditions	Reference
Novozym 435	Oleic acid	Furfuryl alcohol	99% conversion	Furfuryl oleate	Temperature: 60°C; Enzyme amount: 5% (w/w); Acid-to-alcohol molar ratio: 1:1; Reaction time: 6 h; Ping-pong Bi-Bi mechanism.	[73]
Novozym 435	Palmitic acid	2-EtH	93% conversion	2-ethylhexyl palmitate, use in manufacture of water-resistant lubricants	Solvent-free system; Temperature: 70°C; Enzyme amount: 10.5 wt%; Acid-to-alcohol molar ratio: 1:5.5; Reaction time: 6 h; Stirring rate: 150 rpm.	[42]
Novozym 435	Oleic acid	Fusel oil (by-product of distilleries)	99.8% conversion	Oleate esters can be applied as a low-viscosity lubricant, environment friendly with no aquatic toxicity	Solvent-free system; Temperature: 60°C; Acid-to-alcohol molar ratio: 1:2; 0.5% initial water content.	[74]
Lipozyme®	Sunflower oil (high oleic acid, 83.5%)	Butanol-1	77% conversion to ester	Mixture of butyl ester (65%), monoglyceride (26%), diglyceride (6%), and residual triglyceride (3%), which has industrially important properties both as a lubricant and surfactant	Solvent-free system in a continuous plug flow reactor: maintained without any loss in activity for a period of 3 months.	[75]
Candida rugosa lipase	Rapeseed oil methyl ester	TMP	98% total conversion	TMP esters (esterification without additional organic solvent), suitable as biodegradable hydraulic fluids	Temperature: 42°C; Enzyme amount: 40% w/w; TMP and rapeseed oil methyl ester molar ratio: 1:3.5; Pressure: 5.3 kPa.	[32]
Rhizomucor miehei lipase Lipozyme IM 20	Rapeseed oil (fatty acid methyl ester)	TMP	90% conversion to TMP-triester in 66 h	TMP esters (esterification without additional organic solvent), used as biodegradable hydraulic oil and lubricant	Temperature: 58°C; Enzyme amount: 40% w/w; Pressure: 5.3 kPa.	[31]

methyl ester yield at 42°C in 39 h, at methanol-to-oil molar ratio of 6:1 and 5% enzyme feed. The Novozym 435 could be effectively reused. The physicochemical properties of the biolubricant were determined as kinematic viscosity at 40°C (mm²/s): 14; kinematic viscosity at 100°C (mm²/s): 3.7; VI: 165; density at 15°C (g/cm³): 0.925; acid value (mg KOH/g): 0.05; pour point (°C): −27; cloud point (°C): −18; flash point (°C): 186; and refractive index at 20°C: 1.4510. The biolubricant could be used as lubricity enhancer for diesel fuel and viscosity modifier for drilling fluids in petrochemical industries.

In another study, a ping-pong Bi-Bi mechanistic approach was used to explain the kinetics of the synthesis of castor oil-derived biolubricant. Fluidized bed reactor in batch recirculation mode was used for the methanolysis of castor oil catalyzed by Novozym 435. Methyl ester yield of 99.6% was obtained with 1 g enzyme and methanol-to-oil molar ratio of 6:1 at 40°C with flow rate of 30 mL/min. The inhibition constant (K_i methanol: 317.11 mM < K_i oil: 923.23 mM) showed sensitivity of enzyme toward methanol inhibition. Higher affinity of enzyme toward castor oil as substrate was apparent by lower value of K_m (K_m oil: 90.07 mM < K_m methanol: 733.27 mM). The enzyme could be reused until 10 cycles without significant decrease in methanolysis activity [77]. Similar kinetic studies using Michaelis–Menten approach and ping-pong model to study reaction mechanism have been done by many researchers to optimize esterification conditions for biolubricant production [73, 78,79].

Immobilized lipases from different sources, e.g., CALB, *Pseudomomas cepacia*, and *Rhizomucor miehei*, have been tried in synthesis of rapeseed oil-based biolubricants. Rapeseed oil fatty acid methyl ester was transesterified with 2-EtH, NPG, and TMP in a solvent-free medium. CALB expressed as high as 98% conversion. The synthesized esters exhibited good viscosity properties, lower pour point, and higher thermo-oxidative stability [70].

Oleic acid-rich sunflower was used as substrate for biolubricant synthesis by Dossat et al. [75]. Lipozyme RM-IM was used for transesterification in a solvent-free system.

A process has been patented for the production of biolubricant from (i) castor oil methyl biodiesel and (ii) jatropha oil methyl biodiesel, by transesterification with TMP (EP Patent No. 2657324 A1). Conversions of 80% ester from castor oil biodiesel and 99% ester from jatropha oil were attained, which were much higher than that obtained via chemical synthesis (40–50%, respectively). Improved oxidation stability of the biolubricant, easy recovery of enzymatic catalyst, and milder conditions decreased the energy expenditure, leading to economic feasibility of process [80].

Lipases also offer an enzymatic route for the synthesis of estolides, which find use in a wide range of lubricant applications. Estolides are oligomeric esters obtained by linking carboxylic acid functionality of one fatty acid to the site of unsaturation of another fatty acid [81]. Aguieiras et al. [82] used immobilized commercial lipases (Novozym 435, Lipozyme RM-IM, and Lipozyme TL-IM) in a solvent-free medium for estolide synthesis from oleic acid and methyl ricinoleate.

11.6.4 LIPASE-CATALYZED EPOXIDATION OF PLANT OILS

As discussed previously, epoxidation of vegetable oils is one of the widely used methods to render good lubricant properties. Several chemical methods employing different catalysts can be used for epoxidation although Prileshajev epoxidation is applied on industrial scale [83,84]. Lipases can also catalyze epoxidation of fatty acids and thus compete with these conventional chemical routes. Epoxidation using lipase offers many advantages like formation of stable hydroperoxides directly from fatty acids, high conversions, and high regio- and stereoselectivity, and it limits side reactions, is environment friendly, and prevents equipment corrosion [85]. CALB has been widely used to catalyze the epoxidation of olefins, triacylglycerols, FFAs, and their corresponding alkyl esters. The epoxidation reaction involving lipase-catalyzed perhydrolysis step is shown in Figure 11.1 [86].

Klaas and Warwel [87] showed that Novozym 435 catalyzed perhydrolysis and formed peroxyfatty acids to give high epoxide yields via chemoenzymatic self-epoxidation. This chemoenzymatic

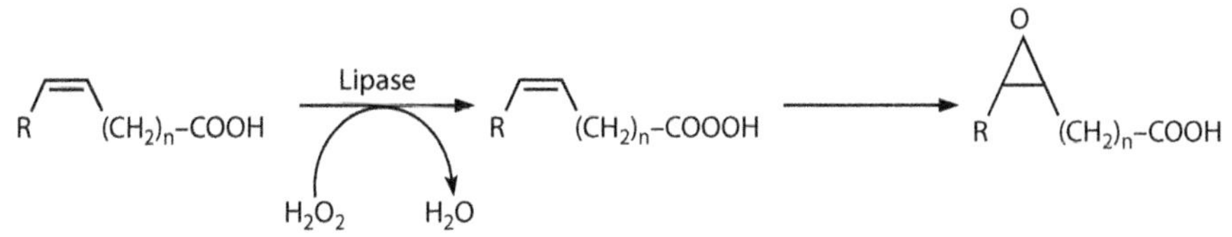

FIGURE 11.1 Reaction scheme of chemoenzymatic epoxidation.

process was highly selective and above 90% conversions were achieved with various plant oils like rapeseed, sunflower, soybean, and linseed oils. The oxirane oxygen content and the iodine values were accurately predicted along with production of tailor-made epoxidized plant oils by this method.

Vlcek and Petrovic [85] reported chemoenzymatic epoxidation of soybean oil by Novozym 435 and showed that the concentration of this biocatalyst greatly influenced the rate of epoxide formation. Over 90% conversion of epoxides was achieved after the optimization of reaction conditions.

Salimon et al. [76] also used Novozym 435 to synthesize monoepoxide biolubricant, i.e., 9(12)-10(13)-monoepoxy 12(9)-octadecanoic acid (MEOA) by chemoenzymatic epoxidation. MEOA showed improved physicochemical characteristics in flash point (128°C) and pour point (−41°C).

Hydrogen peroxide concentration and distribution, temperature, nature of reaction medium, and reactor design are the critical parameters among other various factors that need to be regulated to control reaction rate, degree of epoxidation, and activity of lipases. Although most of the studies have been done using organic solvents like toluene, lipase-mediated epoxidation in a solvent-free medium was carried out by Orellana-Coca et al. [88], making this process more green and environment friendly. Regulation of dosage of hydrogen peroxide is most important as lipase loses its stability in the presence of peroxide and increasing H_2O_2 concentration increases the rate of deactivation [89]. Hagström et al. [90] investigated various operating strategies for effective chemoenzymatic epoxidation but still a lot of work in improving enzymatic activity and stability is required for an economically feasible process.

11.7 FUTURE DIRECTIONS IN ENZYMATIC SYNTHESIS OF BIOLUBRICANTS

There are many limitations in using lipases for biolubricant synthesis in an industrial scale. The cost of using immobilized lipase enzyme for transesterification and epoxidation reactions is much higher than chemical catalysts. Also, loss of enzyme activity occurs in harsh synthesis conditions. Cost-effective immobilization procedures with suitable matrix and optimum amount of enzyme loading can be one of the key areas to be worked on. Various biotechnological innovations like site-directed mutagenesis, protein engineering, and directed evolution can be used to produce tailor-made lipases to selectively and efficiently catalyze hydrolysis, epoxidation, and transesterification of fatty acids.

11.8 CONCLUSION

Increased environmental awareness, stringent environmental regulations, and high cost of disposal of conventional nondegradable lubricants are promoting the development of biobased lubricants. Performance and economic viability remain key issues for increasing their market share. Other challenges like high cost, stability, and reusability of enzymes need to be addressed. Immobilization, site-directed mutagenesis, and protein engineering are being pursued to provide robust biocatalysts for biolubricant production. Interdisciplinary approach of chemical modification, green chemistry, and genetic modification are aimed at developing biobased, biodegradable lubricants from renewable feedstocks using low-energy intensity processes.

ACKNOWLEDGMENT

The financial grant provided by the Department of Biotechnology (Government of India) is gratefully acknowledged.

REFERENCES

1. C. O. Akerman, Y. Gaber, N. A. Ghani, M. Lamsa, and R. Hatti-Kaul, Clean synthesis of biolubricants for low temperature applications using heterogeneous catalysts, *J. Mol. Catal. B Enzym.*, 72, 263–269 (2011).
2. A. K. Jain and A. Suhane, Research approach & prospects of non edible vegetable oil as a potential resource for biolubricant: A review, *Adv. Eng. Appl. Sci.: Int. J.*, 1, 23–32 (2012).
3. J. Salimon, B. M. Abdullah, R. M. Yusop, and N. Salih, Synthesis, reactivity and application studies for different biolubricants, *Chem. Cent. J.*, 8, 16 (2014).
4. N. J. Novick, P. G. Mehta, and P. B. McGoldrick, Assessment of the biodegradability of mineral oil and synthetic ester base stocks, using CO_2 ultimate biodegradability tests and CEC-L-33-T-82, *J. Synth. Lubr.*, 13, 19–30 (1996).
5. N. S. Battersby, The biodegradability and microbial toxicity testing of lubricants—Some recommendations, *Chemosphere*, 41, 1011–1027 (2000).
6. M. Nosonovsky and B. Bhushan, Green tribology: Principles, research areas and challenges, *Phil. Trans. R. Soc. A*, 368, 4677–4694 (2010).
7. S. Asadauskas, J. M. Perez, and J. L. Duda, Oxidative stability and antiwear properties of high oleic vegetable oils, *Lubr. Eng.*, 52, 877–882 (1996).
8. S. Z. Erhan and S. Asadauskas, Lubricant basestocks from vegetable oils, *Ind. Crops Prod.*, 11, 277–282 (2000).
9. P. Nagendramma and S. Kaul, Development of ecofriendly/biodegradable lubricants: An overview, *Renew. Sust. Energy Rev.*, 16, 764–774 (2012).
10. S. J. Randles, Environmentally considerate ester lubricants for the automotive and engineering industries, *J. Synth. Lubr.*, 9, 145–161 (1992).
11. K. R. Sathwik Chatra, N. H. Jayadas, and S. Kailas, Natural oil-based lubricants, in: *Green Tribology*, M. Nosonovsky and B. Bhushan (Eds.), pp. 287–328, Springer, Heidelberg (2012).
12. L. A. T. Honary and E. Richter, Genetic modification and industrial crops, in: *Biobased Lubricants and Greases*, L. A. T. Honary and E. Richter, pp. 63–89, John Wiley & Sons, Chichester (2011).
13. I. Buzás, J. Simon, and J. Holló, Effect of the experimental conditions on the thermooxidative behaviour of vegetable oils, *J. Therm. Anal. Calorim.*, 12, 397–405 (1977).
14. J. Dweck and C. M. S. Sampaio, Analysis of the thermal decomposition of commercial vegetable oils in air by simultaneous TG/DTA, *J. Therm. Anal. Calorim.*, 75, 385–391 (2004).
15. N. J. Fox, B. Tyrer, and G. W. Stachowiak, Boundary lubrication performance of free fatty acids in sunflower oil, *Tribol. Lett.*, 16, 275–281 (2004).
16. J. Santos, I. Santos, M. Conceição, S. Porto, M. Trindade, A. Souza, S. Prasad, V. Fernandes, and A. Araújo, Thermoanalytical, kinetic and rheological parameters of commercial edible vegetable oils, *J. Therm. Anal. Calorim.*, 75, 419–428 (2004).
17. S. Z. Erhan, B. K. Sharma, and J. M. Perez, Oxidation and low temperature stability of vegetable oil-based lubricants, *Ind. Crops Prod.*, 24, 292–299 (2006).
18. A. M. Petlyuk and R. J. Adams, Oxidation stability and tribological behavior of vegetable oil hydraulic fluids, *Tribol. T.*, 47, 182–187 (2004).
19. M. P. Schneider, Plant-oil-based lubricants and hydraulic fluids, *J. Sci. Food. Agric.*, 86, 1769–1780 (2006).
20. B. K. Sharma, J. M. Perez, and S. Z. Erhan, Soybean oil-based lubricants: A search for synergistic antioxidants, *Energ. Fuel*, 21, 2408–2414 (2007).
21. H. Wagner, R. Luther, and T. Mang, Lubricant base fluids based on renewable raw materials: Their catalytic manufacture and modification, *Appl. Catal. A: Gen.*, 221, 429–442 (2001).
22. A. Adhvaryu and S. Z. Erhan, Epoxidized soybean oil as a potential source of high-temperature lubricants, *Ind. Crops Prod.*, 15, 247–254 (2002).
23. B. K. Sharma, Z. Liu, A. Adhvaryu, and S. Z. Erhan, One-pot synthesis of chemically modified vegetable oils, *J. Agric. Food. Chem.*, 56, 3049–3056 (2008).

24. H.-S. Hwang, A. Adhvaryu, and S. Erhan, Preparation and properties of lubricant basestocks from epoxidized soybean oil and 2-ethylhexanol, *J. Am. Oil Chem. Soc.*, 80, 811–815 (2003).

25. C. S. Madankar, A. K. Dalai, and S. N. Naik, Green synthesis of biolubricant base stock from canola oil, *Ind. Crops Prod.*, 44, 139–144 (2013).

26. M. Greaves, E. Zaugg-Hoozemans, N. Khelidj, R. van Voorst, and R. Meertens, Performance properties of oil-soluble synthetic polyalkylene glycols, *Lubr. Sci.*, 24, 251–262 (2012).

27. M. Brown, J. D. Fotheringham, T. J. Hoyes, R. M. Mortier, S. T. Orszulik, S. J. Randles, and P. M. Stroud, Synthetic base fluids, in: *Chemistry and Technology of Lubricants*, R. M. Mortier, M. F. Fox, and S. T. Orszulik (Eds.), pp. 35–74, Springer, Dordrecht (2010).

28. K. Brown, Making good environmental choices for lubricants, *Machinery Lubrication*, http://www .machinerylubrication.com/Read/795/environmental-choices-lubricants (September 2005) (Accessed August 20, 2015).

29. R. R. Leslie and J. B. Wilfried, Comparison of synthetic, mineral oil, and bio-based lubricant fluids, in: *Synthetics, Mineral Oils, and Bio-Based Lubricants*, R. R. Leslie (Ed.), pp. 347–366, CRC Press, Boca Raton, FL (2013).

30. H. M. Mobarak, E. Niza Mohamad, H. H. Masjuki, M. A. Kalam, K. A. H. Al Mahmud, M. Habibullah, and A. M. Ashraful, The prospects of biolubricants as alternatives in automotive applications, *Renew. Sust. Energy Rev.*, 33, 34–43 (2014).

31. Y. Y. Linko, T. Tervakangas, M. Lämsä, and P. Linko, Production of trimethylolpropane esters of rapeseed oil fatty acids by immobilized lipase, *Biotechnol. Tech.*, 11, 889–892 (1997).

32. E. Uosukainen, Y.-Y. Linko, M. Lämsä, T. Tervakangas, and P. Linko, Transesterification of trimethylolpropane and rapeseed oil methyl ester to environmentally acceptable lubricants, *J. Am. Oil Chem. Soc.*, 75, 1557–1563 (1998).

33. S. Gryglewicz, W. Piechocki, and G. Gryglewicz, Preparation of polyol esters based on vegetable and animal fats, *Bioresour. Technol.*, 87, 35–39 (2003).

34. R. Yunus, A. Fakhru'l-Razi, T. L. Ooi, S. E. Iyuke, and J. M. Perez, Lubrication properties of trimethylolpropane esters based on palm oil and palm kernel oils, *Eur. J. Lipid Sci. Technol.*, 106, 52–60 (2004).

35. K. V. Padmaja, B. V. S. K. Rao, R. K. Reddy, P. S. Bhaskar, A. K. Singh, and R. B. N. Prasad, 10-Undecenoic acid-based polyol esters as potential lubricant base stocks, *Ind. Crops Prod.*, 35, 237–240 (2012).

36. K. Kamalakar, A. K. Rajak, R. B. N. Prasad, and M. S. L. Karuna, Rubber seed oil-based biolubricant base stocks: A potential source for hydraulic oils, *Ind. Crops Prod.*, 51, 249–257 (2013).

37. P.-Y. Stergiou, A. Foukis, M. Filippou, M. Koukouritaki, M. Parapouli, L. G. Theodorou, E. Hatziloukas, A. Afendra, A. Pandey, and E. M. Papamichael, Advances in lipase-catalyzed esterification reactions, *Biotechnol. Adv.*, 31, 1846–1859 (2013).

38. F. R. Abreu, D. G. Lima, E. H. Hamú, C. Wolf, and P. A. Z. Suarez, Utilization of metal complexes as catalysts in the transesterification of Brazilian vegetable oils with different alcohols, *J. Mol. Catal. A: Chem.*, 209, 29–33 (2004).

39. J. A. C. da Silva, A. C. Habert, and D. M. G. Freire, A potential biodegradable lubricant from castor biodiesel esters, *Lubr. Sci.*, 25, 53–61 (2013).

40. J. Rocha, M. Gil, and F. Garcia, Optimisation of the enzymatic synthesis of *n*-octyl oleate with immobilised lipase in the absence of solvents, *J. Chem. Technol. Biotechnol.*, 74, 607–612 (1999).

41. M. L. Foresti and M. L. Ferreira, Solvent-free ethyl oleate synthesis mediated by lipase from *Candida antarctica* B adsorbed on polypropylene powder, *Catal. Today*, 107–108, 23–30 (2005).

42. A. Richetti, S. F. Leite, O. C. Antunes, L. Lerin, R. Dallago, D. Emmerich, M. Di Luccio, J. Vladimir Oliveira, H. Treichel, and D. de Oliveira, Assessment of process variables on 2-ethylhexyl palmitate production using Novozym 435 as catalyst in a solvent-free system, *Bioprocess Biosyst. Eng.*, 33, 331–337 (2010).

43. K.-E. Jaeger, B. W. Dijkstra, and M. T. Reetz, Bacterial biocatalysts: Molecular biology, three-dimensional structures, and biotechnological applications of lipases, *Annu. Rev. Microbiol.*, 53, 315–351 (1999).

44. M. Kapoor and M. N. Gupta, Lipase promiscuity and its biochemical applications, *Process Biochem.*, 47, 555–569 (2012).

45. R. D. Schmid and R. Verger, Lipases: Interfacial enzymes with attractive applications, *Angew. Chem. Int. Ed.*, 37, 1608–1633 (1998).

46. P. Adlercreutz, Immobilisation and application of lipases in organic media, *Chem. Soc. Rev.*, 42, 6406–6436 (2013).

47. C. D. Anobom, A. S. Pinheiro, R. A. De-Andrade, E. C. G. Aguieiras, G. C. Andrade, M. V. Moura, R. V. Almeida, and D. M. Freire, From structure to catalysis: Recent developments in the biotechnological applications of lipases, *Biomed. Res. Int.*, Article ID 684506 (2014).

48. S. Hari Krishna and N. G. Karanth, Lipases and lipase-catalyzed esterification reactions in nonaqueous media, *Catal. Rev.*, 44, 499–591 (2002).

49. C. G. Laudani, M. Habulin, Ž. Knez, G. D. Porta, and E. Reverchon, Immobilized lipase-mediated long-chain fatty acid esterification in dense carbon dioxide: Bench-scale packed-bed reactor study, *J. Supercrit. Fluids*, 41, 74–81 (2007).

50. T. Bányai, K. Bélafi-Bakó, N. Nemestóthy, and L. Gubicza, Biolubricant production in ionic liquids by enzymatic esterification, *Hung. J. Ind. Chem.*, 39, 395–399 (2011).

51. F. Hasan, A. A. Shah, and A. Hameed, Industrial applications of microbial lipases, *Enzyme Microb. Tech.*, 39, 235–251 (2006).

52. J. A. C. da Silva, Biodegradable lubricants and their production via chemical catalysis, in: *Tribology—Lubricants and Lubrication*, C.-H. Kuo (Ed.), pp. 185–200, INTECH Open Access (2011).

53. A. Chowdhury, D. Mitra, and D. Biswas, Biolubricant synthesis from waste cooking oil via enzymatic hydrolysis followed by chemical esterification, *J. Chem. Technol. Biotechnol.*, 88, 139–144 (2013).

54. J. Trivedi, M. Aila, C. Sharma, P. Gupta, and S. Kaul, Clean synthesis of biolubricant range esters using novel liquid lipase enzyme in solvent free medium, *SpringerPlus*, 4, 165 (2015).

55. P. Tufvesson, U. Tornvall, J. Carvalho, A. J. Karlsson, and R. Hatti-Kaul, Towards a cost-effective immobilized lipase for the synthesis of specialty chemicals, *J. Mol. Catal. B: Enzym.*, 68, 200–205 (2011).

56. E. M. Anderson, K. M. Larsson, and O. Kirk, One biocatalyst—Many applications: The use of *Candida antarctica* B-lipase in organic synthesis, *Biocatal. Biotransform.*, 16, 181–204 (1998).

57. A. B. Martins, A. M. da Silva, M. F. Schein, C. Garcia-Galan, M. A. Zachia Ayub, R. Fernandez-Lafuente, and R. C. Rodrigues, Comparison of the performance of commercial immobilized lipases in the synthesis of different flavor esters, *J. Mol. Catal. B: Enzym.*, 105, 18–25 (2014).

58. J. Uppenberg, S. Patkar, T. Bergfors, and T. A. Jones, Crystallization and preliminary X-ray studies of lipase b from *Candida antarctica*, *J. Mol. Biol.*, 235, 790–792 (1994).

59. L. Peng, X. Xu, H. Mu, C.-E. Høy, and J. Adler-Nissen, Production of structured phospholipids by lipase-catalyzed acidolysis: Optimization using response surface methodology, *Enzyme Microb. Technol.*, 31, 523–532 (2002).

60. B. Huge-Jensen, D. Galluzzo, and R. Jensen, Partial purification and characterization of free and immobilized lipases from *Mucor miehei*, *Lipids*, 22, 559–565 (1987).

61. J. Oh, S. Yang, C. Kim, I. Choi, J. H. Kim, and H. Lee, Synthesis of biolubricants using sulfated zirconia catalysts, *Appl. Catal. A: Gen.*, 455, 164–171 (2013).

62. V. R. Murty, J. Bhat, and P. K. A. Muniswaran, Hydrolysis of oils by using immobilized lipase enzyme: A review, *Biotechnol. Bioprocess Eng.*, 7, 57–66 (2002).

63. M. T. S. Syaima, K. H. Ong, I. M. Noor, M. I. M. Zamratul, S. A. Brahim, and M. M. Hafizul, The synthesis of bio-lubricant based oil by hydrolysis and non-catalytic of palm oil mill effluent (POME) using lipase, *Renew. Sust. Energy Rev.*, 44, 669–675 (2015).

64. A. Chowdhury, R. Chakraborty, D. Mitra, and D. Biswas, Optimization of the production parameters of octyl ester biolubricant using Taguchi's design method and physico-chemical characterization of the product, *Ind. Crops Prod.*, 52, 783–789 (2014).

65. M. I. M. Zamratul, M. T. S. Syaima, I. M. Noor, and W. M. W. T. Rifdi, Development of bio-lubricant from *Jatropha curcas* oils, *Int. J. Res. Chem. Metall. Civ. Eng.*, 1 (1), 10–12 (2014).

66. J. Madarasz, D. Nemeth, J. Bakos, L. Gubicza, and P. Bakonyi, Solvent-free enzymatic process for biolubricant production in continuous microfluidic reactor, *J. Clean. Prod.*, 93, 140–144 (2015).

67. D. Malhotra, J. Mukherjee, and M. N. Gupta, Lipase catalyzed transesterification of castor oil by straight chain higher alcohols, *J. Biosci. Bioeng.*, 119, 280–283 (2015).

68. E. Kleinaitė, V. Jaška, B. Tvaska, and I. Matijošytė, A cleaner approach for biolubricant production using biodiesel as a starting material, *J. Clean. Prod.*, 75, 40–44 (2014).

69. M. Hajar and F. Vahabzadeh, Artificial neural network modeling of biolubricant production using Novozym 435 and castor oil substrate, *Ind. Crops Prod.*, 52, 430–438 (2014).

70. S. Gryglewicz, M. Muszyński, and J. Nowicki, Enzymatic synthesis of rapeseed oil-based lubricants, *Ind. Crops Prod.*, 45, 25–29 (2013).

71. Y. Tao, B. Chen, L. Liu, and T. Tan, Synthesis of trimethylolpropane esters with immobilized lipase from *Candida* sp. 99–125, *J. Mol. Catal. B: Enzym.*, 74, 151–155 (2012).

72. M. Happe, P. Grand, S. Farquet, S. Aeby, J.-C. Héritier, F. Corthay, E. Mabillard, R. Marti, E. Vanoli, and A.-F. Grogg, Microwave barrel reactor use in trimethylolpropane oleate synthesis by *Candida antarctica* lipase in a biphasic non-solvent process, *Green Chem.*, 14, 2337–2345 (2012).

73. A. Sengupta, T. Dey, M. Ghosh, J. Ghosh, and S. Ghosh, Enzymatic synthesis of furfuryl alcohol ester with oleic acid by *Candida antarctica* lipase b and its kinetic study, *J. Inst. Eng. India Ser. E*, 93, 31–36 (2012).

74. N. Dörmő, K. Bélafi-Bakó, L. Bartha, U. Ehrenstein, and L. Gubicza, Manufacture of an environmental-safe biolubricant from fusel oil by enzymatic esterification in solvent-free system, *Biochem. Eng. J.*, 21, 229–234 (2004).

75. V. Dossat, D. Combes, and A. Marty, Efficient lipase catalyzed production of a lubricant and surfactant formulation using a continuous solvent-free process, *J. Biotechnol.*, 97, 117–124 (2002).

76. J. Salimon, N. Salih, and B. M. Abdullah, Production of chemoenzymatic catalyzed monoepoxide biolubricant: Optimization and physicochemical characteristics, *J. Biomed. Biotechnol.*, 2012, Article ID 693848 (2012).

77. M. Hajar and F. Vahabzadeh, Modeling the kinetics of biolubricant production from castor oil using Novozym 435 in a fluidized-bed reactor, *Ind. Crops Prod.*, 59, 252–259 (2014).

78. A. Zaidi, J. L. Gainer, G. Carta, A. Mrani, T. Kadiri, Y. Belarbi, and A. Mir, Esterification of fatty acids using nylon-immobilized lipase in *n*-hexane: Kinetic parameters and chain-length effects, *J. Biotechnol.*, 93, 209–216 (2002).

79. N. Nemestóthy, L. Gubicza, E. Fehér, and K. Bélafi-Bakó, Biotechnological utilisation of fusel oil, a food industry by-product, *Food Technol. Biotechnol.*, 46, 44–50 (2008).

80. S. J. Da, D. Freire, A. Habert, and V. Soares, Process for the production of bio-lubricant from methyl biodiesel and bio-lubricant obtained by said process, EP 2657324 A1, Patents (2013).

81. S. C. Cermak and T. A. Isbell, Synthesis and physical properties of estolide-based functional fluids, *Ind. Crops Prod.*, 18, 183–196 (2003).

82. E. C. Aguieiras, C. O. Veloso, J. V. Bevilaqua, D. O. Rosas, M. A. da Silva, and M. A. Langone, Estolides synthesis catalyzed by immobilized lipases, *Enzyme Res.*, Article ID 432746 (2011).

83. B. M. Abdullah and S. Jumat, Epoxidation of vegetable oils and fatty acids: Catalysts, methods and advantages, *J. Applied Sci.*, 10, 1545–1553 (2010).

84. S. Tan and W. Chow, Biobased epoxidized vegetable oils and its greener epoxy blends: A review, *Polym.-Plast. Technol.*, 49, 1581–1590 (2010).

85. T. Vlcek and Z. Petrovic, Optimization of the chemoenzymatic epoxidation of soybean oil, *J. Am. Oil Chem. Soc.*, 83, 247–252 (2006).

86. C. Aouf, E. Durand, J. Lecomte, M.-C. Figueroa-Espinoza, E. Dubreucq, H. Fulcrand, and P. Villeneuve, The use of lipases as biocatalysts for the epoxidation of fatty acids and phenolic compounds, *Green Chem.*, 16, 1740–1754 (2014).

87. M. Rüsch gen. Klaas and S. Warwel, Complete and partial epoxidation of plant oils by lipase-catalyzed perhydrolysis1, *Ind. Crops Prod.*, 9, 125–132 (1999).

88. C. Orellana-Coca, U. Törnvall, D. Adlercreutz, B. Mattiasson, R. Hatti-Kaul, Chemo-enzymatic epoxidation of oleic acid and methyl oleate in solvent-free medium, *Biocatal. Biotransfor.*, 23, 431–437 (2005).

89. U. Törnvall, C. Orellana-Coca, R. Hatti-Kaul, and D. Adlercreutz, Stability of immobilized *Candida antarctica* lipase B during chemo-enzymatic epoxidation of fatty acids, *Enzyme Microb. Tech.*, 40, 447–451 (2007).

90. A. E. V. Hagström, U. Törnvall, M. Nordblad, R. Hatti-Kaul, and J. M. Woodley, Chemo-enzymatic epoxidation-process options for improving biocatalytic productivity, *Biotechnol. Progr.*, 27, 67–76 (2011).

12 Synthetic Methodologies of Ester-Based Eco-Friendly, Biodegradable Lubricant Base Stocks for Industrial Applications

Shailesh N. Shah, Sayanti Ghosh, Lalit Kumar, Vivek Rathore, Shivanand M. Pai, and Bharat L. Newalkar

CONTENTS

ABSTRACT

The ever-growing energy demand and strict implementation of environmental norms world-wide has added impetus for the adoption of eco-friendly lubricants in various industrial sectors. It has become the primary driver for the acceptance of ester-based lubricants for major applications such as refrigeration compressor oil, transformer oil, cutting oil, and fire-resistant hydraulic oil. This led to the successful development of vegetable oil- and polyol-based eco-friendly and biodegradable ester lube oil base stocks in the last two decades. This chapter reviews the synthesis chemistry of vegetable oil- and polyol-based ester lubricant oils for meeting the stringent product specifications desired in the targeted application. The synthesis methodologies and the production processes are also discussed. This chapter also stresses future challenges of ester-based lubricant base stocks for meeting the potential demands in various industrial sectors.

12.1 INTRODUCTION

Growing environmental concerns and stringent environmental norms in recent years have led to increased interest in the adoption of biodegradable lubricants in various industrial applications [1]. This has led to the development of biodegradable cutting, transformer, and hydraulic oils in the last decade. Likewise, the adoption of the Kyoto protocol, by developed and developing countries, has added impetus for the adoption of polyolester-based lubricants for their excellent compatibility with hydrofluorocarbon (HFC)-based refrigerants which has been recommended, under the Kyoto protocol, to counter ozone depletion resulting from the use of chlorofluorocarbon (CFC)-based refrigerants.

Typically, vegetable oil- and polyolester-based lubricants offer excellent features with respect to the following:

- High VI > 200
- High flash point
- High degree of biodegradability

Furthermore, fatty acid-based ester lubricants offer a platform to develop base stocks with varied viscosities as per the application requirement. However, poor oxidation stability and poor cold-flow properties of vegetable oil-based base stocks have often limited their potential for high- and low-temperature applications vis-à-vis mineral oil base stocks. Recently Shah et al. have reported glycerol triester derivatives as diluent to improve the low-temperature properties of vegetable oils [2].

Accordingly, numerous research and development efforts have been focused in the past two decades toward the development of technically superior vegetable oil- and polyolester-based lubricant base stocks. In this context, efforts have been made to develop (i) genetically modified varieties of high-oleic vegetable oils, (ii) vegetable oil-based esters with high oxidation stability, and (iii) transesterification pathways [3–9] for fatty acid alkyl ester-based base stocks. However, the worldwide debate about food versus fuel has forced scientists and technologists to pay attention to nonedible-based base stock [10], (iv) polyolester-based esters with excellent low-temperature compatibility. Thus, the present chapter deals with a review of synthesis methodologies developed to produce the aforementioned biodegradable base stocks. The present chapter covers the following synthesis methodologies:

1. Transesterification and esterification of vegetable oil- and biodiesel-based base stocks
2. Selective hydrogenation
3. Epoxidation and ring opening
4. Oligomerization of vegetable oil-derived fatty acids for estolide formation
5. Polyolester-based esterification

In light of developed methodologies, the scale-up operations are also discussed and finally the emerging technical challenges have been pointed out.

12.2 BIODEGRADABLE LUBRICANT BASE STOCKS

Generally, lubricants which are capable of naturally breaking down into common elements and assimilating back into the environment are termed as *biodegradable lubricant base stocks* [11]. The degree of biodegradability is typically measured by standard test protocols as defined under CEC-L-33-A-93 [12], ASTM D6384 [13], and OECD 301B methods [13]. Estimated biodegradabilities of various base stocks are compiled in Table 12.1.

As shown in Table 12.1 [12], it is evident that vegetable oil- and ester-based base stocks are preferred choice for the development of eco-friendly biodegradable lubricants. This is mainly due to the presence of ester groups $-O-C=O$ (Figure 12.1), which is prone toward biodegradation.

Accordingly, the major ester-based lubricant base stocks can be classified in following important categories.

12.2.1 VEGETABLE OIL-BASED AND RELATED BASE STOCKS

Chemically, the vegetable oil-based base stocks constitute a triglyceride moiety (Figure 12.2) which primarily comprises esters of fatty acids ($C_{12}-C_{22}$) linked to a glycerol backbone. Apart from the ester group, it consists of unsaturation ($-C=C-$) and even hydroxyl groups depending on the composition of fatty acids.

The reason for the thermal and oxidative instability of vegetable oils is the β-CH group of the alcoholic component and the presence of double bonds in the fatty acid. Double bonds in alkenyl chains are especially reactive and undergo oxidation in air. The β-hydrogen atom is easily eliminated

TABLE 12.1
Biodegradability Percentages of Different Lubricant Base Stocks

Serial Number	Base Stock	% Biodegradability (as per CEC-L-33-A-93)
1	Mineral oil	20–40
2	Vegetable oil	90–98
3	Esters	75–100
4	Polyolesters	70–100

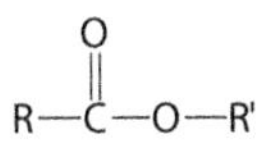

$$\underset{\text{R—C—O—R'}}{\overset{\overset{\textstyle O}{\|}}{}}$$

FIGURE 12.1   Carboxylate ester; R and R′ denotes any alkyl or aryl group.

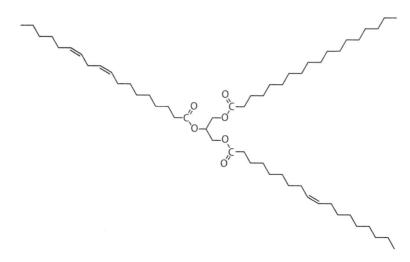

FIGURE 12.2   Molecular structure of a typical triglyceride moiety.

from the molecular structure [14]. This leads to the cleavage of the esters into acid and olefin. Low hydrolytic stability is a further limitation to the natural esters for their direct use as lubricants.

12.2.2   Polyolester-Based Base Stocks

Polyolesters are a class of synthetic lubricants which offer numerous performance advantages over mineral oil-based lubricants [15]. These esters are formed by esterifying polyhydric alcohols, also called *polyols*, like NPG, trimethylol ethane, TMP, PE, and dipentaerythritol (DiPE), with carboxylic acids. The unique feature of the neopentyl structure of polyol alcohols molecules is the fact that there are no hydrogens on the beta-carbon. Since this beta-hydrogen is the first site of thermal attack, eliminating it substantially improves the thermal stability of polyolesters and allows them to be used at much higher temperatures [16]. Mixtures of different polyols and different monocarboxylic acids can also be used to synthesize complex esters. Also, simple esters can be blended to produce a lubricant composition suitable for a particular application. Thus, different polyolesters can be tailored as per the application needs. This makes polyolesters an attractive alternative to mineral oils and gives them a wide scope in the lubrication area [17]. In view of the above, various synthesis methodologies have been evolved in the last two decades to tailor the properties of aforementioned base stocks for various applications. These methodologies are reviewed in the following sections.

12.3   SYNTHESIS OF VEGETABLE OIL-BASED BASE STOCK

Vegetable oil-based lubricants are obtained by modification of vegetable oils to meet the standard specifications for the particular application they are being prepared for. Vegetable oils can be modified by various chemical reaction pathways such as direct transesterification, hydrogenation, epoxidation followed by ring opening of unsaturated bonds, and oligomerization of the fatty acid to form estolides (Figure 12.3).

In view of the poor oxidation stability as well as cold-flow properties of vegetable oil-based base stocks, commonly used chemical modification routes are discussed in this section.

Modification routes are classified into the following groups:

1. Esterification and transesterification
2. Selective hydrogenation
3. Dimerization and oligomerization

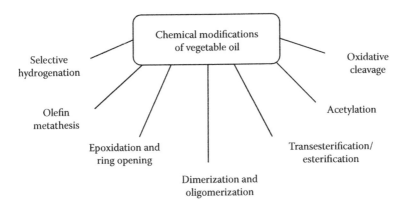

FIGURE 12.3   Various chemical modification routes for vegetable oil-based base stocks.

4. Epoxidation
5. Oxidative cleavage
6. Estolide synthesis

The modifications mentioned above are discussed extensively in the following sections.

12.3.1   ESTERIFICATION AND TRANSESTERIFICATION

One of the most important modifications of the carboxyl group of the fatty acid chain is the transesterification of the triglyceride molecule (Figure 12.4). The transesterification of vegetable oil or vegetable oil-derived fatty acids with short-chain alcohols such as methanol or ethanol leads to biodiesel, while biolubricants are obtained when higher alcohols (C_6–C_8) and polyols are used for transesterification [18]. The transesterification with a triglyceride involves two steps: first, where the triglyceride is converted to a methyl ester usually at temperatures between 50°C and 60°C, followed by the reaction of the fatty acid methyl ester with the higher alcohol at 110–160°C [19]. This is mainly due to the poor yield of the desired product via direct transesterification of vegetable oil with C_6–C_8 alcohols.

A simplified scheme for the combined production of biodiesel and biolubricant is shown in Figure 12.5 [20].

The transesterification reactions are carried out using homogeneous or heterogeneous catalysts. The reaction is faster when a homogeneous acid catalyst such as sulfuric acid (H_2SO_4), p-toluenesulfonic acid, or phosphoric acid is used. However, with homogeneous catalysts, it is difficult to separate the catalyst from the products and the liquid catalysts often cause reactor corrosion [21,22]. On the other hand, heterogeneous acid catalysts are noncorrosive and environmentally benign and present fewer disposal problems. They are more easily separated from the products and can be

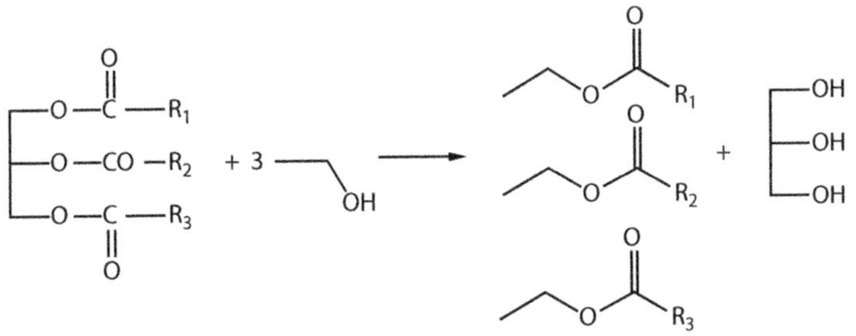

FIGURE 12.4   Synthesis of methyl/ethyl esters via transesterification of vegetable oil.

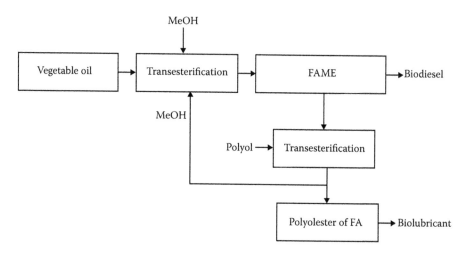

FIGURE 12.5 Simplified flow diagram illustrating the combined production of biodiesel and biolubricant from vegetable oil as starting material.

designed to yield higher activity, selectivity, and longer catalyst lifetime [23]. There have also been several reports on the use of heterogeneous base or acid catalysts [24–29]. Sreeprasanth et al. [30] have synthesized biolubicants by esterification of fatty acids and also by direct transesterification of vegetable oil using Fe–Zn-based double metal cyanide catalysts in the range of 80°C to 170°C. Bokade and Yadav [31] have synthesized biolubricants by transesterification of vegetable oils in the presence of heteropoly acid supported on K-10 clay at 170°C with vegetable oil-to-alcohol ratio of 1:15. They reported a decrease in the oil conversion and an increase in ester selectivity with increase in the carbon chain length of alcohol. This may be due to the reduction in the rate of reaction with increased carbon number of alcohol. Oha et al. [32] have synthesized biolubricants by esterification of oleic acid and transesterification of soybean oil with C_8 alcohol at 140°C using sulfated zirconia catalyst. The soybean oil conversion was 87.8% with the ester yield of 84.6% in the case of direct transesterification. Transesterification reaction of palm oil methyl ester with TMP or palm kernel oil methyl esters produced 98% yield of triester [33]. Transesterification reaction between TMP and palm kernel oil methyl esters has been carried out at 130°C and 20 mbar pressure for 1 hour using 0.8% sodium methoxide catalyst [34]. Another study [35] showed a 99% triester from esterification reaction of RMe with TMP. The reaction was carried out at 110–120°C for 10 hours at 3.3 kPa pressure with 0.5% catalyst sodium methoxide. Recently, alkali-catalyzed production of a jatropha-based lubricant was reported by transesterification at 65°C of oil to a jatropha methyl ester, followed by transesterification (at 150°C, 10 mbar) of jatropha methyl ester with TMP (at jatropha methyl ester/TMP molar ratio of 4:1) at a maximum yield of 50% [36]. Esterification of TMP was carried out with fatty acids from *Jatropha curcas* oil at 150°C for 3 hours at molar ratio of fatty acid/TMP of 4:1, and 2% (w/w) catalyst (based on weight of fatty acid). H_2SO_4 was used as the catalyst in this reaction. The yield of the product was 55% [37]. Furthermore, the effect of various homogeneous catalysts like perchloric acid, sulfuric acid, *p*-toluenesulfonic acid, HCl, and HNO_3 on the esterification of TMP with fatty acids from *Jatropha curcas* oil has been investigated [38]. The results (Table 12.2) show that the yield is a function of the catalyst used for esterification of TMP. It was found that perchloric acid has 70% yield of TMP ester, which was much higher than that of H_2SO_4, *p*-toluenesulfonic acid, hydrochloric acid, or nitric acid. Therefore, perchloric acid is a preferred catalyst for the esterification of TMP ester to increase biolubricant production of the reaction [38].

Lubricant-grade esters have been synthesized by transesterification of sesame oil methyl ester with TMP using base catalyst [20]. Lanthanum- and zinc-incorporated hydrotalcites have been used as base catalysts for production of biodiesel and biolubricant by transesterification of soybean oil

TABLE 12.2
Percentage Yields of TMP Ester Based on Catalyst Used[a]

S. No.	Catalyst	Amount of Catalyst (% w/w)	Yield[b] (%)
1	Perchloric acid	2	70
2	Sulfuric acid	2	46
3	p-Toluenesulfonic acid	2	42
4	Hydrochloric acid	2	41
5	Nitric acid	2	37

Source: Arbain, N. H., and J. Salimon, *E. J. Chem.*, 8, 33–40, 2011. With permission.
[a] Reaction conditions—temperature: 150°C; time: 3 hours; molar ratio of fatty acid/TMP: 4:1; 2% (w/w) catalyst (based on weight of fatty acid).
[b] The percentage yield is based on the weight of the product (% w/w).

with methanol and *n*-octanol [39]. It was concluded that the catalyst doped with 20 mol% lanthanum had the highest transesterification activity for methanol as well as *n*-octanol. Comparable conversions were obtained for both transesterifications at 150°C and 190°C in 4 h at oil-to-methanol molar ratio of 1:15 and oil-to-*n*-octanol molar ratio of 1:9 (*n*-octanol). Transesterification has been reported between a short-chain fatty acid ester (methyl esters and polyolesters) and a vegetable oil in the presence of a suitable catalyst. Such a route results in a random distribution of fatty acids in polyol and glycerol backbones. Attempts to adopt such a route to produce such transesterified vegetable oils have been reported [40–42]. It is worthwhile to note that such an approach enhances the oxidative stability and low-temperature and friction properties of a modified vegetable oil [42].

12.3.2 SELECTIVE HYDROGENATION

Hydrogenation is the conversion of polyunsaturated fatty acid chains, present in vegetable oils, by hydrogen into monounsaturated fatty acids. Complete hydrogenation improves the oxidation stability of the vegetable oils. However, it causes the decrease in the low-temperature fluidity of the modified oils or fatty acids. By partially hydrogenating the fatty acid chains, it is possible to improve the oxidation stabilities of the vegetable oils without affecting their low-temperature fluidity too much. Vegetable oils such as soybean oil, sunflower oil, linseed oil, and rapeseed oil contain polyunsaturated linoleic and linolenic acids and thus have poor thermo-oxidative stability.

Vegetable oils containing polyunsaturated acids can be converted into high monounsaturated oleic acid content oils by selective hydrogenation of the double bonds present in their molecules. Hydrogen atoms attached to the carbon atom connected at its two ends to the carbon atoms with double bonds are *bis*-allylic protons and highly reactive due to poor thermal and oxidation stabilities (Figure 12.6). Linolenic acid chain in soybean oil, as represented in Figure 12.6, contain *bis*-allylic protons [43]. As multiple double bonds of fatty acid chains of vegetable oils are saturated and converted into mono double bond chains, they carry less *bis*-allylic protons, leading to an improvement in the thermo-oxidative stability of the vegetable oils. Also, as the monounsaturated oleic acids are less corrosive in nature compared to linoleic and linolenic fatty acids, selective hydrogenation improves the corrosion resistance of the vegetable oils and makes them suitable for lubricant applications which demand high corrosion-resistant oils [44,45]. However, selective hydrogenation can lead to the formation of *trans*-isomers for monounsaturated fatty acid chain. This is in contrast with the naturally occurring fatty acids which have only *cis*-conformation, due to which they have lower pour point and are liquid at ambient conditions [46]. Therefore, it is of utmost importance to perform partial hydrogenation to achieve *cis*-conformation for fatty acid chains to exploit the treated

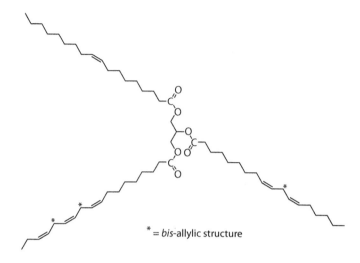

FIGURE 12.6   The *bis*-allylic protons present in soybean oil.

oil for lubricant applications (Table 12.3). Traditionally, vegetable oil hydrogenation at industrial scale is carried out using nickel catalyst on natural silicate–diatomite or silicate support in the range 140°C to 230°C [47]. Performance testing of natural silicate–diatomite supported nickel catalyst for soybean and sunflower oil hydrogenation in a pilot plant and an industrial reactor at between 150°C to 200°C and 0.5 to 2 bar showed high selectivity for monounsaturated oleic acids with 45% (w/w) *trans*-isomers and 55% (w/w) *cis*-isomers [48]. Owing to the higher selectivity of hydrogenating linolenic acid, copper–chromite catalysts at high temperature and pressure are also commercially employed for the selective hydrogenation of soybean oil [48]. Heterogeneous catalyst systems containing noble metals (Pd, Pt, Ru) supported over silica demand gentle hydrogenation conditions. For example, vegetable oil hydrogenations are carried at 40°C under 10 bar hydrogen pressure [49,50]. Precious metal catalysts like platinum, palladium, and ruthenium at low temperatures about 70°C decreased *trans* fatty acids in the hydrogenated vegetable oils compared to conventional nickel catalyst. Hydrogenation of soybean oil at 60°C and 7 bar in the presence of 50 to 65 ppm by palladium catalysts gave the same composition of the hydrogenated oil as obtained from nickel as catalyst. Palladium catalyst did not offer any benefits in reducing the *trans* fatty acids. However, platinum on narrow pore carbon support was found to be more effective in reducing the *trans* fatty acids, but the amounts of saturated acids in hydrogenated oils were much higher than nickel, palladium, and ruthenium catalysts [51]. Boron-modified palladium catalyst (Pd-B/γ-Al$_2$O$_3$) gave better catalytic activity and *trans*-selectivity than Pd/γ-Al$_2$O$_3$ for palm oil hydrogenation reaction at 120°C and 5 bar for 1 h reaction time [52]. Soybean oil was reduced at 5 bar hydrogen pressure and 170–200°C over catalysts (5% Pt/C and 23% Ni/SiO$_2$), which resulted in 97% reduction in linolenic content (three double bonds) and 15% reduction in linoleic content (two double bonds) to increase the oleic content by 26% (w/w) to reach 44% (w/w). However, a remarkable amount of stearic acid was formed over 5% Pt/C, while no stearic acid formation was observed for 23% Ni/SiO$_2$ [53]. Low-temperature (70°C) electrolytic hydrogenation of soybean oil has been reported at atmospheric pressure in the presence of Raney nickel powder cathode. This method resulted in significantly lower transformations of the oil into stearic acid and *trans* fatty acid [54].

Catalytic transfer hydrogen, an alternative method for vegetable oil hydrogenation, has been studied [49]. In this method, an organic molecule such as limonene is used as a hydrogen donor solvent and Pd/C as catalysts at 178°C and ambient pressure. Soybean oil was hydrogenated using homogeneous methyl benzoate–chromium tricarbonyl catalyst in the range of 145°C to 175°C and 35 bar hydrogen for 4–6 h [55]. This method resulted in a decrease of the linoleic of the oil from 55.3% (w/w) to 17.6% (w/w) and linolenic content from 8.2% (w/w) to 0.3% (w/w). On the other hand, the soybean,

TABLE 12.3

Literature Review on Partial Hydrogenation

S. No.	Substrate	Catalyst	Operating Conditions	Conclusion	Ref.
1	Vegetable oil	Nickel on natural silicate–diatomite or silicate support	140–230°C	Commercial process; nickel expressed less selectivity for *cis*-isomers	[47]
2	Soybean and sunflower oil	Natural silicate–diatomite supported nickel catalyst	150–200°C and 0.5–2 bar	Pilot and commercial process; high selectivity for monounsaturated oleic acids with 55% (w/w) *cis*-isomers	[58]
3	Soybean oil	Copper–chromite	High temperature and pressure	Commercial process	[48]
4	Vegetable oil	Noble metals (Pd, Pt, Ru) supported over silica	40°C under 10 bar		[49,50]
5	Vegetable oil	Metal catalysts like platinum, palladium, and ruthenium	70°C	Performance is better than nickel catalyst in reducing the *trans*-isomers	[51]
6	Soybean oil	Precious metals	60°C and 7 bar	Hydrogenated oil composition was same as obtained for nickel as catalyst	[51]
7	Palm oil	Boron-modified palladium catalyst (Pd-B/γ-Al$_2$O$_3$)	120°C and 5 bar	Better catalytic activity and *cis*-selectivity compared to Pd/γ-Al$_2$O$_3$	[52]
8	Soybean oil	5% Pt/C and 23% Ni/SiO$_2$ catalysts	170–200°C and 5 bar	Oleic acid content increased to 44% (w/w) from 18% (w/w) level; negligible *trans*-isomers	[53]
9	Soybean oil	Raney nickel powder cathode	70°C and atmospheric pressure	Significantly lower transformations of the oil into stearic acid and *trans* fatty acids	[54]
10	Vegetable oil	Limonene as hydrogen donor solvent and Pd/C as catalysts	178°C and ambient pressure		[49]
11	Soybean oil	Methyl benzoate-chromium tricarbonyl catalyst	145–175°C and 35 bar hydrogen pressure	Oil oleic content increased from 22.8% (w/w) to 67.2% (w/w), while very low drop in *cis*-isomer was observed from 94.1% (w/w) to 92.9% (w/w)	[55]
12	Vegetable oils	Supercritical hydrogenation using equimolar mixture of CO$_2$ and hydrogen in the presence of nickel catalyst	120–140°C and between 34 and 136 bar	*Trans* acid formation found to be significantly lower between 3.8% (w/w) and 6.4% (w/w) compared to 23.3% (w/w) for nonselective hydrogenation	[56]
13	Sunflower oil	Three phase-catalytic membrane impregnated with palladium	–	No saturation of oleic and elaidic acids monitored in the presence of linoleic acid but selectivity for *trans* fatty acids found to be high	[57]
14	Soybean oil	25% nanocrystalline Ni/SiO$_2$ catalyst	100–120°C and atmospheric pressure	Reduced the *trans* fatty acid formation by 20–25% (w/w) and increased soybean oil reduction up to 55–58%	[59]

oil oleic content increased from 22.8% (w/w) to 67.2% (w/w), while a very low drop in *cis*-isomer was observed from 94.1% (w/w) to 92.9% (w/w). Supercritical hydrogenation of vegetable oils has been performed using a mixture of CO_2 and hydrogen in the range of 120°C to 140°C and a 34 to 136 bar equimolar mixture of CO_2 and H_2 using conventional nickel catalyst [56]. The results were compared with nonselective hydrogenation at 35 bar H_2 pressure for the formation of *trans* acids and hydrogenation activity. *Trans* acid formation for the binary mixtures was significantly lower (3.8–6.4% w/w) compared with nonselective hydrogenation (23.3%) after 4 h of reaction. However, the presence of CO_2 is reported to decrease the hydrogenation activity. Continuous hydrogenation of sunflower oil using a novel three-phase catalytic membrane hydrogenation reactor has been carried out [57]. The membrane was impregnated with palladium, which was the active catalyst. No saturation of oleic and elaidic acids was observed in the presence of linoleic acid. However, the selectivity for *trans* fatty acids was found to be high. Catalytic hydrogenation of soybean oil at atmospheric pressure and at 100°C to 120°C in the presence of 25% nanocrystalline Ni/SiO_2 catalyst was reported [59]. It resulted in good catalytic activity, thermal stability, and selectivity better than those for commercially available nickel catalyst. It reduced *trans* fatty acid formation by 20–25% (w/w) and increased soybean oil reduction by 58%. Partially hydrogenated vegetable oils find applications in margarine and confectionary and also in frying and baking. Selectively hydrogenated vegetable oils into high-oleic content oils find application as electric insulation fluid in oil-filled power and distribution transformers due to their increased thermo-oxidative stability and less corrosive nature [60].

12.3.3 Dimerization and Oligomerization

Dimerization and oligomerization are other commercially feasible modifications of the fatty acid chain molecules that take place at the double bonds of unsaturated fatty acids. Dimerization can result in the formation of cyclic structures by Diels–Alder reaction or it may polymerize the unsaturated fatty acids and their esters into higher-molecular weight products.

12.3.3.1 Cyclic Dimerization/Oligomerization

Cyclization is the formation of cyclic structures by Diels–Alder reaction and is another option to eliminate the unsaturation present in fatty acids and fatty acid esters derived from vegetable oils. Cyclic structures can be formed by dimerizing conjugated fatty acids and their esters or modifying them with another acid or ester carrying a double bond. The conjugated acids or esters act as conjugated diene and other acids or esters acts as substituted alkene or alkyne as for Diels–Alder reaction. Linoleic acid dimers suitable for grease applications with 12% of trimer acids were obtained by heating the linoleic acid at 330°C to 360°C and 6 to 27 bar pressure in the presence of steam atmosphere [61]. Isomerized sunflower methyl esters are reported to be cooligomerized with indene at 40°C to 180°C using acid-activated bleaching earth as catalyst in the presence of hydroquinone [62]. Layered aluminosilicate catalysts (e.g., montmorillonite) also resulted in the cyclic dimerization/oligomerization of C_{18} fatty acids with one or more double bonds at 210–250°C [63,64]. The result was a complex mixture of C_{36} dicarboxylic acids and C_{54} trimer fatty acids. It is assumed that the reaction to the dimeric fatty acids (C_{36}) takes place via a Diels–Alder addition, whereby one fatty acid molecule (after double bond shifting) is the diene and the other is the dienophile [63]. A novel process for oligomerization of polyunsaturated conjugated or conjugatable fatty acids and fatty acid alkyl esters of sunflower seed oil at low temperatures has been reported [65]. It uses Tonsil-activated earth at 130°C to 160°C, for 1–4 hour reaction time and gave 70% formation of dimers and trimers [65]. Brönsted–Lewis acidic ionic liquids, such as (3-sulfonic acid)-propyl-triethylammonium-chlorozincinate, is reported to dimerize fatty acid methyl esters at 240°C during a 5-hour reaction time [66]. Recently, Liu and Shah [67] reported the dimerization of jojoba oil wax ester in the presence of a catalyst in supercritical CO_2—an environmentally friendly green solvent. The authors also reported euphorbia oil (having naturally occurring epoxide ring) polymerization with Lewis acid in carbon dioxide media [68].

12.3.3.2 Noncyclic Dimerization/Oligomerization

This type of dimerization or oligomerization results in the formation of noncyclic dimers, trimers, and higher oligomers from monomeric fatty acid and their esters. Dimerization/oligomerization of monounsaturated fatty acids such as oleic acid, erucic acid, elaidic acid, and undecylenic acid has been carried out at 200°C to 260°C in the presence of acidic catalyst and under steam atmosphere to yield 45–53% (w/w) polymeric acids [69]. The use of alkali to enhance the yield of polymeric acids to 60% has been reported [70]. In order to increase the reaction yield, a two-stage polymerization scheme for unsaturated fatty acids has been presented [71]. In this method, monomeric fatty acids are first polymerized in the presence of acidic montmorillonite clay and glacial acetic acid at 180°C to 260°C. In the second stage, unreacted monomeric fatty acids are polymerized again at 240°C for 4 hours in an autoclave in the presence of water and acid-treated montmorillonite clay. As a result, the concentration of the polymeric acids increased to 71% (w/w) with more than 50% (w/w) dimer acids.

12.3.4 Epoxidation

Epoxidation is an important chemical modification to improve the performance of oil/biodiesel pour point, flash point, viscosity, and oxidative stability [72]. An epoxide is cyclic ether with a three-atom ring system also known as an *oxirane ring*. It is highly strained, which makes it more reactive to form other various chemical compounds by ring opening, as shown in Figure 12.7.

Epoxidations are commonly employed using a peracid formed by reacting a carboxylic acid (usually acetic acid) with hydrogen peroxide (H_2O_2) as shown in Figure 12.8. In this reaction, hydrogen peroxide works as an oxygen donor and carboxylic acid as an active oxygen carrier in the presence of a catalytic amount of an inorganic acid like H_2SO_4 or HCl. Formic acid is preferred over acetic acid as an oxygen carrier due to its high reactivity and does not require catalyst [73].

The degree of epoxidation is commonly estimated by the *oxirane oxygen content* value [74]. Reaction parameters such as concentration of peroxy acid, temperature, time, and stirring speed play an essential role in the synthesis of the product with desired characteristics. Normally, higher-temperature and higher peroxy acid and sulfuric acid concentrations reduce the reaction time and result in higher oxirane content with less hydrolysis. The epoxidation rate increases with increased concentration of H_2O_2.

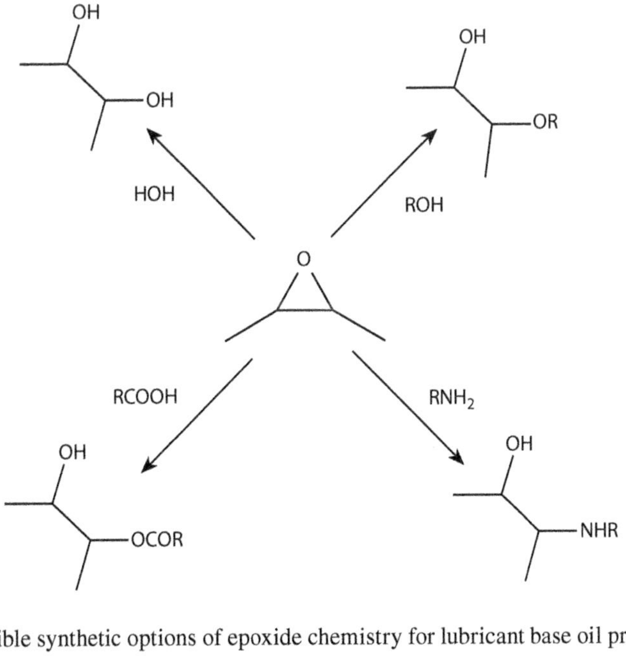

FIGURE 12.7 Possible synthetic options of epoxide chemistry for lubricant base oil preparation.

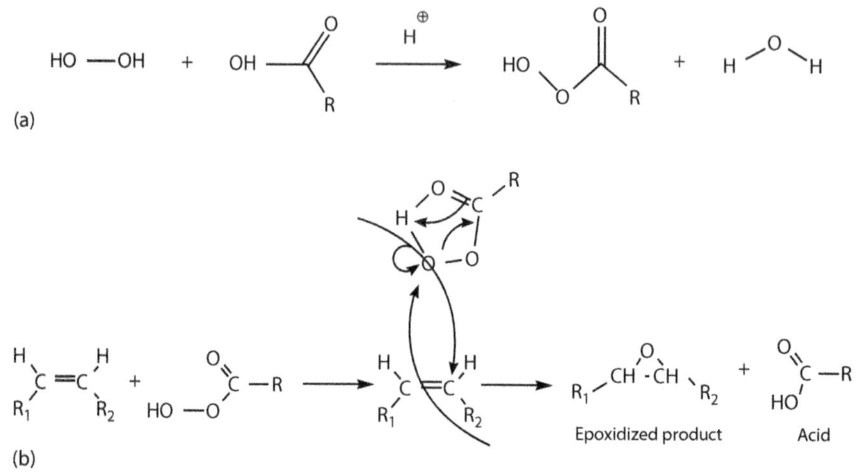

FIGURE 12.8 Generic schematic representation of epoxidation reaction.

TABLE 12.4

Some Reported Epoxidation Reactions of Vegetable Oils and Derivatives

S. No.	Oil	Catalyst	Solvent	Reaction Conditions	Ref.
1	Sunflower oil, soybean oil, and its FAME	Formic acid/30% H_2O_2	Benzene	40–50°C, 20 h	[84]
2	Rubber seed oil	Acetic acid and sulfuric acid/30% H_2O_2	None	50°C, 8 h and 60°C, 3–4 h	[85]
3	Linseed oil	Formic acid/50% H_2O_2	None	50°C, 6 h	[86]
4	Cottonseed oil	Acetic acid/formic acid and H_2SO_4, HCl, H_3PO_4/30% H_2O_2	None	50–60°C, 5–6 h	[75]
5	FAME of palm oil	Formic acid/50% H_2O_2	None	50°C, 3 h	[87]
6	Soybean oil	MoO_2(acac)$_2$/TBHP	Toluene	110°C, 2 h	[80]
7	Poultry-based biodiesel	Formic acid/30% H_2O_2	None	50–60°C, 5 h	[88]
8	High-oleic sunflower oil, coriander oil, castor oil, and soybean oil	Ti-MCM, Ti-SiO$_2$ (aerosil and grace)/TBHP	Ethyl acetate	55–65°C, 24 h	[89]
9	Methyl oleate, methyl linoleate, trioleic sunflower oil, and rapeseed oil	Tungsten-based catalysts/20–50% H_2O_2	None	40°C, 30 min with 50% of H_2O_2 in the presence of oxygen	[90]
10	High oleic acid and sunflower acid	Tungsten and polyoxometalates/35–60% H_2O_2	None	60°C, 50 min	[91]

Note: FAME: fatty acid methyl ester; MCM: Mobil composite material; TBHP: *tert*-butyl hydrogen peroxide.

However, the oxirane ring product is quite stable at low H_2O_2 concentration and very poor at high H_2O_2 concentration. Therefore, there is trade-off about the proper H_2O_2 concentration in the reaction. The oxirane oxygen content value showed a continuous decrease beyond a temperature of 70°C that may lead to severe degradation [75]. The reaction time is also very critical. Beyond the optimum reaction time of 2 h, the oxirane oxygen content value of epoxide rapidly decreases. It is also reported that the value improves with stirring speed up to an optimum level, beyond which it is not substantially affected. This was interpreted as an indication that the reaction was free from mass transfer resistance under the given condition [75]. H_2O_2 is the most promising oxidant for epoxidation reactions because of its low cost and pollution-free effluent [76]. Organic hydrogen peroxides such as *tert*-butyl hydrogen peroxide and urea hydrogen peroxides are also used as epoxidizing agents [77]. Recently, progress has been made with the incorporation of heterogeneous catalyst systems such as tungsten-based catalysts [78], Ti(IV)-grafted silica [79], Mo(IV)-based complex catalysts [80], and ion-exchange resins [81,82]. A brief review of the reports on epoxidation of vegetable oil is compiled in Table 12.4.

The ring opening of epoxy group and subsequent derivatization of epoxy carbons (Figure 12.7) lead to improved oxidative and thermal stability. This reaction is carried out at moderate reaction conditions [83]. Ring-opening reaction is carried out through a cleavage of one of the carbon–oxygen bonds. This initiation can be done by either electrophiles or nucleophiles, or can be acid catalyzed. The rate of the ring opening of epoxidized biodiesel strongly depends on the nature and the structure of the fatty acids. These ring-opening reactions could result in branching at the oxirane ring opening for improving the cold-flow properties. Oxirane ring opening is carried out using different alcohols and carboxylic acids under acid catalysts (acetic acid with catalytic amount of H_2SO_4). Ring opening can also be performed with HCl or HBr, which introduces a halogen to the structure [92]. Other commonly used acid catalysts include acidic ion-exchange resins like Amberlyst-15 (dry) and Amberlite-122R [93]. Acid-catalyzed ring opening of epoxidized oil and biodiesel with low-molecular weight monofunctional acids or alcohols leads to low-molecular weight epoxy esters. Such products find applications as metalworking fluid. Polyfunctional acid or alcohol products form epoxy polyolesters with relatively low viscosities and are used in the production of foams and dispersants. Short-chain alcohol and acid substituent improve pour point better than the long-chain alcohol and acid substituent.

12.3.5 OXIDATIVE CLEAVAGE

Oxidative cleavage of unsaturated fats helps to selectively generate short-chain mono- or dicarboxylic acids from longer-chain carboxylic acids (Figure 12.9). Ozonolysis and ozonation lead to oxidative cleavage of the C=C double bonds. The highly selective ozonolysis of unsaturated fatty acids initially creates ozonides which are transformed to mono- and dicarboxylic acids. Ozonation of oleic acid has been successfully applied on a commercial scale for the production of pelargonic and azelaic acids, used for the manufacture of high-performance lubricants, polyesters, and polyamides [94,95].

Likewise, reductive cleavage leads to the formation of fatty aldehydes which are viscous oils. Azelaic acid is used for the manufacture of high-performance lubricants [96]. Similarly, ozonolytic cleavage of erucic acid in acetic acid followed by oxygen oxidation of the resultant aldehyde at 100°C leads to the saturated brassylic and pelargonic acids (Figure 12.9) [97]. Brassylic acid is used as a plasticizer, biodegradable solvent, and lubricant; whereas pelargonic acid finds application as a herbicide. Highly selective ozonation of unsaturated fatty acids at ambient temperature forms ozonides (1,2,4-trioxolane rings) and peroxy hemiacetals. These are usually further transformed by further ozonolysis. While oxidative cleavage of the products of ozonation gives polyfunctional acids or ketones, reductive decomposition (in a variety of reducing conditions) gives aldehydes and ketones or primary and secondary alcohols [98].

Ozonation may convert all the double bonds to ester groups in a single-step chemical attack in the absence of any catalyst. As the modified vegetable oil products from this process do not contain any tertiary carbons or unsaturation, the modified oil has much higher thermal and oxidative stability. The triglyceride ester linkage is not affected by the ozonolysis process. The fragments of the unsaturated fatty acids beyond the double bonds are also oxidized to form monoesters and diesters. The structure of the

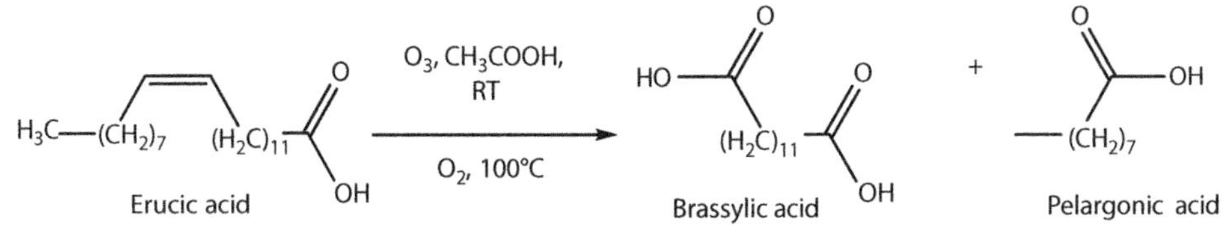

FIGURE 12.9 Schematic representation of oxidative cleavage of erucic acid.

lubricant after the ozonolysis reaction is more uniformed than that of the original, with only three satu-rated fatty acid components (palmitic, stearic, and nonanoic). Ozonation followed by reduction is a very efficient technique for carbon–carbon double bond cleavage in unsaturated oils. The method allows the production of triglyceride polyols and low-molecular weight diols and monols of well-defined structure.

12.3.6 ESTOLIDE SYNTHESIS

Estolides are oligomeric fatty acids (Figure 12.10), formed when the carboxylic acid functionality of one fatty acid molecule forms an ester link at a site of unsaturation of another. More generally, estolides are formed by the condensation of two or more fatty acids [99,100].

Estolide formation may also involve a hydroxy functionality as a site for esterification. The esteri-fication of the residual carboxylic acid group with 2-ethylhexanol (EH) results in the reduction of viscosity and the improvement of the VI and the pour point of the lubricant base stock [99]. Estolides occur in nature [101]. The key benefits of the vegetable oil-based estolides are their higher molecular weight and good lubricity. The viscosity of the estolide can be tailored across a wide range from 30 cSt to >500 cSt (at 40°C) depending on the carboxylic ester functionality and the degree of oligo-merization [102]. Typically, short-chain and branched esters give lower viscosities as well as lower oligomer numbers [102]. Estolides result from a cationic dimerization of two fatty groups in the pres-ence of an acid catalyst, various Brönsted acids can catalyze this oligomerization. Technically, opti-mal conditions, in terms of properties, yields, and cost, for the synthesis of a complex estolide often range from 0.05 to 0.4 equivalents of $HClO_4$ at 60°C for 24 h followed by direct conversion to the cor-responding esters [102,103]. The mechanism of acid-catalyzed estolide formation of a fatty acid with one unsaturation involves carbocation oligomerization across the double bond. This condensation can continue, resulting in polyestolides, where the average degree of oligomerization is designated the estolide number EN (EN = n + 1), which is equal to the average number of fatty acids added to the base fatty acid [103]. When saturated fatty acids are added to an olefin-containing reaction mixture, the oligomerization terminates because the saturate provides no additional reaction site for further reaction. Consequently, the estolide is stopped from further growth and is termed as being *capped* [104]. For instance, coconut oil is 92% saturated; the double bond sites for reaction are very few. For increased estolide formation (via double bonds and hydroxyl groups), castor oil may be added to the coconut oil [105]. The alcohol portion of the primary ester functionality in estolide esters plays a sig-nificant role in pour point reductions as branched chain alcohols dramatically lower the pour point. Different structures and physical properties (e.g., viscosity and density) can be obtained by varying the nature of the base material and reaction conditions. Estolides have certain physical characteristics that could help eliminate common problems associated with vegetable oils as functional fluids [104]. Estolides are potential base stocks for lubricants without the use of additives normally required to improve the lubricating properties of base stocks. The superior properties of estolides are dictated by the combination of the degree of oligomerization, the decrease in the level of unsaturation, and the nature of the specific ester moiety. Estolides of oleic acid [106], meadowfoam oil [107], and castor oil [108] show promise as biolubricants. Estolides from oleic acid have been most intensively explored. Oleic estolide esters have the lowest pour point, −34°C for EH ester. By varying the capping material

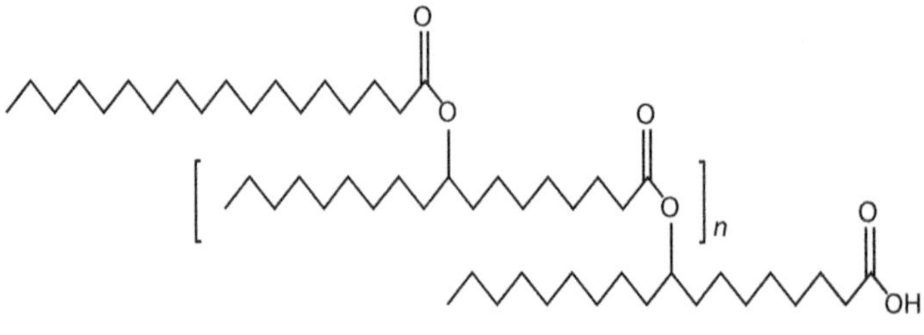

FIGURE 12.10   Typical representation of an estolide moiety.

on the estolide, the crystal lattice structure of the material could be disrupted as it approached its pour point. Saturated fatty acid–capped estolides and their corresponding esters have superior biodegradability and lubricating properties and excellent pour point, cloud point, viscosity, and oxidative stability [109]. Limitations of estolide technology reside in the number of synthetic steps, product work-up, yield, viscosity, and applications. Mono- and polyestolides have a variety of potential applications as lubricants, greases, printing inks, cosmetics, and surfactants.

12.4   SYNTHESIS OF POLYOLESTER-BASED BASE STOCKS

Polyolesters are synthesized by esterification of polyols with monocarboxylic acids in the presence of acid catalyst and azeotrophic solvent for efficient removal of water. The polyols used are DiPE, PE, TMP, and NPG. Usually, the acid is used in 1.5–2.5 molar excess of the amount of polyol used. The reactions are performed in the presence of homogeneous catalysts like p-toluenesulfonic acid and methane sulfonic acid. During esterification reactions, water by-product is continuously removed for reaction completion. The reaction is generally performed at a temperature range of 180–240°C. After the reaction, the excess acid and the catalyst are removed by distillation and alkali treatment, respectively. The synthesis of polyolesters is mostly described in patent literature but scarcely in open literature (Table 12.5).

12.4.1   Applications of Polyolesters

Polyolesters are used in a wide variety of applications including engine oils, compressor oils, aviation lubricants, hydraulic fluids, high-temperature chain oils, automotive gear oils, additive carriers, and metalworking fluids [110]. The last decade has witnessed an adoption of polyolester-based lubricants for refrigeration applications as a replacement for mineral oil-based lubricants. This is in accordance with the Kyoto and Montreal protocols to move away from chemicals that have ozone-depleting and global-warming potentials. The polyolesters are compatible with HFC refrigerants which are being used instead of CFC and hydrochlorofluorocarbons. The HFCs have zero ozone-depleting potential. Various viscosity grades of refrigeration lubricants are used in domestic, automotive, and industrial applications and are synthesized using different combinations of polyols and carboxylic acids of varying carbon lengths (Table 12.5). The list of polyols and acids and their combinations required to obtain esters of various viscosity grades is given in Table 12.6. After synthesis, there are standards which have to be met for the certification of polyolester for use as refrigeration lubricant. There are stringent specifications of TAN < 0.02 and moisture < 30 ppm that must be met. Even after all the specifications have been met at the synthesis level, detailed testing is necessary before the polyolester finds its way in to the compressor industry. A detailed list of required standard tests of polyolesters for refrigeration application is provided in the following section.

12.5   TESTING OF REFRIGERATION LUBRICANTS

Apart from properties like viscosity, pour point, VI, and flash point, there are certain mandatory tests that a fluid should pass so as to be used as a refrigeration lubricant [111].

12.5.1   Thermal Stability

To measure the thermal stability of the ester lubricants with 1,1,1,2-tetrafluoroethane refrigerant gas (referred as R134a hereafter), the sealed tube test was performed at high temperature. The test method involves the following steps: First, 1.5 mL of ester lubricant, 1.5 mL of R134a, and one piece each of iron, copper, and aluminum test pieces (1.7 mm in width, 40 mm in length) were sealed in a glass tube, and the tube was heated to 175°C for 14 days [111]. After the completion of the test, the tube was broken; R134a was released to recover the oil and the metal pieces. The color index of the oil was

TABLE 12.5

Some Open and Patent Literature for the Synthesis of Polyolesters

S. No.	Title	Description	Ref.
1	Process for preparing and purifying complex esters	A polyol such as butyl ethyl propane diol or NPG reacted with mono- or polyvalent acids to obtain a reaction blend with complex esters. Purification of the reaction mixture using tertiary amines. Reaction performed using tin oxide at temperatures of 200–220°C.	[112][a]
2	Synthesis of palm oil-based TMP esters with improved pour points	Preparation of TMP ester of high-oleic palm methyl ester by transesterification to achieve low pour points.	[113]
3	PE-fatty acid ester lubricant composition	Reaction of PE with medium-chain acids for use as lubricants in gas turbines.	[114][a]
4	Cosmetic composition	Preparation of TMP and PE esters of C_8, C_{16}, and C_{17} acids for use in cosmetics. Reaction performed in the presence of sulfuric acid at 200°C to 220°C. Carbon black used as a decolorizing agent.	[115][a]
5	Carboxylic acid esters for PE	Synthesis of complex ester of PE with cyclohexyl carboxylic acid, isostearic acid, and at least one other C_6–C_{16} saturated aliphatic acid, having low volatility, high flash point, and increased viscosity and VI.	[116][a]
6	Method for preparation of polyolesters that are light in color	Process focuses on synthesis of colorless water-clear polyolesters using sodium borohydride.	[117][a]
7	Total and partial erucate of PE	Paper discusses the synthesis of total and partial esters of PE erucate and their infrared spectroscopic study of relationship between structure, reactivity, and thermal properties.	[118]
8	Lubricants and hydraulic fluids	The patent describes synthesis of polyolesters with branch and straight-chain ester oils using p-toluenesulfonic acid as catalyst at a temperature of 125°C.	[119][a]
9	Process for preparing polyolesters	Preparation of triethylene glycol di-2-ethylhexanoate esterification in the presence of activated carbon with fresh acid at a temperature of 225°C.	[120][a]
10	Esterification of TMP with fatty acids	Esterification of TMP with C_5 acids carried out using sodium bisulfate and sulfuric acid as catalysts at 110–120°C.	[121]

[a] Patent literature for synthesis of polyolesters.

measured as compared to that of the oil in the reference tube. The dimensions of the metal pieces were measured to check for the wear and the oil was subjected to chemical analysis to check for the presence of metal ions. A detailed explanation of the above test is given in the test method ASHRAE-97.

12.5.2 PHASE COMPATIBILITY

One milliliter of the lubricant is placed into a thermal shock-resistant volumetrically graduated glass test tube (17 mm in diameter and 145 mm in length) and made up to 10 mL by the addition of refrigerant gas R134a, which is an HFC. The test tube is stoppered and placed into a cooling bath regulated at −29°C ± 0.20°C. After the tube and the contents have equilibrated in the cooling bath, the tube contents are visually examined for evidence of phase separation. If there is any phase separation, the combination is recorded as immiscible. If there is no evidence of phase separation at −29°C, the temperature of the cooling bath is lowered at a rate of 0.3°C/min until phase separation is observed. The temperature of the first observation of phase separation, if within the range of the cooling equipment used, is then noted as the R134a miscibility temperature [122].

TABLE 12.6
Combinations of Polyols and Carboxylic Acids Required for the Synthesis of Various Viscosity Grades of Refrigerating Lubricant

S. No.	Polyol	Polyol (wt%)	Acid	Acid (wt%)	Viscosity at 40°C	Viscosity at 100°C	VI	Ref.
1	NPG	100	i-C_8	100	8.01	2.06	21.82	[111]
2	NPG	82.7	i-C_8	100	10.05	2.46	47.59	[111]
	PE	17.3						
3	NPG	60.6	i-C_8	100	14.94	3.20	61.80	[111]
	PE	39.4						
4	NPG	38	i-C_8	100	22.26	4.08	66.11	[111]
	PE	62						
5	PE	100	nC_5	43.63	30.4	5.74	132.80	[122]
			nC_7	41.00				
			i-C_9	15.37				
6	PE	82.6	nC_5	43.15	30.1	5.7	132.71	[122]
	DiPE	17.4	nC_7	41.38				
			i-C_9	15.47				
7	PE	66	nC_5	70	33.5	6.2	135.96	[123]
	DiPE	34	nC_9	25				
			i-C_8	5				
8	DiPE	100	nC_5	68	52.5	8.7	143.27	[123]
			nC_8	20				
			nC_{10}	12				
9	PE	86	i-C_9	79	69.16	9.011	104.38	[123]
	DiPE	14	nC_5	21				

12.5.3 HYDROLYTIC STABILITY

Hydrolytic stability was measured as follows: Seventy grams of the ester lubricant base stock and 70 g of HFC refrigerant gas R134a containing 1000 ppm water are placed into a tube and sealed. The tube is heated to 175°C and temperature is maintained for 14 days, and the acid value of the products is measured using ASTM D924 procedure. The hydrolytic stability of the ester lubricant is considered good if the measured acid value is equal to or less than 0.05 mg KOH/g [111].

12.5.4 CUSTOMIZED POLYOLESTERS

Polyolesters have to fulfill certain specific structural requirements to successfully pass the above-mentioned tests. Two main types of esters are used:

1. Highly stabilized polyolesters based on completely branched acid esters, for example, 2-ethylhexanoic, *iso*-heptanoic, and 3,5,5-trimethylhexanoic acid.
2. Highly optimized polyolesters based on a mixture of linear and branched esters. For low-viscosity esters (<22 cSt at 40°C), these are based on wholly linear acids.

Generally, highly stabilized polyolesters offer a combination of good miscibility and hydrolytic stability but at the expense of lubricity and thermal stability, while, on the other hand, highly optimized polyolesters offer a combination of excellent wear performance, high thermal stability, acceptable hydrolytic stability, and miscibility. The field trials and the subsequent extensive industrial usage have established highly optimized polyolesters as a commercially important class of polyolesters [124].

12.6 SCALE-UP

12.6.1 Scale-Up of Vegetable Oil-Based Base Stock

For the commercial production of vegetable oil-based biolubricants, it is important to develop and demonstrate the scale-up facilities. Thus, a continuous batch process package can be conceptualized for the vegetable oil-based biolubricant production using transesterified oil and is presented in Figure 12.11.

The proposed process flow diagram consists of three reaction sections: transesterification, epoxidation, and esterification. Transesterification is well known and extensively reported in literature. Henceforth, the biolubricant synthesis is important with respect to the progression of two consecutive reactions, namely, epoxidation followed by esterification. The former reaction is highly exothermic and the latter is endothermic. Thus, a special process scheme is required to optimize the epoxidation reaction wherein the temperature control is essential to the synthesis of the desired quality of product. Epoxidation is followed by ring-opening esterification reaction as discussed in Section 12.6.1.1.

12.6.1.1 Epoxidation Reaction

Typically, an epoxidation reaction is carried out in a batch type of mechanically stirred tank reactor with circulation of reaction mixture to control the exothermicity while addition of oxidizing agent like hydrogen peroxide. The predetermined ratio of transesterified oil and formic acid are

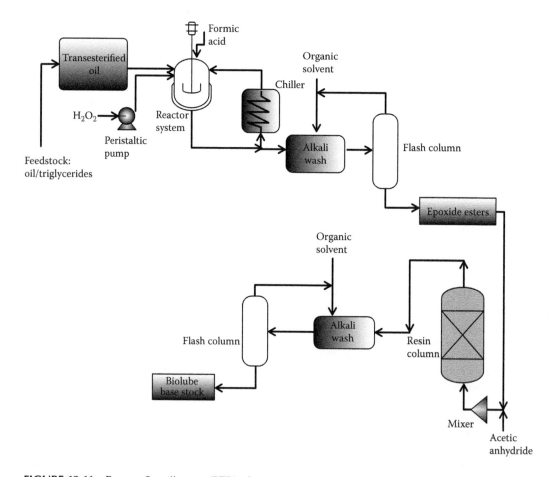

FIGURE 12.11   Process flow diagram (PFD) of vegetable oil-based biolubricant production.

introduced to the reactor and then H_2O_2 is slowly added via a peristaltic pump. Fast or abrupt addition of hydrogen peroxide is not recommended due to exothermic nature of the reaction, which can cause excessive liberation of oxygen due to the decomposition of H_2O_2 at self-attainable high temperatures. Therefore, the continuous recirculation of reaction mixture was recommended to capture the exothermicity and avoid further hydrolysis of the product. On completion of the reaction, the mixture is mixed with organic solvent (such as toluene, benzene, and cyclohexane) and then washed with sodium bicarbonate (5 wt%), distilled water, and sodium chloride (5 wt%) to separate the organic layer from the reaction mixture. Solvent was used to avoid the emulsion formation and was subsequently recovered through flash column and recycled. The epoxide ester product thus formed was analyzed and characterized by epoxy (OOC) value, bromine value, infrared spectroscopy, and kinematic viscosity.

12.6.1.2 Ring-Opening Reaction

The epoxide ester product is mixed with acid anhydride and fed to a solid acid catalyst–filled resin column for esterification. Esterification was carried out in the presence of Amberlyst-15 or its acid-equivalent solid catalyst. These reactants were used in the predetermined molar ratio obtained from optimization studies conducted at lab-scale level. The solid acid catalyst used in the reaction is recovered by regeneration process with strong acids like HCl and H_2SO_4. The reaction mixture was treated similarly to the previously described solvent recovery and its circulation.

The ring-opened esterified product was characterized by infrared spectroscopy and kinematic viscosity as discussed previously. The final ring-opened esterified product thus meets the specification of a lube oil base stock and is termed as *biolube feedstock*.

12.6.2 SCALE-UP OF POLYOLESTER-BASED BASE STOCK

Preparation of polyolesters-based lubricant at lab scale involves a number of unit operations. Based on these unit operations, a complete process for polyolester-based lubricant preparation can be developed. Figure 12.12 is the process flow scheme which can be employed based on the process flow diagram; the development of polyolester lubricants can be divided into reaction, purification, storage and blending, and acid recovery sections. Purification consists of acid extraction, drying, decolorization, acid neutralization, and recovery of the solvent from the ester. A brief discussion about the polyolester reaction and critical unit operations is presented in the following sections.

12.6.2.1 Reaction

Polyolesters are synthesized by the esterification of the polyols with monobasic carboxylic acids in the presence of homogeneous acid catalysts (e.g., sulfuric acid, hydrochloric acid, *p*-toluenesulfonic acid, methane sulfonic acid) and a suitable organic solvent, at 110°C to 175°C. As the reaction proceeds, water is formed as a by-product, which is collected using a Dean–Stark kind of arrangement. The quantity of the water collected is used as a measure of the reaction conversion. The reaction is stopped when all the water has been collected.

12.6.2.2 Purification

Purification is performed to purify the ester in order for it to meet the application requirements. After the completion of reaction, excess reactant acids and acid catalyst from the ester are removed by sequential water washing, which is followed by drying of the esters over inorganic sulfates to remove moisture traces, decolorization, acid neutralization, and solvent distillation.

12.6.2.3 Extraction

Acid extraction is a process by which acid catalyst and excess reactant acids are removed from the product mixture by sequential washings with aqueous sodium bicarbonate, potassium hydroxide, and sodium chloride.

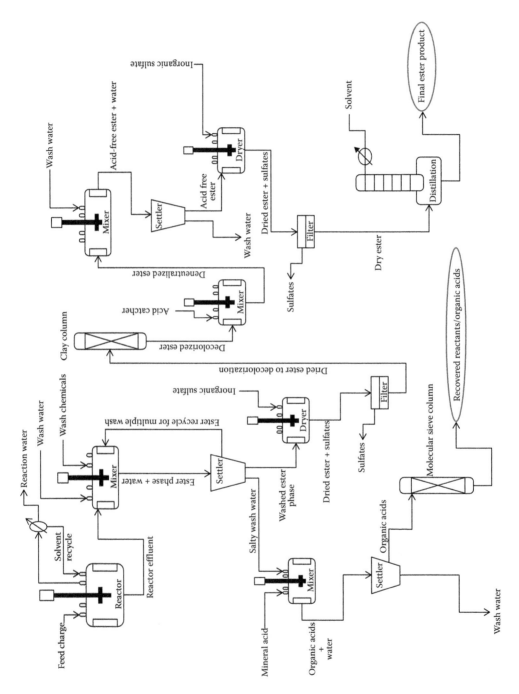

FIGURE 12.12 PFD of polyolester-based biolubricant production.

12.6.2.4 Drying

Drying of the organic phase is carried out to remove the moisture which may be present in the organic phase after water washing. Moisture must be removed or else it would hydrolyze the ester at moderate to high temperature during downstream purification operations. Drying produces purified esters that meet the moisture content specification. Inorganic sulfates (e.g., magnesium sulfate, sodium sulfate) can be employed to remove traces of the moisture from the organic ester phase. Organic product mixture with inorganic sulfates is stirred for 3–4 h and is separated using filtration. During the mixing, the moisture in the organic phase is absorbed by the inorganic sulfates.

12.6.2.5 Decolorization

During synthesis of polyolesters, the color of the reaction mixture changes with increasing temperature and time. It becomes dark brown and may turn black due to thermal degradation of the reactants and ester in the mixture. Decolorization is thus required so the product meets the color specifications. In decolorization, ester is given physical treatment with a decolorization agent to meet the color specifications for the ester application. Decolorization is performed using contact method by employing a batch reactor wherein the oil is agitated with adsorbent material or activated clay for a period and afterward separated from the clay or alternatively, it can be performed by percolation method wherein the oil to be decolorized is percolated through a bed of adsorbent material or activated clay in a column [125].

12.6.2.6 Acid Neutralization

Sodium bicarbonate and potassium hydroxide wash extracts most of the acids present in the product mixture. However, they are not capable of removing minor traces of acids required to make the ester suitable for the applications which demand very low TAN. Decolorization with basic clays improves the color of the ester product and also reduces the TAN but that may still not meet the very low TAN requirements. Therefore, esters are subjected to treatment with acid scavengers like carbodiimides. However, addition of the carbodiimides causes the esters to become turbid. A better method is treating the esters with methoxides of group II elements (e.g., calcium methoxides, magnesium methoxides) in the presence of a solvent to reduce the TAN to significantly lower levels. These methoxides are added in molar excess of 5:1 in the ester, mixed overnight, and then subjected to sodium chloride wash, followed by drying over sodium sulfate.

12.6.2.7 Solvent Recovery

After the acid treatment is over, the ester is subjected to solvent removal. Solvent is used in the esterification reaction in order to dissipate the heat and prevent the degradation of the acids and the esters in the reaction. The solvent is removed by vacuum distillation and the purified ester is obtained as the distillation residue.

12.6.2.8 Storage and Blending

Storage and blending involve storing of the purified esters and blending these to meet the desired property parameters. The blending section includes a mixer wherein two or more esters are blended at 80°C. Heating is required during blending to ensure good mixing of the esters and produce uniform composition. Blended esters are then packed airtight and sealed under nitrogen to prevent diffusion of moisture into the ester.

12.6.2.9 Acid Recovery

Acid recovery is performed to recover and recycle excess reactant acid used in the ester synthesis. From the economic point of view, the recovery of the acids is critical since acids contribute a lot to the cost of the raw materials. Reactant acids are recovered by splitting the organic acid salts in the wash water with mineral acids. Recovered acids are then separated from the water and dried over inorganic sulfates. Additionally, they can be treated with molecular sieves to remove traces of moisture before they are recycled back to the reactor for synthesis.

12.7 FUTURE DRIVERS

As the environmental standards are tightening year by year, the adoption of biodegradable lubricants in various industrial sectors becomes inevitable. Unfortunately, due to their high cost, adoption has not been significant, especially in developing countries. In this context, it is of utmost importance to develop cost-effective routes for biolubricant base stock synthesis. For example, the demand for biobased transformer oils is ever increasing in developing countries. High-oleic vegetable oil offers an excellent platform for transformer oil application. Cost-effective methodologies for the synthesis of oleic acid-rich vegetable oil via a partial hydrogenation route need to be developed. For this purpose, it is essential to develop new catalyst systems for selective hydrogenation of vegetable oils. On similar lines, the increasing demands for the use of biodegradable fire-resistant hydraulic fluids are expected to offer a boost for such developmental activity. Likewise, increased global warming has added impetus for the adoption of high-VI (VI > 100) refrigeration oils in the refrigeration industry. Therefore, to meet the high-VI refrigeration oil demand, it would require developing new cost-effective synthesis pathways for high-VI refrigeration oils. In this context, the development of short-chain fatty acid recovery routes from waste streams derived from caprolactum production would be beneficial. It may be noted that such streams contain C_4 and C_5 short-chain fatty acids, which are reported to be major constituents for refrigeration oils of viscosity grades from 10 to 32 cSt and are currently expensive. In summary, the development of cost-effective pathways is anticipated to offer a platform for the adoption of biodegradable lubricants in various industrial applications.

ACKNOWLEDGMENTS

The authors are thankful to the management of Bharat Petroleum Corporation Limited for granting permission to publish the present chapter. We would also like to thank Grishma S. Patel for preparing the figures.

REFERENCES

1. A. K. Jaina and A. Suhanea, Capability of biolubricants as alternative lubricant in industrial and maintenance applications, *International Journal of Current Engineering and Technology*, 3, 179–183 (2013).
2. S. N. Shah, B. R. Moser, and B. K. Sharma, Glycerol tri-ester derivatives as diluent to improve low temperature properties of vegetable oils, *Journal of ASTM International (JAI)*, 7(3), 102575 (2010).
3. B. R. Moser, S. N. Shah, J. K. Winkler-Moser, S. F. Vaughn, and T. A. Isbell, Composition and physical properties of cress (*Lepidium sativum* L.) and field pennycress (*Thlaspi arvense* L.) oils, *Industrial Crops and Products*, 30, 199–205 (2009).
4. G. N. Jhama, B. R. Moser, S. N. Shah, R. A. Holser, O. D. Dhingra, S. F. Vaughn, R. K. Holloway et al., Wild Brazilian mustard (*Brassica juncea* L.) seed oil methyl esters as a biodiesel fuel, *Journal of the American Oil Chemists' Society*, 86(1), 917–926 (2009).
5. S. N. Shah, B. K. Sharma, B. R. Moser, and S. Z. Erhan, Preparation and evaluation of jojoba oil methyl ester as biodiesel and as blend components in ultra low sulfur diesel fuel, *Bioenergy Research*, 3(2), 214–223 (2009).
6. S. N. Shah, B. K. Sharma, and B. R. Moser, Preparation of biofuel using acetylatation of jojoba fatty alcohols and assessment as a blend component in ultralow sulfur diesel fuel, *Energy & Fuels*, 24(5), 3189–3194 (2010).
7. B. R. Moser, J. K. Winkler-Moser, S. N. Shah, and S. F. Vaughn, Composition and physical properties of arugula, shepherd's purse, and upland cress oil, *European Journal of Lipid Science and Technology*, 112, 734–740 (2010).
8. H. Joshi, B. R. Moser, S. N. Shah, A. Mandalika, and T. Walker, Improvement of fuel properties of cottonseed oil methyl esters with commercial additives, *European Journal of Lipid Science and Technology*, 112(7), 802–809 (2010).

9. S. N. Shah, O. K. Iha, F. C. S. C. Alves, B. K. Sharma, S. Z. Erhan, and P. A. Suarez, Potential application of turnip oil (*Raphanus sativus* L.) for biodiesel production: Physical–chemical properties of neat oil, biofuels and their blends with ultra-low sulphur diesel (ULSD), *Bioenergy Research*, 6(2), 841–850 (2013).

10. N. K. Patel, P. S. Nagar, and S. N. Shah, Identification of non-edible seeds as potential feedstock for the production and application of bio-diesel, *Energy and Power*, 3(4), 67–78 (2013).

11. J. V. Rensselar, The perfect biodegradable lubricant, *Tribology & Lubrication Technology*, 67(6), 44 (2011).

12. N. S. Batters, Biodegradable lubricant: What does biodegradable really mean, *Journal of Synthetic Lubrication*, 22, 3–18 (2005).

13. T. F. Buenemann, S. Boyde, S. Randles, and I. Thompson, Synthetic lubricants—Non aqueous, in *Fuels and Lubricants Handbook: Technology, Properties, Performance and Testing*. George E. Totten (Ed.), West Conshohocken, PA: ASTM International (2003).

14. J. C. J. Bart, E. Gucciardi, and S. Cavallaro, *Biolubricants, Science and Technology*, Cambridge: Woodhead Publishing Limited (2013).

15. A. Jackson, Synthetic versus mineral fluids in lubrication. Keynote address presented at the International Tribology Conference, Melbourne, Australia, December 2–4, 1987, pp. 1–10.

16. S. Gryglewicz, W. Piechocki, and G. Gryglewicz, Preparation of polyol esters based on vegetable and animal fats, *Bioresource Technology*, 87, 35–39 (2003).

17. H. A. Hamid, R. Yunus, U. Rashid, Thomas S. Y. Choong, and A. H. Al-Muhtaseb, Synthesis of palm oil-based trimethylolpropane ester as potential biolubricant: Chemical kinetics modeling, *Chemical Engineering Journal*, 200–202, 532–540 (2012).

18. M. Kotwal, A. Kumar, and S. Darbha, Three-dimensional, mesoporous titanosilicates as catalysts for producing biodiesel and biolubricants, *Journal of Molecular Catalysis A: Chemical*, 377, 65–73 (2013).

19. M. Lamsa, Process for preparing a synthetic ester from a vegetable oil, FI. Patent No. WO 9607632A1 (1995).

20. G. S. Dodos, F. Zannikos, and E. Lois, Utilization of sesame oil for the production of bio-based fuels and lubricants, Proceedings of the Third International CEMEPE & SECOTOX Conference, Skiathos, Greece, June 19–24 (2011).

21. M. A. Harmer and Q. Sun, Solid acid catalysis using ion-exchange resins, *Applied Catalysis A: General*, 221, 45–62 (2001).

22. A. Kleinova, P. Fodran, L. Brncalova, and J. Cvengros, Substituted esters of stearic acid as potential lubricants, *Biomass Bioenergy*, 32, 366–371 (2008).

23. R. Jothiramalingam and M. K. Wang, Review of recent developments in solid acid, base, and enzyme catalysts (heterogeneous) for biodiesel production via transesterification, *Industrial & Engineering Chemistry Research*, 48, 6162–6172 (2009).

24. S. Gryglewicz, Rapeseed oil methyl esters preparation using heterogeneous catalysts, *Bioresource Technology*, 70, 249–253 (1999).

25. G. J. Suppes, K. Bockwinkel, S. Lucas, J. B. Botts, M. H. Mason, and J. A. Heppert, Calcium carbonate catalyzed alcoholysis of fats and oils, *Journal of the American Oil Chemists' Society*, 78, 139–145 (2001).

26. E. Leclercq, A. Finiels, and C. Moreau, Transesterification of rapeseed oil in the presence of basic zeolites and related solid catalysts, *Journal of the American Oil Chemists' Society*, 78, 1161–1165 (2001).

27. L. Bournay, D. Casanave, B. Delfort, G. Hillion, and J. A. Chodorge, New heterogeneous process for biodiesel production: A way to improve the quality and the value of the crude glycerin produced by biodiesel plants, *Catalysis Today*, 106, 190–192 (2005).

28. G. Ondrey, Biodiesel production using a heterogeneous catalyst, *Chemical Engineering*, 10, 13 (2004).

29. S. N. Shah, A. Joshi, A. Patel, and V. P. Brahmkhatri, Synthesis of jatropha oil based biodiesel using environmentally friendly catalyst and their blending studies with diesel, *Energy and Power*, 3(1), 7–11 (2013).

30. P. S. Sreeprasanth, R. Srivastava, D. Srinivas, and P. Ratnasamy, Hydrophobic, solid acid catalysts for production of biofuels and lubricants, *Applied Catalysis A: General*, 314, 148–159 (2006).

31. V. V. Bokade and G. D. Yadav, Synthesis of biodiesel and biolubricants by transesterificattion of vegetable oil with lower and higher alcohols over heteropolyacids supported by clay (K-10), *Process Safety and Environmental Protection Trans IChemE, Part B*, 372–377 (2007).

32. J. Oh, S. Yang, C. Kim, I. Choi, J. H. Kim, and H. Lee, Synthesis of biolubricants using sulfated zirconia catalysts, *Applied Catalysis A: General*, 455, 164–171 (2013).

33. Y. Robiah, A. Fakhru'l-Razi, T. L Ooi, S. E. Iyuke, and A. Idris, Development of optimum synthesis method for transesterification of palm oil methyl esters and trimethylolpropane to environmentally acceptable palm oil-based lubricant, *Journal of Oil Palm Research*, 15(2), 35–41 (2003).

34. Y. Robiah, A. Fakhru'l-Razi, T. L. Ooi, S. E. Iyuke, and A. Idris, Preparation and characterization of trimethylolpropane esters from palm kernel methyl esters, *Journal of Oil Palm Research*, 15(2), 42–49 (2003).

35. E. Uosukainen, Y. Y. Linko, M. Lamsa, T. Tervakangas, and P. Linko, Transesterification of trimethylolpropane and rapeseed oil methyl ester to environmentally acceptable lubricants, *Journal of the American Oil Chemists' Society*, 75 (11), 1557–1563 (1998).

36. T. I. Mohd. Ghazi, M. F. M. Gunam Resul, and A. Idris, Bioenergy II: Production of biodegradable lubricant from *Jatropha curcas* and trimethylolpropane, *International Journal of Chemical Reactor Engineering*, 7, Art. A68 (2009).

37. N. H. Arbain and J. Salimon, Synthesis and characterization of ester trimethylolpropane based *Jatropha curcas* oil as biolubricant base stocks, *Journal of Science and Technology*, 2, 47–58 (2011).

38. N. H. Arbain and J. Salimon, The effects of various acid catalyst on the esterification of *Jatropha curcas* oil based trimethylolpropane ester as biolubricant base stock, *E-Journal of Chemistry*, 8(S1), 33–40 (2011).

39. R. Rahul, J. K. Satyarthi, and D. Srinivas, Utilization of sesame oil for the production of bio-based fuels and lubricants, *Indian Journal of Chemistry*, 50A, 1017–1025 (2011).

40. D. R. Kodali and S. C. Nivens (to Cargill, Inc.), Transesterified oils, US Patent No. 6,278,006 B1 (2001).

41. D. R. Kodali, Z. Fan, and L. R. DeBonte (to Cargill, Inc.), Biodegradable high oxidative stability oils, US Patent No. 6,281,375 B1 (2001).

42. D. R. Kodali and S. C. Nivens (to Cargill, Inc.), Oils with heterogeneous chain lengths, US Patent No. 6,465,401 B1 (2002).

43. M. Ionescu and S. Z. Petrović, Polymerization of soybean oil, Pittsburg State University, Kansas Polymer Research Center, Pittsburg, KS (2011).

44. S. Jain and M. P. Sharma, Long term storage stability of *Jatropha curcas* biodiesel, *Energy* 36, 5409–5415 (2011).

45. S. Jain and M. P. Sharma, Stability of biodiesel and its blends: A review, *Renewable and Sustainable Energy Reviews*, 14, 667–678 (2010).

46. C. Scrimgeour, Chemistry of fatty acids, in *Bailey's Industrial Oil and Fat Products*, Sixth Edition, F. Shahidi (Ed.), Hoboken, NJ: John Wiley & Sons (2005).

47. D. Jovanovic, B. Markovic, M. Stankovic, L. Rozic, T. Novakovic, Z. Vukovic, M. Anic, and S. Petrovic, Partial hydrogenation of edible oils—Synthesis and verification of nickel catalyst, *Hemijska Industrija*, 56 (4), 147–156 (2002).

48. M. A. Tike and V. V. Mahajani, Studies in catalytic transfer hydrogenation of soybean oil using ammonium formate as donor over 5% Pd/C catalyst, *Chemical Engineering Journal*, 123, 31–41 (2006).

49. R. C. S. Schneider, L. R. S. Lara, and M. Martinelli, An alternative process for hydrogenation of sunflower oil, *Orbital—The Electronic Journal of Chemistry*, 2(2), 189–200 (2010).

50. M. B. Fernandez, M. Sanchez, J. F. Tonetto, and G. M. Damini, Hydrogenation of sunflower oil over different palladium supported catalysts: Activity and selectivity, *Chemical Engineering Journal*, 155(3), 941–949 (2009).

51. E. S. Jang, M. Y. Jung, and D. B. Min, Hydrogenation for low trans and high conjugated fatty acids, *Comprehensive Reviews in Food Science and Food Safety*, 4(1), 22–30 (2005).

52. A. M. Alshaibani, Z. Yaakob, A. M. Alsobaai, and M. Sahri, Palladium–boron catalyst for vegetable oils, *Journal of Chemistry*, 5(4), 463–467 (2012).

53. K. W. Lee, B. X. Mei, Q. Bo, Y. W. Kim, K. W. Chung, and Y. Han, Catalytic selective hydrogenation of soybean oil for industrial intermediates, *Journal of Industrial and Engineering Chemistry*, 13 (4), 530–536 (2007).

54. G. Yusem and P. N. Pintauro, The electrocatalytic hydrogenation of soybean oil, *Journal of American Oil Chemists' Society*, 69, 399–404 (1992).

55. E. N. Frankel, Cis-retaining selective hydrogenation, US Patent No. 3,542,821 (1970).

56. J. W. King, R. L. Holliday, G. R. List, and J. M. Snyder, Hydrogenation of vegetable oils using mixtures of supercritical carbon dioxide and hydrogen, *Journal of the American Oil Chemists' Society*, 78(2), 107–113 (2001).

57. J. W. Veldsink, Selective hydrogenation of sunflower seed oil in a three-phase catalytic membrane reactor, *Journal of the American Oil Chemists' Society*, 78, 443–446 (2001).

58. D. Jovanovic, R. Radovic, L. Mares, M. Stankovic, and B. Markovic, Nickel hydrogenation catalyst for tallow hydrogenation and for the selective hydrogenation of sunflower seed oil and soybean oil, *Catalysis Today*, 43, 21–28 (1998).

59. S. Ahmed, K. G. Chandrappa, B. Muhammad, and J. Indurkar, Catalytic hydrogenation of soybean oil promoted by synthesized Ni/SiO nanocatalyst, *Der Pharma Chemica*, 5 (2), 118–126 (2013).

60. G. S. Cannon and L. A. T. Honary, Soybean based transformer oil and transmission line fluid, US Patent No. 6159913 (2000).

61. A. J. Morway, C. Rahway, D. W. Young, and D. L. Cottle, Lubricant grease prepared from the esters of the dimer of linoleic acid, US Patent No. 2673184 (1953).

62. J. Baltes, F. Weghorst, H. Harburg, and O. Wechmann, Process for the co-oligomerization of conjugated unsaturated fatty acids or their esters with indene and/or coumarone, US Patent No. 3441577 (1969).

63. H. Baumann, M. Buhler, H. Fochem, F. Hirsinger, H. Zoebelein, and J. Falbe, Natural fats and oils— Renewable raw materials for the chemical industry, *Angewandte Chemie International Edition*, 27(1), 41–62 (1988).

64. R. M. Koster, M. Bogert, B. de Leeuw, E. K. Poels, and A. Bliek, Active sites in the clay catalysed dimerisation of oleic acid, *Journal of Molecular Catalysis A: Chemical*, 34(1), 159–169 (1998).

65. G. Hillion, R. Stern, and O. L. Borgne, Process for oligomerization of polyunsaturated acids and esters, products obtained and use thereof, US Patent No. 5880298 A (1999).

66. S. Liu, H. Zhou, S. Yu, C. Xie, and F. Liu, Dimerization of fatty acid methyl ester using Brönsted–Lewis acidic ionic liquid as catalyst, *Chemical Engineering Journal*, 174(1), 396–399 (2011).

67. Z. Liu and S. N. Shah, Oligomerization of jojoba oil in super-critical CO_2 for different applications, US Patent No. 8742148 B1 (2014).

68. Z. Liu, S. N. Shah, R. L. Evangelista, and T. A. Isbell, Polymerization of euphorbia oil with Lewis acid in carbon dioxide media, *Industrial Crops and Products*, 41, 10–16 (2013).

69. F. O. Barrett, C. G. Goebel, and R. M. Peters, Process of dimerizing mono unsaturated fatty acids, US Patent No. 2793219 (1957).

70. L. D. Myers, C. G. Goebel, and F. O. Barrett, Polymerization of unsaturated fatty acids, US Patent No. 2955121 (1960).

71. J. E. Milks and N. H. Conroy, Two stage polymerization of unsaturated fatty acids, US Patent No. 3422124 (1969).

72. B. K. Sharma, A. Adhvaryu, Z. Liu, and S. Z. Erhan, Chemical modification of vegetable oils for lubricants applications, *Journal of the American Oil Chemists' Society*, 83, 129–136 (2006).

73. K. D Karlson, R. Kleiman, and M. O. Bagby, Epoxidation of *Lesquerella* and *Limnanthes* (meadowfoam) oils, *Journal of the American Oil Chemists' Society*, 71, 175–182 (1994).

74. D. Derawi and J. Salimon, Epoxidation of palm olein by using performic acid, *E-Journal of Chemistry*, 7(4), 1440–1448 (2010).

75. S. Dinda, A. V. Patwardhan, V. V. Goud, and N. C. Pradhan, Epoxidation of cottonseed oil by aqueous hydrogen peroxide catalysed by liquid inorganic acid, *Bioresource Technology*, 99, 3737–3744 (2008).

76. A. Goti and F. Cardona, Hydrogen peroxide in green oxidation reactions: Recent catalytic processes, *Green Chemical Reactions*, P. Tundo and V. Esposito (Eds.), Dordrecht: Springer, pp. 191–212 (2008).

77. E. Ankudey, H. F. Olivo, and T. L. Peeples, Lipase-mediated epoxidation utilizing urea hydrogen peroxide in ethyl acetate, *Green Chemistry*, 8, 1–5 (2006).

78. Y. Matoba, H. Inoue, J.-I. Akagi, T. Okabayashi, Y. Ishii, and M. Ogawa, Epoxidation of allylic alcohols with hydrogen peroxide catalyzed by $[PMO_{12}O_{40}]^{3-}[C_5H_5N^+(CH_2)_{15}CH_3]_3$, *Synthetic Communications*, 14, 865–873 (1984).

79. M. A. Camblor, A. Corma, P. Esteve, A. Martínez, and S. Valencia, Epoxidation of unsaturated fatty esters over large-pore Ti-containing molecular sieves as catalysts: Important role of the hydrophobic–hydrophilic properties of the molecular sieve, *Chemical Communications*, 8, 795–796 (1997).

80. M. Farias, M. Martinelli, and D. B. Pagliocchi, Epoxidation of soybean oil using a homogeneous catalytic system based on a molybdenum (VI) complex, *Applied Catalysis A: General*, 384, 213–219 (2010).

81. P. D. Meshram, R. G. Puri, and H. V. Patil, Epoxidation of wild safflower oil with peroxy acid in presence of strongly acidic cation exchange resin IR-122 as catalyst, *International Journal of ChemTech Research*, 3, 1152–1163 (2011).

82. S. Sinadinovic, M. Jankovic, and Z. S. Petrovic, Kinetics of in situ epoxidation of soybean oil in bulk catalyzed by ion exchange resin, *Journal of the American Oil Chemists' Society*, 78, 725–731 (2001).

83. J. L. Scala and R. P. Wool, The effect of fatty acid composition on the acrylation kinetics of epoxidized triacylglycerols, *Journal of the American Oil Chemists' Society*, 79, 59–63 (2002).

84. A. Campanella, C. Fontanini, and M. A. Baltanás, High yield epoxidation of fatty acid methyl esters with performic acid generated in situ, *Chemical Engineering Journal*, 144, 466–475 (2008).

85. F. E. Okieimen, O. I. Bakare, and C. O. Okieimen, Studies on the epoxidation of rubber seed oil, *Industrial Crops and Products*, 15, 139–144 (2002).

86. G. Lligadas, J. C. Ronda, M. Galià, and V. Cádiz, Bionanocomposites from renewable resources: Epoxidized linseed oil-polyhedral oligomeric silsesquioxanes hybrid materials, *Biomacromolecules*, 7, 3521–3526 (2006).

87. P. L. Lee, W. M. Z. Wan Yunus, S. K. Yeong, D. K. Abdullah, and W. H. Lim, Optimization of the epoxidation of methyl ester of palm fatty acid distillate, *Journal of Oil Palm Research*, 21, 675–682 (2009).

88. K. Wadumesthrige, S. O. Salley, and K. Y. S. Ng, Effects of partial hydrogenation, epoxidation and hydroxylation on the fuel properties of fatty acid methyl esters, *Fuel Process Technology*, 90, 1292–1299 (2009).

89. M. Guidotti, N. Ravasio, R. Psaro, E. Gianotti, S. Coluccia, and L. Marchese, Epoxidation of unsaturated FAME obtained from vegetable resource over Ti(IV)-grafted silica catalysts: A comparison between ordered and non-ordered mesoporous materials, *Journal of Molecular Catalysis A: Chemical*, 250, 218–225 (2006).

90. E. Poli, J. M. Clacens, J. Barrault, and Y. Pouilloux, Solvent-free selective epoxidation of fatty esters over a tungsten-based catalyst, *Catalysis Today*, 140, 19–22 (2009).

91. I. V. Kozhevnikov, G. P. Mulder, M. C. Steverink-de Zoete, and M. G. Oostwal, Epoxidation of oleic acid catalyzed by peroxo phosphotungstate in a two-phase system, *Journal of Molecular Catalysis A: Chemical*, 134, 223–228 (1998).

92. A. Guo, Y. J. Cho, and Z. S. Petrovic, Structure and properties of halogenated and nonhalogenated soy-based polyols, *Journal of Polymer Science Part A: Polymer Chemistry*, 38, 3900–3910 (2000).

93. P. S. Lathi and B. Mattiasson, Green approach for the preparation of biodegradable lubricant base stock from epoxidized vegetable oil, *Applied Catalysis B: Environmental*, 69, 207–212 (2007).

94. I. Cvetkovic, J. Milic, M. Ionescu, and Z. S. Petrovic, Preparation of 9-hydroxynonanoic acid methyl ester by ozonolysis of vegetable oils and its polycondensation, *Hemijska Industrija*, 62(6), 319–328 (2008).

95. E. H. Pryde, D. E. Anders, H. M. Teeter, and J. C. Cowan, Ozonation of soybean oil: The preparation of some properties of aldehyde oils, *Journal of the American Oil Chemists' Society*, 38, 375–379 (1961).

96. R. G. Fayter, Technical reactions for production of oleochemical monomers, in *Perspectiven Nachwachsender Rohstoffe in der Chemie*, Weinheim, Germany: Wiley-VCH, pp. 107–118 (1996).

97. T. A. Isbell, Derivatives of new oil crops, in *Recent Developments in the Synthesis of Fatty Acid Derivatives*, Urbana, IL: AOCS Press, pp. 44–45 (1999).

98. F. D. Gunstone, Uses of ozone in organic chemistry, *Journal of Chemical Education*, 5(4), 166–170 (1968).

99. T. A. Isbell, R. Kleiman, and B. A. Plattner, Acid-catalyzed condensation of oleic acid into estolides and polyestolides *Journal of the American Oil Chemists' Society*, 71, 169–174 (1994).

100. S. C. Cermak and T. A. Isbell, Synthesis of estolides from oleic and saturated fatty acids, *Journal of the American Oil Chemists' Society*, 78, 557–565 (2001).

101. R. V. Madrigal and C. R. Smith, Estolide triglycerides of *Trewia nudiflora* seed oil, *Lipids*, 17, 650–655 (1982).

102. S. C. Cermak and T. A. Isbell, Synthesis and physical properties of estolide based functional fluids, *Industrial Crops and Products*, 18, 183–196 (2003).

103. T. A. Isbell and R. Kleiman, Mineral acid-catalyzed condensation of meadow foam fatty acids into estolides, *Journal of the American Oil Chemists' Society*, 73, 1097–1107 (1996).

104. S. C. Cermak and T. A. Isbell, Physical properties of saturated estolides and their 2-ethylhexyl esters, *Industrial Crops and Products*, 16, 119–127 (2002).

105. A. Govindapillai, N. H. Jayadas, and M. Bhasi, Analysis of the pour point of coconut oil as a lubricant base stock using differential scanning calorimetry, *Lubrication Science*, 21(1), 13–26 (2009).

106. L. A. García-Zapateiro, M. A. Delgado, J. M. Franco, C. Valencia, M. V. Ruiz-Méndez, R. Garcés, and C. Gallegos, Oleins as a source of estolides for biolubricant applications, *Grasas Aceites*, 61, 171–174 (2010).

107. T. A. Isbell, M. R. Edgcomb, and B. A. Lowery, Physical properties of estolides and their ester derivatives, *Industrial Crops and Products*, 13, 11–20 (2011).

108. C. Yamaguchi, M. Akita, S. Asaoka, and F. Osada, Production of castor oil fatty acid estolide, Jap. Patent No. JP 1,016,591 A (1989).

109. S. C. Cermak and T. A. Isbell, Improved oxidative stability of estolide esters, *Industrial Crops and Products*, 18, 223–230 (2003).
110. S. J. Randles, *Synthetic Lubricants and High-Performance Functional Fluids*, Second Edition, L. R. Rudnick and R. L. Shubkin (Eds.), London: Taylor & Francis Group (2006).
111. N. E. Schnur, Blended polyol ester lubricants for refrigerant heat transfer fluids, US Patent No. 6350392 B1 (2002).
112. J. Koistinen, K. Rissanen, and S. Koskimies, Process for preparing and purifying complex esters, US Patent No. 6362362 B1 (2002).
113. R. Yunus, A. Fakhru'l-Razi, T. L. Ooi, R. Omar, and A. Idris, Synthesis of palm oil based trimethylol-propane esters with improved pour points, *Industrial & Engineering Chemistry Research*, 44, 8178–8183 (2005).
114. R. Yaffe, Pentaerythritol–fatty acid ester lubricant composition, US Patent No. 4216100 (1980).
115. K. Tomita, M. Yanagi, and T. Kobayashi, Cosmetic composition, US Patent No. 3976789 (1976).
116. K. H. Hentschel, R. Dhein, H. Rudolph, K. Nutzel, K. Morche, and W. Kruger, Carboxylic acid esters for pentaerythritol, US Patent No. 4212816 (1980).
117. G. B. Poppe, Method for preparation of polyol esters that are light in colour, US Patent No. 7126018 (2006).
118. V. Eychenne, Z. Mouloungui, and A. Gaset, Total and partial erucate of pentaerythritol, *Journal of the American Oil Chemists' Society*, 75, 293–299 (1998).
119. K. Koch and H. Kroke, Lubricants, hydraulic fluids, US Patent No. 4144183 (1979).
120. M. Adamzik, T. Müller, W. Schulz, and H. J. Schultz, Process for preparing polyol esters, US Patent No. 8524937 (2013).
121. T. M. Itsikson, N. V. Milovidova, and I. B. Rapport, Esterification of trimethylolpropane with fatty acids, *Khimiya i Teknologiya Topliv i Masel*, 6, 14–16 (1967).
122. D. Carr, J. Hutter, R. Kelley, E. T. Hessell, and R. Urrego, Production of polyol ester lubricants for refrigeration systems, US Patent No. 8318647 B2 (2012).
123. H. D. Grasshoff, V. Synek, and H. Kohnz, Synthetic ester lubricants for refrigerator systems, US Patent No. 5830833 (1998).
124. S. J. Randles, Refrigeration lubricants, *Synthetics, Mineral Oils, and Bio-based Lubricants: Chemistry and Technology*, L. R. Rudnick (Ed.), Boca Raton, FL: CRC Press (2006).
125. P. B. Weisz and N. J. Pitman, Decoloriation process, US Patent No. 2687990 (1950).

Section IV

Vegetable Oil-Based Environmentally
Friendly Fluids

13 Engineering and Technology of Environmentally Friendly Lubricants

Carlton J. Reeves, Pradeep L. Menezes,
Michael R. Lovell, and Tien-Chien Jen

CONTENTS

ABSTRACT

Over the last several decades, the lubrication industry has been striving to bring environmentally friendly biobased lubricants, known as biolubricants, to prominence. With an estimated 50% of all lubricants entering the environment, and with much of these being composed of non-biodegradable mineral oils, biolubricants have begun to experience a resurgence. This chapter investigates the various types of eco-friendly biolubricants that are less toxic to the environment, are derived from renewable resources, and provide feasible and economical alternatives to traditional petroleum-based lubricants. Advantages of biolubricants include their higher lubricity, lower volatility, higher shear stability, higher VI, higher load-carrying capacity, and superior detergency and dispersancy when compared to petroleum-based lubricants. Discussions explore the drawbacks of many biolubricants, such as their poor thermal-oxidative stability, solidification at low temperatures, biological deterioration, and hydrolytic instability as well as mechanical and chemical enhancements that seek to rectify these issues. An analysis of the economical and legislative landscape is discussed.

13.1 INTRODUCTION

In the lubrication industry, there are nearly 10,000 different lubricant formulations of oils, greases, and other functional fluids to satisfy more than 90% of all lubricant applications. The primary functions of these lubricants are to control friction, transmit energy, protect against corrosion and wear (attrition), remove heat, disperse wear debris, eliminate foreign contaminants, and act as a sealant [1]. The use of natural organic oils and fats derived from plant- and animal-based raw materials (e.g., soybean, palm, tallow, and lard) for their ability to lower friction and prevent wear dates back to the earliest days of lubrication. In 1859, the first commercial oil well was drilled in Titusville, Pennsylvania, signaling the rise of the modern petroleum oil industry, which would eventually lead to the decline of the use of natural oils as lubricants [2]. The advent of petroleum-based oils produced rapid advances in lubrication technology that quickly dominated other oils, such as natural plant- and animal-based oils. During the mid-1930s, the properties of petroleum-based oils were significantly improved through the use of additives and new chemical synthesis and modification techniques that enhanced the load-carrying capacity, lubricity, corrosive protection, and thermal-oxidative stability. These improvements in the properties of petroleum-based oils often surpassed similar properties of natural oils.

Over the last 150 years, petroleum-based oils have established themselves as the universal lubricant for most industrial, commercial, and personal applications [3–6]. It is estimated that more than 50% of all lubricants used worldwide enter the environment from spills, improper disposal, accidents, volatility, and total loss applications, such as chain saw oils, two-stroke engines, concrete mold release oils, exhaust fumes in engines, and metal-cutting and metal-forming processes [7]. The most problematic of these leakage types are the uncontrolled losses as a result of broken hydraulic hoses or accidents, whereby large quantities of fluids escape into the environment, contaminating soil, surface, ground, and drinking water, as well as the air. It is estimated that 30 to 40 million tons of lubricants is consumed annually, of which 20 million tons enters the environment, amounting to a 55% loss of lubricant [8]. Over 95% of the lubricants entering the environment are petroleum based and harmful to the environment [9]. As a result of their high toxicity and low biodegradability, petroleum-based lubricants and functional fluids (hydraulic fluids) constitute a considerable threat to the environment. More astonishingly, it is estimated by some scientists that over 90% of all petroleum-based lubricants could be replaced by biolubricants [10], which illustrates the substantial potential that biolubricants have to solve our environmental problems caused by toxic petroleum-based lubricants.

13.2 CLASSIFICATION OF LUBRICANTS

Lubricants are often classified into three categories: (i) mineral oils, which are predominantly petro-leum-based lubricants and are the most common lubricants; (ii) natural oils, which are derived from plant-based oils and animal-based fats or tallow; and (iii) synthetic oils, which include PAOs, synthetic esters, PAGs, alkylated aromatics, perfluoroalkylpolyethers (PFPEs), among others. In the last three decades, natural and synthetic oils (not mineral oils) derived from biobased feedstock have seen a resurgence for industrial purposes. The lubrication industry is shifting to becoming more environmentally responsible. Much attention is centered on ecological conservation and sus-tainability through the use of biobased lubricants for industrial purposes to be used as functional fluids. The term *biolubricant* has been ascribed to all lubricants derived from biobased raw materi-als such as plant oils, animal fats, or other environmentally benign hydrocarbons [11]. Biolubricants are intended to be biodegradable and nontoxic to humans and other living organisms, particularly aquatic environments where the impacts are more detrimental [11]. In an effort to curb the use of petroleum-based lubricants, the primary focus has been the development of technologies that incorporate biolubricants as a replacement for traditional petroleum-based industrial lubricants because they are nontoxic, biodegradable, and renewable [12]. In the global lubrication market, the rise of biolubricants is a result of new environmentally friendly initiatives and economic fac-tors, such as protecting the environment from toxic substances; depletion of the world's crude oil reserves; increasing crude oil prices; and increasingly stringent government regulations regarding use, operation, and disposal of petroleum-based oils [13,14]. Furthermore, the emphasis placed on biolubricants is a result of the increase in the demand for environmentally friendly lubricants that are less toxic to the environment, are renewable, and provide feasible and economical alternatives to traditional petroleum-based lubricants [15–17].

Many biolubricants are composed of plant oils, animal fats, or chemical modifications of these oils. They are widely regarded as environmentally benign because of their superior biodegradability and renewable feedstock [11]. Biolubricants are classified according to their base fluid composition into are three groups hydraulic oil environmental triglycerides, also known as natural ester-based or vegetable oils; hydraulic oil environmental ester synthetics; and hydraulic oil environmental polyglycols [8]. Some properties of these groups are listed in Table 13.1. In this chapter, a more rudimentary classification of biolubricants based on the synthesis process is presented. According to this method, biolubricants are classified in two categories: (i) natural oils and (ii) synthetic oils. Natural oils (also known as natural esters) are biolubricants with the base stock consisting of veg-etable oils (plant-based) or animal fats [18]. Many biolubricants have superior lubricity and wear resistance that exceeds those of petroleum-based lubricants resulting in their increased usage as a base stock for industrial oils and functional fluids, thus facilitating the biolubricant resurgence

TABLE 13.1
Suggested Classification of Environmentally Friendly Biolubricants

Fluid	Base Fluid	Saturation	Feedstock
Hydraulic oil environmental triglycerides	Natural ester	Unsaturated	Natural
Hydraulic oil environmental ester synthetics	Synthetic ester	Unsaturated, saturated	Natural synthetic
Hydraulic oil environmental polyglycols	Polyglycol	–	Synthetic

Source: Mang, T., and W. Dresel: *Lubricants and Lubrication.* 2006. Copyright Wiley-VCH Verlag GmbH & Co. KGaA. Reproduced with permission.

[19]. The largest drawback of biolubricants, particularly natural oils, is their poor thermal-oxidative stability and high pour points, which have led to the development of synthetic biolubricants [19]. Ultimately, biolubricants formulated from biobased feedstocks should offer the following advantages over petroleum-based oils [20]:

1. Higher lubricity, lending to lower friction losses, and improved fuel efficiency, affording more power output and better fuel economy
2. Lower volatility, resulting in decreased exhaust emissions
3. Higher VIs
4. Higher shear stability
5. Higher detergency eliminating the need for detergent additives
6. High dispersancy
7. Rapid biodegradation, resulting in decreased environmental and toxicological hazards

The following sections provide a summary of biolubricants, types, vulnerabilities, and enhancement methods. It also examines biodegradability and economy as well as recent advances pertaining to their utilization.

13.3 BIOLUBRICANTS

Biolubricants can be derived from a variety of biobased feedstock. Often, the feedstock is a vegetable oil because it offers natural biodegradability and low toxicity. In other instances, genetically modified natural oils, such as high-oleic sunflower and canola, are being pursued for applications where higher oxidative stability is needed. Synthetic esters derived from natural and artificial resources, i.e., the hydrolysis of solid fats and low-grade waste materials, such as tallow, to produce the constituent ester compounds are also common. Polyglycols are another class of synthetic biolubricants similar to synthetic esters, yet polyglycols are the only type that is water soluble. This property can be advantageous for biological degradation in water. It also poses more of an environmental threat as polyglycol-contaminated water can penetrate more deeply into the soil layers, thus contaminating groundwater. For this reason, polyglycols are not considered environmentally friendly fluids, and their discussion will be limited in this chapter.

Similar to the development of environmentally friendly base fluids, new additives are being developed. Eco-friendly initiatives necessitate the development of additives for two reasons: (i) existing additives used for mineral oils deteriorate the environmental performance properties in many biolubricants and (ii) existing additives contain non-biodegradable substances, leading to a significant deterioration of the biodegradability of the lubricant as a whole [21–25]. Other biolubricants are composed of environmentally benign solid particles such as hexagonal boron nitride mixed with biobased oil such as canola oil to form colloidal suspensions. The formulation of these lubricants means that biodegradable and low-toxicity fluids, i.e., natural oils, are combined with biodegradable and low-toxicity additives, forming two-phase biolubricants that when properly mixed, form colloidal suspensions [10].

13.3.1 NATURAL OILS

The chemical composition of biolubricants derived from vegetable oils and fats affords them the ability to be used as fuels and lubricants for various applications. The sources of biobased oils and fats are numerous and encompass a wide variety of seeds, fruits, nuts, vegetables, animals, and marine life, making them readily available, inexpensive, and environmentally benign [12]. Natural oils derived from plant-based oils are well known because they have viscosity and surface tension properties similar to that of petroleum-based lubricants used in industrial applications for metal stamping and metal forming [26]. Many plant-based oils are obtained by expeller methods and solvent extraction processes [15,27–29]. There is a wide variety of plant-based oils, such as avocado, canola, castor, coconut, corn, cottonseed, olive, palm, palm kernel, peanut, safflower, sesame, soybean, and sunflower oils, among many others, that are native to particular regions of the world. The efficacy of natural oils and fats is determined by their chemical composition, where they predominantly consist of mixtures of fatty acid esters derived from glycerol. Inherently, these oils and fats are naturally occurring organic substances whose properties and utility vary based on biological factors, such as nutrient availability, climate, light, temperature, humidity, and water, as well as being influenced based on different extraction methods [15,27,28,30]. Additionally, it has been shown in the literature that natural oils derived from plants with high-oleic acid (>80%) contents surpass Group I petroleum-based lubricants with respect to lubricity and oxidative stability [31,32]. When comparing natural oils to mineral and synthetic oils, natural oils have a higher lubricity, lower volatility, higher shear stability, higher VI, higher load-carrying capacity, and superior detergency and dispersancy [33].

Many of the accolades associated with natural oils are a result of their molecular structure, which affords superior lubrication properties. The attraction to natural oils is due to their chemical composition of approximately 98% triacylglycerol molecules made up of esters derived from glycerol and long chains of polar fatty acids. Natural oils also contain minor amounts of mono- and diglycerols (0.5%), FFAs (0.1%), sterols (0.3%), and tocopherols (0.1%) [9]. Fatty acids with long unbranched aliphatic chains (4 to 28 carbons) may contain other functional groups and may be saturated or unsaturated. Due to the large amounts of unsaturated fatty acids in natural oils, they tend to suffer from a poor thermal-oxidative stability.

The fatty acids in natural oils are desirable in boundary lubrication for their ability to adsorb to metallic surfaces due to their polar carboxyl group. They form a monolayer film that is effective at reducing friction and wear by minimizing metal-to-metal contact at the tribo-interface [2,29,34]. Figure 13.1 compares the tribological performance of natural oils as a function of sliding distance on pin-on-disk tribometer under ambient conditions. In this study, C101 copper pins were indirectly slid against 2024 aluminum alloy disks under a normal load of 10 N with a sliding velocity of 36 mm/s for a sliding distance of 3000 m. The tests ran for 23 hours using 10 mL of natural oil lubricant in the contact interface [29]. Figure 13.1a shows the coefficient of friction and Figure 13.1b shows the wear volume. These oils are obtained from plants, vegetables, fruits, and nuts that have a diverse fatty acid composition. In Section 13.4.1, the fatty acid compositions will be revealed detailing which fatty acids contribute to the low friction and wear results. Much of the work with natural oils has concentrated on understanding the fundamentals of saturated and unsaturated fatty acids with the bulk of the attention focusing on the use of natural oils as neat lubricants, fatty acids as additives in mineral oils, and biobased feedstock oils for chemically modified lubricants [26,35]. Additionally, new additives are being developed to extend the operable temperature range of natural oils to improve their thermal stability. Recently, natural oils are finding uses as carrier fluids where the oil is being mixed with lamellar particle additives to create colloidal solutions to be used in sliding contacts [36,37].

13.3.2 GENETICALLY MODIFIED VEGETABLE OILS

As their name suggests, genetically modified lubricants are derived from genetically modified oils. They are obtained from plants whose genetic properties have been altered using genetic engineering techniques. In the lubrication industry, genetically modified lubricants are oils obtained from genetically modified plants such as sunflower, soybean, and canola. They are genetically engineered by manipulating the gene sequence within the plants. Genetic modification of vegetable oils often focuses on improving the thermal and oxidative stabilities of natural oils by reducing the linoleic and linolenic acid contents and increasing the oleic acid content [38–40]. Genetic modification of vegetable oils also seeks to improve the cold-flow properties of the lubricants. Higher proportions of short-chain saturated or long-chain monounsaturated fatty acids tend to lower the pour point [41,42].

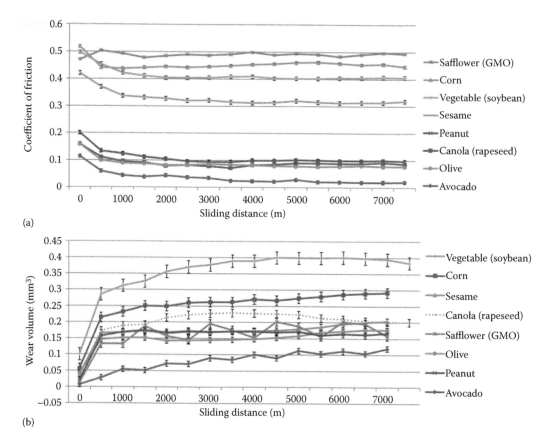

FIGURE 13.1 (a) Coefficient of friction and (b) wear volume for a variety of vegetable oils as a function of sliding distance during ambient conditions. (From Reeves, C. J., P. L. Menezes, T.-C. Jen, and M. R. Lovell, Evaluating the tribological performance of green liquid lubricants and powder additive based green liquid lubricants, STLE Annual Meeting & Exhibition, St. Louis, 2012. With permission.)

The objective of genetic modification is to create oils with higher degrees of saturation whereby the oil is less susceptible to oxidative deterioration by means of exposed double bonds in the fatty acid molecules [43].

13.3.3 Synthetic Esters

Synthetic organic esters are a class of widely used lubricants that were once derived from glycerol molecules found in plant-based oils and animal fats. However, this causes synthetic esters to suffer from similar unsaturated performance issues as those for natural oils. Synthetic esters can be obtained from long-chain alcohols and acids with superior performance than natural ester fluids due to a more uniform molecular structure and use of different alcohols. Completely saturated esters tend to exhibit very stable aging characteristics. New additives specifically for use with synthetic esters have been developed. Early synthetic esters possessed chemical structures similar to those of natural oils. However, due to recent advances, new acids and alcohols are now available for the production of synthetic esters with improved technical performance. Examples of synthetic esters feedstock include C_6–C_{13} alcohols, C_5–C_{18} monoacids, neopentyl polyols, PE, polyolesters, diacids, and various dimer acids [26,32,34,44–49].

Esters provide better low-temperature fluidity and improved thermal-oxidative stabilities when compared to vegetable oils. Similar to natural oils, synthetic esters maintain an affinity for adsorption on metal surfaces due to their high degree of polarity. As a result, they can form monolayers to minimize contact between rubbing surfaces and enhance tribological properties. Esters are inherently sensitive toward hydrolysis and thermal degradation. Their thermal properties can be improved by replacing the glycol with polyols such as NPG, PE, TMP, trimethylolhexane, and trimethyloethane [15,35,50–52]. A subgroup of esters known as complex or oligoesters is derived from mixtures of polyols and mono-, di-, and tricarboxylic acids. Replacing the glycerol molecules with peresters of sugar (i.e., sucrose and sorbital) yields synthetic esters with enhanced oxidative stability, lubricity, and biodegradability [53]. These improved synthetic esters are strong candidates for replacing mineral oils in the food, pharmaceutical, and cosmetic industries. In fact, these esters could be used in many lubricant applications that require low toxicity due to potential contact with animals or humans [53,54]. Esters of NPG were originally developed for the lubrication of aircraft jet engines, whereas PE esters derived from C_5–C_9 carboxylic acids have found use in gas turbines. Still other esters, such as those of TMP and oleic acid, have found widespread use in automotive and marine engine oils, compressor oils (as replacement lubricants for HFC systems), hydraulic fluids, gear oils, and grease formulations [11]. The low toxicity, excellent biodegradability, moderate oxidative stability, and moderate price coupled with the good shear stability make synthetic esters attractive lubricants. Still the use of appropriate additives with ester molecules improves their performance, making them the optimal lubricant of choice for industrial lubricant applications [26,29,35,37,55].

Another type of synthetic ester is the diester, a biolubricant derived partly from renewable resources. Synthetic diesters are derived from dicarboxylic acid and monovalent alcohols. The dicarboxylic acid can be prepared from natural sources, such as azelaic acid (from ozonolysis of oleic acid), sebacic acid, dimeric fatty acid, or from purely petrochemical sources, such as adipic acid or malelic acid. Diesters generally consist of branched alcohols, such as EH (iscooctanol), isodecanol, or guerbet alcohols to offer better low-temperature properties than those of conventional synthetic ester lubricants. Additionally, branched fatty acids, for example, 12-hydroxystearic acid derived from ricinoleic acid, can be utilized to form diester-based lubricants that also exhibit improved low-temperature properties [56,57]. Isostearic acid, a synthetic ester-based lubricant consisting of branched and straight-chain C_{18} fatty acids, has very low levels of unsaturation, resulting in excellent oxidative stability.

13.3.4 PERFLUOROPOLYALKYLETHERS (PFPEs)

High-performance applications that require a combination of high- and low-temperature properties, chemical or oxidative stability, low volatility, material compatibility, inertness, and nonflammability or noncombustibility simultaneously will generally use synthetic lubricants known as PFPEs [58]. PFPEs are a class of lubricants that are undergoing rapid advances to redesign them to be more environmentally friendly [59–61]. PFPEs are known for their superior thermal and oxidative stabilities. Once composed entirely of carbon, fluorine, chlorine, and oxygen to produce a colorless, odorless, and completely inert functional fluid, they are now being derived from more environmentally benign feedstock that reduces their environmental impact. Examples of such are α,ω-dialkoxyfluoropolyethers (DA-FPEs), which are partially fluorinated polyethers that do not contain chlorine atoms and, hence, they do not contribute to ozone depletion [62]. The improvement in these lubricants is due to the incorporation of two alkoxy groups that provide reactive sites, which act to minimize the atmospheric lifetime of these functional fluids [62–64]. Typically, PFPEs enter the environment as an aerosol from exhaust fumes. When compared to PFPEs, DA-FPEs have a lower environmental impact in terms of global-warming potential and DA-FPEs generate less greenhouse gases such as carbon dioxide [62]. Furthermore, perfluorinated ether compounds derived from polymers maintain good properties, such as high thermal and chemical stabilities, no acute toxicity, and excellent heat exchange properties [65]. The presence of the alkoxy groups has allowed DA-FPE based functional fluids to maintain excellent solvent properties in several organic liquids such as ketones and alcohols [66]. These properties of DA-FPEs make them excellent candidates as CFC, perfluocarbon, and halon substitutes in such applications as foaming and fire-extinguishing agents, cleaning agents for electronic devices, heat transfer fluids, and lubricants in extreme conditions [67–69].

13.3.5 IONIC LIQUIDS

Ionic liquids, particularly those that are fluid at room temperature and can be derived from biobased feedstocks, represent a promising new class of lubricants that show potential to improve the limitations associated with both petroleum-based lubricants and natural oils [70]. Room-temperature ionic liquids typically consist of combinations of a bulky asymmetric organic cation paired with an appropriate organic anion with melting points below 100°C and a liquid range above 300°C [71,72]. A schematic of the molecular structure of a generalized ionic liquid is shown in Figure 13.2 [73]. This structure is similar to that of a typical lamellar solid particle, where the anions and the cations form ionic bonds creating layers and these layers are held together by the weak van der Waals force. This structure provides ionic liquids with their liquid lamellar crystal structure [74]. Ionic liquids exhibit a number of unique and useful properties that make them well suited as biolubricants. Ionic liquids have many potential benefits when compared to conventional lubricants including (i) a broad liquid range (low melting and high boiling points), (ii) negligible vapor pressure, (iii) nonflammability and noncombustibility, (iv) superior thermal stability, (v) variable/controlled viscosity, (vi) miscibility and solubility, (vii) environmentally benign/nontoxic, (viii) lamellar-like

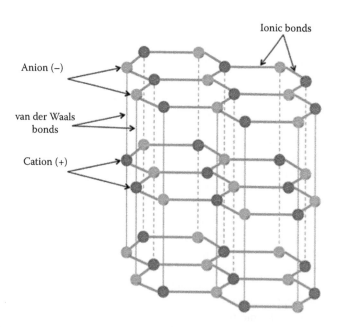

FIGURE 13.2 Molecular structure of generalized ionic liquid. (From Manahan, S. E., *Environmental Chemistry*. Lewis, Boca Raton, FL, 1994; Reeves, C. J., P. L. Menezes, M. R. Lovell, T. C. Jen, S. L. Garvey, and M. L. Dietz, The tribological performance of bio-based room temperature ionic liquid lubricants: A possible next step in biolubricant technology, 5th World Tribology Congress, Torino, Italy, 2013. With permission.)

liquid crystal structure for increased lubricity, (ix) long polar anion–cation molecular chains for monolayer adsorption, and (x) economical costs [58,71,72,75–86].

The use of ionic liquids as lubricants was first reported in 1961, when fluoride-containing molten salts (i.e., LiF and BeF_2) were subjected to high-temperature (650–815°C) bearing tests [87]. Nearly four decades later, low-melting analogues of classical molten salts, room-temperature ionic liquids, were first evaluated as synthetic lubricating fluids [88]. Since that time, considerable attention has been devoted to the utilization of ionic liquids as lubricants. Three main applications have been most extensively explored: the use of ionic liquids as base oils, as additives, and as thin films [80]. When employed as base oils, ionic liquids have been reported to exhibit good tribological performance for steel/steel, steel/copper, steel/aluminum, ceramic/ceramic, and steel/ceramic sliding pairs [84,89–101]. The negligible vapor pressure of ionic liquids makes them good candidates for use under vacuum and in spacecraft applications [80]. Ionic liquids are also effective as additives to the main lubricant (e.g., mineral oils), where because of their tendency to form strong boundary films, the tribological performance of the base lubricant improves [80,102]. Thin-film lubrication employing ionic liquids has been studied by many researchers with the goal of replacing PFPEs [99,100,103–107].

Although the chemical structure of the cationic and anionic substituents of an ionic liquid can greatly vary, the most commonly studied ionic liquids in tribological processes have been those containing a tetrafluoroborate $\left(BF_4^- \right)$ or a hexafluorophosphate $\left(PF_6^- \right)$ anion [108,109], the result of the superior tribological properties that boron- and/or phosphorus-containing compounds often exhibit under the high pressures and elevated temperatures that lubricants can encounter [110–113]. The frequent use of boron- and phosphorus-containing ionic liquids as lubricants does not imply that either of them is optimum. Rather, ionic liquids based on these anions are commonly studied because they are readily available and low cost [76]. Moreover, other hydrophobic anions, such as bis(trifluoromethanesulfonyl)amide and tris(tetrafluoroethyl)trifluorophosphate, actually exhibit better tribological properties for steel–steel contact [76,114,115]. In general, as the hydrophobicity of the anion increases, both the thermal-oxidative stability and the tribological properties improve [76]. Other ionic liquids have been studied with the goal of improving their tribological properties including phosphonium [116–118] and ammonium [98,99,119–121].

Additionally, ionic liquids have a consistent and easily customizable chemical composition that affords them the ability to provide the levels of thermal-oxidative stability and lubricity required for a variety of applications in the aerospace, automotive, manufacturing, and magnetic storage industries [76,80,84,87–101,114,122,123]. The consistent chemical composition of ionic liquids allows them to have physicochemical properties that are readily reproducible. Furthermore, they can be designed to be environmentally friendly by selecting both the cationic and anionic constituents to be nontoxic [124,125]. In many instances, ionic liquids can be prepared from nonpetroleum resources. Lastly, their capacity to overcome the variety of environmental, cost, and performance challenges faced by both petroleum-based and biobased lubricants makes them a potentially attractive alternative biolubricant [85].

The possibility of preparing an ionic liquid capable of functioning as an efficient biolubricant while exhibiting a variety of other useful properties is a result of their physicochemical characteristics, inherent tunability, and structural diversity of the novel compounds. Regarding the latter point, it has been estimated that as many as 10^{18} different combinations of anion and cation moieties are possible [126]. Clearly, this vast assortment of possibilities can pose a significant challenge in the ionic liquid design. As the number of desired properties increases, the number of possible ionic liquid candidates dramatically declines. Here, for example, the desire for an environmentally friendly lubricant means that the use of highly fluorinated anions is unacceptable.

Schematics of the molecular structure of sample environmentally benign anions and cations are depicted in Figure 13.3 [127]. Anions based on common food additives or artificial sweeteners, salicylate and benzoate, are utilized as biolubricants with their structures shown in Figure 13.3a and b [116–118]. These molecules are effective anionic constituents in ionic liquid lubricants due

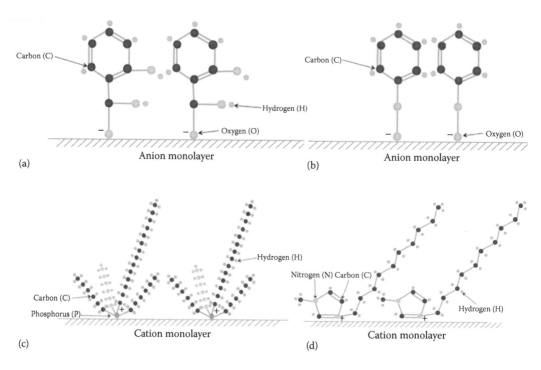

FIGURE 13.3 Environmentally benign ionic liquid anion and cation monolayer molecular structures: (a) salicylate; (b) benzoate; (c) trihexyltetradecylphosphonium; and (d) 1,3-dialkylimidazolium. (From Royal Society of Chemistry, ChemSpider: Search and Share Chemistry, http://www.chemspider.com/, 2014. With permission.)

to their ringlike structure that produce denser monolayers when absorbed onto the charged metal surface [72]. Similar considerations guide the choice of the cation and suggest that trihexyltetradecylphosphonium salts, shown in Figure 13.3c, some of which have been found to exhibit antimicrobial properties, can satisfy many of the desired criteria [128,129]. Trihexyltetradecylphosphonium, abbreviated as $P_{666,14}$, is an effective cationic constituent due to its long alkyl chain of 14 carbon atoms and three additional alkyl chains of 6 carbon atoms, as depicted in Figure 13.3c. Employing renewable feedstocks such as fructose for the use of certain 1,3-dialkylimidazolium cations depicted in Figure 13.3d has also been investigated [78,89,125,130–134]. The imidazolium cation has probably been studied in the most detail, a result of the high thermal stability of imidazole-based rings [122]. Additionally, the chain length of the imidazolium cation can be readily altered. The structure of an imidazolium cation is shown with a 10-carbon chain in Figure 13.3d. In contrast to the improvement in thermal-oxidative stability observed with hydrophobic anions, a decrease in stability is observed with more hydrophobic cations [76]. Nonetheless, ionic liquids with long alkyl chains and lower polarity have been reported to have excellent tribological properties from low to high temperatures (–30°C to 200°C) [74].

13.4 DRAWBACKS OF BIOLUBRICANTS

Natural oils have varying properties, such as thermal-oxidative stability, viscosity, VI, and low-temperature behavior, that are dependent on the structure of the molecules and the triacyclglycerol composition. Despite the many favorable attributes of biolubricants, the largest drawback is their poor thermal-oxidative stability, solidification at low temperatures (high pour points), biological (bacterial) deterioration, hydrolytic instability (aqueous decomposition), and inconsistent chemical composition (biological variations) [72].

13.4.1 Fatty Acid

The structures of the fatty acids affect the properties of the biolubricants in terms of thermal-oxidative stability, viscosity, VI, and low-temperature behavior. By increasing the length of the carbon chain, the fatty acid becomes more oily or fatty and increasingly less water soluble. The short nonbranched fatty acid chains having approximately six carbon atoms are more water soluble due to the presence of the polar –COOH groups [77,114]. If every carbon atom is attached to two hydrogen atoms in the carbon chain, except those at the ends of the chain, which are attached to three hydrogen atoms, then the fatty acid is considered to be a fully saturated, geometrically configured in a linear shape, and referred to as *saturated fatty acid*. When hydrogen atoms are missing from adjacent carbon atoms, the carbons share a double bond and the fatty acid is referred to as *unsaturated fatty acids*. These acids have lower vapor transition temperatures than those of saturated fatty acids as shown in Table 13.2. The fatty acid is polyunsaturated if it contains multiple double bonds. Therefore, the classifications of fatty acids are saturated, monounsaturated, and polyunsaturated with subcategories of diunsaturated and triunsaturated depending on the number of double bonds present [78]. Investigations have revealed that the most common unsaturated fatty acids contained in natural oils are oleic acid (C18:1), linoleic acid (C18:2), and linolenic acid (C18:3). The most common saturated fatty acids are palmitic acid (C16:0) and stearic acid (C18:0) [9]. The fatty acid percentages are shown in Figure 13.4 for several common vegetable oils. These figures include those for genetically modified high-oleic safflower oil. As can be seen from Figure 13.4, natural oils are predominantly composed of oleic acid and linoleic acid with small trace amounts of other acids. Table 13.2 details the characteristics between the most common fatty acids found in the natural oils [135]. It can be seen in Table 13.2 that as the number of double bonds increases, the unsaturation increases and the boiling point decreases. All of these properties influence the tribological performance of a lubricant when used as a base stock or an additive enhancement. Interestingly, comparing Figures 13.1 and 13.4, the oils with low friction and low wear have higher amounts of oleic acid and lower amounts of linoleic acid. The presence of a double bond decreases the density of the fatty acid monolayer [26,29,55]. An effective fatty acid monolayer can be established by saturated and monounsaturated fatty acids, which act to limit the metal-to-metal contact, thereby minimizing the friction force [26,46].

13.4.2 Unsaturation Number (UN)

There is a balance that must be maintained within biolubricants in fatty acid composition. For example, at room temperature, stearic acid is solid wax, whereas oleic acid is liquid. For this reason,

TABLE 13.2
Examples of Common Fatty Acids in Vegetable Oils

Fatty Acid Type	Common Name	Molecular Formula	Lipid Number	Degree of Unsaturation	Vapor Transition Temperature (°C)
Saturated	Palmitic acid	$C_{16}H_{32}O_2$	C16:0	0	352
Saturated	Stearic acid	$C_{18}H_{36}O_2$	C18:0	0	383
Monounsaturated	Oleic acid	$C_{18}H_{34}O_2$	C18:1	1	360
Diunsaturated	Linoleic acid	$C_{18}H_{32}O_2$	C18:2	2	230
Triunsaturated	Linolenic acid	$C_{18}H_{30}O_2$	C18:3	3	228

Source: Reprinted from *Tribology International*, 90, Reeves, C. J., P. L. Menezes, T. C. Jen, and M. R. Lovell, The influence of fatty acids on tribological and thermal properties of natural oils as sustainable biolubricants, 123–134, Copyright (2014), with permission from Elsevier.

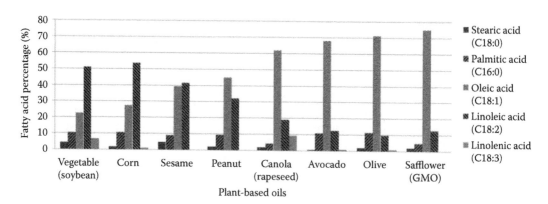

FIGURE 13.4 Fatty acid percentages of common plant-based oils. (From Reeves, C. J., P. L. Menezes, T.-C. Jen., and M. R. Lovell, Evaluating the tribological performance of green liquid lubricants and powder additive based green liquid lubricants, STLE Annual Meeting & Exhibition, St. Louis, 2012. With permission.)

the proportions of saturated and unsaturated acids must be optimal to ensure that oil is liquid at room temperature and can thus serve as a functional fluid. To quantify the degree of saturation, the UN is used. The UN of natural oils refers to the average number of double bonds within a triacyl-glycerol molecule. The greater the UN, the greater the degree of unsaturation. The UN is calculated from the fatty acid composition of the oil as follows:

$$UN = \frac{1}{100}\left[\left(1 \times \sum \text{monounsaturated fatty acids}\right) + \left(2 \times \sum \text{diunsaturated fatty acids}\right)\right.$$
$$\left. + \left(3 \times \sum \text{triunsaturated fatty acids}\right)\right], \tag{13.1}$$

$$UN = \frac{1}{100}\{[1 \times (C18:1 + C20:1 + C22:1)] + [2 \times (C18:2 + C22:2)] + [3 \times (C18:3)]\}. \tag{13.2}$$

In Equation 13.2, the $Cx:y$ is the percentage of fatty acids within the plant-based oil with x carbon in the chain and y double bonds. Similarly, Equation 13.1 could be used to calculate the UNs of unsaturated esters or any other unsaturated hydrocarbon. Table 13.3 shows the calculated UNs of a variety of plant-based oils [135]. Research has shown that natural oils with lower UNs maintain a higher thermal-oxidative stability as well as superior tribological properties.

13.4.3 THERMAL-OXIDATIVE STABILITY

Research indicates that improving the thermal-oxidative stability of natural oils requires a zero or low percentage of polyunsaturated fatty acids (i.e., linoleic) [136–141]. Furthermore, monounsaturated fatty acids, such as oleic acid, improve oxidative stability while simultaneously providing good low-temperature properties. The best compromise between thermal-oxidative stability and low-temperature properties is through the use of high-oleic acid oils, which can be obtained from genetically modified crops. High oleic acid with concentrations above 80% can be commercially obtained from genetically modified safflower, canola, sunflower, or soybean crops [11]. An alternative method would be to perform chemical modifications of the plant-based oil to achieve the desired stability. Fully saturated esters exhibit excellent oxidative stabilities, while partially unsaturated esters require chemical modifications to be useful as engine, transmission, hydraulic, or compressor

TABLE 13.3
Natural Oil UNs

Natural Oil Type	UN
Olive	0.9
Avocado	1.0
High-oleic safflower	1.0
Peanut	1.1
Sesame	1.2
Canola	1.3
Corn	1.4
Soybean	1.5

Source: Reprinted from *Tribology International*, 90, Reeves, C. J., P. L. Menezes, T. C. Jen, and M. R. Lovell, The influence of fatty acids on tribological and thermal properties of natural oils as sustainable biolubricants, 123–134, Copyright (2014), with permission from Elsevier.

lubricants [142]. Moreover, fully saturated diester oils are highly stable toward oxidation and demonstrate a high viscosity–temperature index. By adjusting the chain length of the dicarboxylic acid, the viscosity can be modified, affecting the hydrodynamic capabilities, which will influence the tribological performance.

13.4.4 Viscosity and VI

The viscosity of biolubricants influences the applicability of biolubricants as feasible alternatives to petroleum-based lubricants. Generally, biobased oils are known for their high VIs and can be considered multirange oils. The viscosity and the VI of a biolubricant tend to increase with increasing chain length of the carboxyl acid or with an increase in the molecular weight of the alcohol. In the case of polyols, the viscosity depends on the number of hydroxy functional groups present. The viscosity of polyols with identical fatty acids decreases in the same order as the base fluids with the following series of viscosities (mm²/s): PE > trimethylolhexane > TMP > trimethyloloethane > glycerol > NPG, as illustrated in Figure 13.5 [143–146]. The VI describes

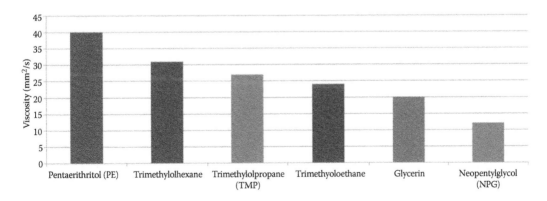

FIGURE 13.5 Viscosities of different polyolesters with the same base acid. (Ullmann, F., and W. Gerhartz, *Ullmann's Encyclopedia of Industrial Chemistry.* 1988. Copyright Wiley-VCH Verlag GmbH & Co. KGaA. Reproduced with permission.)

the dependence of viscosity on temperature. The higher the VI, the smaller the changes in viscosity over a broader temperature range. The VI is also affected by branching where it has similar effects as double bonds. Increasing the branching of either the carboxyl acid or the alcohol functional groups, while maintaining a constant carbon number, will decrease the viscosity and the VI [8]. In contrast, an increase in chain length with no change of structure results in an increase in viscosity. In complex esters and acids where bulk branching and chain length change, the correlation with viscosity is not observed [147].

13.4.5 LOW TEMPERATURE

Low-temperature performance often refers to the pour point and the cloud point of fluids. The pour point is the lowest temperature at which the fluid still flows. Pour points for biobased lubricants exhibit the same dependencies as viscosity. With esters, short-chain branching of the alcohol lowers the pour point. However, this molecular structure also leads to a decrease in oxidative stability of the alcohol. For this reason, neopentyl-polyols are advantageous for the production of synthetic ester lubricants because their molecular structure is primarily composed of branched hydroxyl groups. Natural oils should have fewer saturated fatty acids and shorter chain length and branching for optimal low pour points [148–150], as shown in Figure 13.6. Saturated fatty acids exhibit excellent low-temperature properties, which improves with increased degree of unsaturation (Figure 13.6). However, increased unsaturation leads to high oxidation, which is not desirable. Short-chained saturated fatty acids are optimal for their cold-flow properties; as the chain length increases to 16–18 carbon atoms, these fatty acids become solid at temperatures of 65°C–75°C [74]. Research has shown that oils with high amounts of oleic acid are the best compromise between cold-flow properties and oxidative stability. High–oleic acid oils such as sunflower oil have pour points of –35°C. TMP polyolesters have pour points of –50°C [46,47].

Further investigations into the pour point have shown that the position of the double bonds within the fatty acids has no significant influence on the cold-flow properties. However, slight differences can be observed depending on the degree of distortion imparted by the double bonds on the molecules. Depending on the position of the double bond, the distance between molecules increases or decreases and this can influence the pour point. The influence of the double bond location, for example, at the C_5, C_7, C_9, C_{11}, or C_{13} positions is shown in Figure 13.7 for both the *cis*- and *trans*-configurations [151]. As shown in Figure 13.7, the stereoconfiguration of the fatty acid influences the cold-flow properties, where the *cis*-configuration has a lower pour point than that of the *trans*-configuration [106].

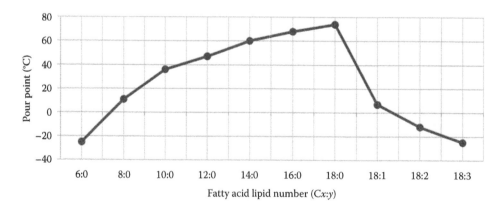

FIGURE 13.6 Dependence of pour point on fatty acid chain length and degree of unsaturation. (From Hart, H., and R. D. Schuetz, *Organic Chemistry: A Short Course*, Houghton Mifflin, Boston, 1978. With permission.)

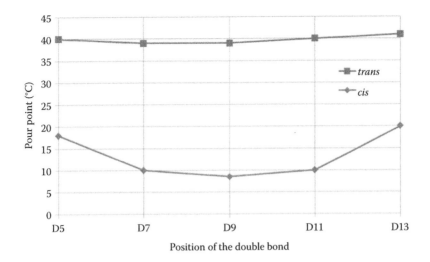

FIGURE 13.7   Influence of double bond position on the pour point. (From Hart, H., and R. D. Schuetz, *Organic Chemistry: A Short Course*, Houghton Mifflin, Boston, 1978. With permission.)

13.4.6   HYDROLYTIC STABILITY

The degree of relative resistance to be attacked or cleavaged by water is known as hydrolytic stability. This property is strongly dependent on the ester structure of biobased fluids, because it is an equilibrium reaction. Biobased fluids cleave at the ester bond, turning into their alcohol and acid components. This process continues until the chemical equilibrium is restored [147]. Biolubricants derived from saturated esters have a higher degree of hydrolytic stability than unsaturated or branched esters.

Steric hindrance is a natural phenomenon that occurs when a large number of methyl groups within a molecule prevent chemical reactions with other molecules. The consequence to the ester bond is improved hydrolytic stability. The methyl groups sterically protect the ester bond against hydrolytic attacks. Depending on the number of methyl groups positioned near the ester group, the reaction rate may be reduced by many orders of magnitude. Table 13.4 illustrates the reaction rate as a function of the number of α-carbon branching of the ester with one substitution when RR = 1 [147]. Biolubricants with linear short-chain monoalcohols display improved hydrolytic stability. Branched alcohols with greater than eight carbon atoms increased their hydrolytic stability to the level of saturated linear monoalcohols [147]. When using glycerol molecules in synthetic esters, the

TABLE 13.4

Influence of α-Carbon Branching of the Oxygen Ester on the Relative Rate of Hydrolytic Reactions as Explained by *ASTM Manual 37*

Alkyl Substitution on α-Carbon	Relative Reaction Rate[a]
CH_3	30
CH_3CH_2	1
$(CH_3)_2CH$	0.3
$(CH_3)_3C$	0

Source: Totten, G. E., S. R. Westbrook, and R. J. Shah, *Fuels and Lubricants Handbook—Technology, Properties, Performance, and Testing*, Vol. 1, ASTM International, West Conshohocken, PA, 2003. With permission.

[a]   Relative to reaction rate of ester with one substitute where RR = 1.

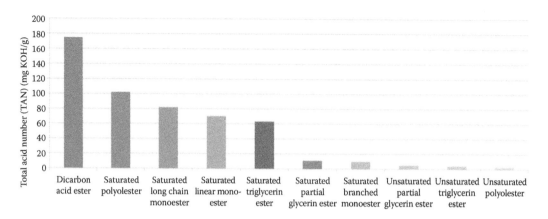

FIGURE 13.8 Increasing hydrolytic stability of different ester structures, as explained by *ASTM Manual 37*. (From Totten, G. E., S. R. Westbrook, and R. J. Shah, *Fuels and Lubricants Handbook—Technology, Properties, Performance, and Testing*, Vol. 1, ASTM International, West Conshohocken, PA, 2003. With permission.)

saturated esters behave with a higher degree of hydrolytic stability than that of stable unsaturated or branched esters. The level of hydrolytic stability is comparable to that of the stable monoesters.

The hydrolytic stability of saturated dicarboxylic esters is almost independent of their chain length and branching, and the structure of the alcohol components. Figure 13.8 depicts the hydrolytic stabilities of different ester structures, where a low hydrolytic stability corresponds to a high acid number [147,152,153]. It can be seen in the figure that saturated partial glycerin ester, saturated branched monoester, unsaturated partial glycerin ester, unsaturated triglycerin ester, and unsaturated polyolester, exhibit superior hydrolytic stability.

Although biolubricants are often susceptible to hydrolytic stability, it is one of the properties that afford them their high biodegradability. Reducing the effects of hydrolysis may reduce the rate of biodegradability. A balance must be maintained when improving hydrolytic stability so as not to impair biodegradability.

13.4.7 Lubrication Mechanisms

Many biolubricants are amphiphilic in nature as they are composed of molecules with polar heads and nonpolar carbon chains [154]. These oils are primarily water insoluble due to the presence of the long hydrocarbon chains in the molecule. Depending on the type of biobased feedstock, other functional groups such as epoxies, hydroxides, and various polar groups might also be present in the molecule, which can impact tribological and other important properties [155]. The amphiphilic properties of biolubricants can affect the boundary as well as the hydrodynamic properties of oils [156].

When lubricants operate in the boundary regime, their performance is often impacted by their adsorption and tribochemical properties with a metal surface [157–163]. Adsorption refers to the ability of lubricant molecules to adsorb to the friction surfaces, minimize asperity contact, and reduce friction and wear at the interface. The ability of a lubricant to adsorb onto a surface is quantified using free energy of adsorption terms [157–163]. Tribochemical reactions often occur as a result of high temperatures, pressures, and shearing of the lubrication process, which can occur in the hydrodynamic regime. This process in the interface is highly volatile and so complex that it is not fully understood. As a result, tribochemical reactions are often mistakenly blamed for mechanical failures resulting from the oil degradation by oxidation and the generation of friction polymers [154].

13.5　PERFORMANCE ENHANCEMENTS

Biolubricant research has focused on solving the deficiencies of natural and synthetic oils while seeking to understand the relationship between chemical composition and molecular structure. In addition, property enhancements through chemical modifications such as epoxidation, metathesis, acylation, estolide formation, transesterification, and selective hydrogenation have been conducted [164,165]. Investigations into the oxidative limitations of biolubricants have been researched and new techniques have been proposed to enhance the oxidative stability of biolubricants [9]. In some instances, through various chemical manipulations and enhancements, biolubricants have been shown to offer higher oxidative stability than traditional petroleum-based lubricants [12,20,41,46,148,166–175]. Still other researchers have investigated the stability of biolubricants when antioxidants are added [176–178]. Nonetheless, the chemical modification to biolubricants by addition reactions to the double bonds constitutes one of the most promising processes for obtaining commercially viable products from renewable raw materials. Thus, to facilitate the use of lubricants derived from biobased materials, additives, chemical modifications, and genetically modified plants will all play critical roles to ensure that adequate functionability and stability as biolubricants and functional fluids grow prominently [179–181].

13.5.1　ADDITIVES

Commercial mineral lubricants can consist of 10% to 25% additives depending on the application [182]. Additives are necessary to impart properties that are application specific and not provided by oil base stock. Examples of additives include antioxidants, metal deactivators, detergents, dispersants, corrosion inhibitors, demulsifiers, rust inhibitors, antiwear additives, extreme pressure additives, viscosity improvers, pour point depressants, and hydrolysis preventers [177,183–185]. Table 13.5 shows some of the examples of additives used with biolubricants [9]. Additives are common practice in lubrication technology, but the toxicity of currently used additives requires research on the development and the use of alternative biobased environmentally benign additives. Thus far, naturally occurring antioxidants, such as tocopherol (vitamin E), L-ascorbic acid (vitamin C), esters of gallic acid (lauric alcohol and dodecanol), citric acid derivatives, or lipid-modified ethylenediaminetetraacetic acid derivatives, serve as synthetic metal scavengers and provide viable alternatives to the currently used toxic antioxidants [11].

TABLE 13.5
Examples of Additives Used in Biolubricants

Additive Type	Compound
Antioxidants	BHT and other phenols
	Alkyl-substituted diphenylamines
Deactivators for Cu, Zn, etc.	Benzotriazoles
Corrosion inhibitors	Ester sulfonates
	Succinic acid esters
Antiwear additives	Zn dithiophosphate
Pour point depressants	Malan styrene copolymers
	Polymethacrylates
Hydrolysis protection	Carbodiimides

Source: Schneider, M. P.: Plant-oil-based lubricants and hydraulic fluids. *Journal of the Science of Food and Agriculture.* 2006. 86. 1769–1780. Copyright Wiley-VCH Verlag GmbH & Co. KGaA. Reproduced with permission.

Other biolubricants function more as antiwear and extreme pressure additives and consist of environmentally benign solid particulate additives. Green solid powders are a class of "solid lubricants" consisting of lamellar crystal structures with low interlayer friction [29,186,187]. These solid powders can function as lubricants themselves in dry applications and as additives when mixed into liquid lubricants such as vegetable oils. Examples of green solid powder additives include boric acid (H_3BO_3) and hexagonal boron nitride (hBN). As shown in Figure 13.9, they have similar properties as graphite, molybdenum disulfide, and tungsten disulfide [186,188,189]. Lamellar powders are superior antiwear additives due to their crystal structure. Atoms lying on the same layer are closely packed and strongly bonded by covalent bonds, whereas the layers are relatively far apart due to weak van der Waals force. This is depicted in Figure 13.9a and b for boric acid and hBN, respectively [187]. Pictures of these solid powders can be seen in Figure 13.9c and d. When entrained between sliding surfaces, these lamellar powders can adhere to the surface, form a protective boundary layer, and minimize contact between opposing surface asperities to prevent wear [190]. The protective boundary acts as a lubricant in sliding contacts by accommodating relative surface velocities. The lamellar powders accomplish lubricity by aligning their layers parallel to the direction of motion and sliding past one another to minimize friction. Moreover, these powders can lubricate in extreme conditions and at high or low temperatures and pressures [27,28,51,191–193].

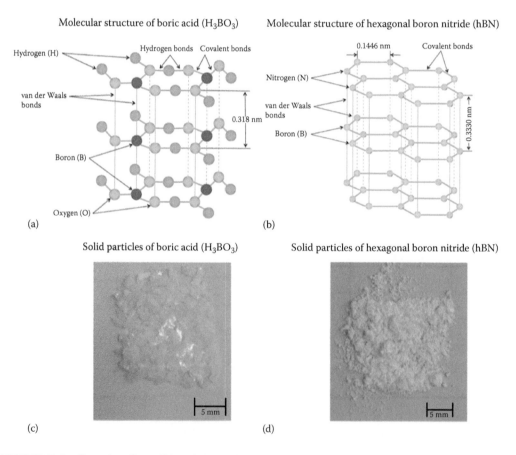

FIGURE 13.9 Green lamellar solid particles: molecular structures of (a) boric acid and (b) hexagonal boron nitride and solid particle photographs of (c) boric acid and (d) hexagonal boron nitride. (With kind permission from Springer Science+Business Media: *Tribology for Scientists and Engineers*, Tribology of solid lubricants, 2013, 447–494, Reeves, C. J., P. L. Menezes, M. R. Lovell, and T.-C. Jen.)

Although there are many solid powder lubricants that can be used as additives, there are far less that are environmentally benign. hBN and boric acid powder represent some of the more environmentally friendly additives that are inert to most chemicals [9]. They have been extensively studied because they are highly refractory materials with physical and chemical properties similar to that of graphite [28,29,194,195]. These powders are extremely lubricious with attractive performance-enhancing attributes similar to that of other lamellar solids. Boric acid is naturally found, whereas hBN is synthesized from boric oxide or boric acid compounds. Generally, boron-based compounds are extremely stable and do not break down to form hazardous materials under normal operation. Thus, they are safe to handle and feasible to use in industrial applications. Furthermore, there are no regulations regarding their use, storage, transport, or disposal, and for these reasons, hBN and boric acid can be considered environmentally benign substances [196].

Research has indicated that powder-based biolubricants can be problematic because they can be forced out of the contact zone during dry sliding contact [196]. In an attempt to remedy this, biolubricant colloidal mixtures composed of powder additives such as boric acid or hBN were combined with natural oils such as canola or soybean oil to create an environmentally friendly lubricant formulation [37]. The formulation remained in the contact zone without degrading over time [197, 198]. These powder-based biolubricants demonstrated improved friction and wear reductions and ecological sustainability [35,51,72,199].

Recently, the effect of hBN particle size in canola oil on friction was investigated using a pin-on-disk apparatus [29,37]. Here, various sizes of 5 wt% hBN particles were formulated into colloidal mixtures with canola oil. The particle sizes investigated were 70 nm, 0.5 μm, 1.5 μm, and 5.0 μm. Figure 13.10a shows the variation of the coefficient of friction with sliding distance for a formulation with a single particle size and Figure 13.10b shows the variation of the coefficient of friction with sliding distance for formulations with multiple particle sizes. It was determined that the nanometer-sized particulate mixtures gave lower friction and a smoother surface finish to the metallic substrates compared to micron- and submicron-sized particulates.

Lovell et al. [35] studied the particle effect of nano- (20 nm), submicron- (600 nm), and micron-sized (4 μm) boric acid powder additives in canola oil on tribological properties. Pure canola oil and canola oil with MoS_2 powder (ranging from 0.5 to 10 μm) were used as reference fluids. Figure 13.11a shows the variation of coefficient of friction with sliding distance for lubricants containing various particulate size additives and Figure 13.11b shows the variation of wear volume in the presence of various sizes of boric acid particles. The results showed that lubricants with the nanosized boric acid particulate additives significantly lowered friction and wear. The nanosized boric acid powder-based lubricants exhibited a wear rate of more than an order of magnitude lower than that of MoS_2 and larger-sized boric acid particles. It was also discovered that the oil mixed with a combination of submicron- and micron-sized boric acid powder additives exhibited better friction and wear performances than the canola oil mixed with either submicron- or micron-sized boric acid additives alone.

13.5.2 Oxidation

Many natural oils and biobased feedstock are composed of fatty acids that are derived from triacylglyceride molecules, which contain glycerol. The presence of glycerol in biobased materials gives rise to a tertiary β-hydrogen attached to the β-carbon of the glycerol, as illustrated in Figure 13.12a. The β-hydrogen present in the glycerol is known to have oxidative instability and is, therefore, a cause for the fast oxidation of natural oils in a process known as autoxidation. Additionally, the *bis*-allylic hydrogen, which attaches to the double-bonded carbon atoms in polyunsaturated fatty acids, are particularly susceptible to free radical attacks, peroxide formation, and production of polar oxidation products similar to that of hydrocarbon mineral oils, except at an expedited rate in natural oils [200–203]. Oxidative instability also arises from the presence of the double bonds present in the triacylglycerol molecule as depicted in Figure 13.12b [15,27,28,204–209]. Figure 13.13 shows how the oxidation stability of vegetable oil worsens with increasing double bonds with the Rancimat

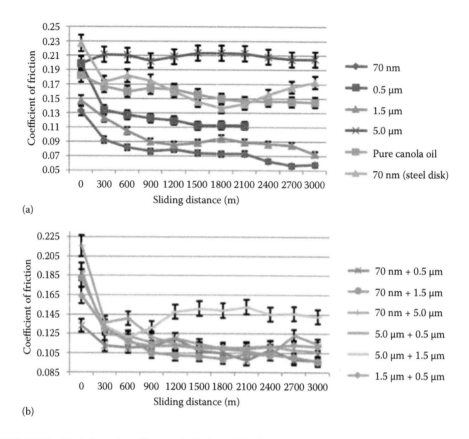

(a)

(b)

FIGURE 13.10 Variation of coefficient of friction with sliding distance for formulation of canola oil with 5% hexagonal boron nitride particles with (a) single particle size and (b) blends of particulates with multiple particle sizes. (With kind permission from Springer Science+Business Media: *Tribology Letters*, The size effect of boron nitride particles on the tribological performance of biolubricants for energy conservation and sustainability, 51, 2013, 437–452, Reeves, C. J., P. L. Menezes, M. R. Lovell, and T.-C. Jen.)

method as explained in *ASTM Manual 37* [147,210]. The effect of the degree of saturation in oxidation of three different polyolesters is conveyed in Figure 13.14. Oxidation stability is estimated by the change in viscosity as a function of aging time according to the Baader test method (DIN 51587) [210]. As a precaution, saturated alcohols and acids are used for the production of lubricants. The oxidation of biobased lubricants can produce insoluble deposits, increase acidity and viscosity, and reduce corrosion protection. Moreover, the existence of unsaturated ester compounds in biolubricants causes hydrolytic degradation, which also increases oxidation [9,40,60,166,211,212]. Oxidation stability can be enhanced by sterically protecting the double bonds with branching. The effects of branching on oxidation stability are similar to the effects of increased hydrolytic stability through branching.

Oxidation instability is negligible in oxygen-free environments since β-hydrogen and *bis*-allylic hydrogen are no longer susceptible to oxidative degradation in the absence of oxygen. The thermal degradation temperatures of biolubricants in oxygen-free atmosphere are significantly higher than that in air [213–215]. Figure 13.15 compares the onset and the maximum thermal degradation temperatures under nitrogen of various natural oils using thermogravimetric analysis method. These values are 20°C to 100°C higher than those reported for fatty acid-based oils under oxygen by Schneider [9], Kodali [60], Birova et al. [166], and Canter [211]; for vegetable oils by Ohkawa et al. [216]; and for synthetic ester base stocks by Jayadas and Nair [41]. To further quantify the susceptibility of the biolubricants to thermal and oxidative vulnerabilities, an unsaturation analysis

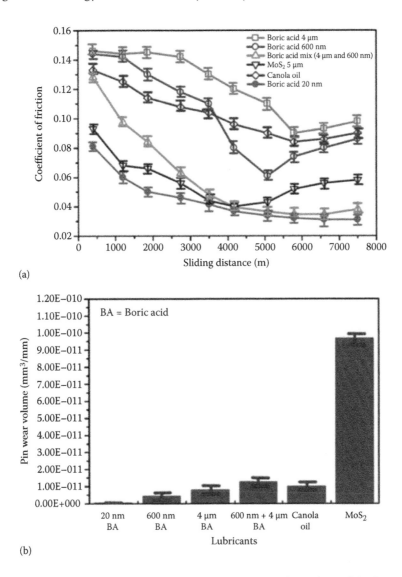

FIGURE 13.11 (a) Coefficient of friction versus sliding distance measured on a pin-on-disk tribometer, steel-on-steel using canola oil (open diamond) without and with 5% particulates of various sizes: open square, boric acid, 4 μm; open circle, boric acid, 600 nm; open triangle, boric acid mix, 4 μm and 600 nm; open inverted triangle, MoS₂, 5 μm; and filled circle, boric acid, 20 nm. (b) Wear rate of the Cu pins slid against Al disks using canola oil with 5% particulates of various sizes of boric acid or MoS₂. (Lovell, M. R. et al., *Philos. Trans. A*, *Math.*, *Phys.*, *Eng. Sci.* 368(1929), 4851–4868, 2010. Reproduced by permission of The Royal Society of Chemistry.)

of biolubricants is often performed. In general, decreasing the unsaturation levels increases the thermal-oxidative stability.

13.5.3 ESTERIFICATION

Esterification or transesterification is the modification of the ester moieties present in the triacylglycerides, which have glycerol as the alcohol. In this process, the glycerol is replaced with simple alcohols such as ethanol, or polyols such as TMP, NPG, and PE, effectively creating a synthetic ester. Results indicate that hydrolytic and oxidative stabilities are considerably increased for biolubricants if the fatty acid portion consists of saturated fatty acids such as stearic acid. High-oleic acid

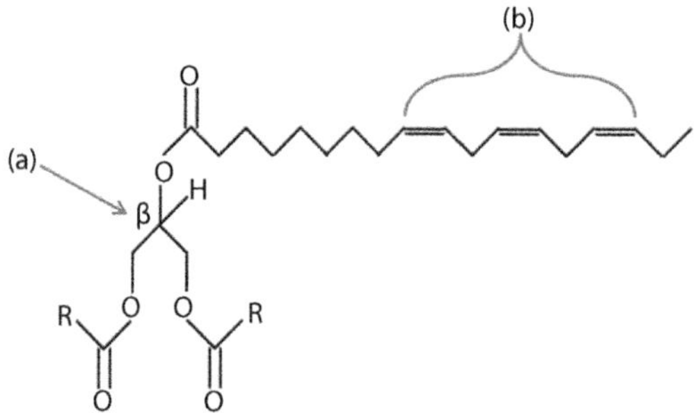

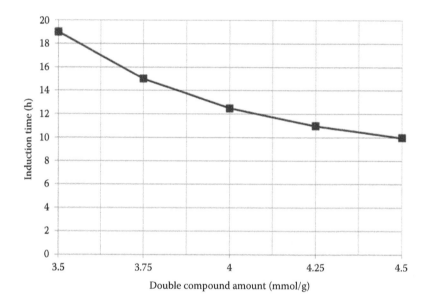

FIGURE 13.12 Carbon–hydrogen bonds in triacylglycerides susceptible to oxidation: (a) carbon–hydrogen bonds on glycerol and (b) allylic and *bis*-allylic carbon–hydrogen bonds on fatty acid. (Schneider, M. P.: Plant-oil-based lubricants and hydraulic fluids. *Journal of the Science of Food and Agriculture*. 2006. 86. 1769–1780. Copyright Wiley-VCH Verlag GmbH & Co. KGaA. Reproduced with permission.)

FIGURE 13.13 Oxidation stability versus concentration of double bonds in vegetable oil by the Rancimat method. (From Totten, G. E., S. R. Westbrook, and R. J. Shah, *Fuels and Lubricants Handbook—Technology, Properties, Performance, and Testing*, Vol. 1, ASTM International, West Conshohocken, 2003; Streitwieser, A., C. H. Heathcock, and E. M. Kosower, *Introduction to Organic Chemistry*, Prentice Hall, Upper Saddle River, 1992. With permission.)

vegetable oils have shown better stability than commodity vegetable oils with lower amounts of oleic acids (68–72%) with TMP [47].

Epoxidation is an important functionalization reaction of double bonds to improve oxidative stability, lubricity, and low-temperature behavior of biolubricants [212]. Epoxidized fatty acids are often subjected to ring-opening reactions as shown in Figure 13.15 [9,60,166,211]. Epoxidation can be carried out using peroxy acids, dioxiranes, peracids, and alkyl hydroperoxides [217]. Other epoxidation techniques include chemoenzymatic self-epoxidation, as well as with in situ performic acid procedures [218]. Previous attempts to chemically modify the fatty acid chain of natural oils lead to the development of diester compounds synthesized from oleic acid and common fatty acids [219]. A multistep process of oleochemical diesters begins with epoxidation, followed by ring opening of epoxidized oleic acid with fatty acids using a *p*-toluenesulfonic acid as a catalyst to yield monoester

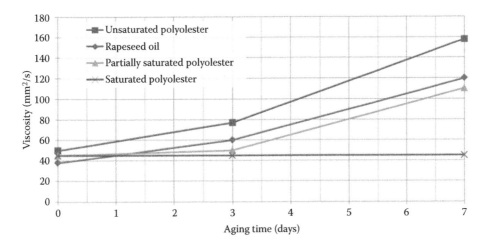

FIGURE 13.14 The oxidation stability of varying degrees of unsaturated carboxylic acids by the viscosity increase after an aging process according to the Baader test (DIN 51587). (Ullmann, F., and W. Gerhartz, *Ullmann's Encyclopedia of Industrial Chemistry*. 1988 Copyright Wiley-VCH Verlag GmbH & Co. KGaA. Reproduced with permission; Streitwieser, A., C. H. Heathcock, and E. M. Kosower, *Introduction to Organic Chemistry*, Prentice Hall, Upper Saddle River, NJ, 1992. With permission.)

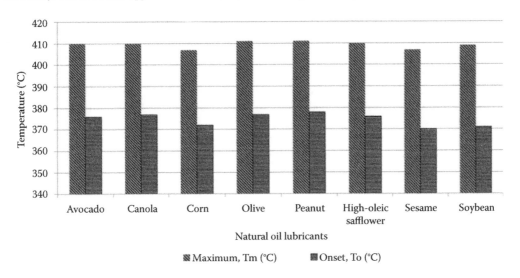

FIGURE 13.15 Thermal degradation temperature of natural oils under a nitrogen environment. (Reprinted from *Tribology International*, 90, Reeves, C. J., P. L. Menezes, T. C. Jen, and M. R. Lovell, The influence of fatty acids on tribological and thermal properties of natural oils as sustainable biolubricants, 123–134, Copyright (2014), with permission from Elsevier.)

compounds. The process ends with the esterification reaction of these compounds, producing the desired diester compounds [220–225]. These diester compounds have demonstrated enhanced low-temperature behavior due to the increased ability of the long-chain esters to disrupt macrocrystalline formation at low temperatures.

13.5.4 ESTOLIDES OF FATTY ACIDS

Estolides are a class of esters derived from natural and synthetic compounds synthesized from fats and oils at the location of the carbon-to-carbon double bond. The estolide structure consists of a secondary ester linkage between one fatty acid acyl molecule and one fatty acid alkyl molecule [150,226–228]. The carbocation on the estolide can undergo nucleophilic attacks by other fatty acids, with or without carbocation migration along the length of the chain, to form estolides [61,229–240]. Estolides can be formed as free acids, esters, or triacylglycerides. Estolides were primarily developed to overcome the thermal-oxidative instabilities and poor low-temperature properties of natural oils [218]. In some instances, the properties of estolides are enhanced with additives, however, at the expense of biodegradability, cost, and toxicity. Recent estolides include saturated monoestolide methyl esters and enriched saturated monoestolide 2-ethylhexyl esters from oleic acid, lauric acid, and free estolides [56]. Results indicate that chain length and estolide number affect low-temperature properties and tribological performance [56].

13.5.5 SELECTIVE HYDROGENATION

The multiple double bonds in fatty acids increase their susceptibility toward oxidation attacks. Monounsaturated fatty acids with only one isolated *cis* double bond, such as oleic acid (C18:1), are considerably less susceptible to thermal-oxidation stability. The ideal solution would be to remove all unsaturation by catalytic hydrogenation within the fatty acids. However, this would "harden" the lubricant, leaving it with saturated fatty acids, such as stearic acid, that would turn the liquid oil to a solid or a semisolid fat, making it harder for use as a lubricant or a hydraulic fluid. For this reason, it is necessary to leave monounsaturated fatty acids in the oil to ensure optimal lubricity, viscosity, and pour point.

Selective hydrogenation is the process of converting polyunsaturated fatty acids to fatty acids with only one double bond. For example, linolenic acid would lose two double bonds, whereas linoleic acid, which has two double bonds, would lose one double bond. Thus, these two acids will be converted to oleic acid with only one double bond. Hydrogenation can be carried out with a variety of catalysts [11]. However, selective hydrogenation is still in its infancy, reactions of linolenic acid and linoleic acid were reduced but not strictly to oleic acid, some of the reaction isomerized to elaidic acid (trans C18:1) which has similarly properties as stearic acid (C18:0), that are undesirable, and thus rendering the process unsubstantiated. Through continued efforts this process does have potential to allow the improvement of readily available natural oils, however more research is needed.

13.5.6 BRANCHED FATTY ACIDS

Biolubricants synthesized from branched fatty acids, such as iostearic acid, exhibit superior pour point, high chemical stability, and high flash point. Isostearic acid is obtained by thermal isomerization of polyunsaturated C18 fatty acids followed by hydrogenation. The branching points are limited to the interior portion of the molecule [70,241]. The 12-hydroxystearic acid derived from ricinoleic acid by hydrogenation can also be used. Compromises are made between tribological performance, oxidative stability, low-temperature performance, and biodegradability. For example, branched fatty acids provide low-temperature capability important in the oils for many industrial applications. However, branched fatty acids also have lower biodegradability than nonbranched, which is a major disadvantage.

13.6   BIODEGRADABILITY AND ECOTOXICITY

In addition to good tribological properties, biolubricants also display low toxicity and high biodegradability. These two properties are discussed in the following sections.

13.6.1 Biodegradability

One of the primary attributes of biobased lubricants is their inherent biodegradability. Biodegradation is the chemical dissolution of organic substances by enzymes of living organisms into CO_2 and H_2O [57,242–245]. Organic material can be biodegraded aerobically in the presence of oxygen or anaerobically without oxygen. Biodegradability of a lubricant can be classified into two types, primary biodegradation and ultimate biodegradation. Primary degradation refers to the disappearance of the original organic compound and may or may not indicate that the substance will biodegrade completely. This method is done by evaluating the infrared bands of the C–H bonds through the standard method CEC-L-33-A-93 [246,247], which has yet to become widely accepted due to the obscurity of the compound disappearance [83]. Ultimate degradation, also known as total degradation, refers to biodegradation of the original organic compound to carbon dioxide and water in 28 days. It is measured by the OECD 301B test method and has gained worldwide acceptance [248]. Currently, there are many other standard test methods for assessing biodegradability [249].

13.6.2 Ecotoxicity

Ecotoxicity is an important property when considering biolubricant usage. With up to 50% of all lubricants entering the environment via waste streams, spills, normal usage, and improper disposal, lubricants create toxicity to plants, animals, and humans. These effects are more severe in aquatic ecosystems due to their high sensitivity. The aquatic toxicity of lubricants to various species is measured using different standard test methods as follows: green algae, *Pseudokirchneriella subcapitata* or *Desmodesmus subspicatus*: OECD 201; freshwater fleas, *Daphina magma*: OECD 202-12; rainbow trout minnows, *Oncorhynchus mykiss*, or zebrafish minnows, *Brachydanio rerio*: OECD 203-13; bacteria, *Pseudomonas putida*: OECD 209; and laboratory rats, Sprague Dawley: OECD 401. A lubricant is classified as biodegradable if it biodegrades by at least 80% within 28 days, CECL-33-A-93 method, or by 60% after 28 days, OECD 301B method. Toxicity measurements are based on the LD_{50} value, the amount required to kill 50% of a given test population. A lubricant is deemed eco-friendly if its LD_{50} is greater than 1000 ppm.

13.6.3 Biolubricant Environmental Definitions

Despite the criteria for evaluating biodegradability and ecotoxicity, many lubricants are still deemed biolubricants that are environmentally acceptable even if they are not properly formulated. For example, if a lubricant consists of a biobased oil in combination with a toxic additive, the environmental friendliness of the lubricant is thus compromised and should be referred to as *not* fully environmentally acceptable. Several European countries have begun eco-labeling lubricants as environmentally acceptable. Examples include Blue Angel in Germany, Nordic Swan in Scandinavia, Euro Margerite in other European countries, and BioPreferred® program in the United States [165,250–253].

13.6.4 Laws and Regulations

Most countries have laws designed to protect water resources, land, workplaces, and air from pollution. In most countries, there are no compulsory legislative measures about the use of biolubricants. Portugal was the first country to institute a mandate requiring outboard two-stroke engines to use two-thirds biodegradable lubricant, the minimum biodegradability requirement according to CECL-33-T-82. In Austria, the use of a plant-derived lubricant for chain saw oils is mandated by federal regulation. Recommendations for the use of biobased lubricants and functional fluids exist in the United Kingdom and Canada [254–256]. In the United States, the USDA has established guidelines for designating and promoting biobased products through two initiatives: (i) product labeling and (ii) the federal procurement preference. It is called the BioPreferred program and its primary focus is to "promote the increased purchase and use of bio-based products" [257]. This program aims at creating economic development by generating new jobs and building new markets for farm commodities. The USDA believes that an increase in the development, purchase, and use of biobased products will reduce the nation's reliance on petroleum. The underlying goal of the BioPreferred program is to increase the use of renewable agricultural resources in hopes of reducing adverse environmental and health impacts of petroleum-based products entering the environment. Currently, many state and local community regulations are trying to convince citizens, contractors, and subcontractors to use environmentally acceptable lubricants and functional fluids.

The price of biolubricants is a major issue that affects their widespread use. On average, biolubricants cost about three times more than traditional petroleum-based lubricants. Vegetable oils and synthetic esters can range anywhere from 1.5 to 5 times more expensive than mineral oils [10]. There must be a balance of the economic versus environmental costs in order to minimize the price difference between petroleum- and biobased lubricants. One way to accomplish this is through government subsidies for manufacturers of biolubricants.

To bring biolubricants to the forefront of the lubrication industry, there need to be economic incentives on protecting the environment. Legislative pressure could promote the use of biolubricants. In order for this to occur, lawmakers would have to quantify and legislate hazardous lubricants at the expense of the environment. Such a decision will provide incentives to consumers, industrial companies, contractors, and other stakeholders to preferentially use biolubricants over petroleum-based lubricants. Currently, many lubrication engineers base their initial decision of which lubricant to use first on price, then on performance, and lastly on environmental consideration.

Currently, the market share of biolubricants is <1% and many manufacturers do not see the immediate value of using a biolubricant because it is more expensive. However, a government subsidy could help to encourage them to create products that operate with biolubricants. This would in turn reduce the price of biolubricants through increased market saturation.

13.7 CONCLUSIONS

Biolubricants can provide economical and feasible alternatives to petroleum-based lubricants while promoting energy conservation and sustainability. They are nontoxic and renewable and provide environmental, health, economic, and performance advantages. Development of biobased lubricants requires maintaining a compromise between environmental, tribological, and price considerations. Although some biolubricants suffer from poor thermal-oxidative stability, high pour points, poor biostability, and poor hydrolytic instability, new techniques are being developed to rectify these drawbacks. In this chapter, biolubricants were examined from the macroscale with their advantages and deficiencies being highlighted. Biolubricants have tremendously progressed over the last 30 years and hopefully in the next 30 years, they can become as prevalent and ubiquitous as mineral oils are now.

REFERENCES

1. Bhushan, B., *Principles and Applications of Tribology*. John Wiley, New York (1999).
2. Menezes, P. L., Reeves, C. J., and Lovell, M. R., Fundamentals of lubrication. In: Menezes, P. L., Nosonovsky, M., Ingole, S. P., Kailas, S. V., and Lovell, M. R. (eds.) *Tribology for Scientists and Engineers*. pp. 295–340. Springer, New York (2013).
3. Reeves, C. J., Menezes, P. L., Lovell, M. R., and Jen, T. C., Macroscale applications in tribology. In: Menezes, P. L., Nosonovsky, M., Ingole, S. P., Kailas, S. V., and Lovell, M. R. (eds.) *Tribology for Scientists and Engineers*. pp. 881–919. Springer, New York (2013).

4. Reeves, C. J., Menezes, P. L., Lovell, M. R., and Jen, T. C., Microscale applications in tribology. In: Menezes, P. L., Nosonovsky, M., Ingole, S. P., Kailas, S. V., and Lovell, M. R. (eds.) *Tribology for Scientists and Engineers.* pp. 921–948. Springer, New York (2013).

5. Menezes, P. L., Reeves, C. J., Kailas, S. V., and Lovell, M. R., Tribology in metal forming. In: Menezes, P. L., Nosonovsky, M., Ingole, S. P., Kailas, S. V., and Lovell, M. R. (eds.) *Tribology for Scientists and Engineers.* pp. 783–818. Springer, New York (2013).

6. Menezes, P. L., Ingole, S., Nosonovsky, M., Kailas, S. V., and Lovell, M. R., *Tribology for Scientists and Engineers.* Springer, New York (2013).

7. Naegely, P. C., Environmentally acceptable lubricants, In: *Seed Oils for the Future.* AOCS Press, Champaign, IL (1993).

8. Mang, T., and Dresel, W., *Lubricants and Lubrication.* Wiley, Weinheim, Germany (2006).

9. Schneider, M. P., Plant oil-based lubricants and hydraulic fluids. *Journal of the Science of Food and Agriculture.* **86**(12), 1769–1780 (2006).

10. IENICA, Biolubricants: Market Data Sheet. http://www.ienica.net/marketdatasheets/biolubricantsmds .pdf (2004). Accessed May 16, 2012.

11. Salimon, J., Salih, N., and Yousif, E., Biolubricants: Raw materials, chemical modifications and environmental benefits. *European Journal of Lipid Science and Technology.* **112**(5), 519–530 (2010).

12. Aluyor, E. O., Obahiagbon, K. O., and Ori-jesu, M., Biodegradation of vegetable oils: A review. *Science Research Essays.* **4**(6), 543–548 (2009).

13. Deffeyes, K. S., *Hubbert's Peak.* Princeton University Press, Princeton, NJ (2009).

14. Goodstein, D. L., *Out of Gas: The End of the Age of Oil.* First ed. W. W. Norton, New York (2004).

15. Lovell, M., Higgs, C. F., Deshmukh, P., and Mobley, A., Increasing formability in sheet metal stamping operations using environmentally friendly lubricants. *Journal of Materials Processing Technology.* **177**(1), 87 (2006).

16. Li, W., Kong, X. H., Ruan, M., Ma, F. M., Jiang, Y. F., Liu, M. Z., Chen, Y., and Zuo, X. H., Green waxes, adhesives and lubricants. *Philosophical Transactions: Series A, Mathematical, Physical, and Engineering Sciences.* **368**(1929), 4869–4890 (2010).

17. Mannekote, J., Menezes, P., Kailas, S., and Sathwik, R. K. C., Tribology of green lubricants. In: Menezes, P. L., Nosonovsky, M., Ingole, S. P., Kailas, S. V., and Lovell, M. R. (eds.) *Tribology for Scientists and Engineers.* pp. 495–521. Springer, New York (2013).

18. Backé, W., The present and future of fluid power: Proceedings of the Institution of Mechanical Engineers, Part I. *Journal of Systems and Control Engineering.* **207**(4), 193–212 (1993).

19. Kumar, A., and Sharma, S., An evaluation of multipurpose oil seed crop for industrial uses (*Jatropha curcas* L.): A review. *Industrial Crops and Products.* **28**(1), 1–10 (2008).

20. Meier, M. A. R., Metzger, J. O., and Schubert, U. S., Plant oil renewable resources as green alternatives in polymer science. *Chemical Society Reviews.* **36**(11), 1788–1802 (2007).

21. Feldmann, D. G., and Remmelmann, A., Biologisch schnell abbaubare Hydraulikfluessigkeiten— Ergebnisse von Pruefstandstests und Folgerungen fuer die Anwendung. *Aachener Fluidtechnisches Kolloquium.* **12**(1), 59–80 (1996).

22. Feldmann, D. G., and Kessler, M., Fluid qualification tests—Evaluation of the lubricating properties of biodegradable fluids. *Industrial Lubrication and Tribology.* **54**, 117–129 (2002).

23. Feldmann, D. G., and Hinrichs, J., Evaluation of the lubrication properties of biodegradable fluids and their potential to replace mineral oil in heavily loaded hydrostatic transmissions. *ASTM Special Technical Publication.* **1310**, 220 (1997).

24. Fessenbecker, A., and Korff, J., Additives for environmentally more friendly lubricant. *Journal of Japanese Society of Tribologists.* **40**(4), 306 (1995).

25. Korff, J., and Fessenbecker, A., Additives for biodegradable lubricants. *NLGI Spokesman.* **57**(3), 19 (1993).

26. Fox, N. J., Tyrer, B., and Stachowiak, G. W., Boundary lubrication performance of free fatty acids in sunflower oil. *Tribology Letters.* **16**(4), 275–281 (2004).

27. Duzcukoglu, H., and Sahin, O., Investigation of wear performance of canola oil containing boric acid under boundary friction condition. *Tribology Transactions.* **54**(1), 57–61 (2011).

28. Erdemir, A., Tribological properties of boric acid and boric acid forming surfaces: Part 1: Crystal chemistry and self-lubricating mechanism of boric acid. Paper presented at the Society of Tribologists Lubrication Engineers Annual Conference, Denver, CO (May 1990).

29. Reeves, C. J., Menezes, P. L., Jen, T.-C., and Lovell, M. R., Evaluating the tribological performance of green liquid lubricants and powder additive based green liquid lubricants. In: STLE Annual Meeting & Exhibition, St. Louis, MO (2012) STLE.

30. Werman, M. J., and Neeman, I., Avocado oil production and chemical characteristics. *Journal of the American Oil Chemists' Society.* **64**(2), 229–232 (1987).

31. Bennion, M., and Scheule, B., *Introductory Foods.* Prentice Hall, Upper Saddle River, NJ (2010).

32. Grushcow, J., High oleic plant oils with hydroxy fatty acids for emission reduction. In: 2005 World Tribology Congress III, Washington, DC (2005) pp. 485–486. American Society of Mechanical Engineers.

33. Biermann, U., and Metzger, J. O., Synthesis of alkyl-branched fatty acids. *European Journal of Lipid Science and Technology.* **110**(9), 805–811 (2008).

34. Grushcow, J., and Smith, M. A., Next generation feedstocks from new frontiers in oilseed engineering. *ASME Conference Proceedings.* **2005**(42010), 487–488 (2005).

35. Lovell, M. R., Menezes, P. L., Kabir, M. A., and Higgs III, C. F., Influence of boric acid additive size on green lubricant performance. *Philosophical Transactions: Series A, Mathematical, Physical, and Engineering Sciences* **368**(1929), 4851–4868 (2010).

36. Randles, S. J., Formulation of environmentally acceptable lubricants. Paper presented at the 49th STLE Annual Meeting, Pittsburgh, PA, May 1–5, 1994 (1994).

37. Reeves, C. J., Menezes, P. L., Lovell, M. R., and Jen, T.-C., The size effect of boron nitride particles on the tribological performance of biolubricants for energy conservation and sustainability. *Tribology Letters.* **51**(3), 437–452 (2013).

38. Smith, S. A., King, R. E., and Min, D. B., Oxidative and thermal stabilities of genetically modified high oleic sunflower oil. *Food Chemistry.* **102**(4), 1208–1213 (2007).

39. Marmesat, S., Morales, A., Velasco, J., and Carmen Dobarganes, M., Influence of fatty acid composition on chemical changes in blends of sunflower oils during thermoxidation and frying. *Food Chemistry.* **135**(4), 2333–2339 (2012).

40. Fox, N. J., and Stachowiak, G. W., Vegetable oil-based lubricants: A review of oxidation. *Tribology International.* **40**(7), 1035–1046 (2007).

41. Jayadas, N. H., and Nair, K. P., Coconut oil as base oil for industrial lubricants—Evaluation and modification of thermal, oxidative and low temperature properties. *Tribology International.* **39**(9), 873–878 (2006).

42. Zeman, A., Sprengel, A., Niedermeier, D., and Späth, M., Biodegradable lubricants studies on thermo-oxidation of metal-working and hydraulic fluids by differential scanning calorimetry (DSC). *Thermochimica Acta.* **268**, 9–15 (1995).

43. Mendoza, G., Igartua A, Fernandez-Diaz, B., Urquiola, F., Vivanco, S., and Arguizoniz, R., Vegetable oils as hydraulic fluids for agricultural applications. *Grasas Aceites.* **62**(1), 29–38 (2011).

44. Lundgren, S. M., Ruths, M., Danerlov, K., and Persson, K., Effects of unsaturation on film structure and friction of fatty acids in a model base oil. *Journal of Colloid and Interface Science.* **326**(2), 530–536 (2008).

45. Hu, Z.-S., Hsu, S. M., and Wang, P. S., Tribochemical reaction of stearic acid on copper surface studied by surface enhanced Raman spectroscopy. *Tribology Transactions.* **35**(3), 417–422 (1992).

46. Salih, N., Salimon, J., and Yousif, E., The physicochemical and tribological properties of oleic acid based triester biolubricants. *Industrial Crops & Products.* **34**(1), 1089–1096 (2011).

47. Erhan, S. Z., Sharma, B. K., and Perez, J. M., Oxidation and low temperature stability of vegetable oil-based lubricants. *Industrial Crops and Products.* **24**(3), 292–299 (2006).

48. Koshima, H., Kamano, H., Hisaeda, Y., Liu, H., and Ye, S., Analyses of the adsorption structures of friction modifiers by means of quantitative structure-property relationship method and sum frequency generation spectroscopy. *Tribology Online Tribology Online.* **5**(3), 165–172 (2010).

49. Bowden, F. P., Leben, L., and Tabor, D., The influence of temperature on the stability of a mineral oil. *Transactions of the Faraday Society.* **35**, 900–904 (1939).

50. Bauccio, M., and American Society for Metals, *ASM Metals Reference Book.* ASM International, Materials Park, OH (1993).

51. Deshmukh, P., Lovell, M., Sawyer, W. G., and Mobley, A., On the friction and wear performance of boric acid lubricant combinations in extended duration operations. *Wear.* **260**(11–12), 1295–1304 (2006).

52. Menezes, P. L., Lovell, M. R., Kabir, M. A., Higgs III, C. F., and Rohatgi, P. K., Green lubricants: Role of additive size. In: Nosonovsky, M., and Bhushan, B. (eds.) *Green Tribology: Green Energy and Technology*, pp. 265–286. Springer, Heidelberg (2012).

53. Havet, L., Blouet, J., Robbe Valloire, F., Brasseur, E., and Slomka, D., Tribological characteristics of some environmentally friendly lubricants. *Wear.* **248**(1–2), 140–146 (2001).

54. Santos, O. D. H., Morais, J. M., Andrade, F. F., Aguiar, T. A., and Rocha Filho, P. A., Development of vegetable oil emulsions with lamellar liquid-crystalline structures. *Journal of Dispersion Science and Technology.* **32**(3), 433–438 (2011).

55. Lundgren, S. M., Persson, K., Mueller, G., Kronberg, B., Clarke, J., Chtaib, M., and Claesson, P. M., Unsaturated fatty acids in alkane solution: Adsorption to steel surfaces. *Langmuir: The ACS Journal of Surfaces and Colloids.* **23**(21), 10598–10602 (2007).
56. Isbell, T. A., Chemistry and physical properties of estolides. *Grasas y Aceites.* **62**(1), 8–20 (2011).
57. Cermak, S. C., and Isbell, T. A., Synthesis and physical properties of mono-estolides with varying chain lengths. *Industrial Crops and Products.* **29**(1), 205–213 (2009).
58. Shubkin, R. L., *Synthetic Lubricants and High-Performance Functional Fluids.* Marcel Dekker, New York (1993).
59. Uosukainen, E., Linko, Y. Y., Lamsa, M., Tervakangas, T., and Linko, P., Transesterification of tri-methylolpropane and rapeseed oil methyl ester to environmentally acceptable lubricants. *Journal of the American Oil Chemists' Society.* **75**(11), 1557–1563 (1998).
60. Kodali, D. R., Biobased lubricants—Chemical modification of vegetable oils. *Inform—Champaign.* **14**, 121–143 (2003).
61. Wagner, H., Luther, R., and Mang, T., Lubricant base fluids based on renewable raw materials—Their catalytic manufacture and modification. *Applied Catalysis A, General.* **221**(1), 429–442 (2002).
62. Andersen, M. P. S., Hurley, M. D., Wallington, T. J., Blandini, F., Jensen, N. R., Librando, V., Hjorth, J., Marchionni, G., Avataneo, M., and Visca, M., Atmospheric chemistry of $CH_3O(CF_2CF_2O)_nCH_3$ (n = 1–3): Kinetics and mechanism of oxidation initiated by Cl atoms and OH radicals, IR spectra, and global warming potentials. *Journal of Physical Chemistry A.* **108**, 1964–1972 (2004).
63. National Oceanic and Atmospheric Administration, National Aeronautics and Space Administration, United Nations Environment Programme, World Meteorological Organization, and European Commission, Scientific Assessment of Ozone Depletion, 2002. Washington, DC; Nairobi, Kenya; Geneva, Switzerland; Brussels, Belgium (2003).
64. Wallington, T. J., Schneider, W. F., Sehested, J., Bilde, M., Platz, J., Nielsen, O. J., Christensen, L. K., Molina, M. J., Molina, L. T., and Wooldridge, P. W., Atmospheric chemistry of HFE-7100 ($C_4F_9OCH_3$): Reaction with OH radicals, UV spectra and kinetic data for $C_4F_9OCH_2\cdot$ and $C_4F_9OCH_2O_2$ radicals, and the atmospheric fate of $C_4F_9OCH_2O$ radicals. *Journal of Physical Chemistry A.* **101**(44), 8264–8274 (1997).
65. Marchionni, G., Avataneo, M., De Patto, U., Maccone, P., and Pezzin, G., Physical properties of four α,ω-dimethoxyfluoropolyethers. *Journal of Fluorine Chemistry.* **126**(4), 463–471 (2005).
66. Avataneo, M., De, P. U., and Marchionni, G., M., Liquid–liquid extraction of polar organic substances from their aqueous solutions with fluorinated extracting liquids. European Patent EP1346757 A1.
67. Marchionni, G., and Visca, M., Perfluoropolyethers (PFPEs) having at least an alkylether end group and respective preparation process. U.S. Patent 20030013923 A1 (20013).
68. Marchionni, G., Petricci, S., Guarda, P. A., Spataro, G., and Pezzin, G., The comparison of thermal stability of some hydrofluoroethers and hydrofluoropolyethers. *Journal of Fluorine Chemistry.* **125**(7), 1081–1086 (2004).
69. Marchionni, G., Maccone, P., and Pezzin, G., Thermodynamic and other physical properties of several hydrofluoro-compounds. *Journal of Fluorine Chemistry.* **118**(1–2), 149–155 (2002).
70. Asadauskas, S., and Erhan, S., Depression of pour points of vegetable oils by blending with diluents used for biodegradable lubricants. *Journal of the American Oil Chemists' Society.* **76**(3), 313–316 (1999).
71. Reeves, C. J., Menezes, P. L., Garvey, S. L., Jen, T. C., Dietz, M. L., and Lovell, M. R., The effect of anion–cation moiety manipulation to characterize the tribological performance of environmentally benign room temperature ionic liquid lubricants. Paper presented at the 2013 STLE Annual Meeting & Exhibition, Detroit, MI (2013).
72. Reeves, C. J., Garvey, S. L., Menezes, P. L., Dietz, M. L., Jen, T. C., and Lovell, M. R., Tribological performance of environmentally friendly ionic liquid lubricants. In: ASME/STLE 2012 International Joint Tribology Conference, Denver, CO (2012) STLE.
73. Manahan, S. E., *Environmental Chemistry.* Lewis, Boca Raton, FL (1994).
74. Jiménez, A.-E., and Bermúdez, M.-D., Ionic liquids as lubricants for steel-aluminum contacts at low and elevated temperatures. *Tribology Letters.* **26**(1), 53–60 (2007).
75. Suisse, J.-M., Bellemin-Laponnaz, S., Douce, L., and Maisse-François, A. W. R., A new liquid crystal compound based on an ionic imidazolium salt. *Tetrahedron Letters.* **46**(25), 4303–4305 (2005).
76. Minami, I., Ionic liquids in tribology. *Molecules (Basel, Switzerland).* **14**(6), 2286–2305 (2009).
77. Bermúdez, M. D., Jiménez, A. E., Sanes, J., and Carrion, F. J., Ionic liquids as advanced lubricant fluids. *Molecules (Basel, Switzerland).* **14**(8), 2888–2908 (2009).
78. Canter, N., Evaluating ionic liquids as potential lubricants. *Tribology and Lubrication Technology.* **61**(9), 15–17 (2005).

79. Liu, W., Ye, C., Gong, Q., Wang, H., and Wang, P., Tribological performance of room-temperature ionic liquids as lubricant. *Tribology Letters.* **13**(2), 81–85 (2002).

80. Zhou, F., Liang, Y., Liu, W., Ionic liquid lubricants: Designed chemistry for engineering applications. *Chemical Society Reviews.* **38**(9), 2590–2599 (2009).

81. Battez, H. A., Alonso, D. B., Rodriguez, R. G., Viesca Rodriguez, J. L., Fernandez-Gonzalez, A., and Garrido, A. H., Lubrication of DLC and tin coatings with two ionic liquids used as neat lubricant and oil additives. In: Proceedings of the STLE/ASME International Joint Tribology Conference 2011, Los Angeles, CA, October 23–26, 2011 (2011) American Society of Mechanical Engineers.

82. Freemantle, M., *An Introduction to Ionic Liquids.* RSC Publishing, Cambridge (2010).

83. Sheldon, R. A., Arends, I., and Hanefeld, U., *Green Chemistry and Catalysis.* Wiley, Weinheim, Germany (2007).

84. Yao, M., Liang, Y., Xia, Y., and Zhou, F., Bisimidazolium ionic liquids as the high-performance anti-wear additives in poly(ethylene glycol) for steel-steel contacts. *ACS Applied Materials and Interfaces.* **1**(2), 467–471 (2009).

85. Reeves, C. J., Menezes, P. L., Lovell, M. R., Jen, T. C., Garvey, S. L., and Dietz, M. L., The tribological performance of bio-based room temperature ionic liquid lubricants: A possible next step in biolubricant technology. In: 5th World Tribology Congress, Torino, Italy (2013) Society of Tribologists & Lubrication Engineers.

86. Reeves, C. J., Jen, T.-C., Garvey, S. L., Dietz, M. L., Menezes, P. L., and Lovell, M. R., The effect of phosphonium- and imidazolium-based ionic liquids as additives in natural oil: An investigation of tribological performance. In: Proceedings of the 2014 STLE Annual Meeting & Exhibition, Lake Buena Vista, FL, May 18–21, 2014 (2014).

87. Smith, P. G., High-temperature molten-salt lubricated hydrodynamic journal bearings. *ASLE Transactions.* **4**(2), 263–274 (1961).

88. Ye, C., Liu, W., Chen, Y., and Yu, L., Room-temperature ionic liquids: A novel versatile lubricant. *Chemical Communications (Cambridge, England).* **2001**(21), 2244–2245 (2001).

89. Phillips, B. S., and Zabinski, J. S., Ionic liquid lubrication effects on ceramics in a water environment. *Tribology Letters.* **17**(3), 533–541 (2004).

90. Wang, H., Lu, Q., Ye, C., Liu, W., and Cui, Z., Friction and wear behaviors of ionic liquid of alkylimidazolium hexafluorophosphates as lubricants for steel/steel contact. *Wear.* **256**(1–2), 44–48 (2004).

91. Liu, W., Ye, C., Chen, Y., Ou, Z., and Sun, D. C., Tribological behavior of sialon ceramics sliding against steel lubricated by fluorine-containing oils. *Tribology International.* **35**(8), 503–509 (2002).

92. Mu, Z., Liu, W., Zhang, S., and Zhou, F., Functional room-temperature ionic liquids as lubricants for an aluminum-on-steel system. *Chemistry Letters.* **33**, 524–525 (2004).

93. Lu, Q., Haizhong, W., Chengfeng, Y., Weimin, L., and Qunji, X., Room temperature ionic liquid 1-ethyl-3-hexylimidazolium-bis(trifluoromethylsulfonyl)-imide as lubricant for steel–steel contact. *Tribology International.* **37**(7), 547–552 (2004).

94. Reich, R. A., Stewart, P. A., Bohaychick, J., and Urbanski, J. A., Base oil properties of ionic liquids. *Lubrication Engineering.* **59**, 16–21 (2003).

95. Mu, Z., Zhou, F., Zhang, S., Liang, Y., and Liu, W., Effect of the functional groups in ionic liquid molecules on the friction and wear behavior of aluminum alloy in lubricated aluminum-on-steel contact. *Tribology International.* **38**(8), 725–731 (2005).

96. Jimenez, A. E., Bermudez, M. D., Iglesias, P., Carrion, F. J., and Martinez-Nicolas, G., 1-*N*-alkyl-3-methylimidazolium ionic liquids as neat lubricants and lubricant additives in steel–aluminium contacts. *Wear.* **260**(7–8), 766–782 (2006).

97. Jimenez, A. E., Bermudez, M. D., Carrion, F. J., and Martinez-Nicolas, G., Room temperature ionic liquids as lubricant additives in steel–aluminium contacts: Influence of sliding velocity, normal load and temperature. *Wear.* **261**(3–4), 347–359 (2006).

98. Omotowa, B. A., Phillips, B. S., Zabinski, J. S., and Shreeve, J. M., Phosphazene-based ionic liquids: Synthesis, temperature-dependent viscosity, and effect as additives in water lubrication of silicon nitride ceramics. *Inorganic Chemistry.* **43**(17), 5466–5471 (2004).

99. Yu, G., Zhou, F., Liu, W., Liang, Y., and Yan, S., Preparation of functional ionic liquids and tribological investigation of their ultra-thin films. *Wear.* **260**(9–10), 1076–1080 (2006).

100. Yu, B., Zhou, F., Mu, Z., Liang, Y., and Liu, W., Tribological properties of ultra-thin ionic liquid films on single-crystal silicon wafers with functionalized surfaces. *Tribology International.* **39**(9), 879–887 (2006).

101. Xia, Y., Wang, S., Zhou, F., Wang, H., Lin, Y., and Xu, T., Tribological properties of plasma nitrided stainless steel against SAE52100 steel under ionic liquid lubrication condition. *Tribology International.* **39**(7), 635–640 (2006).

102. Qu, J., Bansal, D. G., Yu, B., Howe, J. Y., Luo, H., Dai, S., Li, H., Blau, P. J., Bunting, B. G., Mordukhovich, G., and Smolenski, D. J., Antiwear performance and mechanism of an oil-miscible ionic liquid as a lubricant additive. *ACS Applied Materials & Interfaces.* **4**(2), 997–1002 (2012).

103. Mo, Y., Zhao, W., Zhu, M., and Bai, M., Nano/Microtribological properties of ultrathin functionalized imidazolium wear-resistant ionic liquid films on single crystal silicon. *Tribology Letters.* **32**(3), 143–151 (2008).

104. Zhu, M., Yan, J., Mo, Y., and Bai, M., Effect of the anion on the tribological properties of ionic liquid nano-films on surface-modified silicon wafers. *Tribology Letters.* **29**(3), 177–183 (2008).

105. Palacio, M., and Bhushan, B., Ultrathin wear-resistant ionic liquid films for novel MEMS/NEMS applications. *Advanced Materials.* **20**(6), 1194–1198 (2008).

106. Bhushan, B., Palacio, M., and Kinzig, B., AFM-based nanotribological and electrical characterization of ultrathin wear-resistant ionic liquid films. *Journal of Colloid and Interface Science.* **317**(1), 275–287 (2008).

107. Palacio, M., and Bhushan, B., Nanotribological and nanomechanical properties of lubricated PZT thin films for ferroelectric data storage applications. *Journal of Vacuum Science and Technology A: Vacuum, Surfaces and Films.* **26**(4), 768–776 (2008).

108. Minami, I., Inada, T., Sasaki, R., and Nanao, H., Tribo-chemistry of phosphonium-derived ionic liquids. *Tribology Letters.* **40**(2), 225–235 (2010).

109. Zeng, Z., Shreeve, J. M., Phillips, B. S., and Xiao, J. C., Polyfluoroalkyl, polyethylene glycol, 1,4-bismethylenebenzene, or 1,4-bismethylene-2,3,5,6-tetrafluorobenzene bridged functionalized dicationic ionic liquids: Synthesis and properties as high temperature lubricants. *Chemistry of Materials.* **20**(8), 2719–2726 (2008).

110. Shah, F., Glavatskih, S., and Antzutkin, O. N., Synthesis, physicochemical, and tribological characterization of S-Di-n-octoxyboron-O,O'-di-n-octyldithiophosphate. *ACS Applied Materials & Interfaces.* **1**(12), 2835–2842 (2009).

111. Mosey, N. J., Molecular mechanisms for the functionality of lubricant additives. *Science.* **307**(5715), 1612–1615 (2005).

112. Mangolini, F., Rossi, A., and Spencer, N. D., Chemical reactivity of triphenyl phosphorothionate (TPPT) with iron: An ATR/FT-IR and XPS investigation. *Journal of Physical Chemistry C.* **115**(4), 1339–1354 (2011).

113. Shah, F., Glavatskih, S., Höglund, E., Lindberg, M., and Antzutkin, O. N., Interfacial antiwear and physicochemical properties of alkylborate-dithiophosphates. *ACS Applied Materials & Interfaces.* **3**(4), 956–968 (2011).

114. Liu, X., Zhou, F., Liang, Y., and Liu, W., Tribological performance of phosphonium based ionic liquids for an aluminum-on-steel system and opinions on lubrication mechanism. *Wear* **261**(10), 1174–1179 (2006).

115. Itoh, T., Ishioka, A., Hayase, S., Kawatsura, M., Watanabe, N., Inada, K., Minami, I., and Mori, S., Design of alkyl sulfate ionic liquids for lubricants. *Chemistry Letters.* **38**(1), 64–65 (2009).

116. Shah, F. U., Glavatskih, S., MacFarlane, D. R., Somers, A., Forsyth, M., and Antzutkin, O. N., Novel halogen-free chelated orthoborate-phosphonium ionic liquids: Synthesis and tribophysical properties. *Chemical Physics (Incorporating Faraday Transactions).* **13**(28), 12865–12873 (2011).

117. Sun, J., Howlett, P. C., MacFarlane, D. R., Lin, J., and Forsyth, M., Synthesis and physical property characterisation of phosphonium ionic liquids based on $P(O)_2(OR)_2^-$ and $P(O)_2(R)_2^-$ anions with potential application for corrosion mitigation of magnesium alloys. *Electrochimica Acta.* **54**(2), 254–260 (2008).

118. Weng, L., Liu, X., Liang, Y., and Xue, Q., Effect of tetraalkylphosphonium based ionic liquids as lubricants on the tribological performance of a steel-on-steel system. *Tribology Letters.* **26**(1), 11–17 (2007).

119. Minami, I., Kamimura, H., and Mori, S., Thermo-oxidative stability of ionic liquids as lubricating fluids. *Journal of Synthetic Lubrication.* **24**(3), 135–147 (2007).

120. Kamimura, H., Kubo, T., Minami, I., and Mori, S., Effect and mechanism of additives for ionic liquids as new lubricants. *Tribology International.* **40**(4), 620–625 (2007).

121. Zhao W, Mo, Y., Pu, J., and Bai, M., Effect of cation on micro/nano-tribological properties of ultra-thin ionic liquid films. *Tribology International.* **42**(6), 828–835 (2009).

122. Ohtani, H., Ishimura, S., and Kumai, M., Thermal decomposition behaviors of imidazolium-type ionic liquids studied by pyrolysis-gas chromatography. *Analytical Sciences: The International Journal of the Japan Society for Analytical Chemistry.* **24**(10), 1335–1340 (2008).

123. Passerini, S., Alessandrini, F., Appetecchi, G. B., and Conte, M., Ionic liquid based electrolytes for high energy electrochemical storage devices. *ECS Transactions.* **1**(14), 67–71 (2006).

124. Quijano, G., Couvert, A., Amrane, A., Darracq, G., Couriol, C., Le Cloirec, P., Paquin, L., and Carrie, D., Toxicity and biodegradability of ionic liquids: New perspectives towards whole-cell biotechnological applications. *Chemical Engineering Journal.* **174**(1), 27–32 (2011).

125. Gathergood, N., Scammells, P. J., and Garcia, T. M., Biodegradable ionic liquids: Part III. The first readily biodegradable ionic liquids. *Green Chemistry.* **8**(2), 156–160 (2006).

126. Wu, M., Navarrini, W., Spataro, G., Venturini, F., and Sansotera, M., An environmentally friendly class of fluoropolyether: Alpha-omega-dialkoxyfluoropolyethers. *Applied Sciences.* **2**(2), 351–367 (2012).

127. Royal Society of Chemistry, ChemSpider: Search and Share Chemistry. http://www.chemspider.com/(2014). Accessed November 19, 2014.

128. Somers, A. E., Howlett, P. C., Sun, J., MacFarlane, D. R., and Forsyth, M., Transition in wear performance for ionic liquid lubricants under increasing load. *Tribology Letters.* **40**(2), 279–284 (2010).

129. Atefi, F., Garcia, M. T., Singer, R. D., and Scammells, P. J., Phosphonium ionic liquids: Design, synthesis and evaluation of biodegradability. *Green Chemistry.* **11**(10), 1595–1604 (2009).

130. Freire, M. G., Carvalho, P. J., Gardas, R. L., Marrucho, I. M., Santos, L. M., and Coutinho, J. A., Mutual solubilities of water and the $[C_n mim][Tf_2 N]$ hydrophobic ionic liquids. *The Journal of Physical Chemistry: B.* **112**(6), 1604–1610 (2008).

131. Handy, S. T., Greener solvents: Room temperature ionic liquids from biorenewable sources. *European Journal of Chemistry.* **9**(13), 2938–2944 (2003).

132. Gathergood, N., Garcia, T. M., and Scammells, P. J., Biodegradable ionic liquids: Part I: Concept, preliminary targets and evaluation. *Green Chemistry.* **6**(3), 166–175 (2004).

133. Corma, A., Iborra, S., and Velty, A., Chemical routes for the transformation of biomass into chemicals. *ChemInform.* **107**(6), 2411–2502 (2007).

134. Zhang, Z. C., Catalytic transformation of carbohydrates and lignin in ionic liquids. *Wiley Interdisciplinary Reviews: Energy and Environment*, **2**(6), 655–672 (2013).

135. Reeves, C. J., Menezes, P. L., Jen, T. C., and Lovell, M. R., The influence of fatty acids on tribological and thermal properties of natural oils as sustainable biolubricants. *Tribology International.* **90**, 123–134 (2015).

136. Sharma, B. K., Liu, Z., Adhvaryu, A., and Erhan, S. Z., One-pot synthesis of chemically modified vegetable oils. *Journal of Agricultural and Food Chemistry.* **56**(9), 3049–3056 (2008).

137. Hwang, H. S., and Erhan, S. Z., Modification of epoxidized soybean oil for lubricant formulations with improved oxidative stability and low pour point. *Journal of the American Oil Chemists' Society.* **78**, 1179–1184 (2001).

138. Li, S., Blackmon, J., Demange, A., and Jao, T. C., Linear sulphonate detergents as pour point depressants. *Lubrication Science.* **16**(2), 127–137 (2004).

139. Adamczewska, J. Z., and Wilson, D., Development of ecologically responsive lubricants. *Journal of Synthetic Lubrication.* **14**(2), 129–142 (1997).

140. Coscione, A. R., and Artz, W. E., Vegetable oil stability at elevated temperatures in the presence of ferric stearate and ferrous octanoate. *Journal of Agricultural and Food Chemistry.* **53**(6), 2088–2094 (2005).

141. Dunn, R. O., Effect of antioxidants on the oxidative stability of methyl soyate (biodiesel). *Fuel Processing Technology.* **86**(10), 1071–1085 (2005).

142. Adhvaryu, A., Sharma, B. K., Hwang, H. S., Erhan, S. Z., and Perez, J. M., Development of biobased synthetic fluids: Application of molecular modeling to structure–physical property relationship. *Industrial & Engineering Chemistry Research.* **45**(3), 928–933 (2006).

143. Asadauskas, S., Perez, J. M., and Duda, J. L., Oxidative stability and antiwear properties of high oleic vegetable oils. *Lubrication Engineering—Illinois.* **52**(12), 877–882 (1996).

144. Honary, L. A. T., An investigation of the use of soybean oil in hydraulic systems. *Bioresource Technology.* **56**(1), 41–47 (1996).

145. Lal, K., and Carrick, V., Performance testing of lubricants based on high oleic vegetable oils. *Journal of Synthetic Lubrication.* **11**(3), 189–206 (1994).

146. Ullmann, F., and Gerhartz, W., *Ullmann's Encyclopedia of Industrial Chemistry.* Wiley-VCH, Weinheim, Germany (1988).

147. Totten, G. E., Westbrook, S. R., and Shah, R. J., *Fuels and Lubricants Handbook—Technology, Properties, Performance, and Testing*, Vol. 1. ASTM International, West Conshohocken, PA (2003).

148. Jayadas, N. H., Nair, K. P., and Ajithkumar, G., Vegetable oils as base oil for industrial lubricants— Evaluation oxidative and low temperature properties using TGA, DTA and DSC. In: World Tribology Congress, Washington, DC (2005). Proceedings of the World Tribology Congress III—2005, pp. 539–540. American Society of Mechanical Engineers.

149. Quinchia, L. A., Delgado, M. A., Franco, J. M., Spikes, H. A., and Gallegos, C., Low-temperature flow behaviour of vegetable oil-based lubricants. *Industrial Crops and Products*. **37**(1), 383–388 (2012).

150. Salimon, J., and Salih, N., Oleic acid diesters: Synthesis, characterization and low temperature properties. *European Journal of Scientific Research*. **32**(2), 216–222 (2009).

151. Hart, H., and Schuetz, R. D., *Organic Chemistry: A Short Course*. Houghton Mifflin, Boston, (1978).

152. Roehrs, I., and Fessenbecker, A., A new additive for the hydrolytic and oxidative stabilization of ester based lubricants and greases. *NLGI Spokesman*. **61**(3), 10–17 (1997).

153. Boyde, S., Hydrolytic stability of synthetic ester lubricants. *Journal of Synthetic Lubrication*. **16**(4), 297–312 (2000).

154. Adhvaryu, A., Biresaw, G., Sharma, B. K., and Erhan, S. Z., Friction behavior of some seed oils: Biobased lubricant applications. *Industrial and Engineering Chemistry Research*. **45**(10), 3735–3740 (2006).

155. Adhvaryu, A., Erhan, S. Z., Liu, Z. S., and Perez, J. M., Oxidation kinetic studies of oils derived from unmodified and genetically modified vegetables using pressurized differential scanning calorimetry and nuclear magnetic resonance spectroscopy. *Thermochimica Acta*. **364**(1–2), 87–97 (2000).

156. Adhvaryu, A., Erhan, S. Z., and Perez, J. M., Tribological studies of thermally and chemically modified vegetable oils for use as environmentally friendly lubricants. *Wear*. **257**(3–4), 359–367 (2004).

157. Jahanmir, S., and Beltzer, M., An adsorption model for friction in boundary lubrication. *ASLE Transactions*. **29**(3), 423–430 (1986).

158. Jahanmir, S., and Beltzer, M., Effect of additive molecular structure on friction coefficient and adsorption. *Journal of Tribology*. **108**(1), 109–116 (1986).

159. Jahanmir, S., Chain length effects in boundary lubrication. *Wear*. **102**(4), 331–349 (1985).

160. Beltzer, M., and Jahanmir, S., Role of dispersion interactions between hydrocarbon chains in boundary lubrication. *ASLE Transactions*. **30**(1), 47–54 (1987).

161. Beltzer, M., and Jahanmir, S., Effect of additive molecular structure on friction. *Lubrication Science*. **1**(1), 3–26 (1988).

162. Schey, J. A., *Tribology in Metalworking: Friction, Lubrication, and Wear*. American Society for Metals, Metals Park, OH (1983).

163. Reeves, C. J., Menezes, P. L., Jen, T.-C., and Lovell, M. R., The effect of surface roughness on the tribological performance of environmentally friendly bio-based lubricants with varying particle size. In: 2013 STLE Annual Meeting & Exhibition, Detroit, MI, May 5–9, 2013 (2013).

164. Rudnick, L. R., and Shubkin, R. L., *Synthetic Lubricants and High-Performance Functional Fluids*, Second ed. CRC Press, Boca Raton, FL (1999).

165. Eisentraeger, A., Schmidt, M., Murrenhoff, H., Dott, W., and Hahn, S., Biodegradability testing of synthetic ester lubricants—Effects of additives and usage. *Chemosphere*. **48**(1), 89–96 (2002).

166. Birova, A., Pavlovicova, A., and Cvenros, J., Lubricating oils based on chemically modified vegetable oils. *Journal of Synthetic Lubrication*. **18**(4), 291–299 (2002).

167. King, J. W., Holliday, R. L., List, G. R., and Snyder, J. M., Hydrogenation of vegetable oils using mixtures of supercritical carbon dioxide and hydrogen. *Journal of the American Oil Chemists' Society*. **78**(2), 107–113 (2001).

168. Erhan, S. Z., Adhvaryu, A., and Sharma, B. K., Chemically functionalized vegetable oils. *Chemical Industries*. **111**, 361–388 (2006).

169. Doll, K. M., Sharma, B. K., and Erhan, S. Z., Synthesis of branched methyl hydroxy stearates including an ester from bio-based levulinic acid. *Industrial and Engineering Chemistry Research*. **46**(11), 3513–3519 (2007).

170. Yunus, R., Fakhru l-Razi, A., Ooi, T. L., Iyuke, S. E., and Perez, J. M., Lubrication properties of trimethylolpropane esters based on palm oil and palm kernel oils. *European Journal of Lipid Science and Technology*. **106**, 52–60 (2004).

171. Hwang, H.-S., Adhvaryu, A., and Erhan, S. Z., Preparation and properties of lubricant basestocks from epoxidized soybean oil and 2-ethylhexanol. *Journal of the American Oil Chemists' Society*. **80**(8), 811–815 (2003).

172. Verkuijlen, E., Kapteijn, F., Mol, J. C., and Boelhouwer, C., Heterogeneous metathesis of unsaturated fatty acid esters. *Journal of the Chemical Society*. (7), 198–199 (1977).

173. Schmidt, M. A., Dietrich, C. R., and Cahoon, E. B., Biotechnological enhancement of soybean oil for lubricant applications. In: Rudnick, L. R. (ed.) *Synthetics, Mineral Oils, and Bio-Based Lubricants: Chemistry and Technology*, Chemical Industries. pp. 389–398. CRC, Boca Raton, FL (2006).

174. Holser, R., Doll, K., and Erhan, S., Metathesis of methyl soyate with ruthenium catalysts. *Fuel.* **85**(3), 393–395 (2006).

175. Erhan, S. Z., Bagby, M. O., and Nelsen, T. C., Drying properties of metathesized soybean oil. *Journal of American Oil Chemists' Society.* **74**(6), 703–706 (1997).

176. Saad, B., Wai, W. T., and Lim, B. P., Comparative study on oxidative decomposition behavior of vegetable oils and its correlation with iodine value using thermogravimetric analysis. *Journal of Oleo Science.* **57**(4), 257–261 (2008).

177. Yanishlieva, N. V., and Marinova, E. M., Stabilisation of edible oils with natural antioxidants. *European Journal of Lipid Science and Technology.* **103**, 752–767 (2001).

178. Kapilan, N., Reddy, R. P., and Ashok Babu, T. P., Technical aspects of biodiesel and its oxidation stability. *International Journal of ChemTech Research.* **1**(2), 278–282 (2009).

179. Domingos, A. K., Saad, E. B., Vechiatto, W. W. D., Wilhelm, H. M., and Ramos, L. P., The influence of BHA, BHT and TBHQ on the oxidation stability of soybean oil ethyl esters (biodiesel). *Brazilian Chemical Society.* **18**(2), 416–423 (2007).

180. International Organization for Standardization, Animal and vegetable fats and oils: Determination of oxidative stability (accelerated oxidation test). International Organization for Standardization, Geneva (2006).

181. Gertz, C., Klostermann, S., and Kochhar, S. P., Testing and comparing oxidative stability of vegetable oils and fats at frying temperature. *European Journal of Lipid Science and Technology.* **102**, 543–551 (2000).

182. Brimberg, U. I., and Kamal-Eldin, A., On the kinetics of the autoxidation of fats: Influence of pro-oxidants, antioxidants and synergists. *European Journal of Lipid Science and Technology.* **105**(2), 83–91 (2003).

183. Gordon, M. H., and Kourimska, L., The effects of antioxidants on changes in oils during heating and deep frying. *Journal of the Science of Food and Agriculture.* **68**(3), 347–353 (1995).

184. Schober, S., and Mittellbach, M., The impact of antioxidants on biodiesel oxidation stability. *European Journal of Lipid Science and Technology.* **106**(6), 382–389 (2004).

185. Ruger, C. W., Klinker, E. J., and Hammond, E. G., Abilities of some antioxidants to stabilize soybean oil in industrial use conditions. *Journal of the American Oil Chemists' Society.* **79**(7), 733–736 (2002).

186. Menezes, P. L., Reeves, C. J., Rohatgi, P. K., and Lovell, M. R., Self-lubricating behavior of graphite-reinforced composites. In: Menezes, P. L., Nosonovsky, M., Ingole, S. P., Kailas, S. V., and Lovell, M. R. (eds.) *Tribology for Scientists and Engineers.* pp. 341–389. Springer, New York (2013).

187. Reeves, C. J., Menezes, P. L., Lovell, M. R., and Jen, T.-C., Tribology of solid lubricants. In: Menezes, P. L., Nosonovsky, M., Ingole, S. P., Kailas, S. V., and Lovell, M. R. (eds.) *Tribology for Scientists and Engineers.* pp. 447–494. Springer, New York (2013).

188. Farrington, A. M., and Slater, J. M., Monitoring of engine oil degradation by voltammetric methods utilizing disposable solid wire microelectrodes. *The Analyst.* **122**(6), 593–596 (1997).

189. Shankara, A., Menezes, P., Simha, K., and Kailas, S., Study of solid lubrication with MoS_2 coating in the presence of additives using reciprocating ball-on-flat scratch tester. *Sadhana.* **33**(3), 207–220 (2008).

190. Reeves, C. J., Menezes, P. L., Lovell, M. R., and Jen, T. C., The effect of particulate additives on the tribological performance of bio-based and ionic liquid-based lubricants for energy conservation and sustainability. In: STLE (ed.) STLE Annual Meeting & Exhibition, Buena Vista, FL (2014).

191. Clauss, F. J. (ed.), *Solid Lubricants and Self-Lubricating Solids.* Academic Press, New York (1972).

192. Kanakia, M. D., Peterson, M. B., Southwest Research Institute Fuel and Lubricants Research Facility, *Literature Review of Solid Lubrication Mechanisms.* Defense Technical Information Center, Ft. Belvoir, VA (1987).

193. Jamison, W. E., Structure and bonding effects on the lubricating properties of crystalline solids. *ASLE Transactions.* **15**(4), 296–305 (1972).

194. Winer, W., Molybdenum disulfide as a lubricant: A review of the fundamental knowledge. *Wear.* **10**(6), 422–452 (1967).

195. Wornyoh, E. Y. A., Jasti, V. K., and Higgs, C. F., A review of dry particulate lubrication: Powder and granular materials. *Journal of Tribology.* **129**(2), 438–449 (2007).

196. Peng, Q., Ji, W., and De, S., Mechanical properties of the hexagonal boron nitride monolayer: Ab initio study. *Computational Materials Science.* **56**, 11–17 (2012).

197. Lelonis, D. A., Tereshko, J. W., and Andersen, C. M., Boron nitride powder: A high-performance alternative for solid lubrication. In: Vol. QTZ-81506. Copyright 2006–2007 Momentive Performance Materials Inc. (2007, December).

198. Clayton, G. D., Clayton, F. E., Allan, R. E., and Patty, F. A., *Patty's Industrial Hygiene and Toxicology.* Wiley, New York (1991).
199. Nosonovsky, M., and Bhushan, B. (eds.), *Green Tribology: Biomimetics, Energy Conservation and Sustainability.* Springer, New York (2012).
200. Hu, X., On the size effect of molybdenum disulfide particles on tribological performance. *Industrial Lubrication and Tribology.* **57**(6), 255–259 (2005).
201. Huang, H. D., Tu, J. P., Gan, L. P., and Li, C. Z., An investigation on tribological properties of graphite nanosheets as oil additive. *Wear.* **261**(2), 140–144 (2006).
202. Qiu Sunqing, D. J., and Guoxu, C., A review of ultrafine particles as antiwear additives and friction modifiers in lubricating oils. *Lubrication Science.* **11**(3), 217–226 (1999).
203. Xiaodong, Z., Xun, F., Huaqiang, S., and Zhengshui, H., Lubricating properties of Cyanex 302-modified MoS₂ microspheres in base oil 500SN. *Lubrication Science.* **19**(1), 71–79 (2007).
204. Düzcükoğlu, H., and Acaroğlu, M., Lubrication properties of vegetable oils combined with boric acid and determination of their effects on wear. *Energy Sources, Part A: Recovery, Utilization, and Environmental Effects.* **32**(3), 275–285 (2010).
205. Erdemir, A., Bindal, C., and Fenske, G. R., Formation of ultralow friction surface films on boron carbide. *Applied Physics Letters.* **68**(12), 1637–1639 (1996).
206. Erdemir, A., Preparation of ultralow-friction surface films on vanadium diboride. *Wear.* **205**(1–2), 236–239 (1997).
207. Erdemir, A., Eryilmaz, O. L., and Fenske, G. R., Self-replenishing solid lubricant films on boron carbide. *Surface Engineering.* **15**(4), 291–295 (1999).
208. Erdemir, A., Fenske, G. R., Nichols, F. A., Erck, R. A., and Busch, D. E., Self-lubricating boric acid films for tribological applications. Paper presented at the Japan International Tribology Conference, Nagoya, October.
209. Erdemir, A., Lubrication from mixture of boric acid with oils and greases. U.S. Patent 5431830 A (1995).
210. Streitwieser, A., Heathcock, C. H., and Kosower, E. M., *Introduction to Organic Chemistry.* Prentice Hall, Upper Saddle River, NJ (1992).
211. Canter, N., It isn't easy being green—The promise, perils, and progress of environmentally friendly lubricants. *Lubricants World.* **9**(3), 16–21 (2001).
212. Salimon, J., Salih, N., and Yousif, E., Improvement of pour point and oxidative stability of synthetic ester basestocks for biolubricant applications. *Arabian Journal of Chemistry.* **5**(2), 193–200 (2012).
213. Sharma, B. K., and Stipanovic, A. J., Development of a new oxidation stability test method for lubricating oils using high-pressure differential scanning calorimetry. *Thermochimica Acta.* **402**(1–2), 1–18 (2003).
214. Gapinski, R. E., Joseph, I. E., and Layzell, B. D., A vegetable oil based tractor lubricant. SAE Technical Paper 941758, Society of Automotive Engineers, Warrendale, PA (1994).
215. Becker, R., and Knorr, A., An evaluation of antioxidants for vegetable oils at elevated temperatures. *Lubrication Science.* **8**(2), 95–117 (1996).
216. Ohkawa, S., Konishi, A., Hatano, H., and Ishihama, K., Oxidation and corrosion characteristics of vegetable-based biodegradable hydraulic oils. *SAE Transactions.* **104**(4), 737 (1996).
217. Salimon, J., Salih, N., and Yousif, E., Chemically modified biolubricant basestocks from epoxidized oleic acid: Improved low temperature properties and oxidative stability. *Journal of Saudi Chemical Society.* **15**(3), 195–201 (2011).
218. Crivello, J. V., and Fan, M. X., Catalysis of ring-opening and vinyl polymerizations by dicobaltoctacarbonyl. *Journal of Polymer Science: Polymer Chemistry.* **30**(1), 31–39 (1992).
219. Findley, T. W., Swern, D., and Scanlan, J. T. *Journal of the American Chemical Society.* **67**(3), 412–414 (1945).
220. Rangarajan, B., Havey, A., Grulke, E. A., and Dean Culnan, P., Kinetic parameters of a two-phase model for in situ epoxidation of soybean oil. *American Oil Chemists Society.* **72**(10), 1161–1169 (1995).
221. Sonnet, P. E., and Foglia, T. A., Epoxidation of natural triglycerides with ethylmethyldioxirane. *American Oil Chemists Society.* **73**(4), 461–464 (1996).
222. Debal, A., Rafaralahitsimba, G., and Ucciani, E., Epoxidation of fatty acid methyl esters with organic hydroperoxides and molybdenum oxide. *Fat Science Technology.* **95**(6), 236–239 (1993).
223. Ucciani, E., Bonfand, A., Rafaralahitsimba, G., and Cecchi, G., Epoxidation of monoenic fatty esters with cumilhydroperoxide and hexacarbonyl-molybdenum. *Revue Francaise des Corps Gras.* **39**(9–10), 279 (1992).
224. Debal, A., Rafaralahitsimba, G., Bonfand, A., and Ucciani, E., Catalytic epoxidation of methyl linoleate—Cyclisation products of the epoxyacid esters. *Fat Science Technology.* **97**(7/8), 269–273 (1995).

225. Semel, J., and Steiner, R., Renewable raw materials in the chemical industry. *Nachrichten aus Chemie, Technik und Laboratorium.* **31**(8), 632–635 (1983).

226. Salimon, J., and Salih, N., Preparation and characteristic of 9,10-epoxyoleic acid α-hydroxy ester derivatives as biolubricant base oil. *European Journal of Scientific Research.* **31**(2), 265–272 (2009).

227. Salimon, J., and Salih, N., Improved low temperature properties of 2-ethylhexyl 9(10)-hydroxy-10(9)-acyloxystearate derivatives. *European Journal of Scientific Research.* **31**(4), 583–591 (2009).

228. Salimon, J., and Salih, N., Substituted esters of octadecanoic acid as potential biolubricants. *European Journal of Scientific Research.* **31**(2), 273–279 (2009).

229. Biermann, U., Friedt, W., Lang, S., Lühs, W., Machmüller, G., Metzger, U. O., Gen. Klaas, M. R., Schäfer, H. J., and Schneider, M. P., New Syntheses with oils and fats as renewable raw materials for the chemical industry. In: *Biorefineries—Industrial Processes and Products.* pp. 253–289. Wiley-VCH, Weinheim, Germany (2008).

230. Biermann, U., and Metzger, J. O., Friedel-Crafts alkylation of alkenes: Ethylaluminum sesquichloride induced alkylations with alkyl chloroformates. *Angewandte Chemie—International Edition in English.* **38**, 3675–3677 (1999).

231. Metzger, J. O., and Riedner, U., Free radical additions to unsaturated fatty acids. *Fat Science Technology.* **91**, 18–23 (1995).

232. Metzger, J. O., and Linker, U., New results of free radical additions to unsaturated fatty compounds. *Lipid/Fett.* **93**(7), 244–249 (1991).

233. Metzger, J. O., and Biermann, U., Alkylaluminium dichloride induced Friedel–Crafts acylation of unsaturated carboxylic acids and alcohols. *Liebigs Annalen der Chemie.* **1993**(6), 645–650 (1993).

234. Biermann, U., and Metzger, J. O., Lewis acid catalyzed additions to unsaturated fatty compounds—II: Alkylaluminium halide catalyzed ene reactions of unsaturated fatty compounds and formaldehyde. *Lipid/Fett.* **93**(8), 282–284 (1991).

235. Pryde, E. H., Hydroformylation of unsaturated fatty acids. *Journal of the American Oil Chemists' Society.* **61**(2), 419–425 (1984).

236. Frankel, E., and Pryde, E., Catalytic hydroformylation and hydrocarboxylation of unsaturated fatty compounds. *Journal of the American Oil Chemists' Society.* **54**(11), A873–A881 (1977).

237. Xia, Z., Kloeckner, U., and Fell, B., Hydroformylation of mono and multiple unsaturated: Fatty substances with heterogenized cobalt carbonyl and rhodiumcarbonyl catalysts. *Fett Lipid.* **98**, 313–321 (1996).

238. Behr, A., and Laufenberg, A., Synthesis of new branched fatty acids by rhodium catalyzed homogeneous oligomerization. *Fat Science Technology.* **93**, 20–24 (1991).

239. Henkel, K., Al, Dusseldorf, Germany Patent.

240. Keller, U., Fischer, J., and Hoelderich, W. F., New lubricants from renewable resources: Ecotoxicites and oxidative characteristics. *Oelhydraulik und Pneumatik.* **4**, 240–245 (2000).

241. Cermak, S. C., and Isbell, T. A., Improved oxidative stability of estolide esters. *Industrial Crops and Products.* **18**(3), 223–230 (2003).

242. Johansson, L. E., and Lundin, S. T., Copper catalysts in the selective hydrogenation of soybean and rapeseed oils. *Journal of the American Oil Chemists' Society.* **56**(12), 974–980 (1979).

243. Behr, A., Doring, N., Durowitz-Heil, S., Lohr, C., and Schmidtke, H., Selective hydrogenation of multi-unsaturated fatty-acids in the liquid-phase. *Fat Science Technology.* **95**, 2–11 (1993).

244. Behr, A., Homogeneous transition-metal catalysis in oleochemistry. *Fat Science Technology.* **92**, 375–388 (1990).

245. Fell, B., and Schafer, W., Selective hydrogenation of fats and derivatives using Ziegler-type organometallic catalysts—1: Selective hydrogenation of methyllinoleat and other dienic compounds with isolated double bonds. *Fat Science Technology.* **92**, 264–272 (1990).

246. Haase, K., D., Heynen, A. J., and Laane, N. L. M., Composition and application of isostearic acid. *Fat Science Technology.* **91**, 350–353 (1989).

247. Link, W., and Spitellar, G., Products of the dimerization of unsaturated fatty acids—I: The fraction of monomers obtained by dimerization of pure oleic acid. *Fat Science Technology.* **92**, 19–25 (1990).

248. Perin, G., Alvaro, G., Westphal, E., Jacob, R. G., Lenardao, E. J., Viana, L. H., and D'Oca, M. G. M., Transesterification of castor oil assisted by microwave irradiation. *Fuel.* **87**(12), 2838–2841 (2008).

249. Battersby, N. S., and Morgan, P., A note on the use of the CEC L-33-A-93 test to predict the potential biodegradation of mineral oil based lubricants in soil. *Chemosphere.* **35**(8), 1773–1779 (1997).

250. Pagga, U., Testing biodegradability with standardized methods. *Chemosphere.* **35**(12), 2953–2972 (1997).

251. Willing, A., Oleochemical esters—Environmentally compatible raw materials for oils and lubricants from renewable resources. *Lipid–Weinheim.* **101**(6), 192–198 (1999).
252. Remmele, E., and Widmann, B., Hydraulic fluids based on rapeseed oil in agricultural machinery-sustainability and environmental impact during use. Paper presented at the 11th International Colloquium, TA Esslingen, Germany, January 13–15, 1998 (1998).
253. Battersby, N., A correlation between the biodegradability of oil products in the CEC L-33-T-82 and modified Sturm tests. *Chemosphere.* **24**(12), 1989–2000 (1992).
254. Miller, S., Scharf, C., and Miller, M., Utilizing new crops to grow the biobased market: Trends in new crops and new uses. ASHS Press, Alexandria, VA (2002).
255. Willing, A., Lubricants based on renewable resources—An environmentally compatible alternative to mineral oil products. *Chemosphere.* **43**(1), 89–98 (2001).
256. Torbacke, T. N., and Kopp, M., Environmentally adapted lubricants in the Nordic marketplace—Recent developments. *Industrial Lubrication and Tribology.* **54**(3), 109–116 (2002).
257. USDA, BioPreferred Program. http://www.biopreferred.gov/ (2013). Accessed October 23, 2013.

14 Thermo-Oxidative Stability of Base Oils for Green Lubricants

Pattathilchira Varghese Joseph and Deepak Saxena

CONTENTS

ABSTRACT

Environmentally friendly alternatives to mineral oils for formulating green lubricants have been the subject of research for the last few decades. They have become an essential component of an emerging new field, viz., green tribology, which focuses on the scientific and technological principles of tribology toward environmental benefits. Vegetable oils and their modified versions as such lubricants have been the main focus of research in this area. These oils have been finding their way into more applications as eco-friendly lubricant bases with superior characteristics such as very good lubricity and excellent biodegradability.

This chapter deals with the issue of thermo-oxidative stability of vegetable oils in their native state for use as lubricant base oils. A discussion of the different aspects of vegetable oils for designing green lubricants is outlined and a review of the thermo-oxidative stability of vegetable oils for such an application is presented as well. Different approaches to addressing this major drawback and studies on vegetable oils of different sources as lubricants and their thermo-oxidative stability are discussed. The results of studies on jatropha, dilo, and castor oils for application as lubricant base oils are presented.

14.1   INTRODUCTION

Green tribology is being developed as an emerging area, focusing on scientific and technological principles of tribology in order to achieve ecological balance by minimizing environmental impacts and so saving energy and materials as well as enhancing the environment and quality of life. As lubrication is one of the three pillars of tribology, it is essential to make lubricants green in the endeavor toward green tribology. Therefore, the early proponents of green tribology have included green lubricants as one of the three identified areas for green tribology, viz., biomimetic tribology, eco-friendly lubrication, and materials and tribological aspects of sustainable energy applications [1].

14.1.1   Environment-Friendly/Green Lubricants

14.1.1.1   Lubricants and the Environment

In 2014, the global lubricant consumption was 35.4 million metric tons [2]. Waste lubricants produced from usage of this huge volume became a major source of environmental pollution. It is reported that of the total lubricants put under use, 25% is leaked into the environment [3]. In addition to this, a very large proportion of overall lubricant production is dispersed into the environment in the form of spills or leaks of virgin oils or emission of partially combusted oil in automotive exhausts [4]. Thus, lubricants play a major role in introducing millions of tons of oily pollutants into the environment [5,6]. It is estimated that 50% of all lubricants sold worldwide ends up in the environment via total loss applications, evaporation, spills, or major accidents [7,8]. Thus, considerable attention has been given to lubricant biodegradability and persistence in the environment [4]. Because of their high ecotoxicity and low biodegradability, mineral oil-based lubricants constitute a considerable threat to the environment. Several countries have passed legislation to control this situation and some oil companies have started working toward environmentally friendly products [5,6]. The idea of environmentally friendly green lubricants has been in serious discussion for the last three to four decades, and the use of environmentally compatible ingredients from renewable sources has been explored as one of the solutions to this problem [9]. Legal regulations that forbid use of all toxic, nonbiodegradable products in the equipment used in lakes have been around since 1982 [10]. Concerns on the availability of mineral oils from the dwindling source of petroleum have also catalyzed the demand for renewable lubricants [11].

Research has been conducted to develop such lubricants using environmentally friendly and renewable materials from alternate sources. These categories of products are called environmentally acceptable lubricants because of their excellent biodegradability, renewability, and nontoxicity [1]. Such products are developed and approved as environmentally friendly lubricants and allowed to carry labels such as Nordic Swan in Nordic countries, Blue Angel in Germany, NF Environment Mark in France, Ecolabel (earthen pot) in India, etc.

14.1.1.2   Base Oils for Green Lubricants

Lubricants contain around 70–99% base oils with the balance being made up of performance additives. About 35 million tons of lubricant base oils of different viscosities and other characteristics are being used to manufacture more than 500 grades of finished lubricants globally [12]. The different base oils available are mineral oil (solvent neutral and severely hydrotreated), synthetic oils, and vegetable oils. As the base oil is the main component of any lubricant, it should be primarily environmentally friendly or green for the composition of the whole lubricant to be so. Vegetable oils are considered most suitable for application in environmentally friendly/green lubricants. They are renewable, possess high levels of biodegradability, have low aquatic toxicity, and do not accumulate in the environment [8].

14.1.2 Vegetable Oils as Green Base Oil

14.1.2.1 Physicochemical Properties

Green lubricant base oils should meet environmentally friendly characteristics including health, safety, and waste treatment requirements [9]. With their triglyceride structure, vegetable oils display high VI, higher flash points, and low coefficient of friction, which are essential requirements of a lubricant [5,13]. The strong intermolecular interactions in vegetable oils lead to resistance to rapid change in viscosity with increase in temperature, providing high VI to vegetable oils [14]. The low volatility of vegetable oil is beneficial in controlling environmental pollution. Further, vegetable oils have superior power to dissolve contaminants and additive molecules in comparison to severely hydrocracked mineral base oils. Other required lubricant properties such as rust prevention, foaming, and demulsibility, can be improved by incorporation of additives [15]. Low-temperature fluidity and thermo-oxidative stability are the major concerns with vegetable oils for use as lubricant [8,16,17].

14.1.2.2 Tribological Characteristics

Vegetable oils were reported to exhibit excellent lubrication characteristics in laboratory tribological tests and fully formulated lubricants based on vegetable oil displayed a lower coefficient of friction, equivalent load-carrying capacity, and better antiwear in comparison to mineral oil counterparts [14,18–20]. The tribological superiority of vegetable oil base oils was attributed to the triglyceride structure. The polar fatty acid chains provide high-strength lubricant films on metallic surfaces, resulting in reduction of friction and wear [21]. Therefore, vegetable oils are preferred as boundary lubricants [14].

14.1.2.3 Environmental Criteria

For a lubricant to be considered environmentally friendly, the factors to be considered are biodegradability and environmental compatibility in terms of water pollution, toxicity, ecotoxicity, and exhaust emissions. A product which can degrade itself into safe gaseous or liquid products under the biological action of air, water, and bacteria can be considered as biodegradable. Different test methods are being used to establish the extent of biodegradability. Solvent-refined mineral oils are 20–30% biodegradable; hydrocracked mineral oils, 30–55%; diesters and polyolesters, in the range of 50–90%; and vegetable oils, the maximum biodegradability in the range of 90–100% [4].

Mineral oils have poor environmental compatibility because of their negative influence on soil and water along with flora and fauna. Oil contamination in soil markedly affects the chemical, physical, and microbiological properties of the soil [22]. Water containing even 1 to 2 mg/L mineral oil is not suitable for drinking and is considered to be dangerous to the environment. Oil cuts off oxygen entry into water, thereby endangering the lives of organisms [23]. Mineral oils may contain polycyclic aromatic hydrocarbons, which are carcinogenic and mutagenic [24]. Vegetable oils and ester-based oils are regarded the most environmentally compatible [4,9,13]. Vegetable oils are also nontoxic in terms of human contact, which is very essential where contact is unavoidable such as with metalworking lubricants. Shop floor lubricants are also expected to be low fuming, making vegetable oil-based products the ideal choice for meeting this requirement. Because of their lower volatility, vegetable oils give out significantly lower emission compared to mineral oils. The disposal and waste treatment of lubricants is a major concern considering the huge amount of used oils being generated every year. Oils with a higher degree of biodegradability do not accumulate in the environment and their disposal and waste treatment are easier than those of oils with lower biodegradability. Lubricants generating easily recoverable and reusable waste oils have an indirect environmental benefit. Also, those oils that can be converted into other useful products are preferred as environmentally friendly. When mineral oils, synthetics, and vegetable oils are compared from the waste treatment aspect, vegetable oils should be considered superior, because of their ready biodegradability and also since used vegetable oil-based lubricants can be converted into biodiesel by transesterification. These products lessen the quantity of hazardous wastes generated and they also reduce the disposal costs [25].

Even though physicochemical characteristics, biodegradability, and environmental compatibility of vegetable oils make them green lubricants, poor thermo-oxidative stability and cold-flow property are serious drawbacks. Therefore, except in total loss applications such as chain saw lubricants, vegetable oils are generally not preferred as base oils for applications such as automotive, hydraulic, and turbine, where thermo-oxidative stability and cold-flow property are critical requirements.

14.2 THERMO-OXIDATIVE STABILITY AS MAJOR DRAWBACK OF VEGETABLE OILS AS BASE OILS

Lubricants, except total loss lubricants, are supposed to be used under recirculating conditions where exposure to high temperatures and mixing with atmospheric oxygen are involved. Because of this, it is possible for oils to get oxidized under high thermal stress encountered in the stated conditions. It is desirable that the lubricant must retain its original physical properties throughout its lifetime for superior lubrication under the adverse conditions of high temperature and mixing with oxygen. Thermal degradation and oxidation affect fluid physical properties, so a lubricant must have extremely good thermal and oxidative stability to sustain good performance over a longer period. Among the three types of base oils, vegetable oils generally display thermo-oxidative property inferior to that of mineral oils and most of the synthetic esters [15,26,27]. Vegetable oils, being triglycerides of saturated and unsaturated fatty acids, have limited oxidation stability because the unsaturated bonds are active sites for many reactions including oxidation, leading to increases in acidity, viscosity, and volatility [8].

Vegetable oils undergo thermo-oxidation at a faster rate owing to the presence of unsaturation [28,29]. As the unsaturation increases, the oxidation susceptibility also increases, with triply unsaturated fatty acids being the most susceptible to oxidation. The fatty acid composition of vegetable oils is, therefore, the deciding factor of oxidation stability [8]. Oils containing saturated fatty acids are very much resistant to oxidation in comparison to those containing polyunsaturated fatty acids with one or more unsaturation. It was also established that oils with higher content of monounsaturated fatty acids oxidize at higher temperatures, whereas those containing polyunsaturated fatty acids oxidize at lower temperatures [14]. However, saturated vegetable oils are poor in low-temperature fluidity and oils with unsaturation have good flow properties at low temperatures [8]. Therefore, for lubricating oils with optimum oxidation stability and low-temperature flow properties, a high amount of monounsaturation is preferred in vegetable oils.

14.3 EVALUATION TECHNIQUES FOR STUDY OF THE THERMO-OXIDATIVE STABILITY OF VEGETABLE OILS

The test procedures prevalent in the lubricant industry were designed for mineral oil lubricant oils, and because of this reason, it was suggested by some researchers that sometimes those conditions may not be applicable to vegetable oils [14]. It was reported that this is particularly true when water is present in oxidation tests and for those tests where the increase in acidity is measured as the end point. Since triglycerides tend to hydrolyze with increase in acid value, the result of oxidation tests with water and where acid value is used as the indicator would draw a wrong conclusion on the oxidative susceptibility of the vegetable oils [14].

ASTM D943 test is a well-known bench test which is used in the lubricant industry for assessing the overall stability and life of turbine oils. Because of the doubtful test result of oxidation stability of vegetable oils from oxidations tests with added water, some authors considered the ASTM D943 not suitable for triglyceride oxidation studies [30]. Consequently, ASTM D943 without water (dry turbine oil stability test [TOST]) was proposed for ester and vegetable oils' oxidation stability evaluation. Several authors have even found it having very good potential for predicting the field behavior [31–35]. ASTM has adopted the dry TOST method as D7873-13e1 [36]. In this test, six tubes with 360 g sample and coiled iron–copper catalyst are kept at 120°C with oxygen bubbling through the sample. The tubes are removed over time and the samples are examined for rotary pressure vessel oxidation test (RPVOT) induction time and insolubles. The test with six tubes and the insoluble measurement is run up to an RPVOT residual ratio of below 25% or any other specified percentage. The test may also be run also for a specified time(s) with one or many tubes and the insoluble of the sample is determined after that time(s). The amount of insolubles determined at different times or at different RPVOT residual ratios is an indication of the level of oxidation of the oil.

There are many other oxidation tests in the lubricant industry which do not use water, for, e.g., the IP 48 test of the Institute of Petroleum [37] and the Baader test [38].

Some other oxidation test methods for vegetable oil-based formulations are the following:

In the thin-film microoxidation (TFMO) method [39,40], about 25 µL of the test oil kept over a weighed freshly polished cast iron catalyst surface is placed at $(175 \pm 1)°C$ inside a glass-bottomed reactor. The sample is allowed to oxidize by passing compressed air over it at 20 cm^3/min. After the specified time, the catalyst sample along with the oil sample is cooled under a stream of dry nitrogen and transferred to a desiccator. After temperature equilibrium is attained, the catalyst along with the oxidized sample is weighed to determine the loss due to volatilization. The soluble part of the oxidized oil from the catalyst is dissolved in tetrahydrofuran (THF) by soaking in the solvent and the insoluble deposit formed by oil oxidation on the catalyst is determined by again weighing the catalyst after taking it from the solvent and making it free of solvent. The THF extraction of oxidized oil is analyzed to study the nature of the oxidation products. The degree of oxidation of the oil is assessed from the amount of volatile portion, from the deposit formed, and from the analysis of oxidation products.

Differential scanning calorimetry (DSC) ASTM D6186 [41] and E2009 [42] is gaining popularity for studying the oxidation stability of lubricants, including vegetable oils. This test method has been used widely for lubricant studies after the introduction of pressurized differential scanning calorimetry (PDSC) [43]. PDSC is run either in the isothermal or in nonisothermal mode for lubricant evaluation corresponding to ASTM D6186 and E2009, respectively. Thermo-oxidation of vegetable oil lubricants can also be studied by PDSC using either of these methods. In the isothermal mode of the test, ASTM D6186 [41], 3 mg of the oil sample is weighed in an aluminum pan and placed in the cell. The temperature in the cell is rapidly increased from the room temperature to 200°C at 100°C/min and allowed to equilibrate for 2 min. The cell is then pressurized with oxygen at a pressure of 3 MPa, followed by purging at a rate of 100 mL/min. The oxidation induction time (OIT) corresponding to a strong exothermic reaction (oxidation) was then measured considering the starting time from the opening of the oxygen valve. If the OIT is less than 10 min, the temperature is then decreased to 180°C or further down to 155°C. In the nonisothermal method, ASTM E2009 [42], 3 mg of the sample in the pan is made ready in the cell, where the temperature in the cell is increased gradually at a rate of 10°C/min and oxygen flow is maintained at 50 mL/min. From the obtained DSC signal, the oxidation onset temperature is determined. These methods are useful for quick determination of the aging status of oil, but it is not proven to accurately discriminate various compositions with correlation to long-term oxidation [44–46].

RPVOT, ASTM D2272 [47], is a standard oxidation test method widely used in the lubricant industry. As per the standard test method, 50 g of the test oil is taken into a glass container into which 5 mL water is added. Three-meter-long 99.9% copper catalyst coil is placed in the glass vessel with the test oil and the whole assembly is placed in a pressure vessel equipped with a pressure gauge. Five millimeters of water is added into the pressure vessel, which is then filled with oxygen at a pressure of 620 kPa (90 psi) and placed in a constant-temperature bath set at 150°C and rotated axially at 100 rpm. The time taken for a pressure drop of 175 kPa (25.4 psi) in the pressure vessel is noted as the RPVOT service life. The test was successfully used by different authors in studies on a wide range of vegetable oils [28–30,48,49]. A modified version of this test method with less severe conditions was reported to be used in thermo-oxidation studies of vegetable oils and unsaturated esters [50,51]. In the modified test, the temperature is maintained at 120°C and the test is run in the absence of water and copper catalysts. In the glass container, 35 g of the sample is taken and kept in the pressure vessel, which is then filled with oxygen at a pressure of 620 kPa (90 psi). The assembly is kept in a bath maintained at 120°C and rotated axially at 100 rpm. The times taken for pressure drops of 25 psi (1.7 kg/cm^2), 2 × 25 psi (3.4 kg/cm^2), and 3 × 25 psi (5.1 kg/cm^2) in the pressure vessel were recorded as the OITs.

In the Baader test, DIN 51554 Part 3 [38], 60 mL sample with a coiled copper wire catalyst is taken into a tube with a condenser and kept at 95°C for 72 h. The copper wire is allowed to move up and down during the test at a specified speed. At the end of the test, the oil is examined for change in acidity, viscosity, etc. The extent of changes in these properties indicates the degree of aging.

14.4 DIFFERENT VEGETABLE OILS AS LUBRICANT BASE: THERMO-OXIDATIVE STABILITY

Coconut oil, having more than 90% saturation (with 50% lauric acid), has been studied extensively as lubricant base oil. This saturated fatty acid composition imparts favorable oxidation stability to coconut oil [52]. Jayadas and Prabhakaran Nair [53] studied the thermo-oxidative property of coconut oil by using thermogravimetric analysis and found that the oil had lower weight gain in the oxidative environment, indicating better oxidation stability in comparison to that of the other vegetable oil (sunflower) considered in the study. Gas engine oil was reported as one of the potential applications of coconut oil [54]. In a study on coconut oil without additives as engine oil, improved fuel efficiency and emission were reported and the tribological deficiencies were attributed to the oxidation instability due to the absence of additives [55].

Sunflower, abundantly available in North America, was also studied widely. Native sunflower oil contains more than 70% linoleic acid and, therefore, cannot be considered as a thermo-oxidatively stable option. The high-oleic variety (high-oleic sunflower oil [HOSO]) with >70% oleic acid was studied in gear oil application and found to have promising performance in thermo-oxidative stability and tribological properties [56,57]. However, to match with the performance of mineral oils, HOSO needs further improvement. Asadauskas et al. [58] studied HOSO along with other high-oleic varieties of vegetable oils by TFMO test method at high-temperature conditions (200°C and 225°C). It was found that the high-oleic vegetable oils were very poor in oxidation stability at these high temperatures, compared to petroleum mineral oils. In another study, Asadauskas et al. [18] also studied HOSO by TFMO test at a test temperature of 175°C and duration of 30 minutes. Again it was concluded that oxidation stability of HOSO was inferior to that of mineral oil. However, in the presence of antioxidants, HOSO could attain the performance level of mineral oils in oxidation stability [58].

Rapeseed oil with a fairly high amount (72.5%) of monounsaturation (41% erucic acid, 14.5% eicosenoic acid plus 17% oleic) and comparatively lower (22%) polyunsaturation was also studied widely for application in lubricants. In a study conducted, Legisa et al. [59] concluded that formulations based on rapeseed oil showed very poor thermo-oxidative stability above 60°C. Other vegetable oils such as corn oil, cottonseed oil, soybean oil, and groundnut oil are also not favored candidate oils as lubricant base oils because of their high percentage of poly-unsaturation (>30%). Gerbig et al. [60] studied the physicochemical and tribological properties, along with the oxidation stability, of almond oil, castor oil, corn oil, grape seed oil, hazelnut oil, linseed oil, olive oil, peanut oil, pumpkinseed oil, rapeseed oil, safflower oil, sesame oil, soybean oil, sunflower oil, walnut oil, and wheat germ oil. Except for safflower, walnut, sesame, and wheat germ oils, all were found to become solid at the end of the test, showing high susceptibility to oxidation. Mineral oil and wheat germ oil showed similar performance of no change in viscosity.

A recent study conducted by Khemchandani et al. [51] on safflower oil showed low OIT when tested by the modified RPVOT (ASTM D2272) at 120°C without water or catalyst. Addition of synergistic combinations of aminic and phenolic antioxidants plus dilution with TMP and PE esters in 3:1, 1:1, and 1:3 ratios was found to significantly improve the oxidation stability of safflower oil for use as biodegradable lubricant base oil.

A survey of the data [61–66] on various nonedible vegetable oils from Indian sources has revealed a few with suitable fatty acid composition for lubricant base oil application. The screening was carried on the basis of fatty acid distribution, iodine value, and pour point. The authors have identified and studied three suitable oils: *Jatropha curcas* (jatropha) oil, *Calophyllum inophyllum* (dilo or undi) oil, and *Ricinus communis* L., *Euphorbiaceae* (castor) oil. These oils were studied for their tribological and physicochemical properties relative to those of mineral oils, TMP trioleate, PE tetraoleate, and some other edible oils (Table 14.1). The studies showed promising results in terms of tribological and physicochemical characteristics for application as lubricant base oil [67].

These oils were then evaluated for thermo-oxidative stability. In this study, the authors have used RPVOT as per ASTM D2272 and also two modifications of this standard test for evaluating the

TABLE 14.1
Physicochemical and Tribological Characteristics: Vegetable Oils versus Synthetic Esters and Mineral Oils

Test	Method	Jatropha	Dilo	Castor	TMP Ester	PE Ester	Gp. I	Gp. II	Gp. III
Kinematic viscosity at 40°C (cSt)	ASTM D445	35.8	31.2	220.6	49.0	83.9	29.2	29.7	37.6
Kinematic viscosity at 100°C (cSt)	ASTM D445	8.04	7.50	19.70	9.89	14.4	5.14	5.37	6.43
VI	ASTM D2270	208	220	102	193	179	105	116	123
Saponification value (mg KOH/g)	ASTM D94	197	190	180	178	164	0	0	0
Pour point (°C)	ASTM D97	0	+6	<−27	−30	−24	−3	−18	−27
Flash point (°C)	ASTM D93	240	223	>250	>200	>200	224	230	240
Iodine value	ASTM D5554	97	84	87	63.3	80	0	0	0
Coefficient of friction in SRV test, 100 N, 50°C, 50 Hz, 1 mm, 1 h	SRV test	0.08	0.09	0.07	0.09	0.09	0.10	0.20	0.20
References	–	[67,68]	[67,68]	[67,68]	[67,68]	[67,68]	[67,68]	[67,68]	[67,68]

Source: Joseph, P. V. et al., Study of some non-edible vegetable oils of Indian origin for lubricant application. *J. Synth. Lubr.* 2007. 24, 181–197. Copyright Wiley-VCH Verlag GmbH & Co. KGaA. Reproduced with permission.

TABLE 14.2
OITs by Modified RPVOT (I) of Vegetable and Other Oils

Oil	OIT (min) after Reduction of Oxygen Pressure by			References
	25 psi	50 psi	75 psi	
Jatropha	112	123	141	[67,68]
Castor	>300	>300	>300	[67,68]
Dilo	90	105	130	[67,68]
TMP ester	19	33	66	[67,68]
PE ester	<10	25	>300	[67,68]
Group I	>300	>300	>300	[67,68]
Group II	>300	>300	>300	[67,68]
Group III	>300	>300	>300	[67,68]

Note: Test conditions for RPVOT: 120°C in the absence of copper catalyst and water.

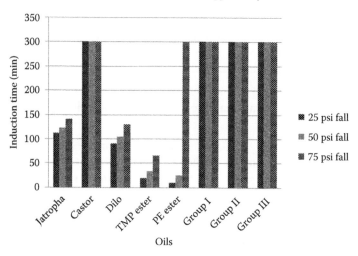

FIGURE 14.1 OITs by modified RPVOT (I): vegetable oils, mineral oils, and synthetic esters. (Joseph, P. V. et al., Study of some non-edible vegetable oils of Indian origin for lubricant application. *J. Synth. Lubr.* 2007. 24. 181–197. Copyright Wiley-VCH Verlag GmbH & Co. KGaA. Reproduced with permission.)

thermo-oxidative stability of vegetable oils and their blends with antioxidants and diluents. The two modifications of the test methods used were as follows:

1. Modified RPVOT test (I): In this method, the RPVOT is run at 120°C without the presence of copper catalyst and water. The induction time in minutes, which gives an indication of the service life of the lubricant in the aforementioned temperature, was recorded after the pressure dropped by 25 psi (1.7 kg/cm^2), 50 psi (3.4 kg/cm^2), and 75 psi (5.1 kg/cm^2).
2. Modified RPVOT test (II): In this, the RPVOT is run at 120°C with copper catalyst and without water. The induction time in minutes, which gives an indication of the service life of the lubricant in the aforementioned temperature, was recorded after the pressure dropped by 25 psi (1.7 kg/cm^2), 50 psi (3.4 kg/cm^2), and 75 psi (5.1 kg/cm^2).

TABLE 14.3
OITs by RPVOT (ASTM D2272) of Vegetable and Other Oils

		OIT (min) after Reduction of Oxygen Pressure by			
S. No.	Candidate Oil	25 psi	50 psi	75 psi	References
1	Jatropha	<5	<5	<5	[67,68]
2	Castor	18	20	–	[67,68]
3	Dilo	<5	<5	<5	[67,68]
4	TMP ester	<5	<5	<5	[67,68]
5	PE ester	<5	<5	<5	[67,68]
8	Group I	40	45	50	[67,68]
9	Group II	35	45	55	[67,68]
10	Group III	38	45	55	[67,68]

Note: RPVOT (ASTM D2272) test conditions: temperature of 150°C with Cu catalyst and water.

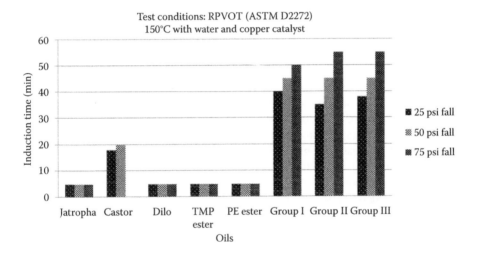

FIGURE 14.2 OITs by RPVOT (ASTM D2272): vegetable oils, mineral oils, and synthetic esters.

Initially the nonedible vegetable oils were studied by modified RPVOT test (I). Table 14.2 provides the results.

As per the modified RPVOT test (I), among the three nonedible vegetable oils, castor oil showed the maximum OIT followed by jatropha and dilo oils (Figure 14.1).

The modified RPVOT test (I) could not differentiate between the oxidation stabilities of castor oil and mineral base stocks. Hence, unmodified RPVOT test was conducted at 150°C in the presence of water and catalyst. In this test, the oxidation stability of castor oil was found to be inferior to that of mineral base stocks (Table 14.3 and Figure 14.2).

14.5 IMPROVING VEGETABLE OILS FOR APPLICATION AS LUBRICANT BASE OIL: STRUCTURAL MODIFICATION, ADDITIVE TREATMENT, AND COMPOUNDING

Different routes are available to improve the oxidation stability and the low-temperature fluidity of vegetable oils, viz., (i) genetic modification, (ii) synthetic modification, (iii) additive treatment, and

(iv) compounding or blending. These approaches have led to considerable success in formulating and introducing locally available vegetable oil-based lubricants into the market.

14.5.1 MODIFICATIONS

Modifications of the acyl chain structure through plant hybridization, genetic alteration, and chemical and physical processing are carried out to convert the unwanted polyunsaturated acids to oleic acid and thus improve oxidation stability [35,69–71].

14.5.1.1 Genetic Modification and Selective Plant Breeding

Fatty acid composition and antioxidant concentration are the two factors responsible for the oxidation stability of vegetable oils. These two factors in turn depend on the genetic property of the plant [72]. For useful lubricating properties, a high proportion of oleic acid component is required. Linolenic and linoleic acids are not desirable in vegetable oils for lubricant application [73]. The fatty acid composition in the plant is altered by genetic engineering, producing transgenic oilseed plants [74]. Such crops have been marketed commercially by DuPont and Monsanto [75]. With this method, oils with more than 75% oleic fatty acids have been obtained [76]. Selective plant breeding can also be adopted to enhance the oleic acid content and decrease the linoleic and linolenic acid content. By combining modified vegetable oils from selective breeding with additive technology, lubricants dominated by the esters can be formulated from vegetable oils. Linoleic acid in sunflower can be reduced from 50–70% and oleic acid increased by cross-pollination [73].

14.5.1.2 Synthetic Modification

Most of the commercially viable routes for upgrading vegetable oils depend on chemical modification and physical separation. Chemical modification of vegetable oils to improve oxidation stability is still under active investigation. More than 90% of chemical modifications have been those occurring at the fatty acid carboxyl group, while less than 10% have involved reactions at the fatty acid hydrocarbon chain [77]. Synthetic methods applied include transesterification, selective hydrogenation, dimerization/oligomerization, C–C and C–O bond formation for branched molecules, hydroformylation, Friedel–Crafts alkylation, Friedel–Crafts acylation, ene reaction, radical addition, acyloxylation, metathesis, and oxidation reactions (such as epoxidation and oxidative cleavage) [77–81]. Vegetable oils with predominantly unsaturated fatty acids can be converted to products with better thermo-oxidative properties and at the same time maintaining the same viscosity temperature characteristics and lubricity [82]. Vegetable oils transesterified with TMP, PE, and neopentyl glycol have been found to possess better lubricant properties including thermo-oxidative stability and low pour points [83]. These esters have been in the market for formulating synthetic and semi-synthetic lubricants with special focus on biodegradability, environmental friendliness, energy efficiency, and long sump life [84]. In epoxidation process the olefinic double bond is saturated by attachment of oxygen between the adjoining carbons of the double bond, forming a three-membered ring. Epoxides are highly viscous oils and are further converted to diesters and other derivatives for lubricant application [85]. Chemical modification in combination with blending additives offers the greatest opportunity for formulating high-performance biolubricants [75].

14.5.2 ADDITIVE TREATMENT

There are a variety of chemical additive structures used to improve the properties and performance of lubricants. Additives are valuable for improving the properties of vegetable oils in lubricants [86,87]. Several investigations were carried out with different vegetable oils incorporated with antioxidants [8]. Hamblin [30] reported that vegetable oils could be stabilized with the use of antioxidants [28–30]. However, higher concentrations of antioxidants are required for vegetable oils. In the study by Hubmann [88], higher acid values and viscosity were observed in rapeseed oil alone

TABLE 14.4

OITs Obtained with the Modified RPVOT (I) Method for Vegetable Oils and Polyolesters with and without Added Antioxidants (Mixture of High–Molecular Weight Hindered Phenol and Alkylated Diphenylamine) by Modified RPVOT (I)

Oil	Antioxidant (Mixture of High–Molecular Weight Hindered Phenol and Alkylated Diphenylamine) (% w/w)	OIT of Blends (min) after Pressure Drop of			Reference
		25 psi	50 psi	75 psi	
Jatropha	1	213	220	230	[68]
	0	112	123	141	
Castor oil	1	>360	>360	>360	[68]
	0	>360	>360	>360	
Dilo oil	1	335	345	355	[68]
	0	90	105	130	
TMP ester	1	>360	>360	>360	[68]
	0	19	33	56	
PE ester	1	272	285	335	[68]
	0	10	25	–	

compared to a blend with 0.3% concentration of phenolic and aminic antioxidants when tested with the Baader test. The response of vegetable oils to oxidation inhibitors increases with decreasing iodine number. It was also reported that the oxidative stability of vegetable oils can be further improved by optimizing the amine and phenol antioxidant types, the amine-to-phenol ratio, and the concentration. Baader oxidation test results on rapeseed oil revealed that the aminic antioxidants studied were insufficient in controlling viscosity increase but effective in controlling acidity, whereas phenolic antioxidants could control bulk viscosity and acidity increase [30]. The observation that phenolic antioxidants outperform aminic antioxidants was also shown by DSC studies [30]. Becker and Knorr [27] reported that synergistic mixtures of selective antioxidants remarkably increased the oxidation stability of HOSO. HOSO with 1% proprietary antioxidant could achieve oxidative stability comparable to that of mineral oil-based lubricants [27]. It was suggested that by means of additives, formulated vegetable oil-based lubricants could make inroads into some of the areas dominated by synthetics [27,58]. Miles [73] has noted that performance of vegetable oil with antioxidants is closer to that of mineral oil than that of synthetic esters despite the disparity in structure. This was attributed to the presence of natural antioxidants in vegetable oils. Minami and Mimura [89] studied the effect of 1% tocopherol, N-phenyl-1-naphthylamine and 2,6-dibutyl-4-methylphenol, and zinc bis(N,N-dibutyl dithiocarbamate) in rapeseed oil by using peroxide value determination at 100°C by air purging at 200 mL/min. Phenolic antioxidants and zinc dialkyldithiocarbamate exhibited good control of the oxidation of vegetable oils.

It was found in the study by the authors that the incorporation of synergistic combination of antioxidants improves oxidation stability of nonedible vegetable oils (Table 14.4) [68]. One percent of a mixture of high-molecular weight hindered phenol with alkylated diphenylamine was selected for the study.

It is seen that incorporation of the right antioxidants improved the thermo-oxidative stability of vegetable oils to a significant level to enhance their usage as base oil.

14.5.3 BLENDING

It was shown earlier that the thermo-oxidation of vegetable oil depends on fatty acid distribution. For the improvement of vegetable oils' properties, the combination of chemical additives, addition of diluents such as PAO, and high-oleic vegetable oils offer the best option for formulating high-performance biodegradable lubricants. These approaches result in superior oxidation stability and improved low-temperature properties compared those of commercially available industrial oils such as biobased hydraulic fluids [75]. Blending is also practiced to lower the viscosity of vegetable oil. PAOs of low molecular weight, adipates, are used for this purpose. A viscosity correlation chart to facilitate finding the viscosity of vegetable oil–synthetic base oil mixtures [90] has been reported. Legisa et al. [59] reported that rapeseed oil blended with complex and other esters showed better oxidative stability. Pop et al. [91] reported that corn oil mixed with di-2-ethylhexyl adipate and di-2-ethylhexyl sebacate provides beneficial viscosity, pour point, and oxidation stability. Blends of vegetable oils and di(2-ethylhexyl) sebacate along with additives offer new base stocks with a wide range of kinematic viscosities, very good low-temperature properties, and lubricity characteristics [92]. The thermogram of the mixture showed weight loss that was specific to each vegetable and synthetic component.

Blends of vegetable oil with mineral oils and antioxidants were found to provide fairly biodegradable base oils with improved thermo-oxidative stability [49]. Jatropha oil was blended with the three ISO VG 32 mineral oils of Groups I, II, and III and synthetic esters (TMP and pentaerythritol ester) in 1:1 ratio with 1% of a mixture of high-molecular weight hindered phenol and alkylated diphenylamine antioxidant and without added antioxidant (Table 14.5).

These blends were first evaluated by modified RPVOT (I) screening test under conditions of 120°C and without copper catalyst and water. The time for pressure drop of 25 psi (1.7 kg/cm^2) was recorded as OIT. The test was conducted for up to 300 minutes maximum without antioxidant additive and for 450 minutes in the presence of antioxidant additive. Four inhibited blends displayed induction time of 450 minutes. To distinguish the oxidation stabilities between these four blends, modified RPVOT (II) (at 120°C with copper catalyst and no added water) was conducted on these blends.

In Figure 14.3, values of modified RPVOT (I) induction time of the five blends are compared with those for jatropha and mineral oils or synthetic ester. Group I, II, and III mineral oils showed induction times of >300 minutes. The method did not differentiate between the mineral oils as the test was terminated after 300 minutes. Blend of jatropha oil with Group I mineral oil displayed induction time higher than those of jatropha oil, whereas blend with Group II and Group III showed induction time lower than those of jatropha oil. Blends of jatropha oil and synthetic esters showed remarkably inferior oxidation stability as shown by the lower induction time. A distinct difference

TABLE 14.5

Blends of Vegetable Oils with Synthetic Esters and Mineral Oils Investigated for Oxidation Stability with and without Antioxidant (Mixture of High–Molecular Weight Hindered Phenol and Alkylated Diphenylamine)

Blend No.	Oil 1—% w/w	Oil 2—% w/w	Antioxidant (Mixture of High–Molecular Weight Hindered Phenol and Alkylated Diphenylamine) (% w/w of Blend)	Reference
1	Jatropha—50	Group I mineral oil—50	0	[49]
6	Jatropha—49.5	Group I mineral oil—49.5	1	[49]
2	Jatropha—50	Group II mineral oil—50	0	[49]
7	Jatropha—49.5	Group II mineral oil—49.5	1	[49]
3	Jatropha—50	Group III mineral oil—50	0	[49]
8	Jatropha—49.5	Group III mineral oil—49.5	1	[49]
4	Jatropha—50	TMP ester—50	0	[49]
9	Jatropha—49.5	TMP ester—49.5	1	[49]
5	Jatropha—50	PE ester—50	0	[49]
10	Jatropha—49.5	PE ester—49.5	1	[49]

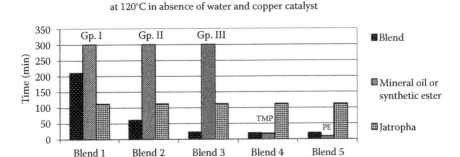

FIGURE 14.3 Modified RPVOT (I) induction times of mineral oils (Groups I, II, and III), synthetic esters, and their blends with jatropha oil with no antioxidant. (Joseph, P. V. et al., Study of some non-edible vegetable oils of Indian origin for lubricant application. *J. Synth. Lubr.* 2007. 24. 181–197; Joseph, P. V., and D. K. Sharma, Improvement of thermooxidative stability of non-edible vegetable oils of Indian origin for biodegradable lubricant application. *Lubr. Sci.* 2010. 22. 149–161. Copyright Wiley-VCH Verlag GmbH & Co. KGaA. Reproduced with permission.)

in induction time was observed among the blends of three mineral oil groups of base oils. The superior oxidation stability displayed by the Group I mineral oil blend was due to the natural antioxidants present in it in the form of sulfur and nitrogen compounds and aromatics. Group II and Group III oils do not contain these antioxidants or contain them in a very small quantity because they are lost during severe refining [93].

Blends with antioxidant were also studied by modified RPVOT (I) and (II) and the results are plotted in Figures 14.4 and 14.5.

In modified RPVOT (I) all the blends with the antioxidant (1% of a mixture of high-molecular weight hindered phenol and alkylated diphenylamine) showed a considerably higher induction time than those of blends without the antioxidant. Jatropha oil when blended with mineral oils or TMP ester and inhibited with antioxidant showed much better thermo-oxidative stability than that of jatropha oil with antioxidant. This study thus clearly shows improved oxidation stability of jatropha oil blends with mineral oil. It also showed that antioxidant addition is essential to maximize the benefit

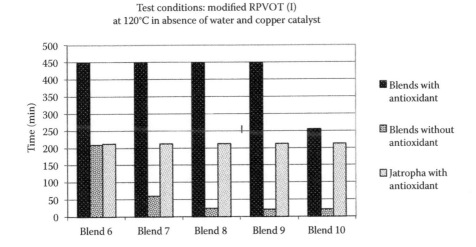

FIGURE 14.4 Modified RPVOT (I) induction times of jatropha oil and its blends with mineral oils (Groups I, II, and III) and synthetic esters (TMP and PE) with and without antioxidant (1% mixture of high-molecular weight hindered phenol and alkylated diphenylamine).

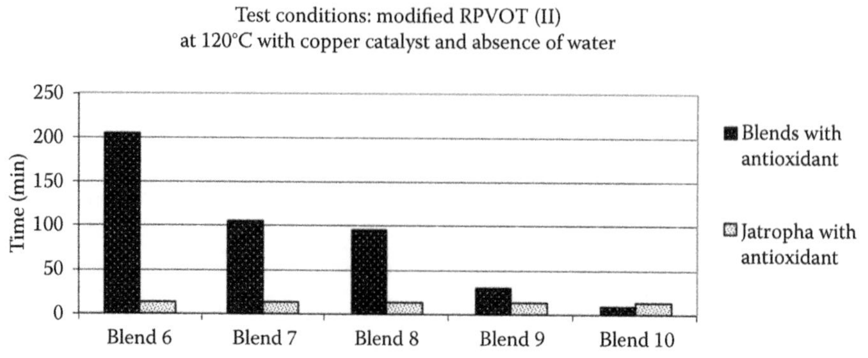

Test conditions: modified RPVOT (II)
at 120°C with copper catalyst and absence of water

FIGURE 14.5 Modified RPVOT (II) induction times of jatropha oil and its blends with mineral oils (Groups I, II, and III) and synthetic esters (TMP and PE) with antioxidant (1% mixture of high-molecular weight hindered phenol and alkylated diphenylamine).

of blending of mineral oil in jatropha oil for achieving better thermo-oxidative stability. The blends with 1% of a mixture of high-molecular weight hindered phenol and alkylated diphenylamine were further studied using modified RPVOT (II). The results showed Group I base oil to be the best for enhancing the oxidation stability of jatropha oil.

Blends with synthetic esters displayed limited improvement, with PE ester blends being the lowest. It can be concluded that the jatropha–mineral oil mixtures possess comparatively much better oxidation stability than that of jatropha oil. The thermo-oxidative stability of blends of jatropha oil with mineral oil and synthetic esters decreases in the order Group I mineral oil > Group II mineral oil > Group III mineral oil > TMP ester > PE ester.

By modifying the locally available vegetable oils through the methods described, different vegetable oils can be put into use for lubricant applications.

REFERENCES

1. P. L. Menezes, M. R. Lovell, M. A. Kabir, C. F. Higgs III, and P. K. Rohatgi, Green lubricants: Role of additive size, in: *Green Tribology*, M. Nosonovsky and B. Bhushan (Eds.), pp. 265–286, Springer, Heidelberg (2012).
2. A. Gosalia, "The Sustainability of the Asian Lubricants Industry," F+L Week 2015, 10–14 March 2015.
3. J. Wilkinson, "Biodegradable Oils—Design, Performance, Environmental Benefits and Applicability," SAE Technical Paper 941077, Proceedings of the 45th Annual Earthmoving Industry Conference, Peoria, Illinois (1994).
4. S. Boyde, Green lubricants: Environmental benefits and impacts of lubrication, *Green Chem.*, 4, 293–307 (2002).
5. S. J. Randles and M. Wright, Environmentally considerate lubricants for the automotive and engineering industries," *J. Synth. Lubr.*, 9 (2), 145–161 (1992).
6. K. Lal and V. Carrick, Performance testing of lubricants based on high oleic vegetable oils, *J. Synth. Lubr.*, 11 (3), 189–206 (1994).
7. D. Horner, Recent trends in environmental friendly lubricants, *J. Synth. Lubr.*, 18 (4), 327–348 (2002).
8. L. R. Rudnick and S. Z. Erhan, Natural oils as lubricants, in: *Synthetics Mineral Oils and Bio-Based Lubricants*, L. R. Rudnick (Ed.), pp. 353–359, CRC Press, Boca Raton, FL (2006).
9. S. P. Carruthers, J. S. Marsh, P. W. Turner, F. B. Ellis, D. J. Murphy, T. Slabas and B. A. Chapman, "Industrial Markets for UK Produced Oilseeds," Proceedings of the Seminar: Lubricants from Oilseeds, 9 May 1996, Ministry of Agriculture, Fisheries and Food, Alternative Crops Unit, London (1996).
10. P. van Broekhuizen, D. Theodori, K. Le Blansch, and S. Ullmer (Eds.), *Lubrication in Inland and Coastal Water Activities*, pp. 3–33, A.A. Balkema Publishers, Amsterdam (2003).
11. S. Z. Erhan, A. Adhvaryu, and B. K. Sharma, Chemically functionalised vegetable oils, in: *Synthetics, Mineral Oils and Bio-Based Lubricants*, L. R. Rudnick (Ed.), pp. 361–388, CRC Press, Boca Raton, FL (2006).

12. S. P. Srivastava, "Modern Lubricant Technology," Technology Publications, Dehradun, India (2007).
13. J. Bhatia and S. Mahanti, "Environment Friendly Lubricants—A Techno-Commercial Audit of Trends and Future Challenges," Proceedings of the 13th LAWPSP Symposium, February 2003, Mumbai, India (2003).
14. N. J. Fox and G. W. Stachowiak, Vegetable oil based lubricants—A review of oxidation, *Tribol. Int.*, 40, 1035–1046 (2007).
15. S. Z. Erhan, Vegetable oils as lubricants, hydraulic fluids and inks, in: *Bailey's Industrial Oil and Fat Products*, Sixth Edition, F. Shahidi (Ed.), John Wiley & Sons, Hoboken, NJ (2005).
16. I. S. Rhee, C. Valez, and K. Bernewitz, "Evaluation of Environmentally Adapted Hydraulic Fluids," TARDEC Tech Report 13640, U.S. Army Tank Automotive Command Research, Development and Engineering Center, Warren, Michigan, pp. 1–15 (1995).
17. E. Kassfeldt and D. Goran, Environmentally adapted hydraulic oils, *Wear*, 207, 41–45 (1997).
18. S. Asadauskas, J. M. Perez, and L. J. Duda, Lubrication properties of castor oil—Potential basestock for biodegradable lubricants, *Lubr. Eng.*, 53 (12), 35–40 (1997).
19. M. Kozma, Investigation into the scuffing load capacity of environmentally-friendly lubricating oils, *J. Synth. Lubr.*, 14 (3), 249–258 (1997).
20. S. Odi-Ovei, Tribological properties of some vegetable oils and fats, *Lubr. Eng.*, 45 (11), 685–690 (1988).
21. A. Adhvaryu, S. Z. Erhan, and J. M. Perez, Tribological studies of thermally and chemically modified vegetable oils for use as environmentally friendly lubricants, *Wear*, 257, 359–367 (2004).
22. B. O. Okonokhua, B. Ikhajiagbe, G. O. Anoliefo, and T. O. Emede, The effects of spent engine oil on soil properties and growth of maize (*Zea mays* L.), *J. Appl. Sci. Environ. Manag.*, 11 (3), 147–152 (2007).
23. R. Musemić and A. Bašić, "The Waste Oil Influence and Risks on Environment in Refer to Bosnia and Herzegovina," 15th International Research/Expert Conference "Trends in the Development of Machinery and Associated Technology" TMT 2011, Prague, Czech Republic, pp. 12–18 (2011).
24. O. P. Abioye, P. Agamuthu, and A. R. Abdul Aziz, Biodegradation of used motor oil in soil using organic waste amendments, *Biotechnol. Res. Int.*, 1–8 (2012).
25. I. Rhee, Evolution of environmentally acceptable hydraulic fluids, *NLGI Spokesm.*, 60 (5), 28–35 (1996).
26. R. E. Gapinski, I. E. Joseph, and B. D. Layzell, A vegetable oil based tractor lubricant, SAE Technical Paper 941758, 1–9 (1994).
27. R. Becker and A. Knorr, Comparative study of the oxidation stability of vegetable oils and the effectiveness of antioxidants, Eurogrease, January–February 1996, 10–19 (1996).
28. D. E. Chasan, "Oxidative Stabilization of Natural Oils," STLE Annual Meeting, May 1994.
29. D. E. Chasan and P. R. Wilson (to Ciba-Geigy Corp.), Stabilized Lubricant Compositions, U.S. Patent No. 5,580,482 (3 December 1996).
30. P. Hamblin, Oxidative stabilisation of synthetic fluids and vegetable oils, *J. Synth. Lubr.*, 16 (2), 157–181 (1999).
31. C. Héry and N. Battersby, Development and applications of environmentally acceptable hydraulic fluids, SAE Technical Paper 981493 (1998).
32. K. D. Erdman, G. H. Kling, and D. E. Tharp, High performance biodegradable fluids requirements for mobile hydraulic systems, SAE Technical Paper 981518 (1998).
33. W. Van Leeuwen, Determination of antioxidant concentration in ester based hydraulic fluids with the use of voltammetric techniques, *Lubr. Eng.*, 56 (10), 30–35 (2000).
34. C. Duncan, J. Reyes-Gavilan, D. Constantini, and S. Oshode, Ashless additive and new polyol ester base oils formulated for use in biodegradable hydraulic fluid applications, *Lubr. Eng.*, 58 (9), 18–28 (2002).
35. A. M. Petlyuk and R. J. Adams, Oxidation stability and tribological behavior of vegetable oil hydraulic fluids, *Tribol. Tran.*, 47, 182–187 (2004).
36. ASTM D7873-13e1, "Standard Test Method for Determination of Oxidation Stability and Insolubles Formation of Inhibited Turbine Oils at 120°C without the Inclusion of Water (Dry TOST Method)," ASTM, West Conshohocken, PA.
37. IP 48, "Determination of Oxidation Characteristics of Lubricating Oil," Standard Methods for the Analysis and Testing of Petroleum Products and British Standard 2000, Energy Institute, London (2012).
38. DIN 51554—Part 3, "Testing of Mineral Oils; Test of Susceptibility to Ageing According to Baader; Testing at 95°C," Deutsches Institut für Normung e.V., Berlin (1978, September).
39. J. M. Perez, F. A. Kelley, E. E. Klaus, and V. Bagrodia, Development and use of the PSU micro oxidation test for diesel engine oils, SAE Fuels and Lubricants Meeting, SAE Technical Paper 872028 (1987).
40. N. Salih, J. Salimon, B. A. Abdullah, and E. Yousif, Thermo-oxidation, friction-reducing and physicochemical properties of ricinoleic acid based diester biolubricants, *Arab. J. Chem.* (2014), doi:10.1016/j .arabjc.2013.08.002.

41. ASTM D6186-98, "Standard Test Method for Oxidation Induction Time of Lubricating Oils by Pressure Differential Scanning Calorimetry (PDSC)," ASTM, West Conshohocken, PA.

42. ASTM E2009-08(2014)e1, "Standard Test Methods for Oxidation Onset Temperature of Hydrocarbons by Differential Scanning Calorimetry," ASTM, West Conshohocken, PA.

43. J. Černý and M. Zelinka, Oxidation stability of lubricants measured by a PDSC technique, *Petroleum and Coal*, 46 (3), 56–62 (2004).

44. T. F. Buneman, M. C. Steverink-de Zoete, and R. P. van Aken, Environmentally acceptable hydraulic fluids based on natural, synthetic esters, in: *Hydraulic Fluids and Alternative Industrial Lubricants*, pp. 1–19, SAE, Warrendale, MI (1998).

45. J. M. Perez, Oxidative properties of lubricants using thermal analysis, *Thermochim. Acta*, 357–358, 47–56 (2000).

46. N. J. Fox, A. K. Simpson, and G. W. Stachowiak, Sealed capsule differential scanning calorimetry—An effective method for screening of the oxidation stability of the vegetable oil formulations, *Lubr. Eng.*, 57 (10), 14–20 (2001).

47. ASTM D2272-14a, "Standard Test Method for Oxidation Stability of Steam Turbine Oils by Rotating Pressure Vessel," ASTM, West Conshohocken, PA.

48. P. Bartl and C. Volkl, Thermooxidative stability of high temperature stability polyol ester jet engine oils—A comparison of test methods, *J. Synth. Lubr.*, 17 (3), 179–189 (2002).

49. P. V. Joseph, P. Bhatnagar, D. Saxena, R. T. Mookken, and R. K. Malhotra, "Designing Green Lubricants for Manufacturing Industry Using Renewable Base Materials," Society of Tribology and Lubrication Engineers Annual Conference & Exhibition, Detroit, MI (2013).

50. X. Zhang, H. Murrenhoff, P. Weckes, and W. Holderich, Effect of temperature on the ageing behavior of unsaturated ester based lubricants, *J. Synth. Lubr.*, 21 (1), 1–11 (2004).

51. B. Khemchandani, A. K. Jaiswal, E. Sayanna, and M. Forsyth, Mixture of safflower oil and synthetic ester as a base stock for biodegradable lubricants, *Lubr. Sci.*, 26 (2), 67–80 (2013).

52. P. Vamsi Krishna, R. R. Srikant, and N. Rao, Experimental investigation on the performance of nano boric acid suspensions in SAE-40 and coconut oil during turning of AISI1040 steel, *Int. J. Mach. Tool Manuf.*, 50, 911–916 (2010).

53. N. H. Jayadas and K. Prabhakaran Nair, Coconut oil as base oil for industrial lubricants—Evaluation and modification of thermal, oxidative and low temperature properties, *Tribol. Int.*, 39 (9), 873–878 (2006).

54. Y. M. Shashidhara and S. R. Jayaram, Vegetable oils as a potential cutting fluid—An evolution, *Tribol. Int.*, 43, 1073–1081 (2010).

55. J. K. Mannekote and S. V. Kailas, Experimental investigation of coconut and palm oils as lubricants in four stroke engines, *Tribol. Online*, 6 (1), 76–82 (2011).

56. K. Chatra, R. Sathwik, N. H. Jayadas, and S. V. Kailas, Natural oil-based lubricants, in: *Green Tribology: Biomimetics, Energy Conservation and Sustainability*, M. Nosonovsky and B. Bhushan (Eds.), p. 287, Springer, Heidelberg (2012).

57. B. Krzan and J. Vizintin, Tribological properties of an environmentally adopted universal tractor transmission oil based on vegetable oil, *Tribol. Int.*, 36, 827–833 (2003).

58. S. Asadauskas, J. M. Perez, and J. L. Duda, Oxidative stability and antiwear properties of high oleic vegetable oils, *Lubr. Eng.*, 52 (12), 877–882 (1996).

59. I. Legisa, M. Picek, and K. Nahal, Some experiences with biodegradable lubricants, *J. Synth. Lubr.*, 13 (4), 347–360 (1997).

60. Y. Gerbig, S. I. U. Ahmed, F. A. Gerbig, and H. Haefke, Suitability of vegetable oils as industrial lubricants, *J. Synth. Lubr.*, 21 (3), 177–191 (2004).

61. O. N. Anand, J. Mehta, and T. S. R. Prasada Rao, Lubricant components from vegetable oils of Indian origin, *J. Synth. Lubr.*, 15 (2), 97–106 (1998).

62. D. K. Salunkhe, J. K. Chavan, R. N. Adsulke, and S. S. Kadam, *World Oilseeds: Chemistry, Technology and Utilization*, pp. 1–8, Van Nostrand Reinhold, New York (1992).

63. E. U. Ikhuoria, A. I. Aigbodion, and F. E. Okieimen, Enhancing the quality of alkyd resins using methyl esters of rubber seed oil, *Trop. J. Pharm. Res.*, 3 (1), 311–317 (2004).

64. S. Khandelwal and R. Y. Chauhan, Biodiesel production from non edible-oils: A review, *J. Chem. Pharm. Res.*, 4 (9), 4219–4230 (2012).

65. A. K. Jain and A. Suhane, Research approach & prospects of non edible vegetable oil as a potential resource for biolubricant—A review, *Adv. Eng. Appl. Sci.*, 1 (1), 23–32 (2012).

66. P. S. Chauhan and V. K. Chhibber, Non-edible oil as a source of bio-lubricant for industrial applications: A review, *Int. J. Eng. Sci. Innov. Technol.*, 2 (1), 299–305 (2013).

67. P. V. Joseph, D. Saxena, and D. K. Sharma, Study of some non-edible vegetable oils of Indian origin for lubricant application, *J. Synth. Lubr.*, 24 (4), 181–197 (2007).
68. P. V. Joseph and D. K. Sharma, Improvement of thermooxidative stability of non-edible vegetable oils of Indian origin for biodegradable lubricant application, *Lubr. Sci.*, 22 (4), 149–161 (2010).
69. N. Canter, It isn't easy being green, *Lubr. World*, 9, 15–21 (2001).
70. A. Birova, A. Pavlovicova, and J. Cvengros, Lubricating oils based on chemically modified vegetable oils, *J. Synth. Lubr.*, 18 (4), 291–299 (2002).
71. D. R. Kodali, Biobased lubricant—Chemical modification of vegetable oils, *Inform Int. News Fats Oils Rel. Mater. (AOCS)*, 14 (3), 121–123 (2003).
72. M. A. Schmidt, C. R. Dietrich, and E. B. Cahoon, Biotechnological enhancement of soybean oil for lubricant, in: Synthetics, Mineral Oils, and Bio-Based Lubricants, L. R. Rudnick (Ed.), pp. 389–398, CRC Press, Boca Raton, FL (2006).
73. P. Miles, Synthetics versus vegetable oils: Applications, options and performance, *J. Synth. Lubr.*, 15 (1), 43–52 (1998).
74. M. J. Hills, "Modification of the Vegetable Oils in the Plant," Proceedings of the Seminar: Lubricants from Oilseeds, 9 May 1996, Ministry of Agriculture, Fisheries and Food, Alternative Crops Unit, London (1996).
75. S. Z. Erhan, B. K. Sharma, and J. M. Perez, Oxidation and low temperature stability of vegetable oil based lubricants, *Ind. Crops Prod.*, 24 (2), 292–299 (2006).
76. B. F. Haumann, Modified oil may be key to sunflower's future, *Inform: Int. News Fats Oils Relat. Mater. (AOCS)*, 5 (11), 1210 (1994).
77. H. Wagner, R. Luther, and T. Mang, Lubricant base fluids based on renewable raw materials: Their catalytic manufacture and modification, *Appl. Catal. A*, 221, 429–442 (2001).
78. W. E. Neff, M. A. El-Agaimy, and T. L. Mounts, Oxidative stability of blends and interesterified blends of soybean oil and palm olein, *J. Am. Oil Chem. Soc.*, 71 (10), 1111–1116 (1994).
79. X. Wu, X. Zhang, S. Yang, H. Chen, and D. Wang, The study of epoxidised rapeseed oil used as potential biodegradable lubricant, *J. Am. Oil Chem. Soc.*, 77 (5), 561–563 (2000).
80. P. S. Sreeprasanth, R. Srivastava, D. Srinivas, and P. Ratnasamy, Hydrophobic solid acid catalysts for production of biofuels and lubricants, *Appl. Catal. A*, 314 (2–9), 148–159 (2006).
81. P. S. Lathi and B. Mattiasson, Green approach for the preparation of biodegradable lubricant base stock from epoxidised vegetable oils, *Appl. Catal. B*, 69 (3–4), 207–212 (2007).
82. P. S. Chauhan, H. C. Joshi, and V. K. Chhibber, Multiple utilization of non-edible seeds for production of eco-friendly products, *Int. J. Chem. Appl.* 4 (4), 319–331 (2012).
83. B. J. Bremmer and L. Plons, "Bio-based Lubricants, a Market Opportunity Study Update," Prepared for United Soybean Board (2008).
84. H. Murrenhoff and A. Remmelmann, Environmentally friendly oils, in: *Fuels and Lubricants Handbook: Technology, Properties, Performance and Testing*, G. E. Totten (Ed.), pp. 267–296, ASTM International, West Conshohocken, PA (2003).
85. S. Z. Erhan, B. K. Sharma, Z. Liu, and A. Adhvaryu, Lubricant base stock potential of chemically modified vegetable oils, *J. Agric. Food Chem.*, 56 (19), 8919–8925 (2008).
86. I. Minami, H. S. Hong, and N. C. Mathur, Lubrication performance of model organic compounds in high oleic sunflower oil, *J. Synth. Lubr.*, 16 (1), 3–12 (1999).
87. I. Minami and S. Mitsumune, Antiwear properties of phosphorous-containing additives in vegetable oils, *Tribol. Lett.*, 13 (2), 95–101 (2002).
88. A. Hubmann, Rapsöl—Ein alternatives Basisöl für Schmierstoffe, *Mineralöltech*, 34, 1–18 (1989).
89. I. Minami and K. Mimura, Synergistic effect of antiwear additives and antioxidants in vegetable oil, *J. Synth. Lubr.*, 21 (3), 193–205 (2004).
90. S. Z. Erhan, S. Asadauskas, and A. Adhvaryu, Correlation of viscosities of vegetable oil blends with selected esters and hydrocarbons, *J. Am. Oil Chem. Soc.*, 79 (11), 1157–1161 (2002).
91. L. Pop, C. Puscas, G. Bandur, G. Vlase, and R. Nutiu, Basestock oils for lubricants from mixtures of corn oil and synthetic diesters, *J. Am. Oil Chem. Soc.*, 85 (1), 71–76 (2008).
92. C. Puscas, G. Bandur, D. Modra, and R. Nutiu, Mixtures of vegetable oils and di-2-ethylhexyl-sebacate as lubricants, *J. Synth. Lubr.*, 23 (4), 185–196 (2006).
93. L. R. Rudnick, *Lubricant Additives: Chemistry and Applications*, Second Edition, p. 477, CRC Press, Boca Raton, FL (2009).

15 Vegetable Oils as Additive in the Formulation of Eco-Friendly Lubricant

Gobinda Karmakar and Pranab Ghosh

CONTENTS

ABSTRACT

The application of vegetable oils as additive in the formulation of biolubricant has attracted considerable interest recently. This is due to their enhanced multifunctional performances and biocompatibility. This chapter is focused on the application of biodegradable additives based on vegetable oils in the formulation of eco-friendly lubricants. The synthesis of some vegetable oil-based additives, particularly soybean oil and sunflower oil; their characterization; and performance evaluation in mineral base stocks are discussed in detail. The additives were synthesized by chemical modification such as polymerization and transesterification, which induce higher thermo-oxidative stability and also enhance their performance. Characterization of the synthesized additives was carried out by spectral techniques (infrared and nuclear magnetic resonance). Molecular weight was determined by gel permeation chromatography–size exclusion chromatography and viscometric analysis. The thermal stability of the additives was determined by thermogravimetric analysis. Finally, performance evaluation of the additives particularly as antiwear, VI improver, and pour point depressant in different base stocks was carried out by ASTM methods. Shear stability was also determined according to the ASTM method. The biodegradability test was carried out by disk diffusion and soil burial test methods. It has been found that the synthesized additives enhance the overall performance of the lubricant base stocks and therefore were successfully used as greener additive in biolubricant formulation.

15.1 BACKGROUND

Modern lubricants are formulated from a range of base fluids and chemical additives. The base fluids have several functions, but they primarily provide a fluid layer, which keeps moving metal surfaces apart and controls friction, removes heat and wear particles, and increases the lifetime of internal combustion engines. Base fluids can be classified as mineral oil (from petroleum), synthetic oil (e.g., PAOs, polyalkylene glycols, synthetic esters, and silicones), and biobased oil (e.g., vegetable oil). Depending upon the nature of the base oils, the lubricants can be categorized as mineral lubricant, synthetic lubricant, or biolubricant. The role of additives in a lubricant composition is very significant. Additives are introduced into the base oil to enhance its performance to the desired level which cannot be achieved by the base fluid alone. The application of additives has a dominant impact on the performance evaluation of the lubricant formulation. In biolubricant formulation the role of additive is also very important. Pour point depressants (PPDs), viscosity index improvers (VIIs) or viscosity modifiers (VMs), dispersants/detergents, antiwear and extreme pressure additives, antioxidants, and corrosion inhibitors are examples of additives generally blended with base stocks. The commercially available additives of such kinds exhibit satisfactory performances but are not biodegradable and create many environmental hazards. Moreover, they have no multifunctional performances. The conventional additives and their environmental impacts are discussed later in this chapter. The application of vegetable oils or their derivatives as biodegradable multifunctional additives in biolubricant composition opens new research horizons in lubrication technology. Although vegetable oils have been widely investigated as base fluids owing to environmental concerns [1–3], the reports of their application as additive in the formulation of eco-friendly lubricant are scarce. Vegetable oils are mainly triglycerides of different long-chain fatty acids having various degrees of unsaturation (Figure 15.1). Their thermo-oxidative stability is very poor. However, to improve their thermal stability and mechanical properties, the triglyceride molecule must be modified chemically [4]. This can be done by polymerizing the vegetable oil or its epoxy derivatives, forming derivatives of the vegetable oil by reacting with suitable reagents, resulting in increasing its molecular weight and cross-link density, since an increase in cross-link density improves its thermal stability and mechanical properties [5]. Li et al. [6] have shown the application of natural garlic oil as a high-performing and environmentally friendly extreme pressure additive in lubricating oils. Biresaw et al. [7] have mentioned the application of biobased polyesters as extreme pressure additive in mineral

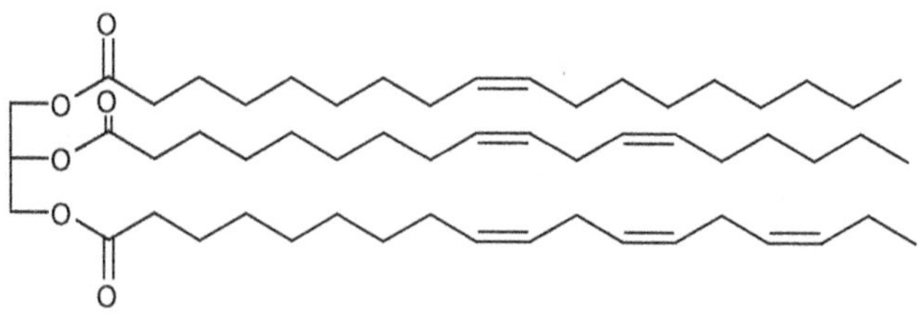

FIGURE 15.1 General structure of triglyceride of long-chain fatty acids, the major constituents of vegetable oil.

oil. Erhan et al. in U.S. Patent 7279448 B2 (2007) [8] describe the use of poly(hydroxy thioether) vegetable oil derivatives as a lubricating oil additive. They have prepared a novel class of chemically modified vegetable oils by performing reaction of epoxidized triglyceride oils with thiols. The resultant poly(hydroxy thioether) derivatives have utility as antiwear/antifriction additives for environmentally friendly industrial oils in automotive applications. Erickson et al. in U.S. Patents 5282989 (1994) [9] and 4925581 (1990) [10] have described the additive properties of triglyceride vegetable oil combined with at least one sulfurized vegetable oil and a phosphite adduct. They have prepared different additive compositions preferably by taking meadowfoam oil as a triglyceride with a sulfurized (20%) and phosphite adduct of the oil and compared its performances as antiwear additive in lubricant base stocks. In U.S. Patent 4152278 [11], the performance of a novel lubricant composition comprising sulfurized derivatives of fatty acid esters (wax esters) derived from vegetable oil resources has been described. The antiwear, extreme pressure, and VI properties of the additives have been investigated by this work. The synthesis and the evaluation of telomerized vegetable oil, sulfurized and phosphorus derivatives of telomerized vegetable oils, and combinations thereof for use as thermal oxidative stability enhancers and viscosity improvers have been discussed in U.S. Patent 5229023 [12]. Telomerization induces vegetable oil which contains no more than 4% polyunsaturated fatty acids, helping enhance its thermo-oxidative stability. The invention is further related to telomerized triglyceride vegetable oils as a lubricating composition base stock substitute. U.S. Patent 4873008 describes the synthesis of lubricant additives made out of jojoba oil, sulfurized jojoba oil, and a phosphite adduct of jojoba oil [13]. The antiwear and frictional properties of the additives obtained by four-ball wear test (FBWT) and Falex test results revealed that a combination of 1% jojoba + 2% sulfurized jojoba + 0.5% phosphite adduct shows better additive performance. Recently, Kumar et al. have discussed the tribological and emission studies on a two-stroke petrol engine lubricated with sunflower methyl ester [14].

Thus, there exists an extensive scope of research on the development of environmentally friendly additives that are compatible with environmental regulations. In this context, we have discussed the synthesis, characterization, and performance evaluation of lubricant additives based on soybean oil, sunflower oil, palm oil, etc., to get additives with improved thermal properties and multifunctional performances.

15.2 BASE OIL CHARACTERISTICS

Petroleum base fluids are nonpolar and generally a mixture of hydrocarbons having different chain lengths. On the other hand, vegetable oils are polar in nature. They have excellent antifrictional properties [14,15], shear stability [16], high VI [2], low evaporative loss, and high flash point [17] compared to mineral oils. Their use in lubricant formulation offers advantages with respect to resource renewability and biodegradability [18,19]. But they have comparatively poor thermo-oxidative stability, poor pour point, and gumming effect [20,21]. Pour oxidation stability is due to the presence of higher degree of unsaturations ($-C=C-$) in the fatty acid chain. Therefore, the use of vegetable oils as lubricant base stock has not been widely spread yet. However, because of increasing environmental concern, much attention has been given to improve their thermo-oxidative stability by means of chemical modification of vegetable oils with a view that they can be used in the lubricant composition both as base fluid and as additive.

In this chapter we focus on the formulation of biolubricant composition by blending vegetable oil-based additives mainly with petroleum-based mineral oils and their performance evaluation. Although the formulation is not environmentally benign, it has the advantage of better performance, easy availability, and lower cost. Petroleum-based mineral oils may be naphthenic or paraffinic. Paraffin mineral oil base stocks have better VI property and biocompatibility than those of naphthenic base oils [22]. In this chapter, the additives are mainly blended with two different mineral base oils of high paraffin content designated as BO1 and BO2. The physical and chemical properties of the base oils are given in Table 15.1. Some examples of bioderived additives added to vegetable base stocks have also been mentioned and discussed.

TABLE 15.1
Physical and Chemical Characteristics of Two
Mineral Oils Used as Base Fluids

Parameter	BO1	BO2
Sulfur	≤0.03%	≤0.03%
Saturates	≤90%	≤90%
Pour point (°C)	−3	−6
Cloud point (°C)	−6	−8
Density (g cm^{-3})	0.84	0.94
Viscosity at 40°C (cSt)	6.7	24.229
Viscosity at 100°C (cSt)	1.77	4.016
VI	80.05	89.02

Source: Reprinted with permission from G. Karmakar and P. Ghosh,
             ACS Sustain. Chem. Eng., 3, 19–25. Copyright 2015 American
             Chemical Society.

15.3   ADDITIVES FOR BASE OILS AND THEIR CHEMISTRY

Nowadays, the application of additives in lubricant base stock to increase their performance up to the desired limit is essential and therefore research in this area is of great interest [23–25]. Additives may be of different types that enhance some specific properties required for lubricant in special conditions. In general, additives improve the lifetime of the machinery parts by reducing friction and providing greater performance, helping achieve fuel economy, carrying away contaminants and debris from the lubricant, reducing pollution, etc. Recently the application of multifunctional additives is increasing rapidly. As already mentioned, bioderived additives are more compatible with the environment compared to commercial ones. Here, the discussion is focused on the application of multifunctional biobased additives that especially act as antiwear, VII, and PPD. Before going into detail on the evaluation of green additives, a brief introduction about conventional additives and their function and drawbacks has been given.

15.3.1   ANTIWEAR AND EXTREME PRESSURE ADDITIVES

Organosulfur and organophosphorus compounds, such as organic polysulfides, phosphates, dithiophosphates, and dithiocarbamates, are the most commonly used antiwear and extreme pressure additives [26]. Zinc dialkyldithiophosphate (ZDDP) (Figure 15.2), after its discovery, became the most widely used antiwear additive in the lubricant industry. As a result, many interesting studies have been undertaken about ZDDP, and many mechanisms have been proposed for its antiwear and extreme pressure action [27,28]. Extreme pressure additives form extremely durable protective films by reacting thermochemically with the metal surfaces. These films can withstand extreme temperatures and mechanical pressures and minimize direct contact between surfaces and thereby protect them from scoring and seizing.

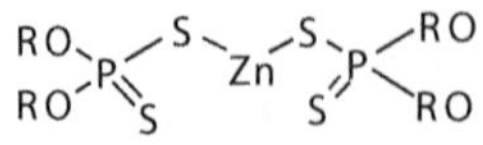

FIGURE 15.2   Structure of ZDDP.

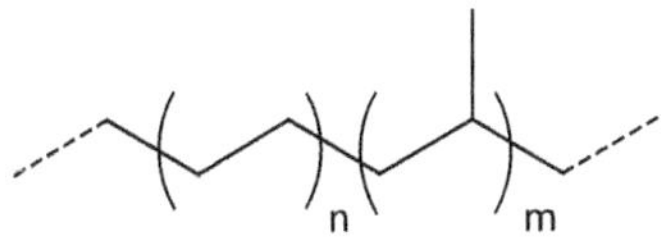

Ethylene-propylene copolymer

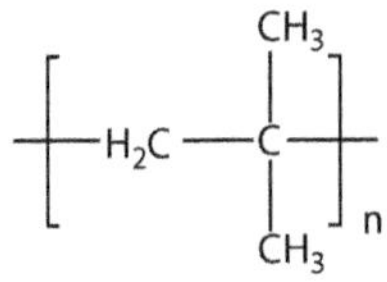

Polyisobutylene

FIGURE 15.3 Structure of commonly used VIIs.

15.3.2 Viscosity Index Improver

The decrease in the viscosity of a fluid with rise in temperature is quantitatively expressed by an empirical term called viscosity index. VIIs or VMs are long-chain, high-molecular weight polymers that function by increasing the relative viscosity of a lubricant more at high temperature than at low temperature. They cause a minimal increase in engine oil viscosity when temperature is low, but prevent considerable decrease when temperature is high. A higher VI value signifies a lesser effect of temperature on viscosity. The viscosity of the lubricant, which increases with increasing volume of the VII/VM additives in the base oil, is related to average molecular weight of the polymer molecules, chain topology, and temperature-dependent solubility [29]. A higher-molecular weight VII/VM polymer will result in a higher VI compared to the lower-molecular weight polymer of the same chemical type. At room temperature the VII molecules exist as random coils. With increasing temperature, the polymer molecules change from tight coils to linear configuration that has larger volume and comparatively better solubility. The increase in the volume of solvated polymers increases the viscosity of the oil, which offsets the reduction in viscosity due to increasing temperature [30]. Commonly used VIIs include polyisobutylene, polyfumarate esters, polymethylacrylates, olefin copolymers such as ethylene–propylene copolymer, hydrogenated styrene–isoprene copolymer, poly(butadiene-styrene), etc. (Figure 15.3).

15.3.3 Pour Point Depressant

The maximum temperature at which the oil stops flowing is known as pour point [31]. Pour point may also be defined as the temperature at which the oil sample no longer flows when subjected to the standardized schedule of quiescent cooling prescribed by ASTM D97. PPDs (also known as low-temperature flow improvers and wax crystal modifiers) are polymeric materials that are added to lubricant base oils to improve their cold-flow properties. Paraffin wax in mineral oils begins to crystallize at a specific low temperature. The needle-shaped crystals form a matrix that will immobilize the oil. Segments of polymeric additive molecules that cocrystallize with the wax modify the shape of the wax and promote the formation of small, free-flowing wax particles which help to flow the oil at lower temperature [31]. Examples of commonly used commercial PPD additives are dialkyl fumarate (C8–C18)/vinyl acetate copolymers, poly(methyl methacrylate) (Figure 15.4), and polyacrylates (Figure 15.5).

The commercially available additives discussed above are synthetic and not eco-friendly. Elements such as zinc, sulfur, and phosphorus in the antiwear and extreme pressure additives and acrylates in the PPD and VII additives generate many harmful materials to the environment. They are not biodegradable and therefore environmentally not benign. On the other hand, vegetable

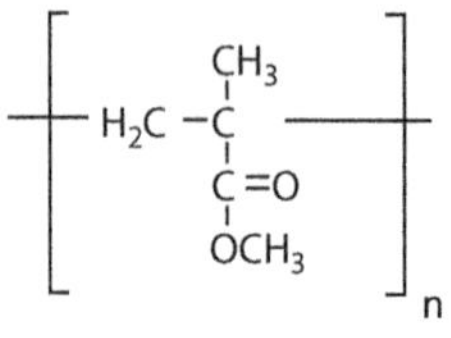

FIGURE 15.4   Structure of poly(methyl methacrylate).

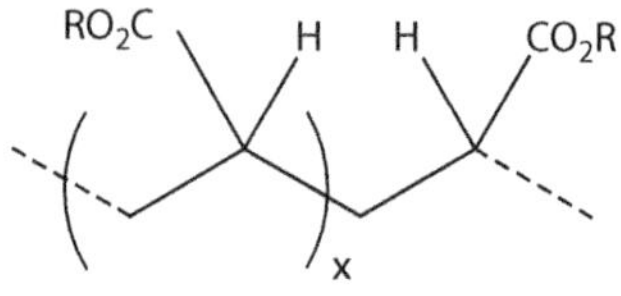

FIGURE 15.5   Structure of polyacrylate.

oil-based additives are biodegradable and show excellent tribological properties [16] and at the same time VII and PPD properties [32]. Thus, research on vegetable oil-based additives as lubricant base stock is increasing recently. The following discussion is made on this emerging topic.

15.4   VEGETABLE OILS USED AS ADDITIVE IN LUBRICANT COMPOSITION

The application of vegetable oils or their derivatives as additive is increasing very recently because of their low toxicity, biocompatibility, and enhanced multifunctional performances in mineral or vegetable base fluids. Vegetable oils are used as antifrictional additive in mineral oils as well as biobased lube oils. Erickson et al. [33] have exposed the application of meadowfoam oil or their derivatives (phosphate or sulfurized) as antiwear additive in paraffinic mineral base oil. The anti-wear test results of the synthesized additives are mentioned in Table 15.2. A significant reduction

TABLE 15.2
Antiwear Test Results of Mineral Base Fluid Blended with Vegetable Oil Derivatives

	Four-Ball Wear		Falex
Lubricant	**WSD (mm)**	**Load (lb)**	**Pounds to Failure**
Base fluid	0.94	120	725
Base fluid + 1% meadowfoam oil	0.82	150	2600
Base fluid + 2% sulfurized meadowfoam oil	0.55	210	4200
Base fluid + 1% meadowfoam phosphite adduct	0.55	210	4200
Base fluid + 2% sulfurized meadowfoam oil[a] + 0.5% meadowfoam phosphite adduct[b]	0.42	280	4500
Base fluid + 1% meadowfoam oil + 1% sulfurized meadowfoam oil + 0.5% meadowfoam phosphite adduct	0.42	280	4500
Base fluid + 1% meadowfoam oil + 0.5% meadowfoam phosphite adduct	0.5	250	4000

Source: Erickson, F. L. et al., U.S. Patent No. 4970010, 1990.
Note: WSD: wear scar diameter.
[a]   The sulfurized meadowfoam oil is a mixture of 10 g of triglyceride meadowfoam oil and 10 g of jojoba oil sulfurized with 4 g of sulfur.
[b]   The meadowfoam phosphite adduct is a monoadduct of dibutyl phosphite with triglyceride meadowfoam oil.

in wear scar diameter (WSD) was observed because of the addition of a mixture of meadowfoam oil and their derivatives in base fluids. Palm oil or its chemically modified adducts also showed significant additive performances in mineral oils as well as biobased base oils. Maleque et al. [34] have mentioned the application of palm oil methyl esters (POMEs) as antiwear additive in mineral base fluids. They reported that four-ball test results for antiwear performance of mineral base oil blended with different percentages of the POME, viz., 0, 3, 5, 7, and 10, showed improvement of performance and 5% additive concentration performed the best. Ossia et al. [35] have shown that eicosanoic and octadecanoic acids, the very long-chain fatty acids in castor and jojoba oil, are used as lubricity additives in biodegradable castor oil and jojoba oil as well as in mineral oil base stocks. Bisht et al. [36] compared the performances of mineral oil base stocks by blending different concentrations of jojoba oil with percentages 5, 10, 20, 30, and 50 and additives such as tricresyl phosphate and zinc dialkyldithiophosphate. It has been found that the use of jojoba oil in lubricant formulation enhances certain desirable properties of the blend such as VI, antirust, antifoam, antiwear, and friction reduction properties. This is due to the straight, high-polarity, and longer-chain molecules of jojoba oil (Figure 15.6), which was strongly adsorbed by the metal surfaces.

Soybean oil (soy oil) and sunflower oil (sun oil) showed very high VI and antiwear properties [16,32]. They contain a higher degree of unsaturation, which is the cause of their poor oxidative stability [37]. The unsaturation, however, can be removed by chemical modification such as polymerization [32] and epoxidation [38], which will improve their thermo-oxidative stability. The fatty acid compositions of soybean and sunflower oil are shown in Table 15.3. Examples where modified soybean and sunflower oil have been successfully applied as additive have been mentioned in our earlier publications [16,17,32]. Balamurugan et al. [2] have reported the wear resistance and the oxidation stability of the base stocks such as SAE 40 (mineral oil) and soybean oil methyl ester (SBME) by blending SBME and a binary blend of POME and castor oil as additives, respectively. The experimental results for the two lubricant compositions (90% SAE 40 + 10% SBME and 85% SBME + 10% castor + 5% POME) showed that mineral oil-based lubricant has comparatively higher

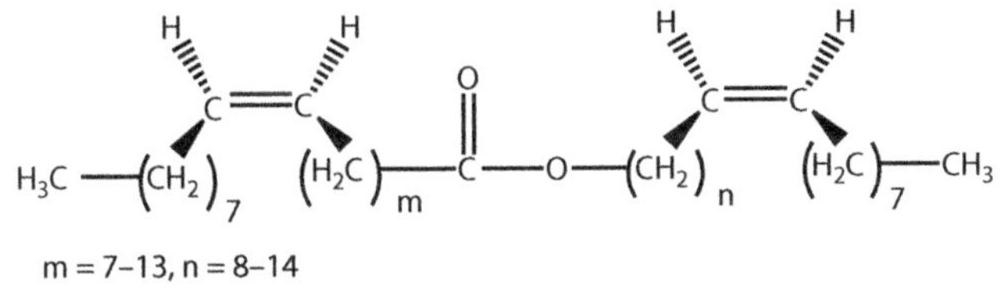

m = 7–13, n = 8–14

FIGURE 15.6 Structure of jojoba oil molecule.

TABLE 15.3
Fatty Acid Profiles of a Few Vegetable Oils

Fatty Acid		% Composition		
		Sunflower Oil	Soybean Oil	Palm Oil
Saturated	C12:0 (lauric)	0.29	0.16	0.1
	C14:0 (myristic)	0.56	0.28	1.0
	C16:0 (palmitic)	7.50	11.00	42.8
	C18:0 (stearic)	4.00	4.10	4.5
	C20:0 (arachidic)	0.41	0.31	–
Unsaturated	16:1 (palmitoleic)	0.30	0.26	–
	18:1 (oleic)	23.00	22.00	40.5
	18:2 (linoleic)	60.00	54.00	10.1
	18:3 (linolenic)	2.90	7.50	0.2

oxidation stability, whereas the biobased lubricant has better wear resistance capacity. Shanta et al. [15] reported that the wear protection power of SAE 15W-40 engine oil was increased by addition of known percentages (5%, 10%, 20%, and 30%) of different biodiesels from vegetable (such as canola oil, peanut oil, and soybean oil) and animal resources (chicken fat) even at small percentages. Thus, extensive works in the field of bioadditive for lubricating oil have already been carried out and are being carried on. In the following, described are the synthesis of few vegetable oil-based additives and their characterization and performance evaluation.

15.4.1 Synthesis

The homopolymer of sunflower and soybean oil used as multifunctional additive can be prepared by conventional thermal or microwave irradiation processes. The microwave method has some advantages over the conventional heating method owing to shorter reaction time and higher yield. Polymerization was carried out by free radical technique using benzoyl peroxide or azobisisobutyronitrile as initiators to the refined oils (Scheme 15.1). In the thermal method [39], 10 g (~11.4 mmol) of each monomer (soybean or sunflower oil) in a four-necked round-bottom flask, fitted with a magnetic stirrer, a condenser, a thermometer, and an inlet for the introduction of nitrogen, was heated at 80°C to 90°C for 30 min in the presence of toluene solvent. Then, 0.1% initiator was added to the monomer and 90°C was maintained while stirring the mixture for 6 hours. The flask was allowed to cool to room temperature, and the reaction mixture was poured into methanol with stirring until the precipitation was completed. The product was then purified by repeated precipitation of its hexane solution with methanol and then dried under vacuum at 40°C. A similar synthetic procedure was applied in the microwave irradiation method [40] using focused monomode microwave oven (CEM Corporation, Matthews, North Carolina) fitted with magnetic stirrer, where both temperature and microwave power were well controlled. The reaction mixture was magnetically stirred at 90°C for 30 min with 300 W applied power without any solvent. The reaction mixture was quenched by rapid cooling to room temperature. A pictorial representation of the probable structure of the homopolymer is shown in Figure 15.7.

The methyl esters of the vegetable oils (palm oil, soybean oil, etc.) used as bioadditives in lubricant base stocks were synthesized by transesterification of the oil with methanol using a base (KOH/NaOH) as catalyst (1% w/w) [41]. The molar ratio of methanol/oil was 6:1 and the reaction was carried out at 60°C for 60 min to obtain the optimum yield [41]. The details of the synthetic procedure was discussed by Komintarachat and Chuepeng [42] and the reaction is shown in Scheme 15.2.

15.4.2 Characterization

Homopolymers of the vegetable oils were characterized using spectroscopy (Fourier transform infrared [FT-IR] and nuclear magnetic resonance [NMR]), thermogravimetric analysis (TGA), shear stability analysis, gel permeation chromatography–size exclusion chromatography (GPC-SEC), and viscometry. In the following the characterization of the polymers of sun oil and soy oil is discussed.

SCHEME 15.1 Synthesis of vegetable oil polymers. BZP is benzoyl peroxide and AIBN is azobisisobutyronitrile.

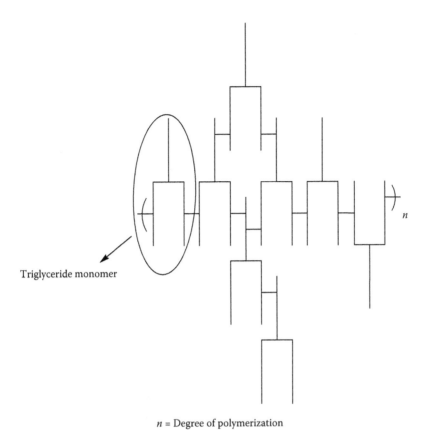

n = Degree of polymerization

FIGURE 15.7 Pictorial diagram of a probable structure of polymer of vegetable oil (sunflower oil/soybean oil).

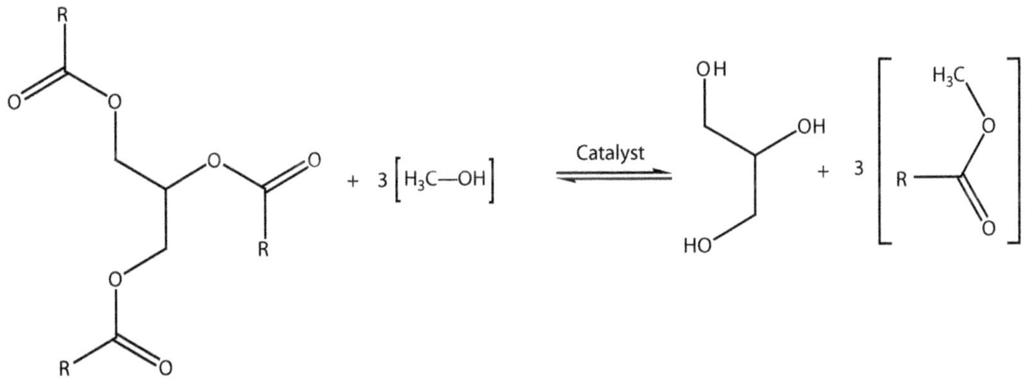

SCHEME 15.2 Transesterification of vegetable oils. R is long-chain hydrocarbon containing saturated and unsaturated carbons.

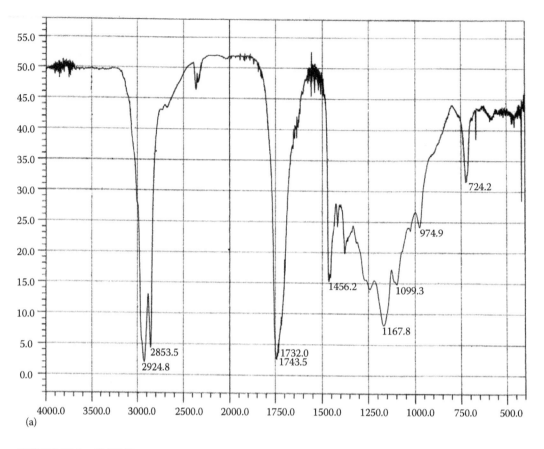

(a)

FIGURE 15.8 (a) FT-IR spectrum of homopolymer of sunflower oil. (*Continued*)

15.4.2.1 Spectroscopy

The infrared (IR) absorption spectra of sunpolymer and soypolymer were recorded by Shimadzu FTIR 8300 (Kyoto, Japan) within the range of 400 to 4000 cm^{-1} using 0.1 mm KBr cells at room temperature and are shown in Figure 15.8a and b, respectively. The peaks appearing at 1732 and 1745.5 cm^{-1} are due to the ester carbonyl groups present in the structures of the polymers. The ester C–O stretching vibration of the two polymers appeared as broad peaks at 1164.9 and 1167.8 cm^{-1}. Peaks that appeared in the ranges of 2853.5 to 2924.8 and 2854.5 to 2923.9 cm^{-1} were for C–H stretching vibrations for sunpolymer and soypolymer, respectively. Peaks due to C–H bending vibrations of the two polymers were found in the region of 1099.3 to 721.3 cm^{-1}. The ^{1}H NMR spectra for the two polymers were recorded by Brucker Avance 300 MHz FT-NMR spectrometer by using 5 mm broadband observe probe, CDCl$_3$ solvent, and tetramethylsilane as reference material. The spectra are shown in Figure 15.9a and b. Broad peaks that appeared at 4.117 and 4.315 ppm were due to ester carbonyl groups present in the triglyceride molecule of sunpolymer and soypolymer, respectively. Methyl and methylene protons appeared in the ranges of 0.857 to 0.911, 1.255 to 1.607, and 2.001 to 2.769 ppm. Peaks in the range of 172.90 to 178.92 ppm in ^{13}C NMR spectra of the two polymers also confirmed the presence of –OCH$_2$ groups in the triester. The formation of the additives also confirmed by the much lower intensity of peaks of around 1600 cm^{-1} in ^{1}H NMR spectra and 130 to 150 ppm in ^{13}C NMR spectra after polymerization.

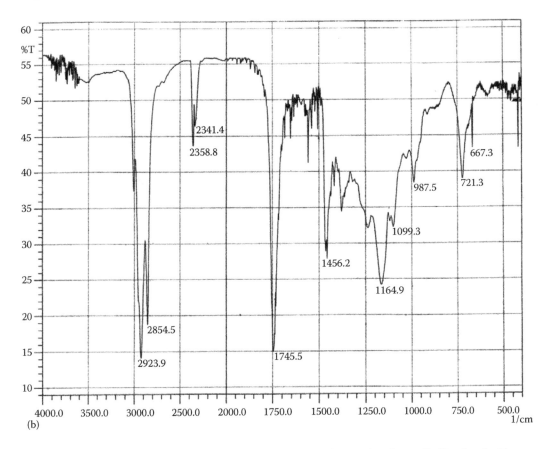

(b)

FIGURE 15.8 (CONTINUED) (b) FT-IR spectrum of homopolymer of soybean oil. (Reprinted with permission from G. Karmakar and P. Ghosh, *ACS Sustain. Chem. Eng.*, 1, 1364–1370. Copyright 2013 American Chemical Society.)

15.4.2.2 Thermal Stability

The thermo-oxidative stability of the polymers was determined by using a thermogravimetric analyzer (Shimadzu TGA-50) in air at a heating rate of 10°C/min. The percentage of weight loss with increasing temperature was calculated. The thermal analysis result for sunpolymer prepared by the thermal method [32] shown in Figure 15.10 indicates that 58% sunpolymer was decomposed at 350°C and 94% was decomposed at 410°C. But it was reported that 95% of sunflower oil was decomposed at 320°C before polymerization [16]. In case of soybean oil, 92% was decomposed at 355°C before polymerization and 93% was decomposed at 440°C after polymerization [16]. Therefore, the thermo-oxidative stability of both soybean oil and sunflower oil could be significantly increased by polymerization.

15.4.2.3 Determination of Molecular Weight by GPC-SEC and Viscometric Methods

The number average molecular weights (M_n), weight average molecular weights (M_w), and polydispersity indices (PDIs) (M_w/M_n) of polymers having molecular weight range of 10^2–5×10^5 g mol^{-1} were determined using a GPC instrument (Waters, Milford, Massachusetts) equipped with refractometer and high-performance liquid chromatography pump plus autosampler. Tetrahydrofuran (0.4% w/v) was used as an eluent at a flow rate of 1.0 mL/min at 40°C. The instrument was calibrated with standard polystyrene before the experiment.

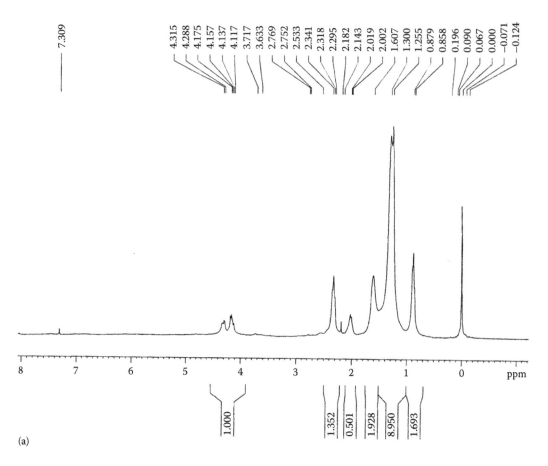

(a)

FIGURE 15.9 (a) ^{1}H NMR spectrum of homopolymer of sunflower oil. (With kind permission from Springer Science+Business Media: *Int. J. Ind. Chem.*, Evaluation of sunflower oil as a multifunctional lubricating oil additive, 5, 2014, Ghosh, P., and G. Karmakar.)   *(Continued)*

Viscosity average molecular weight (M_v) was determined by viscometry technique [43] using the Mark–Houwink–Sakurada relation given in Equation 15.1. The viscometric study was carried out by using an Ubbelohde 0B viscometer (Cannon, India) (with viscometer constant values $K' = 0.00268$ cm^2 s^{-2} and $L = -19.83$ cm^2; volume of the bulb: 3 cm^3; and length of the capillary: 11.3 cm). Viscosity was measured at 40°C by taking different sample concentrations (0.137–0.218 g/mL) of the polymeric additives in toluene. A chronometer was used for recording solution flow times. The intrinsic viscosity values, used in Equation 15.1 for evaluating the average molecular weight, were determined graphically by extrapolation method using the Huggins equation mentioned earlier [43]:

$$[\eta] = KM^a, \tag{15.1}$$

where $[\eta]$ is the intrinsic viscosity (dL g^{-1}) and M is the average molecular weight. K and a are viscometric constants for the given solute–solvent system and temperature. In Equation 15.1, the constants $K = 0.00387$ dL g^{-1} and $a = 0.725$ were used. During this experiment adequate precaution was taken to avoid evaporation of the solvent. The measured average molecular weights of the polymers and the corresponding PDIs are given in Table 15.4. The data show that the average molecular weight of soybean homopolymer determined by both methods was slightly higher than that of sunflower homopolymer. The molecular weight is directly related to the degree of polymerization and cross-link density, which

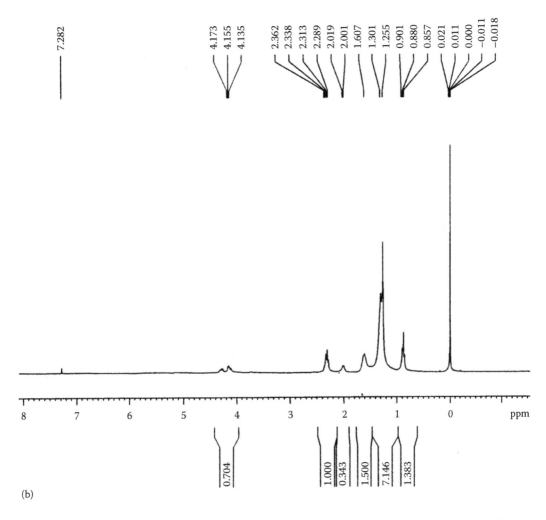

FIGURE 15.9 (CONTINUED) (b) ¹H NMR spectrum of homopolymer of soybean oil. (Reprinted with permission from G. Karmakar and P. Ghosh, *ACS Sustain. Chem. Eng.*, 1, 1364–1370, supplementary material. Copyright 2013 American Chemical Society.)

in turn is largely dependent on the total degree of unsaturation of the fatty acids in the chain. As a result, the average molecular weight of soybean homopolymer is slightly higher than that of sunflower homopolymer because of the higher degree of unsaturation of soybean oil. The PDI values of the homopolymers are almost similar and near unity, indicating that the molecular weight distributions of the homopolymers are uniform and narrow with little branching. This will make the homopolymers highly compatible with the mineral oil base fluids and improve their performance as additives.

15.4.3 PERFORMANCE EVALUATION

15.4.3.1 Permanent Shear Stability Index (PSSI)

Shear stability, which is one of the important criteria, determines the suitability of a lubricant formulation. The shear stability of the additive has strong influence on a multigrade engine oil's ability to retain its viscosity under shearing conditions during the use of the lubricant in engines. The loss of viscosity under shear can be two types: temporary viscosity loss (TVL) and permanent viscosity loss (PVL). PVL is similar to TVL, except that the viscosity loss is measured using kinematic

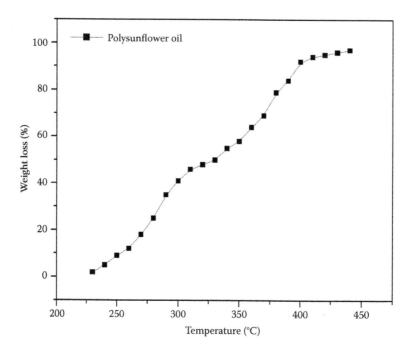

FIGURE 15.10 Percentage weight loss of sun polymer oil with increase in temperature obtained by TGA. (With kind permission from Springer Science+Business Media: *Int. J. Ind. Chem.*, Evaluation of sunflower oil as a multifunctional lubricating oil additive, 5, 2014, Ghosh, P., and G. Karmakar.)

TABLE 15.4

Average Molecular Weight Values Determined by the Mark–Houwink–Sakurada Equation and GPC

| Sample | Viscosity Average M_v | Average Molecular Weight Values (g/mol) | | |
		$M_w \times 10^4$	$M_n \times 10^4$	PDI
		GPC		
S-1	32,849	3.51	3.3	1.06
S-2	47,963	4.85	4.64	1.05

Source: Reprinted with permission from G. Karmakar and P. Ghosh, *ACS Sustain. Chem. Eng.*, 1, 1364–1370. Copyright 2013 American Chemical Society.

viscosity measurement before and after shear at 100°C. The PVL is more frequently expressed in terms of PSSI or simply shear stability index and is determined according to ASTM D6022 [44]. PVL and PSSI are calculated by using the following equation:

$$PVL = (V_i - V_s)/V_i \times 100\% \quad PSSI = (V_i - V_s/V_i - V_0) \times 100\%, \tag{15.2}$$

where V_0 is kinematic viscosity of base fluid before addition of polymer, V_i is the kinematic viscosity of unsheared oil, and V_s is the kinematic viscosity of sheared oil. Kinematic viscosity of fresh toluene and sheared polymer solution in toluene is determined according to ASTM D445 method [45]. Calculated PSSIs of different base oils are shown in Figure 15.11.

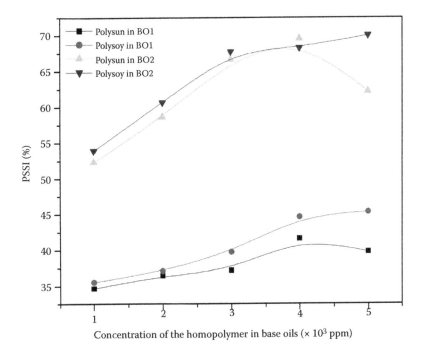

FIGURE 15.11   Shear stability (in terms of PSSI) of the lubricant blended with the vegetable oil-based additives at different concentration levels.

The stability of the polymers against shear decreases with increasing PSSI values [44]. It is observed that (Figure 15.11) with increasing concentration of the additive in the base oils, the PVL values increase and as a result the PSSI also increases. This may be due to an increase in polymer concentration, which causes increase in molecular mass distribution, causing the system to undergo degradation. The higher PSSI values of soypolymer compared to those of sunpolymer reveal that soypolymer is relatively less stable against shear.

15.4.3.2   Evaluation of VI

VI was calculated from measured kinematic viscosity values at 40°C and 100°C of the base oils with different concentrations of additive according to ASTM D2270-10 method. Figure 15.12 indicates that the VI values of the base oils with additives were higher than those of the base oils without additive. The VI values increased with increasing concentration of the additives. Soypolymer showed higher VI improvement in comparison to sunpolymer in both base oils (BO1 and BO2). Significant improvement in VI was observed when methyl acrylate (MA), methyl methacrylate (MMA), or decyl acrylate (DA) was incorporated in the backbone of sunflower oil and soybean oil through copolymerization (Table 15.5) [46]. It has been found that the VI of mineral base stocks has low increases due to addition of telomerized rapeseed oil in different percentages [12].

15.4.3.3   Pour Point

The pour points of base oils at different additive concentrations were evaluated by using the WIL-471 cloud and pour point test apparatus (Wadegati Labequip Pvt. Ltd., Maharashtra, India) in the temperature range of 0–71°C according to ASTM D97 method. As shown in Figure 15.13, the pour points of the lubricant blended with bioadditives showed improvement. It was found that the pour points of the base oils decreased with increasing concentration of the additives. The sunpolymer showed better depression in pour point than soypolymer.

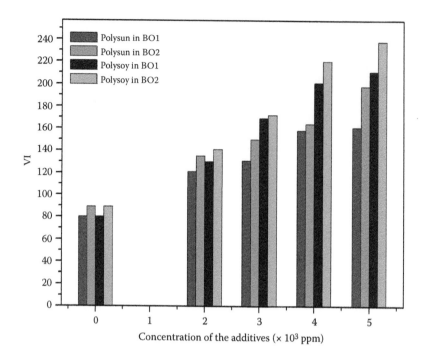

FIGURE 15.12 VI values of the lubricant doped with the bioderived additives at different concentration levels.

TABLE 15.5

VI Values of Copolymers of Sunflower Oil and Soybean Oil

Percentage of Copolymers in Base Oils (w/w)	Sun Oil + 5% MMA		Sun Oil + 5% DA		Soy Oil + 5% MMA		Soy Oil + 5% MA	
	BO1	BO2	BO1	BO2	BO1	BO2	BO1	BO2
0	80	89	80	89	80	89	80	89
2	144	95	109	97	136	142	120	125
3	182	120	184	108	184	201	147	161
4	203	133	230	118	218	232	205	213
5	226	210	262	215	232	255	221	232

15.4.3.4 Antiwear

The antiwear property of lube oil blended with the bioderived additives at different concentration levels was evaluated by a FBWT apparatus according to ASTM D4172-94 method [14]. In this experiment the WSD, a parameter for the determination of antiwear performance of the oils, was measured by applying a definite weld load at 75°C for 60 min [32]. The diameter and the rotating speed of the ball were 12.7 mm and 1200 rpm, respectively. Evaluation of antifrictional property was measured through calculation of coefficient of friction (COF) by multiplying the mean friction torque and spring constant. The frictional torque on the lower balls may be expressed as

$$\mu = \frac{T\sqrt{6}}{3W \times r},$$

(15.3)

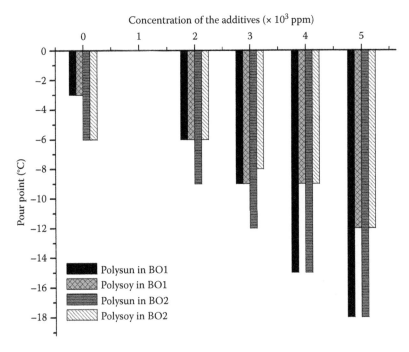

FIGURE 15.13 Pour point values of the lubricant blended with the bioderived additives at different concentration levels.

where μ is COF, T is the frictional torque (kg/mm), W is the applied load (kg), and r is the distance from the center of the contact surfaces on the lower balls to the axis of rotation, which is 3.67 mm. The WSDs of the mineral base oils (BO1 and BO2) blended with sunpolymer were measured at 392 N (40 kgf) weld load and are shown in Figure 15.14. Kalam et al. [47] have measured the WSD and the COF of mineral base fluid (SAE 40) blended with 0.5% amine phosphate and POME of different percentages (1% to 5% w/w) as additive at 70 kg weld load. They observed that the WSD value decreases with increasing concentration of palm oil in the lubricant base fluid and the effect continues up to a certain limit. The positive effect on increasing the wear preventive performances of the lubricating oil blended with the soybean oil-derived additives is also mentioned in our earlier publication [48]. Figures 15.15 and 15.16 show the antifrictional properties of BO1 oil blended with copolymeric additives of soybean oil with MA and MMA, respectively, in different mass fractions. These were determined by the FBWT apparatus at 392 N applied load. In all these cases, a significant improvement of antiwear performance, which was reflected by the decrease in WSD and COF values, was observed when the additives were blended with mineral base stocks. The lower WSD and COF values of the lubricant blended with these additives indicate that the lubricant suffers reduced wear compared to the base oil without additive. It was further observed that both WSD and COF decrease with an increase in concentration of the added polymers in base oils. The additives in lubricant strengthen the film through chemical and physical bonding between the functional groups of the additive molecules and metal atoms on the rubbing zone. Because of the long hydrocarbon chains of the fatty acids and the polar side chains of the ester groups, the copolymers form a stronger layer with metal atoms and perform better as antiwear additive. In Figure 15.16, the maximum reduction in COF that was observed in the case of soy oil–MMA copolymers indicated that the protective film formed between the contacting metal surfaces of the engines during tribochemical process protects the surfaces better from wear.

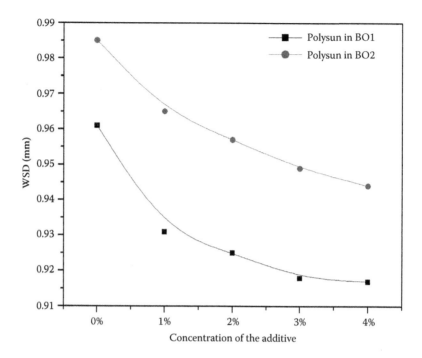

FIGURE 15.14   Antiwear properties (in terms of WSD value) of the base stocks (BO1 and BO2) blended with homopolymer of sun oil at 392 N weld load. (With kind permission from Springer Science+Business Media: *Int. J. Ind. Chem.*, Evaluation of sunflower oil as a multifunctional lubricating oil additive, 5, 2014, Ghosh, P., and G. Karmakar.)

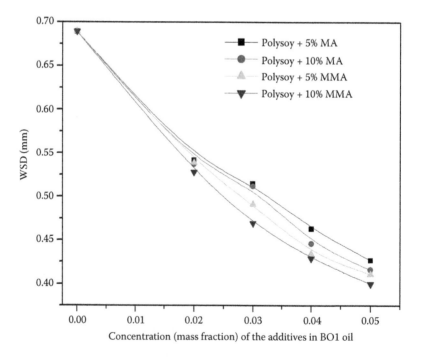

FIGURE 15.15   WSDs of BO1 base oil blended with the copolymers of soybean oil with MA and MMA.

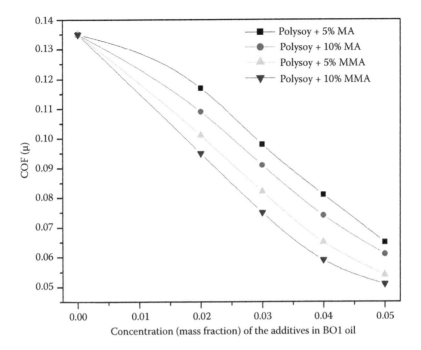

FIGURE 15.16 COFs of BO1 base oil blended with the copolymers of soybean oil with MA and MMA.

15.4.4 BIODEGRADABILITY

Biodegradability is one of the most important properties with regard to the environmental fate of a substance. The main advantage of vegetable oil-based additives over synthetic acrylate-based ones is its excellent biodegradability. The biodegradability was carried out by disk diffusion (DD) [49] and soil burial test (SBT) methods according to ISO 846:1997 [50]. Ultimate biodegradation is achieved when the substance is totally converted into carbon dioxide, water, mineral salts, and biotic mass and confirmed by measuring the shift of IR frequency of the ester carbonyl after the test.

15.4.4.1 Disk Diffusion Degradation

In this process biodegradation of the polymer samples were tested by using five different fungal pathogens, namely, *Colletotrichum camelliae*, *Fusarium equiseti*, *Alternaria alternata*, *Colletotrichum gloeosporioides*, and *Curvularia eragrostidis*. The culture media for fungal strains were prepared by mixing suitable proportions of potato extract, dextrose, and agar powder. The fungal growth was confirmed by a color change from yellow to black. The experiment was performed in petri dishes, which were incubated at 37°C for 30 days after addition of a definite weight of polymer samples. After 30 days, the polymer samples were removed from the fungal media, washed with chloroform, purified by repeated precipitation of their solutions in *n*-hexane with methanol, and dried in an open vessel. The dried samples were weighed and the percentage of weight loss was calculated.

15.4.4.2 Soil Burial Degradation

In this test the polymer film [51] was kept in a bacteriological incubator apparatus (Sigma Scientific Instruments Pvt. Ltd., Chennai, India) and buried for 3 months in soil [52] collected from near landfill with pH of 7.2, moisture of 25%, and relative humidity of 50–60% and thus subjected to the action of microorganisms in which soil is their major habitant. The temperature was kept near 30°C. The buried films were removed for evaluation at regular intervals of 15 days for up to 3 months. Recovered films were washed with chloroform, purified following the mentioned procedure, and dried in a vacuum oven at 50°C to constant weight. The weights of the sample before and after drying were recorded and the percent weight loss of the samples calculated as following:

$$\text{Weight loss} = [(M_0 - M_1)/M_0] \times 100, \tag{15.4}$$

where M_0 is the weight before test and M_1 is the weight after the test.

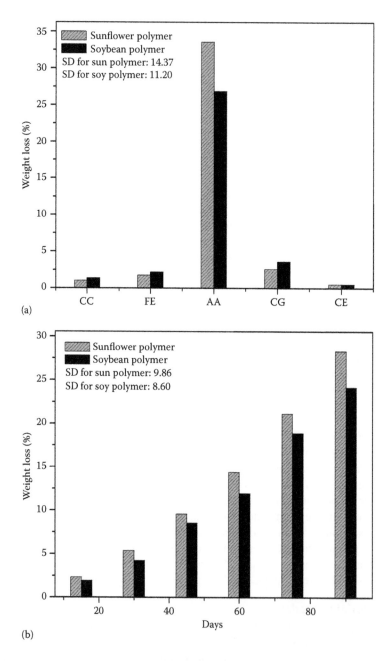

(a)

(b)

FIGURE 15.17 (a) Biodegradability of the synthesized additives by DD test. CC, FE, AA, CG, and CE are the different pathogens used in the test. CC: *Colletotrichum camelliae*; FE: *Fusarium equiseti*; AA: *Alterneria alternata*; CG: *Colletotrichum gloeosporioides*; and CE: *Curvularia eragrostidis*. SD is standard deviation. (b) Degradation of the polymer samples measured after 15- up to 90-day span in SBT.

The results of the biodegradability tests are summarized in Figure 15.17a and b. The result of the test by DD method, shown in Figure 15.17a, indicated a significant degradation against fungal pathogen *Alternaria alternata* for both the polymers. The weight loss of polysun oil is near 33.6%, which is slightly higher than the corresponding weight loss of polysoy oil (26.9%). In SBT (Figure 15.17b), the degradation of recovered samples taken every 15 days increases with increasing time. After 90 days' span the recovered polysun oil and polysoy oil showed 28.33% and 24.18% weight losses, respectively, which are similar. From these results it is evident that the polymers of both vegetable oils showed significant biodegradability and therefore can be successfully used in the formulation of green lubricants.

15.5 SIGNIFICANCE OF ECO-FRIENDLY ADDITIVES IN THE FORMULATION OF BIOLUBRICANTS

The application of vegetable oil or its derivatives as multifunctional additive in mineral base fluid is plausible and, no doubt, a great approach toward the formulation of eco-friendly lubricant at the present environmental situation and demand. Most biolubricant researches are directed toward the development of biobased base fluids. But the application of vegetable oils as base fluid is still not widespread. The major drawbacks of vegetable oils are their lower thermo-oxidative stability (although it can be improved by chemical modification), limited availability, higher cost, and inferior performance in some specific application areas compared to that of petroleum-based mineral oils. Use of vegetable oil-based additives can meet the vast demand for eco-friendly lubricants around the world. Therefore, to get better formulation with minimal cost and protect the environment at the same time, the formulation of the biolubricant with bioderived additives in the mineral base fluid is very significant.

15.6 CONCLUSION

On the basis of the previous discussion, it can be concluded that the potential application of multifunctional vegetable oil-based additives in the formulation of biolubricants is definitely an approach toward green chemistry. The additives perform as excellent antiwear, VII, and PPD. Soypolymer has higher thermo-oxidative stability, higher VI, and higher shear stability index values compared to sunpolymer. However, when pour point properties and biodegradability were considered, the polymer of sunflower oil showed better performance. Therefore, from the previous study it is evident that the application of vegetable oil-based greener additives would definitely be a step toward the formulation of future biolubricants having enhanced performances.

ACKNOWLEDGMENTS

The authors are thankful to University Grants Commission (UGC), New Delhi, India, for financial support. Special thanks to Indian Oil Corporation Limited and Bharat Petroleum Corporation Limited for the supply of base oils.

REFERENCES

1. S.Z. Erhan and S. Asadauskas, Lubricant base stocks from vegetable oils, *Ind. Crops. Prod.*, 11, 277–282 (2000).
2. K. Balamurugan, N. Kanagasabapathy, and K. Mayilsamy, Studies on soybean oil based lubricant for diesel engines, *J. Sci. Ind. Res.*, 69, 794–797 (2010).
3. X. Wu, X. Zhang, S. Yang, H. Chen, and D. Wang, The study of epoxidized rapeseed oil used as a potential biodegradable lubricant, *J. Am. Oil Chem. Soc.*, 77, 561–563 (2000).
4. J. Salimon, B.M. Abdullah, R.M. Yusop, and N. Salih, Synthesis, reactivity, and application studies for different biolubricants, *Chem. Central J.*, 8, 16 (2014).
5. Y. Lu and R.C. Larock, Novel polymeric materials from vegetable oils and vinyl monomers: Preparation, properties, and applications, *ChemSusChem*, 2, 136–147 (2009).

6. W. Li, C. Jiang, M. Chao, and X. Wang, Natural garlic oil as a high-performance, environmentally friendly, extreme pressure additive in lubricating oils, *ACS Sustain. Chem. Eng.*, 2, 798–803 (2014).

7. G. Biresaw, S.J. Asadauskas, and T.G. McClure, Polysulfide and biobased extreme pressure additive performance in vegetable vs paraffinic base oils, *Ind. Eng. Chem. Res.*, 51, 262–273 (2012).

8. S.Z. Erhan, A. Adhvaryu, and B.K. Sharma, "Poly(Hydroxy Thioether) Vegetable Oil Derivatives Useful as Lubricant Additives," U.S. Patent 7279448 B2 (2007).

9. F.L. Erickson, R.E. Anderson, and P.S. Landis, "Vegetable Oil Derivatives as Lubricant Additives," U.S. Patent 5282989 (1994).

10. F.L. Erickson, R.E. Anderson, and P.S. Landis, "Meadowfoam Oil and Meadowfoam Oil Derivatives as Lubricant Additives," U.S. Patent 4925581 (1990).

11. E.W. Bell, "Wax Ester of Vegetable Oil Fatty Acids Useful as Lubricants," U.S. Patent 4152278 (1979).

12. P.S. Landis, "Telomerized Triglyceride Vegetable Oil for Lubricant Additives," U.S. Patent 5229023 (1993).

13. P.S. Landis and F. Erickson, "Jojoba Oil and Jojoba Oil Derivative Lubricant Compositions," U.S. Patent 4873008 (1989).

14. G.S. Kumar, A. Balamurugan, S. Vinu, M. Radhakrishnan, and G. Senthilprabhu, Tribological and emission studies on two stroke petrol engine lubricated with sunflower methyl ester, *J. Sci. Ind. Res.*, 71, 562–565 (2012).

15. S.M. Shanta, G.J. Molina, and V. Soloiu, Tribological effects of mineral-oil lubricant contamination with biofuels: A pin-on-disk tribometry and wear study, *Adv. Tribol.* (2011), Article ID 820795.

16. G. Karmakar and P. Ghosh, Green additives for lubricating oil, *ACS Sustain. Chem. Eng.*, 1, 1364–1370 (2013).

17. B. Wilson, Lubricants and functional fluids from renewable sources, *Ind. Lubr. Tribol.*, 50 (1), 6–15 (1998).

18. E.O. Aluyor, K.O. Obahiagbon, and M. Ori-jesu, Biodegradation of vegetable oils: A review, *Sci. Res. Essays*, 4 (6), 543–548 (2009).

19. N.J. Fox and G.W. Stachowiak, Vegetable oil-based lubricants—A review of oxidation, *Tribol. Int.*, 40 (7), 1035–104 (2007).

20. M. Shahabuddin, H.H. Masjuki, and M.A. Kalam, Experimental investigation into tribological characteristics of biolubricant formulated from Jatropha oil, *Procedia Eng.*, 56, 597–606 (2013).

21. M. Mofijur, H.H. Masjuki, M.A. Kalam, M. Shahabuddin, M.A. Hazrat, and A.M. Liaquat, Palm oil methyl ester and its emulsions effect on lubricant performance and engine components wear, *Energy Procedia*, 14, 1748–1753 (2012).

22. F. Haus, G.A. Junter, and J. German, Viscosity properties of mineral paraffinic base oils as a key factor in their primary biodegradability, *Biodegradation*, 11 (6), 365–369 (2000).

23. M.M. Mohamed, H.H.A.E. Naga, and M.F.E. Meneir, Multifunctional viscosity index improvers, *Chem. Tech. Biotechnol.*, 60, 283–289 (1994).

24. A.A.A. Abdel-Azim, A.M. Nasser, N.S. Ahmed, A.F.E. Kafrawy, and R.S. Kamal, Multifunctional additives viscosity index improvers, pour point depressants and dispersants for lube oil, *Petrol. Sci. Technol.*, 27, 20–32 (2009).

25. P. Ghosh and M. Das, Synthesis, characterization, and performance evaluation of some multifunctional lube oil additives, *J. Chem. Eng. Data*, 58, 510–516 (2013).

26. R.B. Rastogi, J.L. Maurya, V. Jaiswal, and D. Tiwary, Lanthanum dithiocarbamates as potential extreme pressure lubrication additives, *Int. J. Ind. Chem.*, 3, 32 (2012).

27. A.M. Barnes, K.D. Bartel, and V.R.A. Thibon, A review of zinc dialkyldithiophosphates (ZDDPs): Characterization and role in the lubricating oil, *Tribol. Int.*, 34, 389–395 (2001).

28. H. Spikes, The history and mechanisms of ZDDP, *Tribol. Lett.*, 17 (3), 469–489 (2004).

29. J. Wang, Z. Ye, and S. Zhu, Topology-engineered hyperbranched high-molecular-weight polyethylenes as lubricant viscosity-index improvers of high shear stability, *Ind. Eng. Chem. Res.*, 46, 1174–1178 (2007).

30. S. Tanveer, U.C. Sharma, and R. Prasad, Rheology of multigrade engine oils, *Indian J. Chem. Technol.*, 13, 180–184 (2006).

31. K.S. Sedersen, Influence of wax inhibitors on wax appearance temperature, pour point, and viscosity of waxy crude oils, *Energy Fuels*, 17, 321–328 (2003).

32. P. Ghosh and G. Karmakar, Evaluation of sunflower oil as a multifunctional lubricating oil additive, *Int. J. Ind. Chem.*, 5, 7, (2014).

33. F.L. Erickson, R.E. Anderson, and P.S. Landis, "Vegetable Oil Derivatives as Lubricant Additives," U.S. Patent No. 4970010 (1990).

34. M.A. Maleque, H.H. Masjuki, and S.M. Sapuan, Vegetable-based biodegradable lubricating oil additives, *Ind. Lubr. Tribol.*, 55 (3), 137–143 (2003).
35. C.V. Ossia, H.G. Han, and H. Kong, Additive properties of saturated very long chain fatty acids in castor and jojoba oils, *J. Mech. Sci. Technol.*, 22, 1527–1536 (2008).
36. R.P.S. Bisht, G.A. Sivasankaran, and V.K. Bhatia, Additive properties of jojoba oil for lubricating oil formulations, *Wear*, 161, 193–197 (1993).
37. H. Wagner, R. Luther, and T. Mang, Lubricant base fluids based on renewable raw materials: Their catalytic manufacture and modification, *Appl. Catal. A*, 221, 429–442 (2001).
38. V.B. Borugadda and V.V. Goud, Epoxidation of castor oil fatty acid methyl esters (COFAME) as a lubricant base stock using heterogeneous ion-exchange resin (IR-120) as a catalyst, *Energy Procedia*, 54, 75–84 (2014).
39. P. Ghosh, M. Das, M. Upadhyay, T. Das, and A. Mandal, Synthesis and evaluation of acrylate polymers in lubricating oil, *J. Chem. Eng. Data*, 56, 3752–3758 (2011).
40. H. Stange, M. Ishaque, N. Niessner, M. Pepers, and A. Greiner, Microwave-assisted free radical polymerizations and co-polymerizations of styrene and methyl methacrylate, *Macromol. Rapid Commun.*, 27, 156–161 (2006).
41. D. Darnoko and M. Cheryan, Continuous production of palm methyl esters, *J. Am. Oil Chem. Soc.*, 77 (12), 1269–1272 (2000).
42. C. Komintarachat and S. Chuepeng, Methanol-based transesterification optimization of waste used cooking oil over potassium hydroxide catalyst, *Am. J. Appl. Sci.*, 7 (8), 1073–1078 (2010).
43. P. Ghosh, T. Das, and D. Nandi, Synthesis characterization and viscosity studies of homopolymer of methyl methacrylate and copolymer of methyl methacrylate and styrene, *J. Solut. Chem.*, 40, 67–78 (2011).
44. P. Ghosh, T. Das, and D. Nandi, Shear stability and thickening properties of homo and copolymer of methyl methacrylate, *Am. J. Polym. Sci.*, 1(1), 1–5 (2011).
45. B. Gutti, S.S. Bamidele, and I.M. Bugaje, Biodiesel kinematics viscosity analysis of *Balanite aegyptiaca* seed oil, *ARPN J. Eng. Appl. Sci.*, 7 (4), 432–435 (2012).
46. P. Ghosh, T. Das, G. Karmakar, and M. Das, Evaluation of acrylate-sunflower oil copolymer as viscosity index improvers for lube oils, *J. Chem. Pharm. Res.*, 3(3), 547–556 (2011).
47. M.A. Kalam, H.H. Masjuki, M. Shahabuddin, and M. Mofijur, Tribological characteristics of amine phosphate and octylated/butylated diphenylamine additives infused bio-lubricant, *Energy Educ. Sci. Tech. Part A: Energy Sci. Res.*, 30 (1), 123–136 (2012).
48. G. Karmakar and P. Ghosh, Soybean oil as a biocompatible multifunctional additive for lubricating oil, *ACS Sustain. Chem. Eng.*, 3, 19–25 (2015).
49. P. Ghosh, T. Das, D. Nandi, G. Karmakar, and A. Mandal, Synthesis and characterization of biodegradable polymer—Used as a pour point depressant for lubricating oil, *Int. J. Polym. Mater.*, 59, 1008–1017 (2010).
50. A.S. Chandure and S.S. Umare, Synthesis, characterization and biodegradation study of low molecular weight polyesters, *Int. J. Polym. Mater.*, 56, 339–353 (2007).
51. M. Liu, Z. Huang, and Y. Yang, Analysis of biodegradability of three biodegradable mulching films, *J. Polym. Environ.*, 18, 148–154 (2010).
52. N. Lardjane, N. Belhaneche-Bensemra, and V. Massardier, Soil burial degradation of new bio-based additives: Part I—Rigid poly(vinyl chloride) films, *J. Vinyl Addit. Technol.*, 17 (2), 98–104 (2011).

16 Vegetable Oil-Based Lubricant Additives

Padmaja V. Korlipara

CONTENTS

ABSTRACT

Biobased additives are desirable commodities due to their eco-friendly nature. Vegetable oils are suitable raw material for development of bioadditives due to their renewable, biodegradable, and low-toxicity nature. This chapter provides an overview of the synthesis of various vegetable oil-based lubricant additives, their characterization, and evaluation of their additive properties such as antioxidant, antiwear, extreme pressure, viscosity improver, and pour point depressant. Although a significant number of papers and patents are published in this area, there is a lot more to explore.

16.1 INTRODUCTION: DRIVING FORCES FOR VEGETABLE OIL-BASED LUBRICANT ADDITIVES

The environmental impact caused by lubricants polluting air, soil, and water has been raising public concern. Most current lubricants are mineral oil based with chemically derived additives. They can cause harm to the environment because of their low biodegradability and toxicity. It is estimated that 50% of total lubricants consumed worldwide comes into direct contact with the environment because of spillage, machinery failure, leakage, emissions, and careless disposal [1]. Organic compounds containing elements such as S, P, and Cl, as well as metallic dialkyldithiophosphates (zinc dialkyldithiophosphate [ZDDPs]) and organic molybdenum compounds, are widely used in formulating lubricants to enhance their antioxidant, antiwear, extreme pressure (EP), and antifriction properties. However, ash-bearing additives lead to operational problems and the use of some metals and elements is being restricted for environmental reasons [2–4].

To protect the environment, current lubricant additive research should focus on exploring new environment-friendly and multifunctional biobased additives. Renewable oils and fats from vegetable and animal origin offer new feedstock for the chemical industry [5–9]. Structural modification of long-chain fatty acid derivatives such as addition reactions to double bonds or epoxy groups gives compounds with sulfur, phosphorus, nitrogen, oxygen, or a combination of these elements. Such compounds may exhibit good antioxidant, antiwear, or EP characteristics and, in some cases, would function as multifunctional additives [10–13]. The development of biobased additives is attractive for multiple reasons [14–21]. They are compatible with vegetable base oils, which are poor performers and may be incompatible with many currently available additives. If the additives are biobased, it will help the lubricant formulation to meet biopreferred criteria for percentage of biobased carbon [22].

Vegetable oils have great potential as raw material for development of bioadditives owing to their combination of biodegradability and renewability. This chapter provides a review of the development of vegetable oil-based lubricant additives by various structural modifications and their merits, demerits, and challenges.

16.2 SYNTHESIS OF BIOBASED ADDITIVES FROM VEGETABLE OILS

16.2.1 Overview

The structure of vegetable oil offers sites for additional functionalization of additives. The design of environment-friendly additives seeks products that meet commercial, technical, and environmental needs. To a great extent, this requires utilization and sustainment of the natural chemistry of vegetable oils. The structure of vegetable oils/fatty acids offers sites to introduce different groups at the carboxylic acid and at double bonds [23,24]. Chemical, thermal, and enzymatic synthetic methods have been employed to convert vegetable oils to biobased additives. Different techniques have been used to prepare biobased antioxidants, viscosity improvers (VIIs), pour point depressants (PPDs), antiwear, and EP additives from vegetable oils. Some of these methods are summarized in the following.

16.2.2 Sulfurization

Sulfur-containing compounds such as sulfurized oils, fatty acids, esters, and olefins are the earliest known EP additives for lubricants [25]. Sulfurized oils are known to be used for almost 100 years [26,27]. Sulfurized oils are produced commercially by reacting oil with elemental sulfur or hydrogen sulfide under superatmospheric pressures. Sulfurized fats and esters are usually manufactured with 10–15% sulfur and are often good antiwear and mild EP agents. Sufurization causes cross-linking and saturation of double bonds, resulting in a three-dimensional structure of sulfur bridges between chains in the triglyceride or fatty acid ester molecules [28]. Figure 16.1 shows a typical example of sulfurization of methyl oleate with elemental sulfur [29].

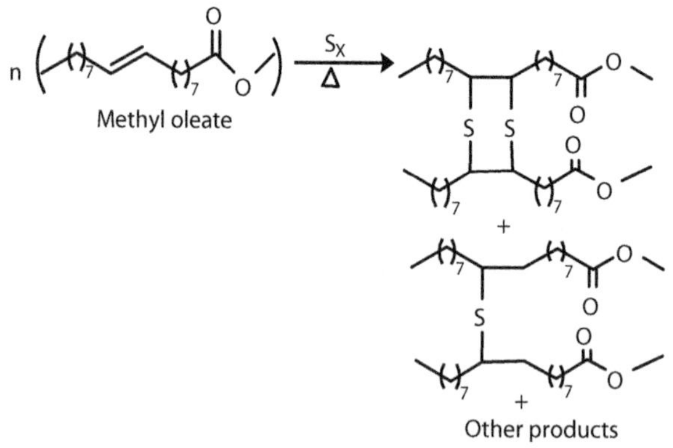

FIGURE 16.1 Reaction scheme for sulfurization of methyl oleate with elemental sulfur. (Adapted from L. O. Farng, "Antiwear Additives and Extreme-Pressure Additives," in: *Lubricant Additives Chemistry and Applications*, L. R. Rudnick (Ed), p. 219.)

Before 1972, lubricant additives were based on raw sperm whale oil, a monoester of monounsaturated fatty acid chains. Sulfurized sperm whale oil was used in many lubricants because it had a combination of properties such as high lubricity, good heat stability, and low freezing point not matched by others. After the inclusion of whales in the endangered species list, their killing was banned. This led to investigation of vegetable oils as substitute for sulfurized sperm whale oil. Sulfurized soybean, sunflower, cottonseed, high-erucic rapeseed, canola, and meadowfoam oils were investigated and proved not to be satisfactory, especially with respect to solubility and stability. When vegetable oils were sulfurized in the presence of large amounts of methyl lardate, the products exhibited potential additive properties [30]. In later studies Kammann [31] tried sulfurization of tranesterified triglycerides and obtained a product with improved solubility in base oil.

Wakim [32] prepared a cosulfurized product from a mixture of triglycerides with an iodine value greater than 80 and nonwax methyl esters of unsaturated fatty acids of 18 to 33 carbon atoms. The product exhibited excellent EP and low-temperature properties and improved solubility in base oils and was a suitable substitute for sulfurized sperm whale oil. Princen and Rothfus [33] prepared sulfurized wax esters of meadowfoam fatty acids with good lubrication properties. However, the product suffered from poor corrosion properties, excessive foaming, and thickening when in use. To improve these properties, the sulfurization method was changed, which gave a product with decreased copper corrosion and similar lubricating properties but produced more foam and failed a thermal stability test.

Erickson et al. [34] disclosed the development of antiwear additives using a mixture of sulfurized meadowfoam oil, sulfurized wax ester, phosphite adducts of meadowfoam oil wax esters, jojoba oil, and triglycerides for automobile engine, gear, spindle, and automatic transmission oils, etc.

Sulfurized vegetable wax esters prepared from hydrogenated vegetable oils and fatty alcohols with properties similar to those of sulfurized sperm whale oil were disclosed in U.S. Patent No. 4152278 [35]. The author claims that the invention is useful as EP additive in lubricant compositions. Landis [36] reported the sulfurization of telomerized rapeseed oil and its use as a VII in a mineral oil lubricant base oil.

Sulfurization of jojoba oil was investigated by several researchers as a suitable substitute for sulfurized sperm whale oil [37–40]. Miwa and Rothfus [41] reported the sulfurization of jojoba oil and its evaluation as an EP additive using laboratory and simulated lubricant tests. They reported that the sulfurized oil prepared from heat-treated and filtered jojoba oil was comparable with or superior to sulfurized sperm whale oil as an EP additive for motor oils, automotive transmission fluids, and gear oils.

Bisht and Bhatia [42] investigated the sulfurization of three triglyceride oils (karanja, rapeseed, and castor) and liquid wax jojoba and evaluated their potential as EP additive for industrial gear oil formulations. Sulfurized rapeseed and castor oil displayed poor solubility in base oil with a tendency to separate out over time. The antiwear and EP properties of the jojoba and karanja oil-based products (sulfurized jojoba oil [SJO] and sulfurized karanja oil [SKO]) were comparable to those of typical commercial EP additive (Table 16.1). Industrial gear oil formulations developed using SJO and SKO at 1% sulfur level in the blend, in the presence of corrosion inhibitor and an emulsifier, met IS 8406-1985 industrial gear oil specifications [43]. Cao et al. [44] prepared odorless sulfurized octadecanoic and docosanoic acids for use as additives in rapeseed oil. The tribological evaluation of these products on four-ball tribometer revealed improved antiwear and EP properties compared with those of unsulfurized acids.

The extensive research for sulfurized sperm whale oil substitute has resulted in several patents. The preferred raw materials turned out to be vegetable oils alone or in combination with synthetic esters or wax ester such as jojoba oil. The largest use of sulfurized oils is in metalworking and grease applications, followed by industrial gear oils. However, sulfurized oils have limited thermal stability, which makes them undesirable for high-temperature applications, where fast decomposition of the EP additive is not desired. Corrosion toward yellow metals is another drawback of sulfurized compounds. At high temperatures, the active sulfur reacts with copper to form copper sulfide.

TABLE 16.1

Antiwear[a] and EP[b] Properties of SJO and SKO

Sulfurized Oil/Commercial Additive	WSD (mm)	Weld Point (kgf)
Jojoba oil (low S)	0.332	501
Jojoba oil (high S)	0.375	562
Karanja oil (low S)	0.300	355
Karanja oil (high S)	0.287	398
SP-type commercial additive	0.223	316

Source: Adapted from Bisht, R. P. S., and V. K. Bhatia: Sulphurised vegetable oils as EP additives for industrial gear oil formulations. *J. Synth. Lubr.* 1997. 14. 23–33.

[a] Four-ball antiwear test; ASTM D4172 (rotating speed of 1200 rpm, 40 kg load, 60 min duration).

[b] Four-ball EP test; IP 239/ASTM D2783 (1475 rpm, ambient temperature at variable load, test duration for 1 min or until the balls welded).

Thus, improved thermal stability, low odor, low copper corrosion, and excellent solubility in base fluids are some of the challenges for sulfurized products for application in the new generation of industrial lubricants.

Epithios are another class of sulfur compounds that are very effective lubricant additives. In particular their EP, antiwear, and antioxidant properties are superior to those of other sulfur compounds [45]. The synthesis of C_{18} fatty acids containing a thiophene nucleus has been reported [46]. Synthesis of a number of epithio C_{18} fatty esters from methyl oleate, linoleate, ricinoleate, and derivatives has been reported by Marcel et al. [47]. Epoxy fatty esters were treated with dimethylthioformamide and trifluoroacetic acid in dichloromethane and converted to corresponding 2,3-epithio fatty esters. 2,2′-Epoxy fatty esters were transformed to 2,2′-epithio fatty esters by a similar reaction and characterized by ^1H and ^{13}C NMR spectroscopy. Evaluation of these epithio fatty esters for their additive properties is not reported.

Magne et al. [48] synthesized a series of substituted stearates of various mono- and polyhydric alcohols as additives for paraffin and diester base oils. The effect of substituents such as epithio, chloro, phosphato, phosphorodithio, thio, mercapto, and sulfuryl, individually and in mixtures, on antiwear and EP performance was investigated. The epithio substituent was found to be the most effective on EP and antiwear performance. Sulfurized and chloro-sulfone derivatives performed well also, but were less effective than epithio derivatives.

Recently novel alkyl-10-undecenoate acid-based epithio compounds (alkyl, n-C_1, n-C_4, n-C_6, i-C_3, i-C_4, and i-C_8) were synthesized by Geethanjali et al. [49]. Epoxy alkyl esters of undecenoic acid were reacted with ammonium thiocyanate in ionic liquid 1-methylimidazolium tetrafluoroborate–H_2O

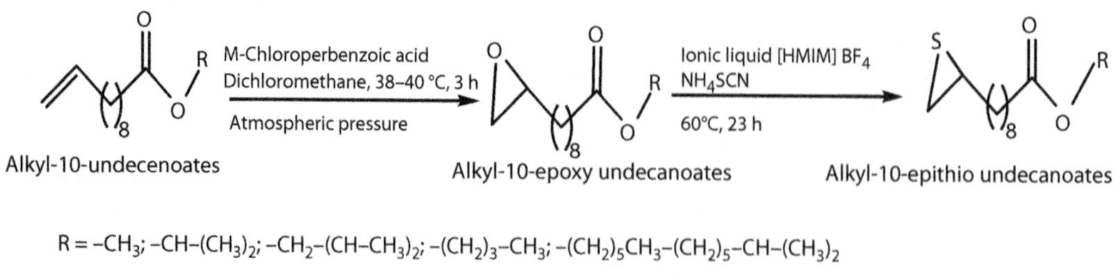

$R = -CH_3; -CH-(CH_3)_2; -CH_2-(CH-CH_3)_2; -(CH_2)_3-CH_3; -(CH_2)_5CH_3-(CH_2)_5-CH-(CH_3)_2$

FIGURE 16.2 Synthesis of alkyl epithio undecanoates. (Adapted from G. Geethanjali et al., *J. Agric. Food Chem.*, 62, 11505–11511.)

TABLE 16.2

Effect of 2 wt% Epithio Undecanoate Additives on Four-Ball Weld Point in Three Different Base Oils

	Weld Point (kgf)		
Additive	EJB	DOS	S-105
None	160	120	120
Methyl 10-epithio undecanoate	210	170	160
1-(2-Propyl)-10-epithio undecanoate	180	140	130
2-Methyl-1-propyl 10-epithio undecanoate	180	140	130
1-Butyl 10-epithio undecanoate	210	170	160
1-Hexyl 10-epithio undecanoate	170	130	120
2-Ethylhexyl 10-epithio undecanoate	180	140	130

Source: Adapted from G. Geethanjali et al., *J. Agric. Food Chem.*, 62, 11505–11511.

Note: DOS: di-2-ethylhexyl sebacate; EJB: epoxy jatropha fatty acid *n*-butyl ester; S-105: mineral oil.

(2:1) solvent system (Figure 16.2). The products were evaluated for antioxidant, EP, and antiwear properties in the following base oils: epoxy jatropha fatty acid *n*-butyl esters (EJBs), di-2-ethylhexyl sebacate (DOS), and mineral oil (S-105). The results showed that *n*-butyl epithio undecanoate additive had superior antioxidant properties, with an onset temperature of 229.2°C, compared with butylated hydroxytoluene (BHT, 193.8°C), in DOS base oil but had comparable performance with EJB and S-105 base oils. All the epithio derivatives exhibited significant improvement in weld point at 2 wt% level in the base oils EJB and DOS and moderately in S-105 base oil in EP test as per ASTM D2783 test method (Table 16.2). Methyl epithio undecanoate at 0.6% in DOS base oil improved its wear scar diameter (WSD) by more than 26%.

16.2.3 THERMAL MODIFICATION

Thermal polymerization of vegetable oils involves heating the oil under nitrogen to produce telomerized or heat-bodied oils. It is a free radical process and produces high-molecular weight oils. The use of heat-bodied oils as additives was investigated by many researchers. Telomers are polymerized triglyceride oils with high viscosity and enhanced thermo-oxidation stability for use in a variety of lubricant formulations.

Telomerized vegetable oils prepared by heating oil for 5 h at 200–400°C in nonoxidizing atmosphere are reported to be applied as thermal-oxidative stability enhancers and VIIs [36,50]. These telomerized vegetable oils and their derivatives, such as phosphite adduct and sulfurized derivatives, were projected to be cheaper substitutes for wax ester derivatives that are used in lubricant additives.

Shahanan and Landis [51] prepared improved telomerized oils using a process that required lower reaction temperature (150°C to about 400°C versus 200°C to 400°C) and shorter reaction time (3–10 h versus 6–40 h) than those for the original process described in U.S. Patent 5222023. The improved telomerized oils had a lighter color, lower acid value (varying from 4.68 to 5.79 versus from 7.65 to 23.99), improved viscosity (6848–41,328 Saybolt universal second [sus] versus 646–2636 sus at 40°C), and enhanced antiwear properties (89 versus 57.7% reduction in teeth wear) when determined using a Falex Pin and Vee Block method (ASTM D2670) compared with previous telomerized oils. The improved telomerized vegetable oil exhibited comparable thickening effect in an MVI Shell base oil with a commercially available VII (Acryloid 1019).

TABLE 16.3

VIs and Pour Points of Base Oils Doped with Homopolymers of Sunflower and Soybean Oils

		Concentration of Additive-Doped Base Oils ($\times 10^3$ ppm)							
		2		3		4		5	
Sample	Base Oil	VI	Pour Point (°C)	VI	Pour Point (°C)	VI	Pour Point (°C)	VI	Pour Point (°C)
S-1	BO1	121	−6	131	−9	158	−15	161	−18
	BO2	135	−9	150	−12	164	−15	198	−18
S-2	BO1	130	−6	169	−9	201	−9	211	−12
	BO2	141	−6	172	−8	220	−9	238	−12

Source: Adapted from G. Karmakar and P. Ghosh, *ACS Sustain. Chem. Eng.*, 3, 19–25.
Note: BO1: base oil 1; BO2: base oil 2; S-1: homopolymer of sunflower oil; S-2: homopolymer of soybean oil.

In another disclosure [52] preparation of lower-viscosity telomers (5,000–12,000 sus versus 41,328 sus at 40°C) from a mixture consisting of 5–15% conjugated triglyceride oils and 85–95% of nonconjugated triglyceride oils with low acid value (>20 versus 50) was reported. These oils were used as antiwear agents, VIIs, and high-performance thickeners without having corrosion or oxidation problems.

U.S. Patent 20110302825 [53] describes the preparation of biobased lubricant additives containing 50–100% biobased carbon by transesterification of heat-bodied oils with biobased or petrobased alcohols followed by hydrotreatment. Biresaw et al. [54] synthesized different grades of heat-bodied oils by thermal treatment of soybean oil under inert atmosphere by varying reaction time. The film-forming properties of these oils under elastohydrodynamic conditions were investigated. The film thickness of these oils was measured as a function of load, temperature, and entrainment speed.

Ghosh et al. [55] synthesized homo- and copolymers of sunflower oil with different mass fractions of methyl methacrylate, decyl acrylate, and styrene. The copolymers were evaluated for their potential as PPDs in two different base oils. The copolymers exhibited superior additive performance, reducing the pour point of base oils (from −3°C to −18°C) compared with that of homopolymers (−3°C to −6°C). Karmakar and Ghosh [56] synthesized homopolymers of soybean and sunflower oil by a thermal method using benzoyl peroxide as a radical initiator and evaluated their PPD and VII properties. Polymer-doped base oils displayed improved VI and pour points (Table 16.3).

16.2.4 EPOXY RING-OPENING REACTIONS

Epoxidation of oleochemicals has been known for many years. Epoxidized fatty components can be reacted to synthesize a number of interesting products. Ring opening of epoxidized oils with organic acids or alcohols produce hydroxy esters and hydroxy polyethers. The cleavage of the epoxy ring also allows the introduction of heteroatoms and other functional groups. Thus, a whole series of oleochemical products can be obtained from epoxidized oleochemicals for use as lubricant base fluids, additives, etc. [57]. The additives developed from epoxidized fatty acid derivatives display high solubility in biodegradable vegetable oils and synthetic esters. As a result, additives derived from epoxidized natural raw materials will be considerably better than conventional petroleum-based additives.

Epoxidized vegetable oils and their esters are also useful as lubricant additives imparting anticorrosive, antiwear, and antifriction properties. The epoxidized ester additives impart protection to the lubricated system against the corrosive tendencies of other additives in the lubricant. U.S. Patent 5368776 [58] discloses the development of corrosion protection additives from epoxidized methyl esters of unsaturated fatty acids. Preparation of epoxidized fatty acid esters as additives for gasoline, fuel oil, greases, industrial oils, gear oils, and lubricants for engines and other equipment with moving parts operating under boundary lubricating conditions is disclosed in U.S. Patent 4244829 [59].

Various synthetic approaches are applied to develop products with improved thermo-oxidative and low temperature stability, antiwear, and EP properties by nucleophilic epoxy ring-opening reactions with alcohols, acids, amines, thiols, borates, etc. Preparation of hydroxylphenoxy stearic acid from epoxidized crude palm olein and its performance as a rust and oxidation inhibitor agent in a synthetic base fluid is reported [60]. Reacting 70% epoxidized palm olein with 39% phenol at 2 bar pressure using 1% zinc oxide as catalyst resulted in a product with 0% oxirane value and 110.4 hydroxyl value (Figure 16.3). This product was found to be effective as an antioxidant and rust inhibitor.

Biswas et al. [61] synthesized amine-functionalized soybean oil by ring opening of epoxy groups of soybean oil with amines using different catalysts. The reaction did not cause cross-linking or disruption of the ester linkage in the oil (Figure 16.4). Biswas et al. [62] in their efforts to explore new and green reaction pathways for development of biobased additives, developed a straightforward and environmentally friendly reaction to functionalize methyl oleate with aniline using a catalytic amount of ionic liquid to produce an oleate–aniline adduct, without the formation of fatty amide. Blends of PAO base oil (PAO 8) with the oleate–aniline adduct additive were evaluated for oxidation

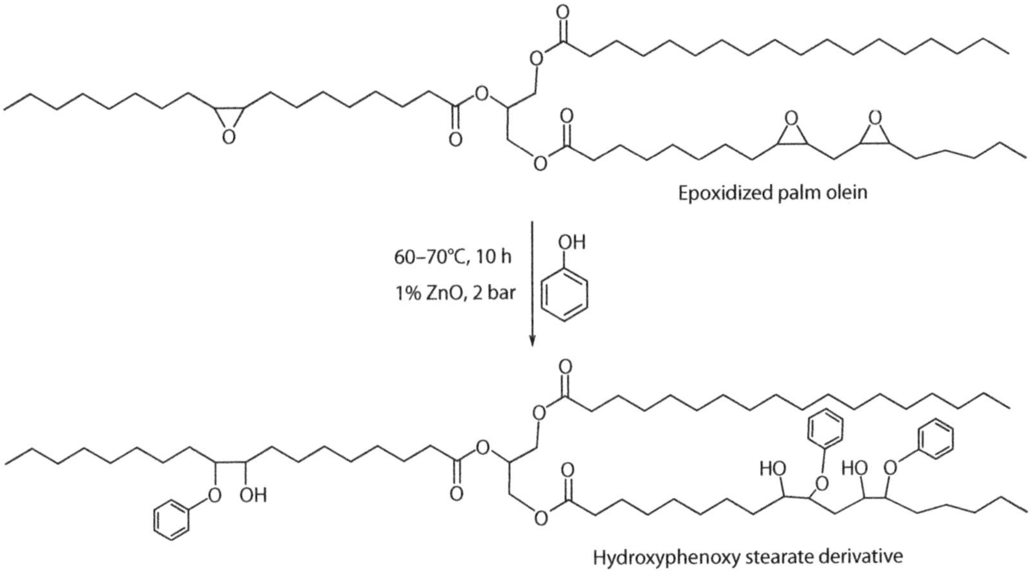

FIGURE 16.3 Preparation of hydroxylphenoxy stearate derivative from epoxidized crude palm olein. (Adapted from Sadi, S. et al., Preparation and performance of lubricating oil additive derived from epoxidized crude palm olein, *Proceedings of International Symposium on Advances in Alternative Renewable Energy [ISAAC]*, Johor Bahru, Malaysia, 137–153, 1998.)

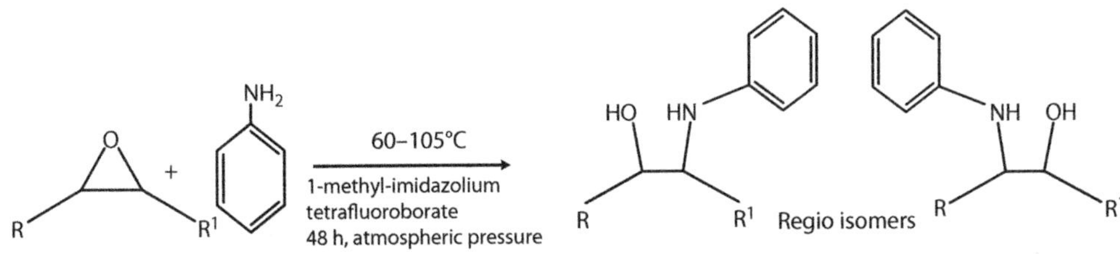

FIGURE 16.4 Preparation of amine–oleate derivative from methyl oleate. (Adapted from A. Biswas et al., *J. Agric. Food Chem.*, 57, 8136–8141.)

property using pressure differential scanning calorimetry (DSC). The results indicated that the additive increased the onset temperature of PAO 8 by more than 32°C. Lubricity tests on a four-ball tribometer using ASTM D4172 method with 1200 rpm and a normal load of 20 kg applied at 75°C for 60 minutes lowered scar diameter in the presence of the oleate–aniline adduct.

Synthesis and characterization of ß-amino alcohols from methyl epoxy stearate and terminal epoxy fatty acid methyl ester using zinc(II) perchlorate hexahydrate as catalyst were reported [63]. However, evaluation of these products for their potential as additives is not reported. 10-Undecenoic acid-based alkyl 11-anilino-10-hydroxy undecanoates were prepared by ring opening of alkyl epoxy undecanoates with aniline without catalyst [64,65]. The products (Figure 16.5) were evaluated for their antioxidant properties in a biolubricant base oil, epoxy karanja fatty acid 2-ethylhexyl esters. All compounds exhibited good antioxidant activity by improving the base oil oxidation induction period (IP). At 0.25% concentration, the products 1-(2-propyl)-11-anilino-10-hydroxy undecano-ate and 1-butyl-11-anilino-10-hydroxy undecanoate exhibited better performance with IP values of 66.5 and 78.6 h, respectively, compared with commercial antioxidant BHT (42.3 h). When the additive concentration was increased to 0.5%, all products at 110°C and 140°C exhibited enhanced antioxidant activity. Additive 1-butyl-11-anilino-10-hydroxy undecanoate enhanced the IP value of base oil from 78 to 127 h at 110°C with increase in concentration, and additive methyl-11-anilino-10-hydroxy undecanoate also prolonged the IP value from 44.7 to 96.2 h. The rest of the additives exhibited marginal improvement in IP value. The results obtained from isothermal DSC correlated well with those based on Rancimat method (Table 16.4).

9(10)-hydroxy-10(9)-ester derivatives of methyl oleate were prepared by Sharma et al. [66] from commercially available methyl oleate and common organic acids (Figure 16.6) and their properties were investigated. These derivatives exhibited improved low-temperature properties and increased thermo-oxidative stability relative to those of olefinic oleochemicals. The lubrication enhancement properties of these branched oleochemicals were investigated in hexadecane, soybean oil, and PAO base oils at 5 wt% using four-ball geometry (Falex MultiSpecimen apparatus) under a normal load of 40 kg at room temperature for 15 min at both 1200 and 2400 rpm. Addition of the hydroxyl ester products as additives in PAO lowered the WSD by more than 0.1 mm in all cases compared with that of PAO alone at 1200 rpm, and a similar trend was observed even at the high speed of 2400 rpm in hexadecane. However, the additives' effect was much less evident in soybean oil. This may be due to the good lubricity of soybean oil in its natural state. The synthesis retained the triglyceride struc-ture with its inherent high flash point, high VI, biodegradability, and lubricity. It also eliminated the polyunsaturation in the fatty acid chains. Addition of polar functional groups improved the tribo-logical properties of hydroxy ester products. Hydroxy and ester functionalities at C_9 and C_{10} posi-tions help the compounds to adsorb to the metal surface and reduce friction under boundary regime.

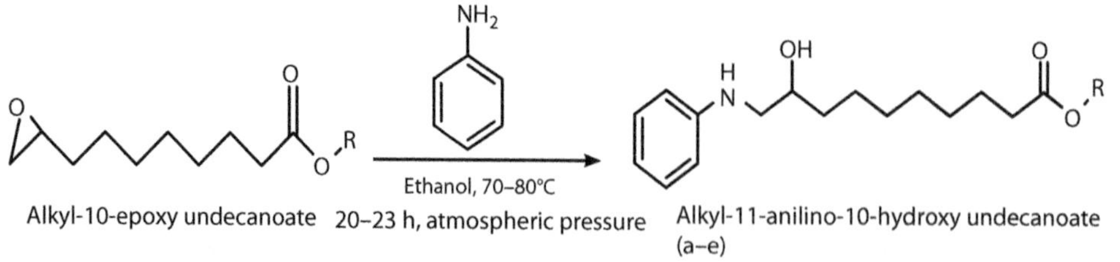

Alkyl-10-epoxy undecanoate   Ethanol, 70–80°C   Alkyl-11-anilino-10-hydroxy undecanoate
20–23 h, atmospheric pressure   (a–e)

R: a, –CH$_3$; b, –CH–(CH$_3$)$_2$; c, –(CH$_2$)$_3$–CH$_3$; d, –CH$_2$–CH–(CH$_3$)$_2$; e, –(CH$_2$)$_5$–CH–(CH$_3$)$_2$

FIGURE 16.5 Reaction scheme for synthesis of 10-undecenoic acid-based amino alcohols. (Adapted from Geethanjali, G. et al.: Preparation of alkyl 11-anilino-10-hydroxy undecanoates and evaluation of their anti-oxidant activity. *Eur. J. Lipid Sci. Technol.* 2013. 115. 740–746.)

TABLE 16.4

Effect of Antioxidants on Epoxy Karanja Fatty Acid 2-Ethylhexyl Esters as Determined by Rancimat* and DSC** Methods

Sample (Structure in Figure 16.5)	Rancimat IP Value (h)				DSC IP Value (h)
	0.25%		0.5%		0.5%
Additive Concentration	110°C	140°C	110°C	140°C	140°C
Epoxy karanja fatty acid 2-ethylhexyl esters	45.78	5.7	–	–	1.78
Base oil + BHT	42.3	5.3	87.7	10.9	1.8
Base oil + methyl-11-anilino-10-hydroxy undecanoate (R = a)	44.7	5.6	96.2	12.0	3.2
Base oil + 1-(2-propyl)-11-anilino-10-hydroxy undecanoate (R = b)	66.5	8.3	74.4	9.3	2.2
Base oil + 1-butyl-11-anilino-10-hydroxy undecanoate (R = c)	78.6	9.8	127.2	15.9	3.3
Base oil + 2-methyl-1-propyl-11-anilino-10-hydroxy undecanoate (R = d)	44.6	5.6	68.4	8.5	2.6
Base oil + 2-ethylhexyl-11-anilino-10-hydroxy undecanoate (R = e)	44.0	5.5	53.4	6.7	2.5

Source: Adapted from Geethanjali, G. et al.: Preparation of alkyl 11-anilino-10-hydroxy undecanoates and evaluation of their antioxidant activity. *Eur. J. Lipid Sci. Technol.* 2013. 115. 740–746.

* AOCS cd 12b-92.
** ASTM E 2009-08.

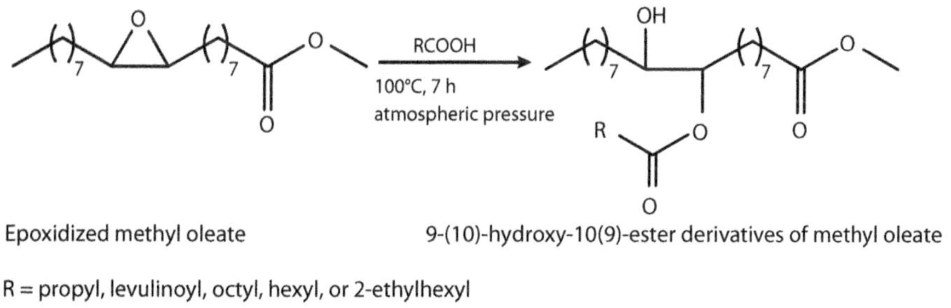

Epoxidized methyl oleate   9-(10)-hydroxy-10(9)-ester derivatives of methyl oleate

R = propyl, levulinoyl, octyl, hexyl, or 2-ethylhexyl

FIGURE 16.6 Ring-opening reaction of epoxidized methyl oleate to prepare ester hydroxyl derivatives. (Adapted from Sharma, B. K. et al.: Ester hydroxyl derivatives of methyl oleate: Tribological, oxidation and low temperature properties, *Bioresour. Technol.*, 99, 7333–7340.)

Poly(hydroxy thioether) derivatives were prepared [67] by reaction of epoxidized soybean oil with common thiols such as 1-butanethiol, 1-decanethiol, 1-octadecanethiol, and cyclohexyl mercaptan. Figure 16.7 gives the reaction scheme for the synthesis of 1-butanethiol–ether derivative of soybean oil whose antifriction and antiwear properties were investigated in toluene solution relative to those of three commercial antiwear additive packages using a four-ball tester. The friction coefficient and WSD were measured at 1200 rpm speed, 40 kg load, for 15 minutes at room temperature. The results indicated that this derivative significantly lowered the coefficient of friction at 5 wt% concentration compared with that of commercial antiwear additive packages. Scar diameter measurements revealed that the antiwear performance of the 1-butanethiol–ether soybean oil derivative was comparable with those of other multicomponent additives [17].

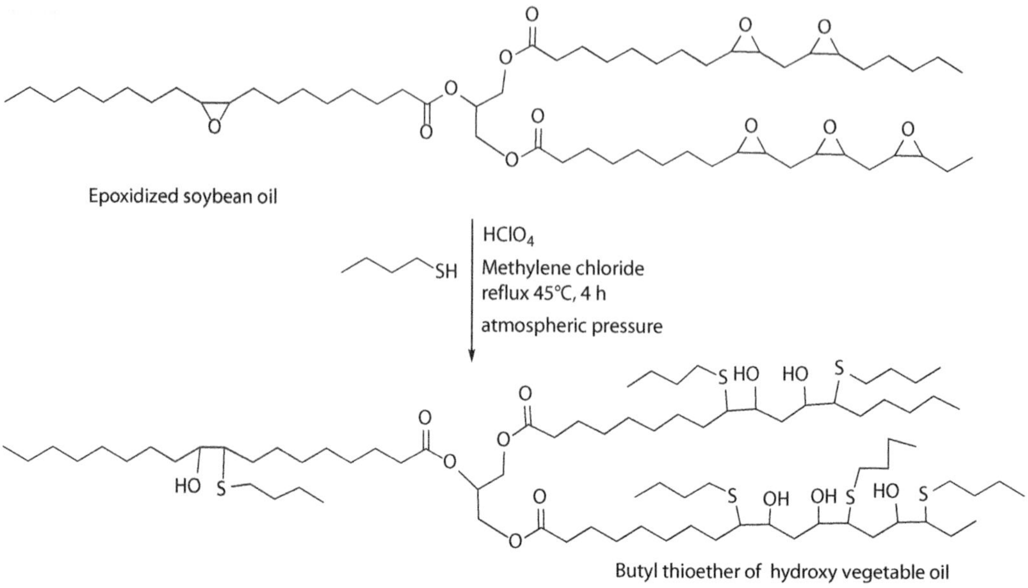

FIGURE 16.7 Ring-opening reaction of epoxidized soybean oil with butanethiol. (Adapted from B.K. Sharma et al., *J. Agric. Food Chem.*, 54, 9866–9872.)

Sharma et al. [22] also synthesized antiwear additives by a ring-opening reaction of the epoxidized soybean with mono and difunctional borates using titanium isopropoxide as catalyst. The additive properties of the borate derivatives were evaluated in soybean oil, PAO, Group III mineral oil, and hexadecane. The aromatic additive when used at 2 wt% increased the oxidation onset temperature of the base stocks by 14°C, 52°C, 48°C, and 49°C, respectively. Monofunctional borate additive at 1% concentration reduced the wear scar of soybean oil from 0.72 to 0.52 mm. However, both the additives did not reduce the wear scar of mineral oil. The aromatic additive when tested in a gear oil blend at 4 wt% reduced wear by 0.2 mm. Both the additives were compatible with a commercial additive zinc dialkyl dithiophosphate.

16.2.5 THIOL-ENE OR H-PHOSPHONATE-ENE REACTIONS

The functionalization of fatty acids and their esters via the addition of sulfur and phosphorus radicals onto C=C bonds has been known for long time. Biobased materials prepared from unsaturated lipids reacted with H-E, where E is either a sulfur- or phosphorus-centered group, by free radical mechanism were found to be good antioxidant, antiwear, and EP additives [68].

Recently the radical addition of thiols to plant oil derivatives received increased interest as this is considered as an efficient and green procedure for the production of biomaterials for different applications. Sulfide-modified corn, canola, and castor–lauric estolide oils were prepared [69,70] by direct photochemical reaction of butanethiol with the double bonds on the hydrocarbon chains (Figure 16.8). These oils, along with commercial mono- and polysulfide additives, were investigated as EP additives at 0.6% (w/w) S concentration, in corn and PAO base fluids. In both the fluids, the additives resulted in minor changes in pour point, cloud point, coefficient of friction, WSD, and weld point. The additives when tested for oxidation stability using rotary pressure vessel oxidation test (RPVOT) method did not improve the RPVOT time of corn oil, but made big improvement (from 22 to 249–344 min) in the oxidation stability of PAO (Table 16.5).

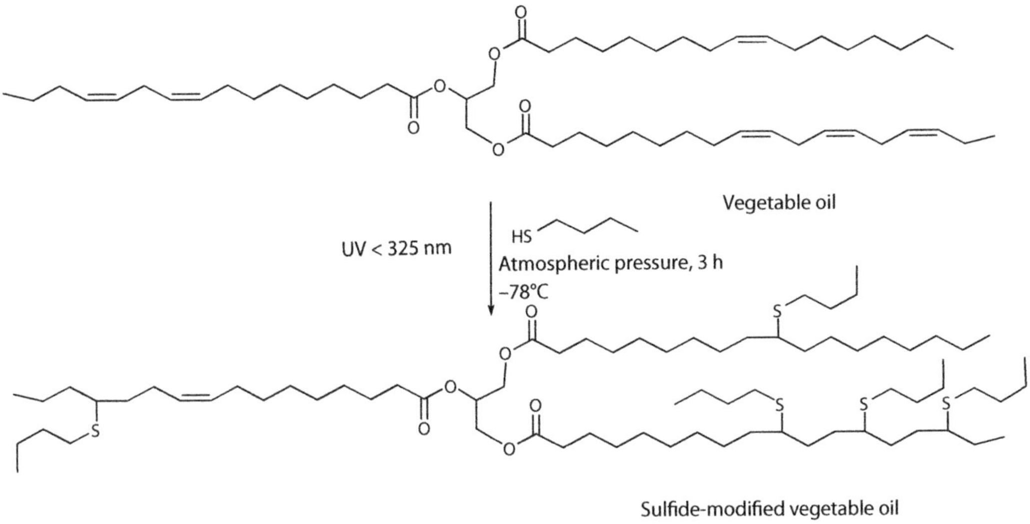

Vegetable oil

UV < 325 nm

HS

Atmospheric pressure, 3 h
−78°C

Sulfide-modified vegetable oil

FIGURE 16.8 Reaction of butanethiol with double bonds in vegetable oil under photochemical conditions. (Adapted from ACS: *J. Agric. Food Chem.*, Free radical addition of butanethiol to vegetable oil double bonds, 57, 2009, 1282–1290, Bantchev G.B. et al.)

Phosphonates are well known for their excellent lubricity properties and are used as EP additives and carbon dispersants [28]. The addition of dialkyl phosphonates to unsaturated compounds under free radical conditions is a general reaction of wide applicability. Literature reports indicate that phosphonated oils have been investigated for their tribological properties by themselves [71–76] or in combination with sulfur- and/or halogen-containing groups [77–79]. Sasin et al. [80] reported preparation of trialkyl ω-phosphono undecanoates and trialkyl 9,(10)-phosphono stearates by radical-catalyzed addition of dialkyl phosphonates (such as dimethyl, diethyl, di-*n*-butyl,

TABLE 16.5
Effect of Sulfide-Modified Oils on Oxidation and Cold-Flow Properties of Corn and PAO Base Oils

Oil	RPVOT[a] Value (min)	Pour Point[b] (°C)	Cloud Point[c] (°C)
Corn oil	12.5 ± 0.7	−12	−1
Corn oil + 6.2% (w/w) SM corn oil	13.0 ± 0.0	−12	−5
Corn oil + 6.3% (w/w) SM canola oil	13	−12	−3
Corn oil + 13.4% (w/w) SM castor lauric ester	13	−12	−5
PAO base oil	22.0 ± 0.0	<−54	<−45
PAO base oil + 6.2% (w/w) SM corn oil	311 ± 44	<−54	<−33
PAO base oil + 6.3% (w/w) SM canola oil	249	<−54	<−54
PAO base oil + 13.4% (w/w) SM castor lauric ester	344	<−54	<−54

Source: Adapted from Springer Science+Business Media: *Tribol. Lett.*, Tribological properties of sulfur modified vegetable oils, 43, 2011, 17–32, Biresaw, G. et al.

Note: SM: sulfide-modified.

[a] RPVOT: rotating pressure vessel oxidation test; ASTM D2272-98.

[b] ASTM D97-96a.

[c] ASTM D2500-99.

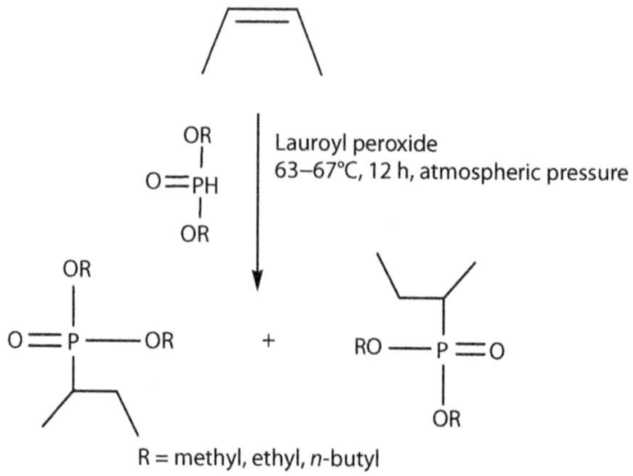

FIGURE 16.9   Reaction scheme for the preparation of phosphonated methyl oleates. (Adapted from Biresaw, B. and Bantchev, G.B., Tribological properties of biobased ester phosphonates, *J. Am. Oil Chem. Soc.*, 90, 891–902, 2013.)

and di-2-ethylhexyl phosphonates) to a series of alkyl undecanoates and alkyl oleates. These products were characterized by IR spectroscopy but were not evaluated for their additive properties.

Alkyl phosphonate derivatives of polyunsaturated acid alkyl esters and vegetable oils (such as soybean, cottonseed, safflower, corn, sunflower, tung, and linseed oils) prepared by free radical addition of dialkyl phosphonates were reported by Knight et al. [81]. The phosphonated triglycerides and esters were evaluated for their antiwear properties using a four-ball tester at applied load of 50 kg at 120°C for 1 h at 5 wt% in paraffin and synthetic diester base fluids. Phosphonated cottonseed and oleo oils, lard, and soybean oil butyl esters reduced wear of paraffin oil (0.40–0.48 mm from 0.63 mm) and synthetic fluid (0.84 to 0.43–0.50 mm). Other products were only moderately effective as antiwear additives.

Recently Bantchev et al. [82] synthesized three phosphonate derivatives of methyl oleate (MeOPs) in a radical chain reaction and characterized their physical properties. The reaction scheme for synthesis of MeOPs is given in Figure 16.9. In a later study the authors investigated the additive properties of these phosphonate blends in soybean oil and PAO 6 base oils at various treat rates relative to those of ZDDP. MeOP additives reduced the wear properties of soybean oil by more than 25%, but compared with ZDDP, they were less effective. At 5 wt%, additives had no effect on EP weld point in petroleum-based oils, but exhibited smaller improvement in soybean oil relative to ZDDP (Table 16.6) [83].

Thiol-ene or H-phosphonate-ene reactions are green alternatives for introducing sulfur or phosphorus in fatty acid chains. Such derivatives of vegetable oils and fatty esters were investigated for their potential as lubricant additives. Some of the derivatives were found to be suitable as antiwear and EP additives. But these reactions are not well studied and their true potential cannot be fully estimated.

16.2.6   Enzymatic Modification

Enzymatic catalysts are preferred in lipid chemistry for modifications of sensitive compounds. They can be carried out under mild reaction conditions (low temperature) and avoid unwanted side reactions and make the syntheses more environment friendly. In an enzymatic modification reported by Biresaw et al. [84], lipoic acid, a naturally occurring cellular cofactor with a cyclic disulfide moiety, was transesterified with high-oleic sunflower oil in 2-methyl-2-butanol solvent. The reaction gave a crude product mixture comprising unreacted lipoic acid, free fatty acids, and several lipoyl

TABLE 16.6

Weld Point Values of Base Oils with and without Phosphonates and ZDDP

| | | Weld Point (kgf) | | |
| | | Blend with 5% Additive | | |
Base Oil	Base Oil with No Additives	Etop[a]	n-Buop[b]	ZDDP
PAO	120	120	120	120
150 N (mineral oil)	120	–	–	150
Soybean oil	120	140	140	250

Source: Adapted from Springer Science+Business Media: *J. Am. Oil Chem. Soc.*, Tribological properties of bio-based ester phosphonates, 90, 2013, 891–902, Biresaw, G., and G. B. Bantchev.

[a] Etop: ethoxy phosphonated derivative:

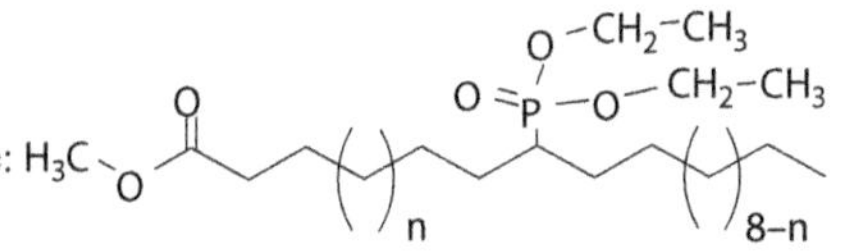

[b] n-Buop: n-butoxy phosphonated derivative:

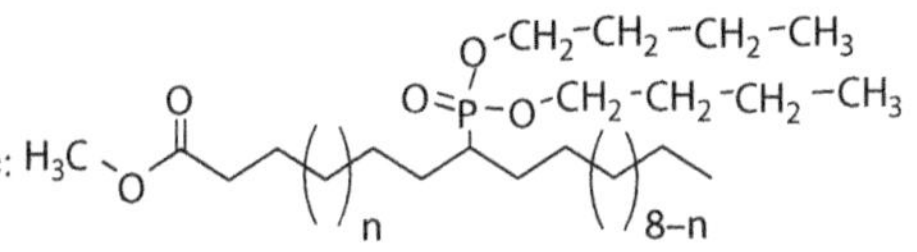

glyceride structures of varying lipoic acid substitutions. A highly purified product mixture was prepared by extraction of polar components with methanol (Figure 16.10). Both these product mixtures were evaluated for their EP properties in soybean and high-oleic sunflower oils. Both the product mixtures displayed four- to sixfold improvement in weld point values over the value of that of the neat vegetable oils, indicating the beneficial EP characteristics of lipoyl derivatives.

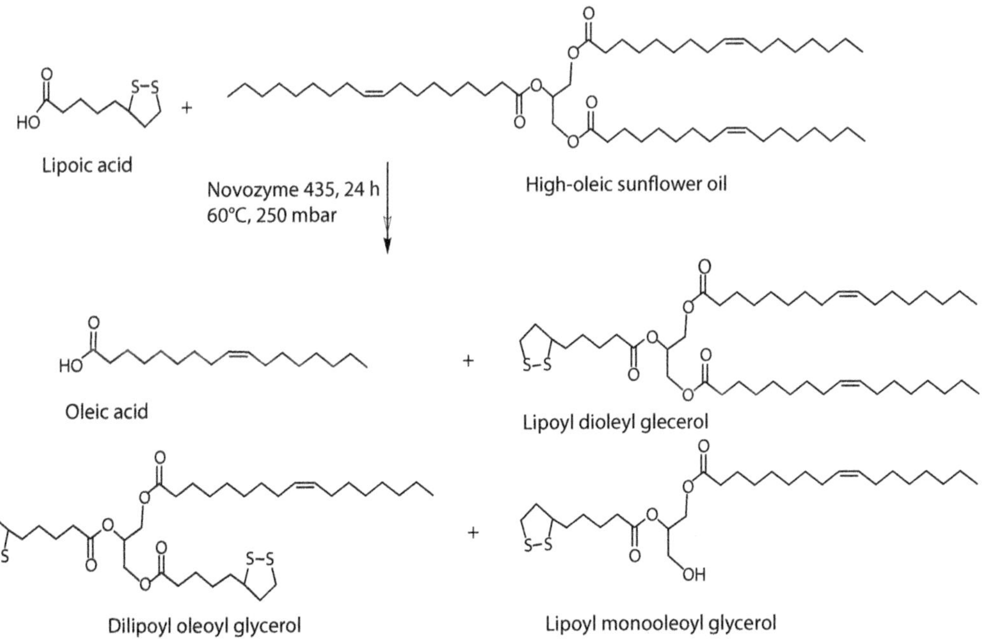

FIGURE 16.10 Preparation of high-oleic sunflower oil-based lipoyl derivatives. (Adapted from G. Biresaw et al., *J. Agric. Food Chem.*, 62, 2233–2243.)

16.3 CONCLUSIONS

Increasing environmental awareness has increased investigation into new biobased lubricant additive chemistry. Additives such as antioxidants, PPDs, viscosity improvers, and antiwear and EP additives can be synthesized from vegetable oils through chemical, thermal, photochemical, and enzymatic processes. Soybean, rapeseed, meadowfoam, sunflower, and rapeseed oils and monoesters such as sperm whale and jojoba oils were used as feedstocks to synthesize lubricant additives. Different reactions used include sulfurization, thermal, enzymatic, photochemical, and epoxy ring-opening reactions. Some of the products from the syntheses are far from perfect and will require further research to achieve desired performance. One of the major challenges in the development of vegetable-based additives is to produce bioadditives compatible with biolubricants, retaining their biodegradability, high lubricity, and low toxicity. Review of the previous works shows that it is possible to produce predominantly biobased lubricant additives from vegetable oils by using different chemical reactions.

REFERENCES

1. M.P. Schneider, Plant-oil-based lubricants and hydraulic fluids, *J. Sci. Food Agric.*, 86, 1769–1780 (2006).
2. K.G. Allum and J.F. Ford, The influence of chemical structure of the load-carrying properties of certain organosulfur compounds, *J. Int. Petrol.*, 51, 145–152 (1965).
3. A. Mammen, C.V. Agarwal, and V.K. Verma, The load-carrying properties of certain S-alkylisothioamides and their effectiveness as extreme pressure lubricant additives, *Wear*, 71, 355–361 (1981).
4. A. Bhattacharya, T. Singh, and V.K. Verma, The role of certain substituted 2-amino benzothiazolyl benzoyl thiocarbamides as additives in extreme pressure lubrication of steel bearing balls, *Wear*, 136, 345–357 (1990).
5. W.J. Bartz, Ecotribology: Environmentally acceptable tribological practices, *Tribol. Int.*, 39, 728–733 (2006).
6. J. Crawford, A. Psaila, and S.T. Orszulik, Miscellaneous additives and vegetable oils, in: *Chemistry and Technology of Lubricants*, M. Roy, R.M. Mortier, M.F. Fox, and S.T. Orszulik (Eds.), pp. 189–211, Springer, Heidelberg (2010).
7. C.I. Betton, Lubricants and their environmental impact, in: *Chemistry and Technology of Lubricants*, M. Roy, R.M. Mortier, M.F. Fox, and S.T. Orszulik (Eds.), pp. 435–445, Springer, Heidelberg (2010).
8. S. Sasaki, Environmentally friendly tribology (eco-tribology), *J. Mech. Sci. Technol.*, 24, 67–71 (2010).
9. S. Plaza and L. Margielewski, Surfactants and lubricating oil additives, in: *Advanced Tribology*, J. Luo, Y. Meng, T. Shao, and Q. Zhao (Eds.), pp. 920–921, Springer, Heidelberg (2010).
10. R.J. Gunderson and A.H. Hart, *Synthetic Lubricants*, Reinhold Publishing, New York (1962).
11. F.C. Magne, R.R. Mod, G. Sumrell, R.E Koos, and W.E. Parker, Lubricants and lubricant additives: II. Performance characteristics of some substituted fatty acid esters, *J. Am. Oil Chem. Soc.*, 52, 494–497 (1975).
12. R.A. Mod, F.C. Magne, G. Sumrell, and R.E. Koos, Lubricants and lubricant additives: III. Performance characteristics of some thioacetate, phosphorodithioate, and hexachlorocyclopentadiene derivatives of stearic acid amides and esters, *J. Am. Oil Chem. Soc.*, 54, 589–591 (1977).
13. V.K. Chibber, R.B. Chaudhary, O.S. Tyagi, and O.N. Anand, Antiwear and antifriction characteristics of tribochemical films of alkyl octa-decenoates and their derivatives, *Lubr. Sci.*, 18, 63–76 (2006).
14. J. Filley, New lubricants from vegetable oil: Cyclic acetals of methyl 9,10-dihydroxy stearate, *Bioresour. Technol.*, 96, 551–555 (2005).
15. N.N. Tupotilov, V.V. Ostrikov, and A.Y. Kornev, Plant oil derivatives as additives for lubricants, *Chem. Technol. Fuels Oils*, 42, 192–199 (2006).
16. K.M. Doll, B.K. Sharma, and S.Z. Erhan, Synthesis of branched methyl hydroxy stearates including an ester from bio-based levulinic acid, *Ind. Eng. Chem. Res.*, 46, 3513–3519 (2007).
17. S.Z. Erhan, A. Adhvaryu, and B.K. Sharma, "Poly(Hydroxy Thioether) Vegetable Oil Derivatives Useful as Lubricant Additives," U.S. Patent 7279448 (2007).
18. S.C. Cermak, G. Biresaw, and T. Isbell, Comparison of a new estolide oxidative stability package, *J. Am. Oil Chem. Soc.*, 85, 879–885 (2008).

19. B.R. Moser and S.Z. Erhan, Synthesis and evaluation of a series of α-hydroxyethers derived from iso-propyl oleate, *J. Am. Oil Chem. Soc.*, 83, 959–963 (2006).
20. B.R. Moser and S.Z. Erhan, Preparation and evaluation of a series of α-hydroxy ethers from 9,10-epoxy-stearates, *Eur. J. Lipid Sci. Technol.*, 109, 206–213 (2007).
21. F. Shah, S. Glavatskih, and O. Antzutkin, Novel alkyl borate dithiocarbamate lubricant additives: Synthesis and tribophysical characterization, *Tribol. Lett.*, 45, 67–78 (2012).
22. B.K. Sharma, K.M. Doll, G.L. Heise, M. Myslinska, and S.Z. Erhan, Antiwear additive derived from soybean oil and boron utilized in a gear oil formulation, *Ind. Eng. Chem. Res.*, 51, 11941–11945 (2012).
23. R. Becker and A. Knorr, An evaluation of antioxidants for vegetable oils at elevated temperatures, *Lubr. Sci.*, 8, 95–117 (1996).
24. A. Adhvaryu, S.Z. Erhan, Z.S. Liu, and J.M. Perez, Oxidation kinetic studies of oils derived from unmodified and genetically modified vegetables using pressurized differential scanning calorimetry and nuclear magnetic resonance spectroscopy, *Thermochim. Acta*, 364, 87–97 (2000).
25. S.O. Jones and E.E. Reid, The addition of sulfur, hydrogen sulfide and mercaptans to unsaturated hydro-carbons, *J. Am. Oil Chem. Soc.*, 60, 2452–2455 (1938).
26. K.E. Davis, "Sulfurized Compositions," U.S. Patent 4191659 (1980).
27. D.E. Johnson, "Sulphurised Olefins, Extreme pressure, Antiwear Additives and Compositions," U.S. Patent 5135670 (1992).
28. A.S. Patil, V.A. Pattanshetti, and M.C. Dwivedi, Functional fluids and additives based on vegetable oils and natural products: A review of potential, *J. Synth. Lubr.*, 15, 193–211 (1997).
29. L.O. Farng, Antiwear additives and extreme-pressure additives, in: *Lubricant Additives Chemistry and Applications*, L.R. Rudnick (Ed.), p. 219, Taylor and Francis, Boca Raton, FL (2009).
30. K.P. Kammann and A. Phillips, Sulfurized vegetable oil products as lubricant additives, *J. Am. Oil Chem. Soc.*, 62, 917–923 (1985).
31. K.P. Kammann, "Modified Fatty Amides and Sulfurized Fatty Oils as Lubricant Additives," U.S. Patent 4921624 (1990).
32. J.M. Wakim, "Cosulfurized Products of High Iodine Value Triglyceride and Non-wax Ester of Monoethenoid Fatty Acid as Lubricant Additives," U.S. Patent 3986966 (1976).
33. L.H. Princen and J.A. Rothfus, Development of new crops for industrial raw materials, *J. Am. Oil Chem. Soc.*, 61, 281–289 (1984).
34. F.L. Erickson, R.E. Anderson, and P.S. Landis, "Meadowfoam Oil and Meadowfoam Oil Derivatives as Lubricant Additives," U.S. Patent 4970010 (1990).
35. E.W. Bell, "Wax Esters of Vegetable Oil Fatty Acids Useful as Lubricants," U.S. Patent 4152278 (1979).
36. P.S. Landis, "Telomerized Vegetable Oil for Lubricant Additives," Patent WO1992007051 (1992).
37. J. Wisniak and H. Benajahu, Sulfurization of jojoba oil, *Ind. Eng. Chem. Prod. Res. Dev.*, 14, 247–258 (1975).
38. H. Gisser, D. Messina, and D. Chasan, Jojoba oil as a sperm oil substitute, *Wear*, 34, 53–63 (1975).
39. C.L. Herman, Discussions on comparisons of sperm oil based EP additives with their replacement prod-ucts, *NLGI Spokesm.*, 36, 214–217 (1972).
40. V.K. Bhatia, A. Chaudhry, A. Masohan, R.P.S. Bisht, and G.A. Sivasankaran, Sulphurization of jojoba oil for application as extreme pressure additive, *J. Am. Oil Chem. Soc.*, 65, 1502–1507 (1988).
41. T.K. Miwa and J.A. Rothfus, Extreme-pressure lubricant tests on jojoba and sperm whale oils, *J. Am. Oil Chem. Soc.*, 56, 765–770 (1979).
42. R.P.S. Bisht and V.K. Bhatia, Sulphurised vegetable oils as EP additives for industrial gear oil formula-tions, *J. Synth. Lubr.*, 14, 23–33 (1997).
43. IS 8406-1985: Indian standard test method for gear lubricants for enclosed industrial gear drives, *Annual Book of Bureau of Indian Standards*, New Delhi, India, Doc: No PCD 4 (1128) (1993).
44. Y. Cao, L. Yu, and W. Liu, Study of the tribological behavior of sulfurized fatty acids as additives in rapeseed oil, *Wear*, 244, 126–131 (2000).
45. P. Michaelis, "Epithio Compounds as Additives for Lubricants," U.S. Patent 4217233 (1980).
46. F.D. Gunstone, M.G. Husain, and D.M. Smith, The preparation and some properties of methyl monomer-capto stearates, some related thiols, and some methyl epithiostearates, *Chem. Phys. Lipids*, 13, 71–91 (1974).
47. S.F. Marcel, L.K. Jie, and Y.F. Zheng, Synthesis and physical properties of some 2,3- and 2,2′-epithio C_{18} fatty acid derivatives, *Chem. Phys. Lipids*, 49, 167–178 (1988).
48. F.C. Magne, R.R. Mod, G. Sumrell, R.E. Koos, and W.E. Parker, Lubricants and lubricant additives: II. Performance characteristics of some substituted fatty acid esters, *J. Am. Oil Chem. Soc.*, 52, 494–497 (1975).

49. G. Geethanjali, K.V. Padmaja, A. Sammaiah, and R.B.N. Prasad, Synthesis, characterization, and evaluation of 10-undecenoic acid-based epithio derivatives as multifunctional additives, *J. Agric. Food Chem.*, 62, 11505–11511 (2014).

50. P.S. Landis, "Telomerised Triglyceride Oil for Lubricant Additives," U.S. Patent 5229023 (1993).

51. A. Shahanan and P.S. Landis, "Telomerized Triglyceride Oil Product," U.S. Patent 5454965 (1995).

52. P.S. Landis, "Lower Viscosity Telomer Oil," U.S. Patent 5567345 (1996).

53. P. Bloom, "Lubricant Additives," U.S. Patent 20110302825 (2011).

54. G. Biresaw, B.K. Sharma, G.B. Bantchev, T.L. Kurth, K.M. Doll, S.Z. Erhan, B. Kunwar, and J. Scott, Elastohydrodynamic properties of biobased heat-bodied oils, *Ind. Eng. Chem. Res.*, 53, 16183–16195 (2014).

55. P. Ghosh, T. Das, D. Nandi, G. Karmakar, and A. Mandal, Synthesis and characterization of biodegradable polymer—Used as a pour point depressant for lubricating oil, *Int. J. Polym. Mater.*, 59, 1008–1017 (2010).

56. G. Karmakar and P. Ghosh, Green additives for lubricating oil, *ACS Sustain. Chem. Eng.*, 3, 19–25 (2013).

57. H. Wagner, R. Luther, and T. Mang, Lubricant base fluids based on renewable raw materials, their catalytic manufacture and modification, *Appl. Catal. A: Gen.*, 221, 429–442 (2001).

58. V. Schafer, R. Kohler, A. Pauli, and A. Fessenbecker, "Corrosion Protection Additives Based on Epoxides," U.S. Patent 5368776 (1994).

59. K. Coupland, "Hydrocarbon-Soluble Epoxidized Fatty Acid Esters as Lubricity Modifiers for Lubricating Oils," U.S. Patent 4244829 (1981).

60. S. Sadi, R.A. Rahman, M. Abu Baker, and A.H. Khadum, Preparation and performance of lubricating oil additive derived from epoxidized crude palm olein, Proceedings of International Symposium on Advances in Alternative Renewable Energy (ISAAC), Johor Bahru, Malaysia, 137–153 (1998).

61. A. Biswas, A. Adhvaryu, S.H. Gordon, S.Z. Erhan, and J.L. Willet, Synthesis of diethylene amine-functionalized soybean oil, *J. Agric. Food Chem.*, 53, 9485–9490 (2005).

62. A. Biswas, B.K. Sharma, K.M. Doll, S.Z. Erhan, J.L. Willet, and H.N. Cheng, Synthesis of amine-oleate derivative using ionic-liquid catalyst, *J. Agric. Food Chem.*, 57, 8136–8141 (2009).

63. S. Singh and R. Kamboj, Synthesis of amino alcohols from methyl epoxy stearate, *Ind. Eng. Chem. Res.*, 49, 3106–3111 (2010).

64. K.V. Padmaja and R.B.N. Prasad, "Alkyl 11-Anilino-10-hydroxy Undecanoates as Potential Antioxidant Additives for Biolubricant Basestocks," Indian Patent 1397/DEL/2010 (2010).

65. G. Geethanjali, K.V. Padmaja, B. Sreedhar, and R.B.N. Prasad, Preparation of alkyl 11-anilino-10-hydroxy undecanoates and evaluation of their antioxidant activity, *Eur. J. Lipid Sci. Technol.*, 115, 740–746 (2013).

66. B.K. Sharma, K.M. Doll, and S.Z. Erhan, Ester hydroxyl derivatives of methyl oleate: Tribological, oxidation and low temperature properties, *Bioresour. Technol.*, 99, 7333–7340 (2008).

67. B.K. Sharma, A. Adhvaryu, and S.Z. Erhan, Synthesis of hydroxy thio-ether derivatives of vegetable oil, *J. Agric. Food Chem.*, 54, 9866–9872 (2006).

68. G. Bantchev, S.C. Cermak, G. Biresaw, M. Appell, J.A. Kenar, and R.E. Murray, Thiol–ene and H-phosphonate–ene reactions for lipid modification, in: *Green Materials from Plant Oils*, Z. Liu and G. Krauss (Eds.), pp. 59–60, Royal Society of Chemistry, Cambridge (2015).

69. G.B. Bantchev, J.A. Kenar, G. Biresaw and M.G. Han, Free radical addition of butanthiol to vegetable oil double bonds, *J. Agric. Food Chem.*, 57, 1282–1290 (2009).

70. G. Biresaw, G.B. Bantchev, and S.C. Cermak, Tribological properties of sulfur modified vegetable oils, *Tribol. Lett.*, 43, 17–32 (2011).

71. L.P. Peale, J. Messina, B. Ackerman, R. Sasin, and D. Swern, Evaluation of long-chain phosphorus compounds as lubricity additives, *ASLE Trans.*, 3, 48–54 (1960).

72. D. Swern, W.E. Palm, R. Sasin, and L.P. Witnauer, Viscosity characteristics of long chain phosphorus compounds, *J. Chem. Eng. Data*, 5, 486–488 (1960).

73. E.S. Forbes and H.B. Silver, The effect of chemical structure on the load-carrying properties of organophosphorus compounds, *J. Inst. Pet.*, 56, 90–98 (1970).

74. A. Berman, S. Steinberg, S. Campbell, A. Ulman, and J. Israelachvili, Controlled microtribology of a metal oxide surface, *Tribol. Lett.*, 4, 43–48 (1998).

75. K. Hirao, T. Hasegawa, Y. Kidera, M. Memita, and I. Minami, Antiwear additives for neopentyl type esters: Design and performance evaluation of amine salts of phosphonic acid, *Toraibarojisuto/J. Jpn. Soc. Tribol.*, 48, 734–740 (2003).

76. H. Nadano, M. Nakasako, M. Kohno, H. Yamaguchi, and I. Minami, Nihon kikai gakkai ronbunshu, *C. Hen. Trans. Jpn. Soc. Mech. Eng. Part C*, 71(8), 2657–2664 (2005).

77. P.I. Sanin, E.S. Shepeleva, A.V. Ulyanova, and B.V. Kleimenov, The effect of synthetic additives in lubricating oil on wear under friction, *Wear*, 3, 200–218 (1960).

78. Y.A. Lozovoi, Y.S. Shepeleva, G.V. Shipilov, and P.I. Sanin, Comparative study of aryl phosphonic acid esters as anti-wear additives, *Neftekhimiya*, 12, 901–906 (1972).

79. R.I. Barber, The preparation of some phosphorus compounds and their comparison as load carrying additives by the four-ball machine, *ASLE Trans.*, 19, 319–328 (1976).

80. R. Sasin, W.F. Olszewski, J.R. Russell, and D. Swern, Phosphorus derivatives of fatty acids: VII.[2] Addition of dialkyl phosphonates to unsaturated compounds, *J. Am. Chem. Soc.*, 81, 6275–6277 (1959).

81. H. Knight, B. Ambler, and D. Swern, "Phosphonated Oils and Phosphonated Esters," U.S. Patent 3189628 (1965).

82. G.B. Bantchev, G. Biresaw, K.E. Vermillion, and M. Appell, Synthesis and spectral characterization of methyl 9(10)-dialkylphosphono stearates, *Spectrochim. Acta Part A: Molec. Biomolec. Spectrosc.*, 110, 81–91 (2013).

83. G. Biresaw and G.B. Bantchev, Tribological properties of biobased ester phosphonates, *J. Am. Oil Chem. Soc.*, 90, 891–902 (2013).

84. G. Biresaw, J.A. Laszlo, K.O. Evans, D.L. Compton, and G.B. Bantchev, Synthesis and tribological investigation of lipoyl glycerides, *J. Agric. Food Chem.*, 62, 2233–2243 (2014).

Section V

Additives for Environmentally Friendly Fluids

17 Friction of Fatty Acids in Nanoscale Contacts

Marina Ruths, Sarah M. Lundgren, and Karin Persson

CONTENTS

ABSTRACT

Boundary films and loose-packed self-assembled monolayers of fatty acids (stearic, oleic, and linoleic acids) and a rosin acid (dehydroabietic acid) were formed on mica and steel surfaces from *n*-hexadecane solutions and studied with the surface force apparatus technique and atomic force microscopy in friction mode. When the sliding contact was immersed in adsorption solution or in ethanol, the friction force increased linearly with load ($F_f = \mu F_n$), whereas the friction force in an adhesive contact (in N_2 gas) showed a sublinear increase with load and an apparent contact-area dependence ($F_f = S_c A$). The friction coefficient μ and the critical shear stress S_c were found to increase with increasing degree of unsaturation of the fatty acid.

17.1 INTRODUCTION

17.1.1 Background

Mixtures of unsaturated fatty acids such as oleic, linoleic, and linolenic acids, which are obtained as by-products of the papermaking process, are commonly used in formulations of friction modifiers for conventional diesel fuel. There is a renewed interest in the details of the lubricating properties of these compounds because of their occurrence in blends of biodiesel with ultralow sulfur petroleum-based fuel [1,2], and because of increasing restrictions on the components allowed in conventional friction-modifying additives [3].

Although the lubricating properties of saturated and unsaturated fatty acids have been utilized since ancient times [4] and have been subjects of scientific investigation since the beginning of the 20th century, our fundamental understanding of the influence of chain unsaturation on friction is still limited. The commercial importance of fatty acids has inspired a large number of studies at the macroscopic level, where saturated chains have been more extensively investigated than unsaturated ones [2,5–9]. Thus far, rather few investigations have been concerned with the boundary friction at the nanoscale in these systems, and they have mostly focused on saturated fatty acids in close-packed layers formed by Langmuir–Blodgett (LB) deposition or from the dilute solution of an evaporating solvent [10–13]. Much less is known about the lubricating properties of molecularly thin films of unsaturated fatty acids, in particular, in systems where they are formed at adsorption conditions similar to real lubrication applications, i.e., from a base oil, where it is known that the adsorbed amount is typically less than a full monolayer [14]. A few systematic investigations have been published, revealing that the adsorbed amount on steel surfaces increases with increasing unsaturation [15,16] and increasing chain length [17]. Adsorption from a nonaqueous solution typically results in adsorbed amounts of only one-half to one-third of that in a close-packed monolayer [14,15], with the exception of linoleic acid, which has recently been shown [15] to adsorb on steel from n-hexadecane with a molecular area similar to that of its maximum close packing in LB films.

Fatty acids may form chemical bonds with metal oxide surfaces [5,14,17]. Early work on lubrication with various functionalized alkanes showed that the friction behavior of alcohols, ketones, and amines changed from smooth to stick-slip sliding at their melting point [5,18]. For fatty acids, this transition coincided with the melting point of the metal carboxylate instead of that of the fatty acid, and this was in some cases interpreted as that a chemical reaction had taken place on the surface. However, later investigations revealed that a more likely reason for the transition was the changes in the orientation of physisorbed molecules [19]. It was shown that the amount that was chemisorbed on steel from an alkane solution was similar for oleic, linoleic, and linolenic acids [15]. The different total adsorbed amounts [15] therefore arose from additional physisorption due to interactions between the hydrocarbon parts of the molecules, and only a fraction of the molecules on the surface was chemisorbed. In an alkane solution, coadsorption of the solvent is also expected [20,21], and chain matching is important for both the tribological performance and bulk properties of the mixtures of fatty acids and alkanes [6,22]. For example, the load-carrying ability of a fatty acid peaks when its chain length is matched by that of the solvent [6].

17.1.2 Effects of Unsaturation on Lubricant Film Formation from Model Base Oil

Investigations at the nanoscopic and microscopic scales have shown that the friction of stearic, oleic, and linoleic acids on steel and other surfaces increased with increasing unsaturation [12,15,16]. The same trend has also been seen at the macroscopic scale, where the friction coefficient measured in a four-ball test apparatus increased [23] and the load-carrying abilities decreased [6] when going from stearic to oleic acid (in n-hexadecane), and it was suggested that the double bond prevented the oleic acid from forming a close-packed layer on the surface. In ball-on-plate experiments, stearic acid (dissolved in sunflower oil) gave a lower friction coefficient and better wear protection than oleic and linoleic acids [24]. However, another investigation of the friction and wear of steel surfaces in the presence of oleic, linoleic, and linolenic acids in *n*-hexadecane showed that the performance of the three systems was similar, possibly due to the severe test conditions of very high load and low sliding velocity [25]. In a different study, the friction coefficient was found to decrease with increasing degree of unsaturation [26].

The friction and wear of a surface in the presence of an adsorbed layer of fatty acid is thus expected to depend not only on the structure of the hydrocarbon chain, but also on the attachment of the layer to the surface. In order to understand the influence of double bonds, without the complication of the wear typically observed in macroscopic systems or the presence of a fraction of chemisorbed material, we have studied well-defined single-asperity contacts in a surface force apparatus (SFA) with attachments for tribological measurements [27]. In Sections 17.3.1 and 17.3.2, we report on the layer thickness, the friction, and the load-carrying capabilities of one saturated and two unsaturated fatty acids and a rosin acid, physisorbed on mica from a model base oil, *n*-hexadecane. One of the unsaturated fatty acids, linoleic acid, formed a different film structure with good stability at high loads and gave a more complex friction response than the other systems.

17.1.3 EFFECTS OF ADHESION ON FRICTION OF MONOLAYERS ON STEEL

Many previous studies have indicated that the magnitude and the shape of the friction force versus the load curve obtained for a single-asperity contact depend on the strength of the adhesion. For a variety of systems, the friction force F_f in an adhesive contact increased in proportion to the contact area at low loads (adhesion-controlled friction) [28–33], and there was a finite friction force at zero applied external load ($F_n = 0$). In contrast, systems with low or no adhesion typically showed a linear increase in friction with load (load-controlled friction) at low loads, with $F_f = 0$ at $F_n = 0$ [31,34–36]. Systems that are adhesion controlled at low loads often became load controlled at high loads. These experimental observations have been generally expressed as [31]

$$F_f = S_c A + \mu F_n, \tag{17.1}$$

where S_c is the critical shear stress (a constant) and μ is the friction coefficient. This empirical equation can also be written as *shear strength* [13] by dividing it by the contact area A to obtain $S = S_c + \mu p$, where p is the pressure F_n/A. The friction behavior expressed in the empirical Equation 17.1 is not yet well understood despite having been observed and modeled for a long time (cf. [28,31,34–36]). In one of the simplest models, only S_c depends on the adhesion, so that the area dependence of the friction force would be strongly reduced, even to the point of complete removal, if the adhesion (or interfacial energy γ) was sufficiently reduced. As a result, only the term μF_n in Equation 17.1 would remain, and such a linear load-dependent friction has been experimentally found in systems where the interaction between the surfaces was purely repulsive [31,34]. It has also been seen in systems with low adhesion, $\gamma \approx 1$–4 mJ/m^2 [35,36], where very similar friction coefficients were obtained in atomic force microscopy (AFM) and SFA experiments, although the radius of curvature R differed by about six orders of magnitude [35].

Other models assume that F_f is always proportional to A [37,38], and since the area of an elastically deforming single-asperity contact increases nonlinearly with F_n [39–46], a linear load dependence may be explained as resulting from a pressure-dependent shear stress. While a pressure dependence of the shear stress is possible, it is not likely to be the complete explanation for the linear load dependence seen in many different types of systems (cf. [31,35,36,47,48]). Recent computer simulations [49–51] have suggested that within the nominal contact area, the size of the real molecular contact area as well as its increase with increasing load are adhesion dependent, but in a different manner than that observed for nominal or macroscopic contacts.

The work presented in Sections 17.3.3 through 17.3.6 aims to address these commonly observed phenomena by studying the formation of self-assembled fatty acid films and their friction both at dry conditions and when immersed in a liquid [16]. In those sections, we focus on systems where only one surface carries a fatty acid layer, formed by adsorption from n-hexadecane solution on a steel-coated silicon wafer. The steel surface was of a type previously used in quartz crystal microbalance (QCM) measurements of the adsorbed amount, allowing us to work on surfaces with a known coverage [15]. The sample was removed from the adsorption solution, rinsed with water and dried, and probed with an opposing surface of unmodified silicon oxide. The technique used in this case is friction force microscopy [52–54], which is based on AFM.

To investigate the dependence of friction on the adhesion, the adsorbed layers were studied at two conditions: in dry N_2 gas (high adhesion) and immersed in ethanol (very low adhesion). When the adhesion was kept low, we observed load-dependent friction ($F_f = \mu F_n$), whereas at higher adhesion, F_f sublinearly increased with F_n. The F_f versus F data obtained at low F_n in dry N_2 were compared to a function of the form $F_f = S_c A$, where the contact area A was calculated from the thin-coating contact mechanics (TCCM) model [45,46]. This recently developed model for a macroscopic contact describes the deformation of a thin film confined between rigid substrates. We found that once the area dependence in N_2 was accounted for by using the TCCM model, the lowest friction (expressed as μ and S_c) was obtained in the most close-packed system, stearic acid, whereas the unsaturated systems showed significantly higher friction coefficients and critical shear stresses.

17.2 MATERIALS AND METHODS

17.2.1 MONOLAYER FORMATION

Monolayers were formed from solutions of stearic acid (C18:0, octadecanoic acid, ≥99.5%), oleic acid (C18:1, cis-9,10-octadecenoic acid, 99.5%), and linoleic acid (C18:2, cis,cis-9,12-octadecadienoic acid, 99%) in n-hexadecane (99%). All were obtained from Sigma-Aldrich and used without further purification. The fatty acids were stored at −15°C and care was taken to avoid oxidation by adding an N_2 atmosphere to opened bottles and, in the case of linoleic acid, by working from newly opened bottles. Monolayers were also formed from a rosin acid, dehydroabietic acid (8,11,13-abietatrien-18-oic acid) with a purity of 98% ± 1.5% as determined by gas chromatography, synthesized by us from disproportioned rosin (Arizona Chemical) according to the procedure described by Hedman [55].

The concentrations, 0.05 wt% (stearic acid), 1.0 wt% (oleic and linoleic acids), and 0.1 wt% (dehydroabietic acid) solutions in n-hexadecane, were chosen based on previous observations of plateaus in the adsorption of fatty acids on various types of surfaces from nonaqueous solutions [7, 14,15,23,56–59]. Since previous work has shown that the adsorption of fatty acids is sensitive to the presence of water [7,15], the solutions for the SFA experiments were dried over molecular sieves (3 Å, 8–12 mesh, Tamro Med-Lab) for 24 h. The solutions were then filtered through an Acrodisc CR syringe filter with polytetrafluoroethylene membrane (pore size 0.2 μm, Pall [Life Sciences]) when injected between the mica surfaces in the SFA. The muscovite mica surfaces were prepared according to standard procedures for SFA work [60].

The steel surfaces were prepared by QuartzPRO (Järfälla, Sweden) by sputtering Swedish standard steel no. 2343 in an argon atmosphere to form a 500 Å thick layer on the top electrode of quartz crystals for a QCM for experiments described by Lundgren et al. [15], and on silicon wafers for the AFM measurements [16]. The composition and properties of the resulting steel surfaces have been described in ref. [15]: The surface of the evaporated steel on silicon wafer contains 65% oxygen as oxides and 16% iron (other elements being silicon, nickel, chromium, and molybdenum), as determined by X-ray photoelectron spectroscopy (spot size ca. 1 mm^2). Its roughness was 0.3 nm rms over 1 μm × 1 μm, as determined from AFM images, and the diameter of individual grains on the steel surface on silicon wafer was ca. 20 nm (slightly larger than the estimated diameter of

TABLE 17.1

Adsorbed Amounts on Steel, Molecular Areas, Layer Thicknesses *h*, and Contact Angles

System	Conc. (wt%)	Ads. Amount (mg/m^2)	Molec. Area (nm^2)	*h* (nm)	Ads. Time (h)	θ_{adv} (°)	θ_{rec} (°)
Bare steel						54 ± 2	24 ± 6
Stearic acid	0.05	1.2	0.4	1.3	1	82 ± 3	50 ± 7
					24	81 ± 3	48 ± 5
Oleic acid	1.0	0.4[a]	1.2[a]	0.4	1	84 ± 2	57 ± 5
					25	89 ± 1	56 ± 7
Linoleic acid	1.0	1.1[a]	0.4[a]	1.2	1	78 ± 2	60 ± 5
					24	85 ± 1	60 ± 3
Dehydroabietic acid	0.10	0.9	0.6	0.8	1	72 ± 2	43 ± 4
					24	75 ± 4	52 ± 6

[a] From S. M. Lundgren et al., *Langmuir*, 23, 10598–10602, 2007; *Langmuir*, 24, 9922, 2008.

the contact area between the AFM tip and the surface). The different metals may form different domains of metal oxides inside a grain and at its boundaries. However, our friction measurements involve sliding a distance of 1 μm across a quite smooth surface composed of many such grains, and no distinct effects were seen at the boundaries on this scale. The steel-covered substrates were cleaned in 2% (v/v) Hellmanex for 1 h, rinsed with water and ethanol, and blown dry with N_2. Right before immersion in the adsorption solution, they were cleaned with air plasma for 5 min on low setting (6.8 W) in a Harrick PDC-32G plasma cleaner. The monolayers were formed by immersion for 1 h or 24 h in adsorption solutions (cf. Table 17.1) that had been dried over molecular sieves for a minimum of 12 h. The samples were removed from the solution, and the excess solution was removed by touching the edge of the substrate with a paper towel. The samples were then rinsed with distilled water and dried with a stream of N_2 gas. For the AFM experiments, the samples were mounted in the instrument and dried in an N_2 atmosphere for at least 1 h, or immersed in ethanol. Layers prepared in this manner may still contain small amounts of hexadecane, which may dissolve in ethanol. However, the polar group of the fatty acids adsorbs on steel, and it has been shown that covalently bound fatty acids are not removed from steel surfaces by ethanol [15]. Advancing and receding contact angles of water (Table 17.1) were measured with a ramé-hart contact angle goniometer (model 100-00 230) (Mountain Lakes, New Jersey).

17.2.2 NORMAL FORCE AND FRICTION FORCE MEASUREMENTS WITH SFA

The measurements of normal and lateral (friction) forces were performed with an SFA3 apparatus [61] (SurForce Corp., Santa Barbara, California) with bimorph and friction device attachments [62]. In this technique, which has been described in detail elsewhere [61–63], the separation distance or film thickness *D* between two atomically smooth mica substrates is measured with an accuracy of 1–2 Å by multiple-beam interferometry [64]. The substrates are placed in a crossed-cylinder configuration, which at small *D* is equivalent to a sphere interacting with a flat surface. The optical interference fringes can also be used to measure the radius of curvature *R*, which in this study was in the range 0.7–2.2 cm. The normal force F_n, which is normalized by *R* to allow comparison between different experiments, can be measured to better than 10^{-7} N by monitoring the deflection of a double cantilever leaf spring (spring constant 260 N/m) on which the lower surface is mounted. For the measurements of normal force, the lower surface is moved vertically toward or away from the upper by moving the base of the leaf spring. The rate of approach and separation of the surfaces was constant, ca. 3 Å/s, in all experiments. To induce sliding, the lower surface is moved laterally at a constant speed by applying a triangular waveform to the bimorph device [62]. The friction force F_f between the lower and upper surfaces is obtained by measuring the deflection of a double cantilever leaf spring (spring constant 3200 N/m) of a friction device on which the upper surface is mounted. The sensitivity of this device, which is equipped with semiconductor strain gauges, is approximately 50 μN. The electronic signals were collected using a program written in LabVIEW.

The interior of the instrument and the surfaces were dried by allowing the instrument chamber to equilibrate with P_2O_5, which was frequently replaced during the course of the experiment. The wavelength positions of the interference fringes with the mica surfaces in contact with dry air were taken as $D = 0$. After this measurement, the surfaces were separated and a large drop of filtered, dry solution was injected. The fatty acids were allowed to adsorb for 1 or 24 h at a temperature of 25–26°C. Slight differences were seen in the normal forces of stearic and linoleic acids at 1 and 24 h of adsorption, and in the linoleic acid system, the friction force was slightly higher at the longer adsorption time. The friction experiments were done at two different sliding speeds, 0.4 and 4 μm/s. These were chosen since at speeds of 0.4 μm/s and higher, only smooth sliding with no stick-slip motion has been observed for n-hexadecane [65], the solvent in our systems.

17.2.3 NORMAL FORCE AND FRICTION FORCE MEASUREMENTS WITH AFM

The friction force F_f was measured with a NanoScope IIIa Multimode AFM (Bruker), using unfunctionalized Si tips with a native silicon oxide surface (MikroMasch). The tips were plasma cleaned for 20 s at medium setting before use. The normal and lateral spring constants k_N and k_l, respectively, of the rectangular cantilevers were calculated from their dimensions as measured with scanning electron microscopy. The values of k_N were in the range 0.06–0.09 N/m and k_l in the range 8–10 N/m. The tip radii R were obtained by reverse imaging of a calibration sample (TGT01, MikroMasch), and were 21 and 58 nm. The calculations of the spring constants [66,67] and the calibration of the signal from the AFM have been previously described in detail [35,36]. The friction measurements were done either in dry N_2 gas (Air Liquide, purity 99.9%), in a Plexiglass environmental chamber enclosing the AFM, or in ethanol (Primalco Etax, Finland, 99.7%, used as received) in a fluid cell. The relative humidity in N_2, measured with a Vaisala HMI 31 humidity meter, was <1.5% after purging for 1 h. No layering or oscillatory forces were seen in the ethanol, in agreement with previous studies [35,36]. In this study, the F_f versus F_n data were taken over a scan size of 1 μm with a sliding velocity $v = 3$ μm/s.

17.2.4 THE TCCM MODEL

In AFM friction experiments on self-assembled monolayers, the radius of the contact area at low loads can be comparable to the thickness of the monolayers, whose elastic modulus is expected to be 10–50 times lower than the bulk modulus of the confining surfaces (cf. Ref. [16]). The initial deformations in such a system are thus expected to occur in the soft, thin film. At increased loads, the effective stiffness of the system is still lower than that of the substrates, and the situation is more complex than that assumed in contact mechanics models for homogeneous bodies [39–41,68]. Several models [42–46] have been developed for the contact mechanics of a thin, elastic film confined between stiffer substrates. In nanometer-sized contacts, where the atomic structure of the substrate is expected to affect the pressure distribution [69], it is not yet established how well such models apply, although the influence of the substrate is reduced when a molecularly thin film is introduced [51].

In the TCCM model [45,46], the indenter and the flat substrate are assumed to be rigid; i.e., the deformation is allowed to occur only in the confined, thin film. The model describes two limiting cases [45]: a Derjaguin–Muller–Toporov (DMT)-like response, in systems where the range of adhesion is large compared to the elastic deformations [41]; and a Johnson–Kendall–Roberts (JKR)-like case, which describes deformations that are large compared to the range of the adhesion [40]. JKR-like deformations of a thin film have also been previously described for elastically deforming probe and substrate [42–44]. For the DMT-like contact, finite element simulations indicate that the rigid indenter and the substrate model can be used when the uniaxial compression modulus E_u of the thin film is <5% of the indenter and the substrate Young's modulus [45]. A recent extension [46] of the model describes a continuous transition between the DMT- and JKR-like cases. This model has been successfully applied to AFM friction force data [70] and compared with a molecular dynamics simulation [71] of confined alkylsilane monolayer.

We compare the F_f versus F_n data in our adhesive systems (discussed in Section 17.3.5) to $F_f = S_c A$, where S_c is a constant, the critical shear stress, and A is the contact area at a given load F_n. The relationship between F_n, radius of the contact area (a), and work of adhesion (W) in the extended TCCM model is given in nondimensional form by [45,46]

$$\overline{F}_n = \frac{\pi}{4}\overline{a}^4 - \zeta^{1/2}\pi\overline{a}^2(2\overline{W})^{1/2} - 2\pi\overline{W}(1-\zeta), \tag{17.2}$$

where $\overline{F}_n = F_n/(E_u Rh)$, $\overline{a} = a/\left(\sqrt{Rh}\right)$, and $\overline{W} = W/(E_u h)$. The uniaxial strain modulus is $E_u = E(1-\nu)/[(1+\nu)(1-2\nu)]$, where E is Young's modulus (in this case, 0.1 and 0.6 GPa) and ν is Poisson's ratio (0.4). The film thickness h is the thickness of the monolayer (cf. Table 17.1). The transition parameter is ζ ($0 \leq \zeta \leq 1$), a measure of the ratio of the elastic deformation to the effective range of the surface forces, with $\zeta = 0$ corresponding to the DMT-like and $\zeta = 1$ to the JKR-like limit of the TCCM model [45]. Values of the interfacial energies γ_{TCCM} ($= W/2$) will be given in Section 17.3.5. The propagation of uncertainties to the values of S_c and ζ is discussed by Yang and Ruths [72]. In the current work, the uncertainty is ca. 20% in S_c and 25% in ζ.

17.3 RESULTS

17.3.1 NORMAL FORCES ACROSS FATTY ACID FILMS IN HEXADECANE

The force normal to the surfaces (normalized by the radius of curvature R) F_n/R, measured with the SFA for stearic, oleic, and linoleic acid in n-hexadecane, is shown as a function of separation distance or film thickness D in Figure 17.1. In the stearic acid system (Figure 17.1a), there was an onset of a weak repulsive force at a film thickness of $D \approx 75$ Å, followed by a small transition (marked with arrows pointing left in Figure 17.1a) from $D = 60$ Å to 54–55 Å. The force at this transition was $F_n/R \approx 5$ mN/m after 1 h of adsorption and lower, $F_n/R \approx 1$ mN/m, after 24 h. The same film thickness of 54 Å was gradually reached on compression in the oleic acid system (Figure 17.1b), without the transition seen for stearic acid. On the separation of the surfaces, the compressed oleic acid film was seen to expand back as the pressure was released, so that the jump apart of the surfaces occurred from $D = 63–69$ Å, whereas the stearic acid system separated from the hard-wall film thickness of 54 Å. The adhesion was weak, $F_n/R \approx -1$ to -2 mN/m in both systems, and the overall appearance of the force curves was similar to that observed in previous work on palmitic acid and hexadecylamine adsorbed on mica from n-tetradecane [73,74], and several previous studies of surfactant monolayers interacting across organic solvents [75–77]. The film thicknesses will be discussed in detail below. The forces were different from the ones in the pure, dry n-hexadecane, which are known to be oscillatory with a period corresponding to the cross-sectional diameter (4–5 Å) of the alkane chain [78]. The normal forces in pure n-hexadecane were not determined in detail in this study, but the characteristic molecular film thicknesses in this system were observed in the friction measurements (cf. Figure 17.2a).

In the linoleic acid system (Figure 17.1c), a weak repulsion was also observed starting at $D \approx 70$ Å, i.e., at a film thickness similar to the onset of forces in stearic and oleic acids. However, as the compression was increased to $F_n/R = 1–6$ mN/m (Hertzian pressure at midpoint of contact: $p_0 = 0.4–0.7$ MPa), there was a transition to a film thickness of 34 Å, and at slightly higher compression, a very stable film thickness of 28–32 Å was found. The transition from the outer maximum occurred at $F_n/R = 4–6$ mN/m after an adsorption time of 1 h, and at a lower force, $F_n/R = 1–1.5$ mN/m, after

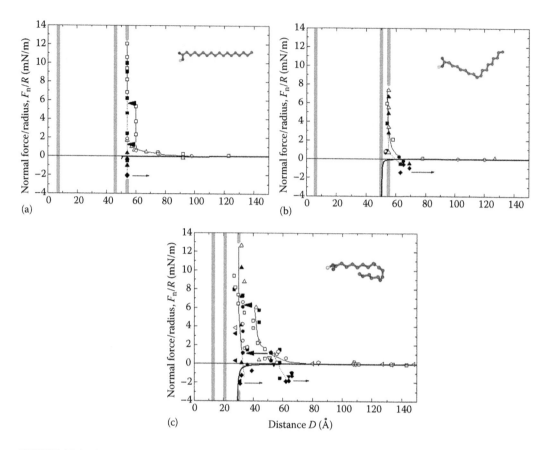

FIGURE 17.1 SFA measurements of the normal force divided by the radius of curvature, F_n/R, as a function of the separation distance or film thickness D between mica surfaces separated by dry n-hexadecane solutions of (a) stearic, (b) oleic, and (c) linoleic acid. Open and solid symbols indicate compression and separation, respectively, and their shapes indicate the adsorption time: 1 h ($\square$, $\triangle$) or 24 h ($\bigcirc$, $\triangledown$, $\triangleleft$, $\bigcirc$). Thin black curves are intended as guides for the eye. Film thickness transitions on compression are indicated by arrows pointing to the left, and jumps apart of the surfaces from adhesive minima are indicated by arrows pointing to the right (from jump-off points indicated with $\blacklozenge$). Thick gray vertical lines indicate other stable film thicknesses observed in the friction measurements at higher pressure (cf. text and Figure 17.2). Thick black curves indicate van der Waals–Lifshitz forces calculated for five-layer systems as described in the text. (Reprinted from *J. Colloid Interface Sci.*, 326, Lundgren, S. M., M. Ruths, K. Danerlöv, and K. Persson, Effects of unsaturation on film structure and friction of fatty acids in a model base oil, 530–536, Copyright (2008), with permission from Elsevier.)

adsorption for 24 h. The outer minimum in the force curve at $D = 62$–66 Å was quite similar in magnitude to the inner one at $D = 31$–39 Å.

At the higher pressures reached at large normal forces (loads) in the friction experiments discussed below, smaller film thicknesses were seen in all systems. These film thicknesses are indicated with vertical gray lines in Figure 17.1. In the stearic and oleic acid systems, there was a gradual compression from $D = 54$–55 Å to a very stable film thickness of 45 and 50 Å, respectively, and at very high loads (pressure $p_0 \geq 3.5$ MPa), an abrupt transition to a film thickness of ca. 6 Å, accompanied by a change in the friction response (cf. Figure 17.2a). In the linoleic acid system, there was a gradual transition from $D \approx 30$ to ca. 22 Å, followed by an abrupt transition to a final film with a thickness of 12–15 Å, which was not removed even at the highest loads in our experiment ($p_0 \geq 5$ MPa). No discernible differences were found in the normal forces measured before and after lateral sliding, as long as the surfaces remained undamaged.

17.3.2 Friction Forces across Fatty Acid Films in Hexadecane

The friction forces F_f measured with the SFA as a function of load (normal force) F_n are shown in Figure 17.2. The film thicknesses in different ranges of load are indicated with gray vertical lines in Figure 17.1 and also marked in Figure 17.2. The film thickness was always uniform throughout the whole contact area, and all systems showed smooth sliding in the investigated range of load. Wear of the mica surface was sometimes observed in the stearic and oleic acid systems at very high loads, after the transition to a film thickness of ca. 6 Å ($p_0 \geq 3.5$ MPa). The applied energy at this transition (calculated from F_n/R, with a molecular area of 50 Å^2) is 6 kcal/mol (2×10^{-20} J/molecule), which is within the typical range expected for physisorbed fatty acid molecules [14,19]. All the data shown in Figure 17.2 represent wearless friction, i.e., before any damage occurred.

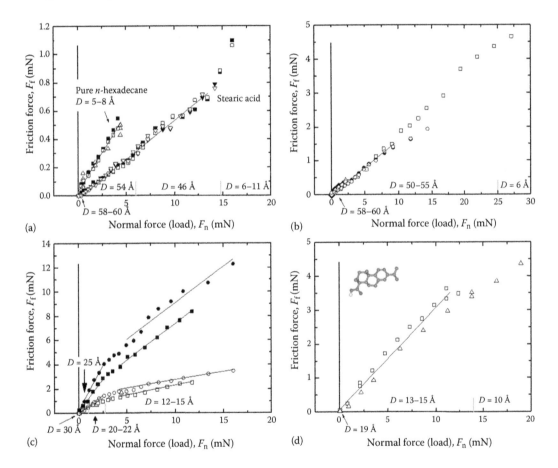

FIGURE 17.2 SFA measurements of the friction force F_f as a function of normal force or load F_n, for (a) pure *n*-hexadecane and stearic acid, (b) oleic acid, and (c) linoleic acid, i.e., the systems in Figure 17.1, and (d) a rosin acid, dehydroabietic acid, adsorbed from dry *n*-hexadecane solution on mica. The measured film thicknesses are indicated by arrows or given between gray tick marks on the load axes, and the corresponding pressures are given in the text. The friction force of pure, dry *n*-hexadecane with a film thickness of 5–8 Å is also shown in panel a. All systems show linear friction force versus load, with no dependence on the radius of curvature R. Open and solid symbols indicate sliding speeds of 0.4 and 4 μm/s, respectively, and their shapes indicate the adsorption time: 1 h (□, △, ◇) or 24 h (○, ▽, ◁, ◯). Note the different scales on the *y*-axes. The friction coefficients μ = F_f/F_n (or $\Delta F_f/\Delta F_n$, in cases where the fitted line does not go through the origin) are listed in Table 17.2. (Reprinted from *J. Colloid Interface Sci.*, 326, Lundgren, S. M., M. Ruths, K. Danerlöv, and K. Persson, Effects of unsaturation on film structure and friction of fatty acids in a model base oil, 530–536, Copyright (2008), with permission from Elsevier.)

All these systems showed load-controlled friction, i.e., a linear dependence of F_f on F_n, and F_f was very small at $F_n = 0$, except for in the case of n-hexadecane, where a small but finite friction force was measured at $F_n = 0$. This was expected [47], since the adhesion [78] in this system at the film thickness investigated in Figure 17.2a was larger than in the fatty acid systems. In pure n-hexadecane (Figure 17.2a), there was no effect of equilibration time (1 or 24 h) or of sliding velocity (0.4 or 4 µm/s, shown as open and solid symbols). Our measured value of the friction coefficient µ in n-hexadecane (Table 17.2) was in good agreement with a previous result on kinetic friction in SFA experiments at a sliding speed of 0.4 µm/s (µ = 0.12) [65,79]. Higher friction coefficients have been observed in n-tetradecane and n-hexadecane at different experimental conditions [80–82]. The low friction coefficient we observed in the stearic acid system (Table 17.2) was in good agreement with previous SFA work on palmitic acid adsorbed from n-tetradecane, where a friction coefficient of 0.05 ± 0.01 can be estimated from the two loads investigated ($F_n \leq 3$ mN) [73]. At the transition to a film thickness of 6–11 Å at high loads in the stearic acid system (Figure 17.2a), the friction coefficient increased to a value similar to that for pure n-hexadecane. Oleic and linoleic acids gave higher friction coefficients than n-hexadecane and stearic acid (Table 17.2). The linoleic acid system (Figure 17.2c) showed a complex friction response with a decrease in the friction coefficient at the film thickness transition to $D = 12$–15 Å, a higher friction at the larger sliding speed, and also a slightly higher friction after adsorption for 24 h. These features were not observed in the other systems.

Fatty acid blends for lubrication applications also contain components other than the unsaturated fatty acids discussed previously. Commonly, they contain rosin acids, which are by-products of the papermaking process. One typical component is dehydroabietic acid, which we studied in a manner similar to that done for the fatty acids. No normal forces were observed in this system until a repulsion at $D = 19$–20 Å (not shown). A weak adhesion was seen on separation, similar to the one in the stearic and oleic acid systems. A small increase in pressure to $p_0 \geq 1$ MPa reduced the total film thickness to 13–15 Å, which was stable up to a pressure of 5–7 MPa, i.e., to higher pressure than the stearic and oleic acid films. Comparison of the measured film thickness to the shape of the dehydroabietic acid molecule (cf. Figure 17.2d) indicated that only a monolayer formed on each surface. We expect the monolayers of this molecule to be loose packed because of the irregular shape of the molecule. The friction of this compound was higher than that of the fatty acids at comparable conditions (Figure 17.2d and Table 17.2).

TABLE 17.2

SFA Friction Measurements: Typical Film Thicknesses D, Adsorption Times, and Friction Coefficients µ

System	D (Å)	Ads. Time (h)	μ^a $v = 0.4$ µm/s	$v = 4$ µm/s
n-Hexadecane	5–8	1, 24	0.11 ± 0.01[b]	
Stearic acid	46–54	1, 24	0.055 ± 0.005	
Oleic acid	50–55	1, 24	0.16 ± 0.01	
Linoleic acid	20–25	1	0.50 ± 0.02	1.1 ± 0.1
		24	0.58 ± 0.03	1.5 ± 0.1
	12–15	1	0.16 ± 0.01[b]	0.60 ± 0.02[b]
		24	0.14 ± 0.01[b]	0.60 ± 0.05[b]
Dehydroabietic acid	13–15	1	0.30 ± 0.01	

[a] The standard deviations are from the linear fits in Figure 17.2.
[b] In these cases, F_f does not pass through the origin and µ was therefore defined as $\Delta F_f / \Delta F_n$ instead of F_f / F_n.

17.3.3 Adsorbed Amount and Monolayer Structure on Steel

The amounts of fatty acid and rosin acid adsorbed from *n*-hexadecane onto evaporated steel surfaces were determined by QCM measurements as described by Lundgren et al. [15]. The solution concentration, the adsorbed amount at adsorption equilibrium, and the corresponding area per molecule are given in Table 17.1. Also listed are the calculated film thicknesses h of such adsorbed layers after removal of the solvent by rinsing and drying (cf. Section 17.2), which are the relevant thicknesses of the dry layers in the contact mechanics calculations discussed later.

The amounts adsorbed from *n*-hexadecane (Table 17.1) are not large enough to form close-packed layer structures on steel, except in the case of the linoleic acid. This agrees with other reports on adsorption of fatty acids from nonaqueous solutions [14]. When comparing our molecular areas and calculated film thicknesses to the closest packing [83,84] obtained in LB films, 0.24, 0.48, and 0.47 nm^2 [83], for stearic, oleic, and linoleic acids, respectively, the adsorbed amounts of stearic and oleic acids are about one-half to one-third of that in a close-packed monolayer. In the dry state, the molecules must be strongly tilted, so that the stearic acid would have a tilt angle of about 60° from the surface normal. The oleic acid molecule is bent and cannot form as close-packed a structure as stearic acid. Its calculated dry layer thickness is particularly low and similar to the cross-sectional diameter of an alkane chain (0.4–0.5 nm). Linoleic acid, with double bonds at the 9 and 12 positions, can attain a variety of folded structures [85], where the U-shaped ones would have a molecular area of about twice that of a straight chain.

Advancing and receding contact angles θ of water were measured on layers formed after different adsorption times on the steel-covered Si wafer. The advancing contact angles of water on these monolayers on steel (Table 17.1) were lower than the values above 100° seen on close-packed structures, e.g., on stearic acid adsorbed from hexadecane on Ag, Cu, and Al surfaces [86], and on palmitic acid on stainless steel where a voltage was applied to facilitate the adsorption of a close-packed layer [87]. A contact angle of 80° was reported on stearic acid on Cu [88], where it was noted that the monolayer thickness 2.3 nm did not correspond to maximum close packing. The large contact angle hysteresis in our systems suggests low coverage and rearrangements in the layers during the contact with water. AFM height traces and friction loops measured in 1 μm scans at different positions on the surfaces to obtain the friction results presented below did not reveal any domains of different coverage, suggesting that although the adsorption density is low, the fatty acid and rosin acid layers may evenly cover the steel substrate when adsorbed from *n*-hexadecane, which contrasts with many LB-deposited layers that form physisorbed, close-packed domains on various substrates.

17.3.4 Tip–Monolayer Adhesion in Ethanol and in Dry N₂

In ethanol, the interaction between the AFM tip and the adsorbed monolayers was very weak. For simplicity, we determined the interfacial energy γ from the pull-off forces $F_{n,adh}$ in normal force versus separation curves obtained with the AFM (not shown). Assuming $\gamma_{JKR} = -F_{n,adh}/3\pi R$ (based on JKR contact mechanics, a possible overestimation of at most 20%), we obtained experimental values of $\gamma < 1$ mN/m (Table 17.3). The low values of γ were, as expected, similar to the data obtained in investigations of smoother surfaces carrying CH_3-terminated alkanethiol and alkylsilane monolayers in ethanol, $\gamma = 3.5$ mN/m [35,89], and previous investigations of aromatic thiols and octadecanethiol in ethanol, $\gamma < 1$–4 mN/m [36].

The measured values were compared to γ_{vdW} based on van der Waals–Lifshitz theory [90]. Our γ_{JKR} values were lower than that of γ_{vdW} at contact (Table 17.3) calculated for three-layer systems (steel/ethanol/amorphous silicon dioxide or monolayer/ethanol/amorphous silicon dioxide) from the dielectric properties of the interacting materials [91,92] and using a cutoff distance of 0.165 nm [90]. Several previous studies, including ones on highly close-packed systems where the monolayer

TABLE 17.3
AFM Friction Experiments in Ethanol (Low Adhesion), R = 58 nm[a]

System	$F_{n,adh}$ (nN)	γ_{JKR} (mN/m)	γ_{vdW} (mN/m)	μ[b]
Bare steel	−0.5	0.9	4.7	0.60 ± 0.02
Stearic acid	−0.4	0.7	1.8	0.22 ± 0.01
Oleic acid	−0.4	0.7	2.1	0.48 ± 0.01
Linoleic acid	−0.5	0.9	2.2	0.72 ± 0.01
Dehydroabietic acid	−0.5	0.9	3.5	0.72 ± 0.03

[a] R, tip radius; $F_{n,adh}$, pull-off force from AFM force curves; γ_{JKR}, surface energy according to the JKR model; γ_{vdW}, surface energy according to van der Waals–Lifshitz theory for three-layer systems; μ, friction coefficient.

[b] The standard deviations are from the linear fits in Figure 17.3a.

structure could be imaged at different conditions, have suggested that the adsorbed monolayers and a sharp AFM tip are not separated by a complete monomolecular film of ethanol [35,36].

In the experiments performed in dry N_2 gas (relative humidity < 1.5%), the adhesion was always significantly larger than in ethanol, as expected from the difference in dielectric properties. Values of the pull-off forces obtained from the normal force versus separation curves ($F_{n,adh}$) and the jump-off loads in the friction measurements ($F_{n,adh,fr}$, lowest loads in each data set in Figure 17.4) are listed in Table 17.4. In most cases, we found a close agreement between these values, although premature jumps off can occur in the friction measurements, for example, if the tip is scanned at high velocity over a small asperity or dip on the surface. The values of the interfacial energy in these systems, γ_{TCCM}, were obtained from fits of the contact mechanics model (Table 17.4 and the curves in Figure 17.4). There was a good agreement between these values and γ_{vdW} calculated for a three-layer system (monolayer/dry N_2 atmosphere/amorphous silicon dioxide) for most of the tip-and-monolayer combinations (Table 17.4).

17.3.5 FRICTION FORCES BETWEEN FATTY ACID MONOLAYERS AND A SILICA TIP

Different types of friction responses were observed at low and at moderately high adhesion (Figures 17.3 and 17.4, respectively). The friction at low load (before the transition or plateau) was very reproducible in each system, whereas the region after the transition (at high loads) would occasionally differ between samples of the same composition. The discussion of friction below concerns only the very reproducible regions at low loads, before the transition or plateau.

In ethanol, all systems showed load-dependent friction (Figure 17.3), in good agreement with many previous investigations (see Section 17.1). The friction coefficients obtained from linear fits to the low-load regimes in Figure 17.3a are listed in Table 17.3. At higher adhesion (Figure 17.4), the term S_cA in Equation 17.1 is expected to dominate at low loads, causing the friction force to initially increase less strongly than linearly [31,34]. Areas were calculated with the TCCM model for two values of the elastic modulus (see Section 17.2) and multiplied by a constant S_c (Table 17.4) to obtain curves $F_f = S_cA$ to fit the low-load regimes in Figure 17.4. The dry layer thickness of the monolayer h in Table 17.1 was used. The results obtained with the two tip radii gave very similar S_c values, suggesting that the data in N_2 gas are examples of area-dependent friction. Note that since one actually calculates load as a function of contact radius a, similarly to in the JKR model [40], the S_cA versus F_n curve has a parabolic shape and there is an experimentally inaccessible region of parameter space (below the experimental data points in Figure 17.4).

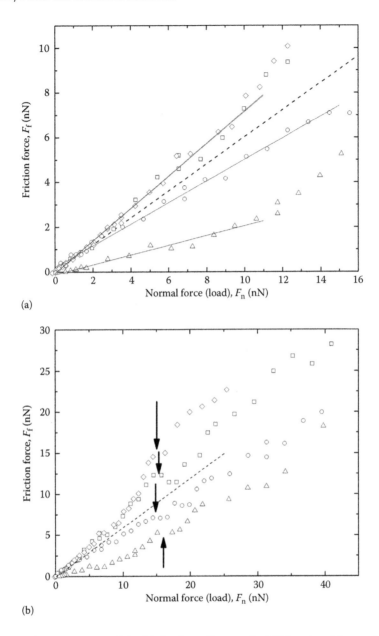

FIGURE 17.3 AFM measurements of the friction force F_f versus load F_n in ethanol for stearic ($\triangle$), oleic ($\bigcirc$), linoleic ($\diamond$), and dehydroabietic acid ($\square$) monolayers adsorbed on evaporated steel from n-hexadecane solution. The data were obtained with a Si tip with radius $R = 58$ nm. The dashed line indicates the friction force of an uncoated evaporated steel sample in ethanol, measured with the same tip. (a) Linear fits to the data at low load (μ given in Table 17.3). (b) Full range of investigated loads. The arrows indicate the onsets of plateaus or transition regimes in the data. (Reprinted with permission from M. Ruths et al., *Langmuir*, 24, 1509–1516. Copyright 2008 American Chemical Society.)

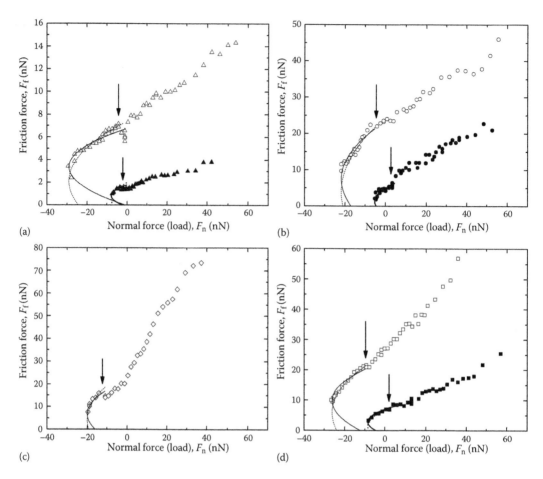

FIGURE 17.4 Friction force F_f versus load F_n in dry N_2 (relative humidity < 1.5%) for (a) stearic, (b) oleic, (c) linoleic, and (d) dehydroabietic acid on steel substrates. The data were obtained with two different Si tips: $R = 58$ nm (open symbols) and $R = 21$ nm (solid symbols). Arrows indicate the onset of transition regimes or plateaus in the F_f versus F_n data. The curves show $F_f = S_c A$ where the contact area A was calculated according to the TCCM model using a Young's modulus of $E = 0.1$ GPa (solid curves) or 0.6 GPa (dotted curves). The critical shear stress S_c (Table 17.4) was adjusted to obtain the best agreement with the data at lowest load, below the point indicated by the arrows. (Data reproduced with permission from M. Ruths et al., *Langmuir*, 24, 1509–1516. Copyright 2008 American Chemical Society.)

17.3.6 Monolayer Transitions at Larger Loads

In each data set in Figures 17.3b and 17.4, there is a transition regime or plateau. The onsets of the transitions are indicated with arrows. Such plateaus have been previously observed for alkanethiol monolayers on gold [93] and ascribed to a reversible displacement of the monolayer at high pressure so that the tip comes in contact with the substrate. Similar behavior has also been observed in mixed and single-component aromatic monolayers [35,36]. No plateau or transition regime was observed for the bare steel surface in ethanol (Figure 17.3); instead, the friction force linearly increased with load of up to over 20 nN. In all of the monolayer systems, the friction at high load is different from the one for a bare steel surface (Figure 17.3) and it is therefore unlikely that all of the molecules have been removed from the contact region. It should be noted that no damage to the monolayers could be detected after scanning within the investigated load range.

TABLE 17.4

AFM Friction Experiments in Dry N$_2$ (Higher Adhesion), TCCM Parameters

System	R (nm)	$F_{n,adh}$ (nN)	$F_{n,adh,fr}$ (nN)	E = 0.1 GPa		E = 0.6 GPa		γ_{vdW} (mN/m)
				γ_{TCCM} (mN/m)	S_c (MPa)	γ_{TCCM} (mN/m)	S_c (MPa)	
Stearic acid	21	−15.1	−7.5	30	8.5	30	32	28
	58	−31.1	−29.5	40	9.5	40	36	
Oleic acid	21	−5.2	−5.2	20	70	20	200	29
	58	−20.4	−21.8	30	92	30	290	
Linoleic acid	58	−15.7	−19.7	28	40	28	150	30
Dehydro-	21	−17.3	−8.2	32	45	32	140	37
abietic acid	58	−26.5	−26.5	36	50	36	180	

Note: R, tip radius; $F_{n,adh}$, pull-off force from AFM force curves; $F_{n,adh,fr}$, pull-off force from AFM friction curves; E = Young's modulus of monolayer; γ_{TCCM}, surface energy according to the TCCM model; γ_{vdW}, surface energy according to van der Waals–Lifshitz theory for three-layer systems.

In AFM measurements on LB-deposited monolayers of stearic acid and Cd stearate, the layer thickness decreased with increasing load, probably so that a tilt of the molecules was induced [10]. This gradual change started before any distinct transition regime was seen in the friction force [10]. It can be expected that such a gradual compression occurs in our systems as well. Since a constant value of *h* is used to calculate the TCCM curves, this compression is not taken into account, and for this reason, the curves in Figure 17.4 should be regarded as fits only to the data at the lowest load. LB-deposited layers showed an elastic response to about 35% compression before the transition point [10], although only the initial compression of about 10–12% appeared to be linearly elastic. Computer simulations [94] have suggested elastic compression of up to 20% for very close-packed alkyl monolayers. It is likely that the regime of linear elastic response is exceeded before the transition in the monolayer is reached. A changing region before the plateau may also be present at low adhesion (in ethanol), but cannot be identified in our data in Figure 17.3.

17.4 DISCUSSION

17.4.1 EXPECTED MOLECULAR CONFORMATIONS

Molecular dynamics simulations [85] at different temperatures have shown that the extended all-*trans* conformation of stearic acid found in the crystal is the state with the lowest energy also at higher temperatures in vacuum. Other low-energy conformers have few bonds in the gauche conformation. In the crystalline state, oleic [85,95] and linoleic acid [85,95,96] have a stretched (extended) conformation; i.e., the oleic acid is slightly bent with a *cis*-configuration at the double bond, and the central region of the linoleic acid has an S-like shape. However, above the melting temperature, low-energy conformers of linoleic and oleic acids were mostly U-shaped, although some low-energy structures with an extended conformation were also observed [85]. Rich [85] shows two U-shaped conformers of linoleic acid with energy of ca. 5 kcal/mol below that of the extended structure found in the crystal state. One U-shaped and one extended structure of oleic acid are also shown as low-energy states. Schematic drawings of the extended all-trans conformation of stearic acid, the extended structure of oleic acid, and a U-shaped conformation of linoleic acid (based on bond angles and lengths from Rich [85]) are shown as insets in Figure 17.1.

The extended length of stearic acid (calculated from bond lengths, bond angles, and van der Waals radii) is 26–27 Å, and the projected length of oleic acid is 24 Å [84]. X-ray measurements are typically reported as the size of a unit cell containing more than one molecule (stearic, oleic and linoleic acid all form some crystal structures containing dimers), and the measured spacings therefore include not only the chain lengths, but also a distance of ca. 5.6 Å representing the combined length of the hydrogen bond between dimers and space at the terminal –CH$_3$ groups [95]. X-ray data therefore indicate that each stearic acid molecule occupies a length of 24–25 Å in different crystal forms [97,98] and 26 Å in the neat liquid [99], and of this length, 21–23 Å is the end-to-end length of the stearic acid molecule itself. Similarly, oleic acid occupies 21–22 Å per molecule in the crystal [100], and 24 Å in the neat liquid [99], and the projected length of the molecule is 18–19 Å. The two U-shaped low-energy conformers shown for linoleic acid by Rich [85] have lengths of approximately 12 and 14 Å (based on bond lengths and angles from Rich's study [85]). X-ray data [95,101] on linoleic acid crystal polymorphs suggest that the extended molecular conformation occupies 21 Å (projected length of molecule 18 Å) and a smaller peak was visible at a spacing of 13.9 Å [101]. Calculations based on bond lengths and angles and semiempirical relationships suggest a length of *n*-hexadecane of 19–22 Å [65,90,96].

17.4.2 FILM STRUCTURE ON MICA SURFACES

Previous force and friction measurements on palmitic acid in *n*-tetradecane revealed a stable film thickness larger than the typical thickness of monolayers formed from surfactants with all-*trans* chains of 16 carbon atoms [73]. It is known that at moderate pressure, an alkane layer with a thickness of about 10 Å remains between surfaces covered with well-characterized surfactant layers with low-packing density [75]. Ruths et al. [73] concluded that such a layer (consisting of fatty acid and solvent) separated the loose-packed palmitic acid monolayers and penetrated into them to some extent. At the low range of loads investigated ($F_n \le 3$ mN), this extra layer was not removed [73]. The loose-packed monolayers themselves contained alkane molecules coadsorbed with the fatty acid, oriented along the fatty acid chain direction [20,21]. This is the likely structure of our stearic acid and oleic acid monolayers as well. Experiments on more close-packed monolayers containing all-*trans* chains of 18 carbon atoms have shown that the molecules typically have a 27° to 35° tilt with respect to the surface normal [10,102]. The resulting monolayer thickness is 19–22 Å, which is an upper limit for the thickness of our monolayers. Although it has been argued that a quite densely packed structure is formed when stearic acid is matched by *n*-hexadecane [6], the adsorbed amount of stearic acid is much lower than that of the systems in the works of Tsukruk et al. [10] and Laibinis et al. [102]. Our loose-packed monolayers were separated by a layer of fatty acid and *n*-hexadecane molecules with a thickness of about 10 Å, oriented parallel to the surfaces. Thus the transition (from $D = 60$ to 54 Å) in the stearic acid system was a removal of part of this layer. At higher pressures ($p_0 \approx 2$–3.5 MPa), this additional layer was gradually squeezed out or a larger tilt of the molecules in the adsorbed layer was induced until a very stable film with a thickness of ca. 45–50 Å was formed (gray vertical lines in Figure 17.1a and b). At the very highest loads in Figure 17.2a and b ($p_0 \ge 3.5$ MPa), there was a sudden transition to a film thickness corresponding to one or two cross-sectional diameters of an alkane chain (innermost gray vertical line at $D \approx 6$ Å); i.e., all remaining fatty acids or solvent molecules were then oriented parallel to the surfaces.

More complex film thickness transitions were seen in the linoleic acid system (Figure 17.1c). Two adhesive minima were readily detected at low normal forces, and the outer repulsive barrier showed a higher compressibility than the repulsion in the stearic and oleic acid systems. The origin of these forces was considered in some more detail: This type of force curve could arise from an inherently layered structure, of a type observed for dry neat 1-octanol and 1-undecanol confined between mica surfaces [103], or from a bending of extended (S-shaped) linoleic acid molecules into a U-shape as a result of the applied normal forces. The latter case was excluded, based on the following considerations: The energy needed to change the conformation of one linoleic acid molecule from the extended to the bent (U-shaped) state was assumed to be 5 kcal/mol (3.5×10^{-20} J/molecule) [85]. The externally applied energy at the transition was calculated for interacting flat surfaces from the force measured between curved surfaces ($F_n/R = 1$–6 mN/m; cf. Figure 17.1c) using the Derjaguin approximation [90], $F_{n,curved}/R = 2\pi E_{flat}$. With an adsorption density of ca. 2×10^{18} molecules/m^2 (molecular area, 50 Å^2, based on studies by Mathieson [20] and Bailey [21]), the energy applied at the transition was $(1-5) \times 10^{-22}$ J/molecule, which was much smaller than the energy needed to bend one molecule. In dilute solution at temperatures well above its melting temperature, linoleic acid is not expected to be in the extended conformation found in the crystal, but in one of the U-shaped conformations (length ca. 14 Å) shown in the study by Rich [85] (cf. Figure 17.1c). The outer force barrier represented the compression of a structure, four U-shaped molecules thick, and the transition (from $D = 60$ to 30 Å) showed the removal of a bilayer, so that one monolayer remained on each surface. The mechanisms leading to the formation of this layered structure are discussed below.

17.4.3 INTERACTION ENERGY BETWEEN LINOLEIC ACID MOLECULES

The conclusion that the linoleic acid molecules formed multilayers on the surfaces was supported by calorimetric work on the interaction of n-alkenes and n-alkadienes with n-alkanes (six to eight carbon atoms) [104,105]. As a measure of the interaction, we used the excess enthalpy of the mixtures of n-hexane with different hexenes and with 1,5-hexadiene and 1,3-hexadiene. At a mole fraction of 0.5, the excess enthalpies of mixing alkane with alkene and alkadiene were $0.03RT$ (1×10^{-22} J/molecule) and $0.1RT$ (4×10^{-22} J/molecule), respectively [104,105]. The alkadiene contact with alkane was thus energetically unfavorable. The higher excess enthalpy implied a lack of hexadecane molecules in the vicinity of the diene parts of the linoleic acid, i.e., between the layers. Here, it was also interesting to note that recent measurements with a QCM [15] indicated the formation of a close-packed, mainly physisorbed layer of linoleic acid on steel without coadsorbed n-hexadecane molecules. Although the layered structure in our confined film was stabilized by interactions between the exposed diene groups on the monolayer-covered surfaces and the ones on the intervening bilayer, this attraction was overcome by the compression discussed above, which amounted to an applied energy of $(1-5) \times 10^{-22}$ J/molecule.

Many carboxylic acids form hydrogen-bonded dimers in the gas phase [106], in the neat liquid [99], in the crystalline state [95,97,98,100], and when dissolved in nonpolar solvents [106–110]. In the gas phase [106], and in liquid stearic acid [106], each of the two hydrogen bonds between the carboxylic acid groups has a binding energy of ca. 7 kcal/mol (5×10^{-20} J/bond, i.e., 10×10^{-20} J/dimer), which has also been found in direct measurements of the hydrogen bond strength with AFM [111]. The enthalpy of association is approximately 14 kcal/mol (9×10^{-20} J/dimer) for dimer formation of stearic, dodecanoic and benzoic acid in carbon tetrachloride [108,109], hexane, and benzene [109]. A lower value of 4 kcal/mol (1.4×10^{-20} J/dimer) has also been reported for dimer formation of dodecanoic acid in benzene [107], and values in this lower range have been reported [106] for many short-chain alkanoic acids in nonpolar solvents.

The hydrogen bonding in linoleic acid was thus stronger than the diene–diene interaction and kept the intervening bilayer of linoleic acid together. This bilayer was entirely removed when the interactions between its diene regions and those of the monolayers on the surfaces were overcome by the compressive force (transition from $D = 60$ to 30 Å in Figure 17.1c). A similar structure has been suggested for confined layers of neat 1-octanol and 1-undecanol between mica surfaces [103]. No bilayer formation was detected in the stearic and oleic acid systems, where the confined films were expected to contain a significant amount of solvent.

17.4.4 ADHESION ACROSS FATTY ACID FILMS IN HEXADECANE

The interfacial energy γ can be determined from the pull-off force by assuming that $\gamma = -F_{n,pull-off}/3\pi R$ [90]. The values obtained from the force curves in Figure 17.1 were $\gamma = 0.1-0.2$ mN/m. Although this was a low interfacial energy, it was larger than what can be accounted for by the van der Waals forces in the stearic and oleic acid systems, where the dielectric properties of the monolayer compounds and the solvent were very similar. For these two systems, the van der Waals–Lifshitz force was calculated for a five-layer system consisting of mica/adsorbed layer/confined hexadecane/ adsorbed layer/mica [90], with 25 Å thick adsorbed layers of a 1:1 mix of stearic or oleic acid and hexadecane. The resulting attraction is shown as thick black curves in Figure 17.1a and b. For linoleic acid, the van der Waals–Lifshitz force for 13 Å thick adsorbed layers of pure linoleic acid is shown as a thick black curve in Figure 17.1c.

A broad adhesive minimum of the shape and magnitude seen in the oleic acid system has been observed between various surfactant-coated surfaces interacting across hydrocarbon liquids [73–76], and was believed to be of entropic origin [75,112,113]. The first few molecular layers of the intervening film (i.e., the molecules closest to the adsorbed layer on each surface) were oriented mainly parallel to the surfaces and therefore lost some of their orientational entropy. As the two surfaces were brought closer, some of these molecules moved out into the bulk solution, where they could attain other orientations and a net attraction was observed. High compression of the film can cause stronger orientation of remaining, trapped molecules, which would lead to a repulsive force such as the hard-wall repulsion seen at small distances.

17.4.5 Friction Forces between Fatty Acids in Hexadecane

In the stearic and oleic acid systems, the friction force did not depend on the adsorption time or on the sliding velocity within the investigated range. The observed film thicknesses (cf. Figure 17.2 and Table 17.2) indicated that some of the molecules oriented parallel to the surfaces were still present between the adsorbed layers at the lowest loads. Oleic acid formed more loose-packed and thus less well-ordered mixed layers with n-hexadecane compared to stearic acid.

Only the stearic acid system showed a lower friction than that of pure n-hexadecane, and its friction coefficient was ca. three times smaller than that in the oleic acid system (Table 17.2). Such trends have been previously observed in experiments at both the microscopic and macroscopic scales [6,12,19–24]. Studies of macroscopic friction report $\mu = 0.08$–0.09 [19,23] or 0.10 [56] when both interacting steel surfaces were replenished with stearic acid from n-hexadecane, the solvent used for the adsorption in our study. Similar values have also been obtained in macroscopic systems containing LB-deposited monolayers [114]. Under similar sliding conditions, pure n-hexadecane gave $\mu = 0.2$–0.6 [23,56].

The friction of stearic acid and calcium stearate has also been measured in SFA experiments at conditions quite different from the ones in our study [13,115,116]. Very close-packed monolayers (each 24–25 Å thick [13,115]) were formed on mica by LB deposition, and the experiments were done under dry conditions (i.e., no solvent was present in the system), where the surfaces strongly adhered and the friction nonlinearly increased with load (adhesion-controlled, area-dependent friction). Furthermore, the investigated loads were very high, and stick-slip sliding was observed in some cases. If the nonlinear friction force is interpreted as consisting of additive area-dependent and load-dependent terms, the friction coefficient of close-packed stearic acid in the study done by Briscoe et al. [116] was 0.038 at a sliding velocity of 0.36 μm/s.

The friction in our linoleic acid system showed several features not observed in the other systems. Two different load regimes were seen, and the one at the lower load (larger D) had a higher friction coefficient (cf. Table 17.2). Based on the measurements of normal force, the film structure in this low-load regime consisted of monolayers of adsorbed, bent linoleic acid molecules in an upright orientation on each surface. In contrast to the stearic and oleic acid systems, there did not appear to be an intervening film with molecules in an orientation parallel to the surfaces present at low loads (cf. discussion in Sections 17.4.1 through 17.4.3 of interactions between the double-bond rich regions). At loads above 3–4 mN ($p_0 = 2.5$–3 MPa), there was an abrupt film thickness transition to 12–15 Å, accompanied by a decrease in the friction coefficient (Figure 17.2c). We interpret this as a pressure-induced strongly tilted or parallel orientation of the bent molecules between the surfaces, with less resistance to sliding than the upright structure. In contrast to the stearic and oleic acid systems, a film thickness lower than 12–15 Å was not reached at high loads, and the surfaces were not easily damaged. This higher film thickness suggested that not all linoleic acid molecules were removed from the contact, possibly because of interactions of their double bond–rich regions within the film or with the mica in the tilted or parallel orientation obtained at high load. The molecules were not completely extended from their U-shape by the shearing, since this would allow a different packing and result in a lower film thickness. The friction coefficients in both load regimes were higher than in the other systems, and there was also a strong dependence of the friction force on sliding velocity (Table 17.2). The velocity dependence was interpreted as a longer time allowed for the system to relax and possibly disentangle during slower sliding speeds, and could reflect a rearrangement/ change of orientation of the U-shaped molecules in the sliding direction. There was a slight increase in the friction coefficient with adsorption time, which can be due to a slight difference in packing, or to the oxidation of the linoleic acid because of exposure to oxygen dissolved in the solution.

17.4.6 FRICTION COEFFICIENT AND CRITICAL SHEAR STRESS IN MONOLAYER SYSTEMS

Most previous works on lubrication with fatty acids or metal stearates and oleates have been done on systems in contact with a bulk reservoir of base oil containing fatty acid [116,117], or monolayers formed ex situ by LB deposition [10,11,13,114,116,118,119], giving higher packing density than in our current study. Several studies have been done with an opposing surface carrying the same fatty acid monolayer as the substrate, or an alkanethiol monolayer. All these conditions will likely result in a lower friction than that of in our systems with poorly packed layers and an unfunctionalized tip.

Stearic acid and Cd stearate LB monolayers on Si wafer have been probed with a Si AFM tip in ambient air [10]. Stearic acid transferred at a compressed solid state gave very low friction with $\mu = 0.02$, and a slightly lower μ was measured on Cd stearate transferred at a liquid-expanded state. The Cd stearate film consisted of domains with a height of 1.9 nm, implying a tilt angle of the molecules from the surface normal of ca. 35°, i.e., more close packed than our systems, where no domains were detected. AFM has also been used to measure the friction between functionalized tips and stearic and oleic acid monolayers, adsorbed on cellulose from an evaporated acetone solution [12]. On stearic acid, a friction coefficient of approximately 0.05 was obtained with $-CH_3$ functionalized tips. The oleic acid monolayer did not reduce the friction coefficient of the cellulose surface from $\mu = 0.7$ [12]. Friction force microscopy on domains of LB-deposited Cd arachidate showed that an intact patch of bilayer could be laterally translated, suggesting a lower limit of the critical shear stress between the two monolayers of $S_c = 1.0$ MPa [119]. Molecular dynamics simulations suggested $S_c = 15.6$ MPa for close-packed hydrocarbon chains that were six units long [120]. Measurements in dry N_2 gas on a system where both the tip and the substrate carried highly packed octadecanethiol monolayers gave $S_c = 12$ or 33 MPa (with another contact mechanics model, and depending on the choice of monolayer modulus), which was similar to the values we find for our stearic acid monolayers (Table 17.4) for very similar scan sizes, load ranges, tip radii, and interfacial energies [36]. Measurements in ethanol on the octadecanethiol system gave linear F_f versus F_n with $\mu = 0.08$ [36].

Measurements at larger length scales have been performed with the colloidal probe technique and with the SFA. The friction coefficient measured between a bare iron particle and a bare polished iron surface in air was $\mu = 0.5$ [121], and decreased to $\mu = 0.3$ when both surfaces were covered with Zn stearate. LB-deposited monolayers on mica have been studied with the SFA [13,116]. With this technique, the contact area can be directly measured and is large compared to the monolayer film thickness so that the JKR model can be applied. The friction between two close-packed, adhering monolayers was found to be area dependent with $S_c = 0.6$ MPa for myristic (C_{14}) and stearic acids, 2.2 MPa for behenic acid (C_{22}), and 1.0 MPa for Cd stearate [13]. Other SFA experiments on Cd stearate show $S_c = 2.5$ GPa [122] and 3–4 GPa [115]. SFA experiments on Cd arachidate (C_{20}) monolayers adhering in dry N_2 gas suggest area-dependent friction with $S_c \approx 0.6 \pm 0.1$ MPa at 25°C and 0.8 ± 0.2 MPa at 35°C [123,124]. These values are lower than the S_c values in Table 17.4 for our loose-packed stearic acid at $E = 0.1$ GPa, the modulus that corresponds most closely to the ones determined for confined monolayers in the SFA [16].

Bare surfaces and LB-deposited films have also been studied on the macroscopic scale. The friction coefficient between a steel ball slider and a flat steel surface in air was $\mu = 0.6$–0.8, and when the flat surface carried a monolayer of stearic acid, μ was reduced to 0.09–0.10 [114]. When replacing the flat surface with a glass substrate covered with stearic acid, the friction coefficient was slightly lower, 0.05–0.08 [114]. In a ball-on-plate sliding friction test in air at a relative humidity of 55%, a friction coefficient of $\mu = 0.10$ was measured between a glass lens and a stearic acid monolayer deposited on an Al plate coated with NiP film [11]. The macroscopic friction in a glass/stearic acid/Al system suggested $S_c = 5.8$ MPa [118], and the low-load range in a glass/stearic acid/glass agreed well with results of the study done by Briscoe et al. [125] which gave $S_c \approx 0.5$ MPa.

The studies found in the literature span a large range of substrate compositions, roughnesses, and lateral scales. Some studies [12,19] have indicated that stearic acid monolayers reduce the friction between surfaces more than monolayers of oleic acid or of metal oleates do, at both the nanoscopic and macroscopic scales. The same was found in our studies. Only the stearic acid had low μ and S_c, and the values we obtained were larger than for more close-packed LB monolayers and octadecanethiol. There was an increase in μ with increasing unsaturation. The trend was not as clear for the critical shear stress S_c, where a higher value was found for oleic acid than for linoleic acid. The calculation of S_c was very sensitive to the choice of h, and the particularly low experimental value of h for oleic acid might cause this discrepancy.

17.5 SUMMARY

When adsorbed on mica from dry n-hexadecane solution, stearic and oleic acids formed upright, single layers containing coadsorbed solvent oriented in the direction of the fatty acid hydrocarbon chains. When the surfaces were brought in contact in an SFA, these fatty acid layers were separated by a thin film of solvent and fatty acid molecules oriented mainly parallel to the surfaces, similarly to in systems of preformed surfactant monolayers in contact with alkane solvent. Linoleic acid formed a more complex, multilayered structure of U-shaped molecules, where a dimer layer was stabilized between adsorbed monolayers by interactions between the exposed double bond regions of the molecules. Only the stearic acid system showed low friction, and the friction increased with increasing unsaturation.

Fatty acids adsorbed from n-hexadecane on a model steel surface formed loose-packed layers where no distinct domains were observed in AFM in friction mode. In ethanol, the friction between the monolayers and an unfunctionalized Si tip was load-dependent. The friction coefficient μ in the stearic acid system, which had the lowest friction, was 3–10 times higher than that found in the literature for more close-packed systems. In dry N_2, where the adhesion was higher, the friction force appeared to be area-dependent. The contact area for the two different probe radii used in these systems was calculated using a contact mechanics model for layered systems, and similar critical shear stresses S_c for the two radii were found in each system. Also, when measured with this method, the friction of the unsaturated fatty acids and a rosin acid, dehydroabietic acid, was higher than for stearic acid.

ACKNOWLEDGMENTS

We would like to thank K. Danerlöv for many helpful discussions and assistance with the instrumentation used in this work. We also thank P. Claesson and his research group for discussions and for access to his laboratory for the sample preparation and SFA work. B. Kronberg is thanked for valuable discussions on diene interactions, J. Clarke and D. Broere for enabling the determination of the purity of the dehydroabietic acid, and T. Pettersson for providing the software for analyzing the AFM friction data. Financial support from the Academy of Finland (Stipend No. 204951) and the VINNOVA Competence Centre, "Surfactants Based on Natural Products, SNAP," is gratefully acknowledged.

REFERENCES

1. G. Knothe and K. R. Steidley, Lubricity of components of biodiesel and petrodiesel: The origin of bio-diesel lubricity. *Energy Fuels*, 19, 1192–1200 (2005).
2. J. M. Martin, C. Matta, M.-I. De Barros Bouchet, C. Forest, T. Le Mogne, T. Dubois, and M. Mazarin, Mechanism of friction reduction of unsaturated fatty acids as additives in diesel fuels. *Friction*, 1, 252–258 (2013).
3. C. Bovington and R. Castle, Lubricant chemistry including the impact of legislation. *Tribol. Ser.*, 40(Boundary and Mixed Lubrication), 141–146 (2002).
4. D. Dowson, *History of Tribology*, Second ed., Professional Engineering Publishing, London, UK (1998).
5. J. J. Frewing, The influence of temperature on boundary lubrication. *Proc. R. Soc. London, Ser. A*, 181, 23–42 (1942).
6. T. C. Askwith, A. Cameron, and R. F. Crouch, Chain length of additives in relation to lubricants in thin film and boundary lubrication. *Proc. R. Soc. London, Ser. A*, 291, 500–519 (1966).
7. M. Ratoi, V. Anghel, C. Bovington, and H. A. Spikes, Mechanisms of oiliness additives. *Tribol. Int.*, 33, 241–247 (2000).
8. R. Simič and M. Kalin, Comparison of alcohol and fatty acid adsorption on hydrogenated DLC coatings studied by AFM and tribological tests. *Stroj. Vestn.—J. Mech. Eng.*, 59, 707–718 (2013).
9. K. Nakano and H. A. Spikes, Process of boundary film formation from fatty acid solution. *Tribol. Online*, 7, 1–7 (2012).
10. V. V. Tsukruk, V. N. Bliznyuk, J. Hazel, D. Visser, and M. P. Everson, Organic molecular films under shear forces: Fluid and solid Langmuir monolayers. *Langmuir*, 12, 4840–4849 (1996).
11. P. Cong, T. Igari, and S. Mori, Effects of film characteristics on frictional properties of carboxylic acid monolayers. *Tribol. Lett.*, 9, 175–179 (2000).
12. N. Garoff and S. Zauscher, The influence of fatty acids and humidity on friction and adhesion of hydrophilic polymer surfaces. *Langmuir*, 18, 6921–6927 (2002).
13. B. J. Briscoe and D. C. B. Evans, The shear properties of Langmuir–Blodgett layers. *Proc. R. Soc. London, Ser. A*, 380, 389–407 (1982).
14. D. H. Wheeler, D. Potente, and H. Wittcoff, Adsorption of dimer, trimer, stearic, oleic, linoleic, nonanoic, and azelaic acids on ferric oxide. *J. Am. Oil Chem. Soc.*, 48, 125–128 (1971).
15. S. M. Lundgren, K. Persson, G. Mueller, B. Kronberg, J. Clarke, M. Chtaib, and P. M. Claesson, Unsaturated fatty acids in alkane solution: Adsorption to steel surfaces. *Langmuir*, 23, 10598–10602 (2007); Erratum, *Langmuir*, 24, 9922 (2008).
16. M. Ruths, S. Lundgren, K. Danerlöv, and K. Persson, Friction of fatty acids in nanometer-sized contacts of different adhesive strength. *Langmuir*, 24, 1509–1516 (2008).
17. E. L. Cook and N. Hackerman, Adsorption of polar organic compounds on steel. *J. Phys. Colloid Chem.*, 55, 549–557 (1951).
18. D. Tabor, Desorption or "surface melting" of lubricant films. *Nature*, 147, 609–610 (1941).
19. S. Jahanmir, Chain length effects in boundary lubrication. *Wear*, 102, 331–349 (1985), and references therein.
20. R. T. Mathieson, Electron microscopy of oleophobic monolayers. *Nature*, 183, 1803–1804 (1959).
21. A. I. Bailey, Friction and adhesion of clean and contaminated mica surfaces. *J. Appl. Phys.*, 32, 1407–1412 (1961).
22. F. Hirano, T. Sakai, N. Kuwano, and N. Ohno, Chain matching between hydrocarbon and fatty acid as interfacial phenomena. *Tribol. Int.*, 20, 186–204 (1987).
23. S. Jahanmir and M. Beltzer, Effect of additive molecular structure on friction coefficient and adsorption. *J. Tribol.*, 108, 109–116 (1986).
24. N. J. Fox, B. Tyrer, and G. W. Stachowiak, Boundary lubrication performance of free fatty acids in sunflower oil. *Tribol. Lett.*, 16, 275–281 (2004).
25. S. Lundgren, Unsaturated fatty acids in alkane solution: Adsorption and tribological properties. Doctoral thesis, Royal Institute of Technology, Stockholm, Sweden (2008).
26. R. C. Castle and C. H. Bovington, The behavior of friction modifiers under boundary and mixed EHD conditions. *Lubr. Sci.*, 15, 253–263 (2003).
27. S. M. Lundgren, M. Ruths, K. Danerlöv, and K. Persson. Effects of unsaturation on film structure and friction of fatty acids in a model base oil. *J. Colloid Interface Sci.*, 326, 530–536 (2008).
28. R. W. Carpick and M. Salmeron, Scratching the surface: Fundamental investigations of tribology with atomic force microscopy. *Chem. Rev.*, 97, 1163–1194 (1997).

29. R. W. Carpick, N. Agraït, D. F. Ogletree, and M. Salmeron, Variation of the interfacial shear strength and adhesion of a nanometer-sized contact. *Langmuir*, 12, 3334–3340 (1996).

30. M. Enachescu, R. J. A. van den Oetelaar, R. W. Carpick, D. F. Ogletree, C. F. Flipse, and M. Salmeron, Observation of proportionality between friction and contact area at the nanometer scale. *Tribol. Lett.*, 7, 73–78 (1999).

31. J. N. Israelachvili and A. D. Berman, Surface forces and microrheology of molecularly thin liquid films, in *Handbook of Micro/Nanotribology*, Second ed., Bhushan, B. (Ed.), pp. 371–432, CRC Press, Boca Raton, FL (1999).

32. G. Bogdanovic, F. Tiberg, and M. W. Rutland, Sliding friction between cellulose and silica surfaces. *Langmuir*, 17, 5911–5916 (2001).

33. S. Ecke and H.-J. Butt, Friction between individual microcontacts. *J. Colloid Interface Sci.*, 244, 432–435 (2001).

34. A. Berman, C. Drummond, and J. Israelachvili, Amontons' law at the molecular level. *Tribol. Lett.*, 4, 95–101 (1998).

35. M. Ruths, N. A. Alcantar, and J. N. Israelachvili, Boundary friction of aromatic silane self-assembled monolayers measured with the surface forces apparatus and friction force microscopy. *J. Phys. Chem. B*, 107, 11149–11157 (2003).

36. M. Ruths, Friction of mixed and single-component aromatic monolayers in contacts of different adhesive strength. *J. Phys. Chem. B*, 110, 2209–2218 (2006).

37. F. P. Bowden and D. Tabor, The area of contact between stationary and between moving surfaces. *Proc. R. Soc. London, Ser. A*, 169, 391–413 (1939).

38. F. P. Bowden and D. Tabor, *An Introduction to Tribology*, Anchor Press/Doubleday, Garden City, NY (1973).

39. H. Hertz, Über die Berührung fester elastischer Körper, *J. Reine Angew. Math.*, 92, 156–171 (1881).

40. K. L. Johnson, K. Kendall, and A. D. Roberts, Surface energy and the contact of elastic solid. *Proc. R. Soc. London, Ser. A*, 324, 301–313 (1971).

41. B. V. Derjaguin, V. M. Muller, and Y. P. Toporov, Effect of contact deformations on the adhesion of particles. *J. Colloid Interface Sci.*, 53, 314–326 (1975).

42. I. Sridhar, K. L. Johnson, and N. A. Fleck, Adhesion mechanics of the surface forces apparatus. *J. Phys. D: Appl. Phys.*, 30, 1710–1719 (1997).

43. K. L. Johnson and I. Sridhar, Adhesion between a spherical indenter and an elastic solid with a compliant elastic coating. *J. Phys. D: Appl. Phys.*, 34, 683–689 (2001).

44. I. Sridhar, Z. W. Zheng, and K. L. Johnson, A detailed analysis of adhesion mechanics between a compliant elastic coating and a spherical probe. *J. Phys. D: Appl. Phys.*, 37, 2886–2895 (2004).

45. E. D. Reedy, Jr., Thin-coating contact mechanics with adhesion. *J. Mater. Res.*, 21, 2660–2668 (2006).

46. E. D. Reedy, Jr., Contact mechanics for coated spheres that includes the transition from weak to strong adhesion. *J. Mater. Res.*, 22, 2617–2622 (2007).

47. M. Ruths and J. N. Israelachvili, Surface forces and nanorheology of molecularly thin films, in *Handbook of Nanotechnology*, Third ed., Bhushan, B. (Ed.), pp. 857–922, Springer-Verlag, Berlin and Heidelberg (2010).

48. J. Gao, W. D. Luedtke, D. Gourdon, M. Ruths, J. N. Israelachvili, and U. Landman, Frictional forces and Amontons' law: From the molecular to the macroscopic scale. *J. Phys. Chem. B*, 108, 3410–3425 (2004).

49. Y. Mo, K. T. Turner, and I. Szlufarska, Friction laws at the nanoscale. *Nature*, 457, 1116–1119 (2009).

50. Y. Mo and I. Szlufarska, Roughness picture of friction in dry nanoscale contacts. *Phys. Rev. B*, 81, 035405/1–17 (2010).

51. S. Cheng, B. Luan, and M. Robbins, Contact and friction of nanoasperities: Effects of adsorbed monolayers. *Phys. Rev. E*, 81, 016102/1–17 (2010).

52. C. M. Mate, G. M. McClelland, R. Erlandsson, and S. Chiang, Atomic-scale friction of a tungsten tip on a graphite surface. *Phys. Rev. Lett.*, 59, 1942–1945 (1987).

53. G. Meyer and N. M. Amer, Simultaneous measurement of lateral and normal forces with an optical-beam-deflection atomic force microscope. *Appl. Phys. Lett.*, 57, 2089–2091 (1990).

54. E. Meyer, R. M. Overney, K. Dransfeldt, and T. Gyalog, *Nanoscience: Friction and Rheology on the Nanometer Scale*, World Scientific, Singapore (1998).

55. B. E. O. Hedman, Tall oil products as raw material for surfactant synthesis. Licentiate thesis, Royal Institute of Technology, Stockholm (2000).

56. P. Studt, Boundary lubrication: Adsorption of oil additives on steel and ceramic surfaces and its influence on friction and wear. *Tribol. Int.*, 22, 111–119 (1989).

57. S. H. Chen and C. W. Frank, *n*-Alkanoic acid self-assembled monolayers: Adsorption kinetics. *ACS Symp. Ser.*, 447, 160–176 (1990).

58. V. V. Korolev, A. G. Ramazanova, V. I. Yashkova, O. V. Balmasova, and A. V. Blinov, Adsorption of fatty acids from solutions in organic solvents on the surface of finely dispersed magnetite: 1: Isotherms of adsorption of oleic, linoleic, and linolenic acid from carbon tetrachloride and hexane. *Colloid J.*, 66, 700–704 (2004).

59. S. M. Lundgren, K. Persson, B. Kronberg, and P. M. Claesson, Adsorption of fatty acids from alkane solution studied with quartz crystal microbalance. *Tribol. Lett.*, 22, 15–20 (2006).

60. J. N. Israelachvili, N. A. Alcantar, N. Maeda, T. E. Mates, and M. Ruths, Preparing contamination-free mica substrates for surface characterization, force measurements, and imaging. *Langmuir*, 20, 3616–3622 (2004).

61. J. N. Israelachvili and P. M. McGuiggan, Adhesion and short-range forces between surfaces: Part I: New apparatus for surface force measurements. *J. Mater. Res.*, 5, 2223–2231 (1990).

62. G. Luengo, F.-J. Schmitt, R. Hill, and J. Israelachvili, Thin film rheology and tribology of confined polymer melts: Contrasts with bulk properties. *Macromolecules*, 30, 2482–2494 (1997).

63. J. N. Israelachvili and G. E. Adams, Measurement of forces between two mica surfaces in aqueous electrolyte solutions in the range 0–100 nm. *J. Chem. Soc. Faraday Trans.*, 1, 74, 975–1001 (1978).

64. J. N. Israelachvili, Thin film studies using multiple-beam interferometry. *J. Colloid Interface Sci.*, 44, 259–272 (1973).

65. H. Yoshizawa and J. Israelachvili, Fundamental mechanisms of interfacial friction: 2. Stick-slip friction of spherical and chain molecules. *J. Phys. Chem.*, 97, 11300–11313 (1993).

66. Y. Liu, T. Wu, and D. F. Evans, Lateral force microscopy study on the shear properties of self-assembled monolayers of dialkylammonium surfactant on mica. *Langmuir*, 10, 2241–2245 (1994).

67. Y. Liu, D. F. Evans, Q. Song, and D. W. Grainger, Structure and frictional properties of self-assembled surfactant monolayers. *Langmuir*, 12, 1235–1244 (1996).

68. R. W. Carpick, D. F. Ogletree, and M. Salmeron, A general equation for fitting contact area and friction vs load measurements. *J. Colloid Interface Sci.*, 211, 395–400 (1999).

69. B. Luan and M. O. Robbins, The breakdown of continuum models for mechanical contacts. *Nature*, 435, 929–932 (2005).

70. E. D. Reedy, Jr., M. J. Starr, R. E. Jones, E. E. Flater, and R. W. Carpick, Contact modeling of SAM-coated polysilicon asperities. In Proceedings of the 28th Annual Meeting of the Adhesion Society (Mobile, AL), pp. 366–368 (2005).

71. M. Chandross, C. D. Lorenz, M. J. Stevens, and G. S. Grest, Simulations of nanotribology with realistic probe tip models. *Langmuir*, 24, 1240–1246 (2008).

72. Y. Yang and M. Ruths, Friction of polyaromatic thiol monolayers in adhesive and nonadhesive contacts. *Langmuir*, 25, 12151–12159 (2009).

73. M. Ruths, H. Ohtani, M. L. Greenfield, and S. Granick, Exploring the "friction modifier" phenomenon. Nanorheology of *n*-alkane chains with polar terminus dissolved in *n*-alkane solvent. *Tribol. Lett.*, 6, 207–214 (1999).

74. Y. Zhu, H. Ohtani, M. L. Greenfield, M. Ruths, and S. Granick, Modification of boundary lubrication by oil-soluble friction modifier additives. *Tribol. Lett.*, 15, 127–134 (2003).

75. M. L. Gee and J. N. Israelachvili, Interactions of surfactant monolayers across hydrocarbon liquids. *J. Chem. Soc., Faraday Trans.*, 86, 4049–4058 (1990).

76. C. E. Herder, B. W. Ninham, and H. K. Christenson, Interaction of hydrocarbon monolayer surfaces across *n*-alkanes: A steric repulsion. *J. Chem. Phys.*, 90, 5801–5805 (1989).

77. R. Tadmor, R. E. Rosensweig, J. Frey, and J. Klein, Resolving the puzzle of ferrofluid dispersants. *Langmuir*, 16, 9117–9120 (2000).

78. H. K. Christenson, D. W. R. Gruen, R. G. Horn, and J. N. Israelachvili, Structuring in liquid alkanes between solid surfaces: Force measurements and mean-field theory. *J. Chem. Phys.*, 87, 1834–1841 (1987).

79. H. Yoshizawa, P. McGuiggan, and J. Israelachvili, Identification of a second dynamic state during stick-slip motion. *Science*, 259, 1305–1308 (1993).

80. M. L. Gee, P. M. McGuiggan, J. N. Israelachvili, and A. M. Homola, Liquid to solidlike transitions of molecularly thin films under shear. *J. Chem. Phys.*, 93, 1895–1906 (1990).

81. L.-M. Qian, G. Luengo, and E. Perez, Thermally activated lubrication with alkanes: The effect of chain length. *Europhys. Lett.*, 61, 268–274 (2003).

82. L. Qian, G. Luengo, D. Douillet, M. Charlot, X. Dollat, and E. Perez, New two-dimensional friction force apparatus design for measuring shear forces at the nanometer scale. *Rev. Sci. Instrum.*, 72, 4171–4177 (2001).

83. I. Langmuir, The shapes of group molecules forming the surfaces of liquids. *Proc. Natl. Acad. Sci. U.S.A.*, 3, 251–257 (1917).

84. M. Tomoaia-Contisel, J. Zsako, A. Mocanu, M. Lupea, and E. Chifu, Insoluble mixed monolayers: III: The ionization characteristics of some fatty acids at the air/water interface. *J. Colloid Interface Sci.*, 117, 464–476 (1987).

85. M. R. Rich, Conformational analysis of arachidonic and related fatty acids using molecular dynamics simulations. *Biochim. Biophys. Acta*, 1178, 87–96 (1993).

86. Y.-T. Tao, Structural comparison of self-assembled monolayers of *n*-alkanoic acids on the surfaces of silver, copper, and aluminum. *J. Am. Chem. Soc.*, 115, 4350–4358 (1993).

87. G. Shustak, A. J. Domb, and D. Mandler, Preparation and characterization of *n*-alkanoic acid self-assembled monolayers adsorbed on 316L stainless steel. *Langmuir*, 20, 7499–7506 (2004).

88. M. E. Tadros, P. Hu, and A. W. Adamson, Adsorption and contact angle studies: I: Water on smooth carbon, linear polyethylene, and stearic acid coated copper. *J. Colloid Interface Sci.*, 49, 184–195 (1974).

89. S. C. Clear and P. F. Nealey, Chemical force microscopy study of adhesion and friction between surfaces functionalized with self-assembled monolayers and immersed in solvents. *J. Colloid Interface Sci.*, 213, 238–250 (1999).

90. J. N. Israelachvili, *Intermolecular and Surface Forces*, Third ed., Academic Press, Amsterdam (2011).

91. J. Vissner, Hamaker constants: Comparison between Hamaker constants and Lifshitz–van der Waals constants. *Adv. Colloid Interface Sci.*, 3, 331–363 (1972).

92. *CRC Handbook of Chemistry and Physics*, 66th ed., CRC Press, Boca Raton, FL (1985).

93. G.-y. Liu and M. Salmeron, Reversible displacement of chemisorbed *n*-alkanethiol molecules on Au(111) surface: An atomic force microscopy study. *Langmuir*, 10, 367–370 (1994).

94. M. Chandross, G. S. Grest, and M. J. Stevens, Friction between alkylsilane monolayers: Molecular simulations of ordered monolayers. *Langmuir*, 18, 8392–8399 (2002), and references therein.

95. J. Ernst, W. S. Sheldrick, and J.-H. Fuhrhop, The structures of the essential unsaturated fatty acids: Crystal structure of linoleic acid and evidence for the crystal structures of α-linolenic and arachidonic acid. *Z. Naturforsch. B*, 34, 706–711 (1979).

96. A. L. Rabinovich and P. O. Ripatti, On the conformational, physical properties and functions of polyunsaturated acyl chains. *Biochim. Biophys. Acta*, 1085, 53–62 (1991).

97. A. Müller, An X-ray investigation of certain long-chain compounds. *Proc. R. Soc. London, Ser. A*, 114, 542–561 (1927).

98. K. Sato and M. Okada, Growth of large single crystals of stearic acid from solution. *J. Cryst. Growth*, 42, 259–263 (1977).

99. M. Iwahashi, Y. Kasahara, H. Matsuzawa, K. Yagi, K. Nomura, H. Terauchi, Y. Ozaki, and M. Suzuki, Self-diffusion, dynamical molecular conformation, and liquid structures of *n*-saturated and unsaturated fatty acids. *J. Phys. Chem. B*, 104, 6186–6194 (2000).

100. S. Abrahamsson and I. Ryderstedt-Nahringbauer, The crystal structure of the low-melting form of oleic acid. *Acta Cryst.*, 15, 1261–1268 (1962).

101. S. Ueno, A. Miyazaki, J. Yano, Y. Furukawa, M. Suzuki, and K. Sato, Polymorphism of linoleic acid (*cis*-9, *cis*-12-octadecadienoic acid) and α-linolenic acid (*cis*-9, cis-12, *cis*-15-octadecatrienoic acid). *Chem. Phys. Lipids*, 107, 169–178 (2000).

102. P. E. Laibinis, G. M. Whitesides, D. L. Allara, Y.-T. Tao, A. N. Parikh, and R. G. Nuzzo, Comparison of the structure and wetting properties of self-assembled monolayers of *n*-alkanethiols on the coinage metal surfaces, copper, silver, and gold. *J. Amer. Chem. Soc.*, 113, 7152–7167 (1991).

103. F. Mugele, S. Baldelli, G. A. Somorjai, and M. Salmeron, Structure of confined films of chain alcohols. *J. Phys. Chem. B*, 104, 3140–3144 (2000).

104. W. Wóycicki, Excess enthalpies of binary mixtures containing unsaturated aliphatic hydrocarbons: 4. *n*-Alkane + *n*-alkane. *J. Chem. Thermodyn.*, 7, 77–81 (1975).

105. W. Wóycicki, Excess enthalpies of binary mixtures containing unsaturated aliphatic hydrocarbons: 1. *n*-Diene + *n*-alkane and + cyclohexane. *J. Chem. Thermodyn.*, 12, 165–171 (1980).

106. G. C. Pimentel and A. L. McClellan, *The Hydrogen Bond*, W.H. Freeman, San Francisco, CA (1960).

107. O. Levy, G. Y. Markovits, and I. Perry, Thermodynamics of aggregation of long chain carboxylic acids in benzene. *J. Phys. Chem.*, 79, 239–242 (1975).

108. V. P. Tikhonov, G. I. Fuks, and N. A. Kuznetsova, Infrared spectroscopic study of the association of fatty acid molecules in carbon tetrachloride. *Kolloidn. Zh.*, 36, 998–1002 (1974).

109. G. I. Fuks, V. P. Tikhonov, I. G. Fuks, and G. V. Rakaeva, Effect of the chemical nature of nonpolar solvents on the primary association and micelle formation of surfactants. *Kolloidn. Zh.*, 46, 976–979 (1984).

110. E. E. Tucker and E. Lippert, High resolution nuclear magnetic resonance studies of hydrogen bonding, in *The Hydrogen Bond: Recent Development in Theory and Experiment; II. Structure and Spectroscopy*, Schuster, P., Zundel, G., and Sandorfy, C. (Eds.), pp. 806–809, North-Holland, Amsterdam (1976).

111. T. Han, J. M. Williams, and T. P. Beebe, Jr., Chemical bonds studied with functionalized atomic force microscopy tips. *Anal. Chim. Acta*, 307, 365–376 (1995).

112. J. N. Israelachvili, S. J. Kott, M. L. Gee, and T. A. Witten, Forces between mica surfaces across hydrocarbon liquids: Effects of branching and polydispersity. *Macromolecules*, 22, 4247–4253 (1989).

113. G. Ten Brinke, D. Ausserre, and G. Hadziioannou, Interaction between plates in a polymer melt. *J. Chem. Phys.*, 89, 4374–4380 (1988).

114. D. D. Dominguez, R. L. Mowery, and N. H. Turner, Friction and durabilities of well-ordered, close-packed carboxylic acid monolayers deposited on glass and steel surfaces by the Langmuir–Blodgett technique. *Tribol. Trans.*, 37, 59–66 (1994).

115. J. N. Israelachvili and D. Tabor, Shear properties of molecular films. *Wear*, 24, 386–390 (1973).

116. B. J. Briscoe, D. C. B. Evans, and D. Tabor, The influence of contact pressure and saponification on the sliding behavior of stearic acid monolayers. *J. Colloid Interface Sci.*, 61, 9–13 (1977).

117. F. P. Bowden and D. Tabor, *The Friction and Lubrication of Solids*. Clarendon, Oxford (1954).

118. R. S. Timsit, Effect of surface reactivity on tribological properties of a boundary lubricant, in *Fundamentals of Friction*; Singer, I. L, and Pollock, H. M. (Eds.), pp. 287–298, Kluwer, Dordrecht (1992).

119. E. Meyer, R. Overney, D. Brodbeck, L. Howald, R. Lüthi, J. Frommer, and H. Güntherodt, Friction and wear of Langmuir–Blodgett films observed by friction force microscopy. *Phys. Rev. Lett.*, 69, 1777–1780 (1992).

120. J. N. Glosli and G. M. McClelland, Molecular-dynamics study of sliding friction of ordered organic monolayers. *Phys. Rev. Lett.*, 70, 1960–1963 (1993).

121. A. Meurk, I. Larson, and L. Bergström, Tribological properties of iron powder subjected to various surface treatments. In *Fundamentals of Nanoindentation and Nanotribology, Materials Research Society Symposium Proceedings*, Vol. 522. Moody, N. R., Gerberich, W. W., Baker, S. P., and Burnham, N. (Eds.), pp. 427–432, Materials Research Society, Pittsburgh, PA (1998).

122. A. I. Bailey and J. S. Courtney-Pratt, The area of real contact and the shear strength of unimolecular layers of a boundary lubricant. *Proc. R. Soc. London, Ser. A*, 227, 500–515 (1955).

123. M. Ruths, S. Steinberg, and J. N. Israelachvili, Effects of confinement and shear on the properties of thin films of thermotropic liquid crystal. *Langmuir*, 12, 6637–6650 (1996).

124. M. Ruths, S. Steinberg, and J. N. Israelachvili, unpublished data.

125. B. J. Briscoe, B. Scruton, and F. R. Willis, Shear strength of thin lubricant films. *Proc. R. Soc. London, Ser. A*, 333, 99–114 (1973).

18 Additives for Biodegradable Lubricants

Daniel M. Vargo and Brian M. Lipowski

CONTENTS

ABSTRACT

This chapter focuses on chemical additives that enhance the performance characteristics of a biodegradable lubricant. In general, the same types of additives commonly used in mineral oil-based lubricants can be utilized in biodegradable lubricants. However, special emphasis must be placed on additive compatibility and on environmental toxicity. The topics discussed include vegetable oils or natural oils as base oils and performance additives such as detergents and dispersants, rust inhibitors, corrosion inhibitors, antioxidants, and pour point depressants, among others. Lesser emphasis is given to synthetic biodegradable esters as base oils.

18.1 INTRODUCTION

According to the Office of the Federal Environmental Executive, biobased products are commercial or industrial products (other than food or feed) composed in whole or in significant part of biological products or renewable domestic agricultural materials (including plant, animal, and marine materials) or forestry materials [1]. The Office of the Federal Environmental Executive is responsible for promoting sustainability and environmental stewardship throughout federal government operations. The office works to implement executive orders on federal environmental performance, including achieving federal goals for greenhouse gas emissions reduction, energy efficiency, and water conservation and documenting the economic benefits of environmental performance.

Biodegradable and biobased lubricants are becoming increasingly popular as alternatives to petroleum-based products. The introduction of petroleum oils into oceans and lakes is hazardous to affected wildlife and ecosystems. Federal and state agency fines and cleanup costs are escalating and equipment operators must find a cost-effective, environmentally safe lubricant to meet their needs. One gallon of petroleum oil can contaminate one million gallons of water. The BP oil spill in 2010 released an estimated 210 million gallons into the Gulf of Mexico [2]. Another example is the IXTOC I exploratory well in the Bay of Campeche, which blew out on June 3, 1979, causing a major oil spill in the Gulf of Mexico. By the time the well was brought under control in 1980, an estimated 140 million gallons of oil had spilled into the bay. IXTOC I is currently number two on the list of largest oil spills of all time [3]. Readily available, affordable biobased lubricant products which are environmentally acceptable and perform equivalently to their petroleum-based counterparts must be produced.

Since 2010 research on biobased oils and lubricant additives has escalated. A useful report, "Technical and Market Survey of Biobased Lubricants January–July 2012," compiles biobased lubricant patents for this period [4].

18.2 HISTORY

Over the past two decades, a renewed interest in vegetable oil-based lubricants has occurred as environmental concerns, sustainability concerns, and the impact of the carbon footprint have increased. In Europe during the 1980s, various mandates and regulations were placed on petroleum products necessitating the use of biodegradable lubricants [5]. During the 1990s, many American companies began research to develop biodegradable products. A prime example is the introduction of Mobil Corporation's Environmental Awareness Lubricants line of hydraulic fluids. Lubrizol Corporation also developed biodegradable additive base stocks derived from sunflower oil. These endeavors, however, lacked appropriate regulatory mandates in the United States and made biodegradable oils too expensive to compete.

18.3 VEGETABLE OILS

Vegetable oils have been used as lubricants without modification to their chemical structures. They have several advantages and disadvantages when considered for industrial and machinery lubrication. Vegetable oils have excellent lubricity, far superior to that of mineral oils. Lubricity is so high that in some applications, such as tractor transmissions, materials to increase friction need to be added to reduce clutch slippage. Some vegetable oils have passed hydraulic pump/wear tests, such as ASTM D2882 and ASTM D2271, without additional additives [5].

Vegetable oils have a very high VI, typically over 200 for canola oil, compared to about 100 for most petroleum oils. A high-VI vegetable oil does not reduce in viscosity as much when exposed to high temperatures (oil thinning), and does not increase in viscosity as much when exposed to cool temperatures (oil thickening) when compared to a petroleum oil.

Another important advantage of vegetable oils is their high flash and fire points, typically greater than 300°C, compared to a flash point of approximately 200°C for mineral oils. This gives the formulated lubricant a greater margin of safety at high operating temperatures.

Most importantly, vegetable oils are biodegradable, are generally less toxic, are renewable, and reduce our dependency on petroleum oils [6].

Disadvantages of unmodified, unadditized vegetable oils include insufficient oxidative stability for lubricant use. If untreated, the oil will quickly oxidize during use, becoming more viscous and more acidic, and acquiring an unpleasant odor, thus reducing lubricant life and necessitating frequent changing of the oil. The chemical modification of vegetable oils and/or the use of antioxidants can address this problem. Chemical modification could involve a partial hydrogenation of the vegetable oil and/or a modification of its fatty acid profile. The challenge with hydrogenation is to determine at what point the optimum oil could be produced. A high level of hydrogenation will greatly improve its oxidative stability but will also increase the pour point of the oil to the point that it may become solid at ambient temperatures.

Advances in biotechnology have led to the development of genetically modified oilseeds that do not require chemical modification and/or use of antioxidants to enhance their oxidative stability. A soybean seed developed through DuPont technology, for example, presents more than 83% oleic acid as compared to only 20% oleic acid content in conventional soybean oil [7]. High-oleic varieties of canola, rapeseed, sunflower, and soybean oils are becoming standard base oils for biodegradable lubricants and greases [7,8].

Another disadvantage of vegetable oils is their high pour point. This problem can be addressed by winterization, chemical additives such as PPDs, and/or blending with other fluids with lower pour points [5].

Vegetable oils are triglycerides formed from the biochemical reaction of fatty acids and glycerol in the plant. The fatty acid chain length and the degree of unsaturation vary. A broad range of fatty acid profiles in the plant is possible. The most commonly used vegetable oils have a fatty acid chain length distribution centering around 18 carbons with various degrees of unsaturation. The relative proportion of these long-chain fatty acids depends on the vegetable type. In general, the higher the oleic acid content (C_{18} with 1 double bond) and the lower the polyunsaturation (2 or more double bonds per chain), the better the oxidative stability and the higher the pour point [9,10].

Standard petroleum products typically contain aromatic hydrocarbons, usually cyclic ring structures, that cause the appearance of rainbow sheen on the surface of water. Biobased products do not contain aromatics and therefore do not produce rainbow sheen on water if spilled. A biobased lubricant may not produce sheen on water; however, the oil will persist on the water surface and may harm aquatic life and the ecosystem.

The widespread application of biobased oils is limited due to performance deficiencies when compared to petroleum products. Chief among these concerns are hydrolytic and oxidative stabilities and cold temperature flow properties. Additives need to be used to improve the areas in which performance is lacking to take full advantage of the natural benefits of biobased oils such as lubricity and solvency.

Oxidative stability can be improved through use of additives or by chemical modification such as alkylation, epoxidation, or hydrogenation [11]. Cloud and pour points can be improved by additives such as PPDs and by blending with low-temperature biodegradable fluids such as low–molecular weight esters and PAOs [12]. Further information on these and other types of additives for biobased systems is presented in the following sections.

18.4 ADDITIVES USED WITH BIODEGRADABLE OILS

18.4.1 DETERGENTS AND DISPERSANTS

Vegetable oils have a high level of solvency due to polar ester structures present in the molecule and typically act as a detergent; additional detergent additives are generally not required. If a detergent must be used in a formulation, it is necessary to avoid metallic phenates and sulfonate chemistries which may be harmful to the environment. Likewise, components such as sulfurized phenates and salicylates are typically not used in biobased oils.

Typically, dispersants are designed for and used in petroleum-based engine oils. Recently, dispersants with a high total base number have found use in biobased hydraulic oils. These dispersants tend to neutralize and solubilize acidic products formed upon oxidation of the oil over its service life. Oxidation may lead to deposits and sludge formation. Sludge in a hydraulic fluid will clog filters, leading to eventual pump failure if fluid flow is not restored. A test which measures acid number (AN) and sludge buildup over time under given oxidative conditions is the turbine oil stability test (TOST) [13].

18.4.2 RUST INHIBITORS—FERROUS METALS

Rusting is a process where iron metal reacts with water to produce iron oxides. These oxides are typically reddish brown and have a brittle structure. Rusting can result in the following:

- The wearing away of the metal surface
- The formation of (micro)pits and (micro)fractures of the surface, leading to failure
- The suspension of debris in the lubricant accelerating wear

Typical rust inhibitors for biobased lubricants are those that are surface active; these additives bond with the metal surface through, for example, an oxygen moiety, including those in carboxylic acids, borates, amine phosphates, or sulfonates.

18.4.3 CORROSION INHIBITORS—NONFERROUS METALS

A common mechanism for inhibiting corrosion involves formation of a coating, often a passivating layer, which prevents the corrosive substance from contacting the vulnerable portion of the metal. Benzotriazole, as shown in Figure 18.1, inhibits the corrosion and staining of copper surfaces by forming an inert layer of polymer on the metal surface [14].

A disadvantage of corrosion inhibitors is that the same mechanism of action is involved for other functional additives such as friction modifiers and antiwear agents. As a result, multiple additives compete for bonding sites on the metal surface and the compound that has the greatest affinity will be present on the surface at the highest concentration, resulting in some additive components becoming less effective than anticipated. The goal of the formulator is to select and balance all additive components in a formulation for optimum performance.

18.4.4 ANTIOXIDANTS

One of the major disadvantages of vegetable oils is their poor oxidative stability. Vegetable oils can contain one or more double bonds per chain of the triglyceride molecule, depending on the type of oil. Free radicals are produced over the course of the service lifetime of the fluid through heat and contact with the air. These ions will react with the double bond, producing oxidation by-products such as acids, aldehydes, ketones, and ether linkages. A higher level of unsaturation leads to greater oxidative instability [15]. Table 18.1 shows that as the degree of polyunsaturation decreases, the oil becomes more resistant to oxidation.

For most applications, the antioxidant will be the major component of the formulation. The antioxidant will prevent discoloration, rancidification, sludge buildup, and AN increase, greatly extending the useful life of the lubricant or grease made from vegetable oils [9].

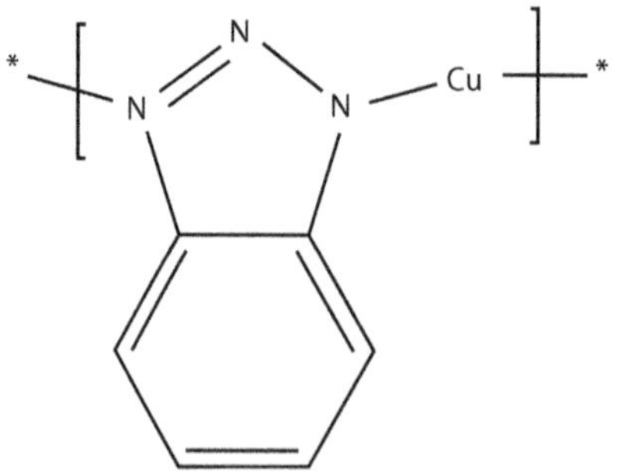

FIGURE 18.1 Copper benzotriazole polymer film formed when used as a corrosion inhibitor.

TABLE 18.1

Relationship between Modified PDSC Onset Temperature and Degree of Unsaturation in Soybean Oil

	Average Number of Double Bonds per Chain	PDSC Onset Temperature (Modified CEC L-85-T-99) (°C)
Regular soybean	1.5	173
High-linoleic soybean	1.4	179
Mid-oleic soybean	1.1	190
High-oleic soybean	0.9	198

Source: Rudnick, L., *Lubricant Additives: Chemistry and Applications*, CRC Press, Boca Raton, p. 452, 2008; Mookken, R. T. et al. Dependence of oxidation stability of steam turbine oil on base oil composition, *Lubr. Eng.*, 53, 19–24, 1997. With permission.

Note: PDSC: pressure differential scanning calorimetry.

Excellent antioxidant and antiwear agents include ZDDPs. These compounds have the ability to scavenge free radical ions, preventing the double bond in the oil from reacting. These compounds, however, are hazardous to the environment and, as a result, their use in vegetable oils is extremely limited.

Another group of antioxidant compounds used in biobased oils is arylamines and hindered phenols. These compounds also act as free radical scavengers are usually used together as they can form a synergistic mixture. The most widely used compounds of the arylamines used to control oxidation are C_4- to C_9-alkylated diphenylamine compounds since they are generally not hazardous. Hindered phenols work in a similar manner and are generally not hazardous. One disadvantage of hindered phenols is high volatility. Various substituents such as esters or amides are usually inserted to the 4-position to lower their volatility and to maintain antioxidant concentration during the lubricant lifetime.

Two tests to determine oxidative stability are currently widely used in the industry. One is the TOST (ASTM D943). This test runs at 95°C to represent thermo-oxidative conditions of steam turbines and for hydraulic fluids in running applications. The number of hours to reach an AN of 2.0 for the fluid is determined. For biobased oils, the test procedure has been modified to exclude water, thus preventing premature hydrolysis of the oil; this is called the dry TOST test. Even so, unmodified vegetable oil-based fluids generally perform poorly in this test.

The RPVOT (ASTM D2272) is also frequently used. This test runs at 150°C under pressurized oxygen. While this test is quick and less costly to run, it can give a fair indication of the stability of a fluid. It does not replicate the normal fluid conditions a typical turbine oil or a hydraulic fluid will encounter [16].

A formulator will determine whether a fluid will encounter relatively high operating temperatures (120–150°C) or relatively low temperatures (50–90°C) and balance the antioxidant components and the necessary tests accordingly. Work done by Petlyuk and Adams has shown that good TOST performance is obtained when the hindered phenolic component is present in the formulation at a higher concentration than the arylamine component, typically at a 3:1 ratio by weight [17]. Conversely, to obtain good RPVOT performance, the arylamine component is present at a higher concentration than the hindered phenolic component.

18.4.5 ANTIWEAR AGENTS

Unlike petroleum oils, vegetable oils and biodegradable esters are polar and have affinity to metal surfaces. Therefore, vegetable oils can form a stronger film, providing better antiwear performance than a petroleum oil of the same viscosity grade.

Commonly used antiwear additives include ZDDPs and molybdenum dialkyldithiophosphates, phosphites, phosphorothionates, and amine phosphates. Sulfonamides also possess antiwear properties [18].

Amine phosphates are unique in that they are very polar salts. Therefore, they have very good solubility in biobased ester oils. Amine phosphates can have dual functions, serving as a ferrous corrosion inhibitor due to their ability to be adsorbed to a metal surface and as a mild antiwear additive due to the phosphate moiety. It is known, however, that amine phosphates are detrimental to the oxidative stability of oil, especially in combination with ZDDPs and phenolic antioxidants [15].

18.4.6 Viscosity Modifiers

Natural oils have excellent lubricity but they are available in limited viscosity grades. Typical viscosities tend to fall in the ISO 32 range. The biobased products these oils are suited for use in tend to fall in the following ISO grades: hydraulic fluids, ISO 32 to ISO 68; bar and chain oils, from ISO 68 to ISO 200; and rock drilling oils, from ISO 46 to ISO 320. Depending on the application, addition of viscosity modifier is therefore required to attain the desired viscosity.

Polyisobutylene and olefin copolymers, traditionally used as viscosity modifiers in petroleum-based oils, are not soluble in vegetable oils. However, polyisoprene, polybutadiene, and styrene–butadiene rubber are generally soluble due to the unsaturation in the structures of these polymers. Additionally, ethylene–vinyl acetate copolymers are generally soluble in biobased and vegetable oils.

18.4.7 Tackifiers

Tackifiers are generally added to increase the overall cohesive energy of a lubricant. Cohesive energy imparts stringiness (tackiness) or viscoelasticity to a fluid, resulting in resistance to removal from the surface. This property is most widely used in way oils, open gear lubricants, and bar and chain oils where the lubricant can be exposed to mechanical action and water impingement and must remain in place.

Traditionally, ultrahigh-molecular weight polyisobutylenes (one to five million daltons) are used as tackifiers in petroleum oils. Polyisobutylene is not sufficiently soluble in biobased fluids to impart tackiness but other polymers such as those discussed in Section 18.4.6 are soluble and may impart tackiness depending on the molecular weight and the base fluid used.

18.4.8 Pour Point Depressants

Another negative of vegetable oils is their high pour point. This problem can be addressed by winterization, addition of chemical additives (pour point suppressants), and/or blending with other fluids possessing lower pour points. Various synthetic oils can be used for this purpose. With a combination of these techniques, UNI-ABIL has developed hydraulic fluids with a pour point of −32.8°F (−36°C) for use in snowblowers used by the Iowa Department of Transportation. Another experimental hydraulic fluid using genetically enhanced oils meets military specifications for a pour point of −65.2°F (−54°C). While the use of genetically modified seed oils alleviates the problem of oxidation stability, the cold-temperature properties must be enhanced by the addition of chemical PPDs and/or the addition of other liquids with much lower pour points [5].

Traditional polymethacrylate PPDs that perform well in mineral oils may also perform well in vegetable oil-based lubricants. However, in certain oils, higher levels of PPD may be required.

Vegetable oils or biobased oils that have been highly refined or chemically modified do not generally respond to PPDs to the same degree that typical commodity vegetable oils do at the same treat level. In general, the higher the oleic content and the lower the level of polyunsaturates, the more difficult it is to lower the pour point of the oil [19], but these oils respond to a much greater degree to oxidation inhibitors, which improves the oxidative stability more than in the case of unmodified oil.

18.4.9 ANTIFOAM AGENTS/DEMULSIFIERS

Typical antifoam agents that work in petroleum-based lubricants also work in vegetable oil-based or biobased lubricants but the level of efficacy may vary. For example, polymethacrylates are successfully used in petroleum oil lubricants as defoamers but they have limited success in biobased lubricants. Silicon-based defoamers and ethylene oxide/propylene oxide block copolymers are the most effective in vegetable oils, functioning as a surfactant that lowers the interfacial tension between oil and air or water.

ACKNOWLEDGMENT

We are grateful to Functional Products Inc. for supporting this work.

REFERENCES

1. Office of Federal Procurement Policy Office of Management and Budget, *Report to Congress on Implementation of the Resource Conservation and Recovery Act (RCRA), Farm Security and Rural Investment Act of 2002 and the Food, Conservation, and Energy Act of 2008*, Washington, D.C.: United States Government Printing Office, 2011.
2. Deepwater horizon oil spill, last modified December 9, 2015, http://en.wikipedia.org/wiki/Deepwater _Horizon_oil_spill, 2015.
3. IXTOXC I, http://incidentnews.noaa.gov/incident/6250, 2013 (Accessed November 20, 2014).
4. J. Roiz, Biolubricants: Technical and market survey, http://www.valbiom.be/files/library/Docs /Biolubrifiants/valbiom_biolubricants_technicalandmarketsurvey_dec2011_final1326186412.pdf, 2011 (Accessed November 20, 2014).
5. L. A. T. Honary, Biodegradable/Bio-based lubricants and greases, *Mach. Lubr.* 1 (September 2001).
6. S. J. Randles et al., Synthetic base fluids, in: *Chemistry and Technology of Lubricants*, R. M. Mortier and S. T. Orszulik (Eds.), pp. 34–74, London: Blackie Academic and Professional–Chapman and Hall, 1996.
7. Plenish™ high oleic soy: Product overview, http://www2.dupont.com/Biotechnology/en_US/products /plant_biotech_products/plenish_overview.html (Accessed November 20, 2014).
8. E. O. Aluyor and M. Ori-Jesu, The use of antioxidants in vegetable oils—A review, *Afr. J. Biotechnol.*, 25, 4836–4842 (2008).
9. R. Begstra, Green means go, *Lubes'n'Greases*, 10, 40–41 (November 2004).
10. R. Scarth and P. B. E. McVetty, Designer oil canola—A review of new food-grade brassica oils with focus on high oleic, low linolenic types (Paper presented at the annual meeting for the International Consultative Group for Research on Rapeseed, Canberra, Australia, September 26–29, 1999).
11. M. A. Maleque, H. H. Masjuki, and S. M. Sapuan, Vegetable-based biodegradable lubricating oil additives, *Ind. Lubr. Tribol.*, 55, 137–143 (2003).
12. S. Z. Erhan and J. M. Perez, *Biobased Industrial Fluids and Lubricants*, L. Rudnick (Ed.), Champaign, IL: AOCS Press, 2002.
13. A. S. Yano et al., Study on sludge formation during the oxidation process of turbine oils, *Tribol. Trans.*, 47, 111–122 (2004).
14. M. Finšgarand and I. Milošev, Inhibition of copper corrosion by 1,2,3-benzotriazole: A review, *Corros. Sci.*, 52, 2737–2749 (2010).
15. S. Z. Erhan, Oxidative Stability of Mid-Oleic Soybean Oil: Synergistic Effect of Antioxidant-Antiwear Additives, Peoria, IL: National Center for Agricultural Utilization Research, USDA/ARS, 2006.
16. R. T. Mookken et al., Dependence of oxidation stability of steam turbine oil on base oil composition, *Lubr. Eng.*, 53, 19–24 (1997).
17. A. M. Petlyuk and R. J. Adams, Oxidation stability and tribological behavior of vegetable oil hydraulic fluids, *Tribol. Trans.*, 47, 182–187 (2004).
18. H. Wu and T. H. Ren, Tribological performance of sulfonamide derivatives as lubricating oil additives in the diester, *J. Synth. Lubr.*, 23, 211–221 (2006).
19. J. K. Mannekote et al., Structure and composition of vegetable oils, in: *Tribology for Scientists and Engineers: From Basics to Advanced Concepts*, P. Menezes et al. (Eds.), pp. 498–501, New York: Springer Science and Business Media, 2013.

19 Micro- and Nano-TiO$_2$, a Lubricant Additive for Environmentally Friendly Lubricants

Sudeep Ingole, Archana Charanpahari, and Suresh Umare

CONTENTS

ABSTRACT

With stringent environmental regulations, there is a need for better lubricant ingredients (stock and additives) with minimal adverse effects on the environment. This chapter reviews nano-TiO$_2$ as an additive to solid and liquid lubricants. Its versatility, nontoxicity, low cost of synthesis, and biocompatibility promote further investigation on the performance of nano-TiO$_2$ as a lubricant additive. Synthetic methods and approaches, properties, and tribological performance are discussed.

19.1 INTRODUCTION: LUBRICATION ADDITIVES

A typical lubricant consists of additives such as antiwear, friction modifier, and EP for its effective functioning. Antiwear and EP additives are certain types of chemical compounds which provide good boundary lubrication and have the ability to build strong boundary lubricant layers during extreme events (i.e., severe loading conditions). They protect sliding surfaces from indenting by the asperities of opposite surface. From the mechanical properties point of view, antiwear and EP additives are semi-plastic deposits with high shear strength and show moderate to high COF [1]. ZDDPs are antiwear/EP additives. Friction modifiers, on the other hand, are composed of orderly and closely packed layered structure of multimolecules. They will have the polar head attached to the metallic surface and inter-molecular layers are loosely adhered to each other. This structure of friction modifier produces lower COF [1]. Molybdenum disulfide (MoS$_2$) is a friction modifier. Other additives include VI improvers, rust and oxidation inhibitors, anticorrosion agents, and antifoaming agents.

MoS_2 is widely used as a solid lubrication additive in composite materials, lubricating oils, and greases. It has excellent lubricity produced by weak van der Waals forces, which allow easy sliding between two S–Mo–S layers. It is insoluble in lubricant base oil, and it is used in disperse form. Oil-soluble compounds such as molybdenum dialkyldithiophosphates and molybdenum dithiocarbamates are used to disperse MoS_2 in base oils [2].

Surface-capped nano-MoS_3 and ZDDP together have demonstrated friction-modifying and anti-wear properties. Nano-MoS_3 does not exhibit antiwear properties in the absence of ZDDP and provides small wear reduction at elevated temperature. ZDDP stabilizes (i.e., inhibits the oxidation of nano-MoS_2) the tribofilm formation on the sliding surfaces and thus reduces the COF. Compared to bulk MoS_3, nanosized MoS_3 usually has better COF reduction and antiwear properties [2,3], but fails at higher temperature.

19.2 INTRODUCTION TO TiO_2

TiO_2 is used in many engineering applications such as photovoltaic, electrochromic, fuel cells, self-cleaning surfaces, hydrogen sensors, photocatalyst for detoxification of pollutants, and hydrogen generation from water [4,5]. TiO_2 has potential application in tribology. Nanosized TiO_2, an n-type semiconductor, is a candidate for lubricant additive. TiO_2 is nontoxic, has low cost of synthesis, is biocompatible [6–8], and has unique optical and electronic properties [4,9].

Nano-TiO_2 has high surface-to-volume ratio, good refractive index, excellent load-bearing capacity, and high mobility of electron–hole pairs due to quantum confinement. As its size decreases from bulk to nano, there are some unsatisfied valencies of surface atoms, known as dangling bonds, resulting in high surface energy. In order to reduce surface energy, surface relaxation or surface adsorption or agglomeration of particles occurs [10]. Surface properties play an important role in determining the adsorption of various molecules (like water molecules) or groups like the hydroxyl group. Surface properties may be subdivided into (i) chemical effects, e.g., coordination structure of surfaces that controls adsorption of molecules; (ii) electronic structure of the clean surface or defects and adsorbate (e.g., hydroxyl)-induced states that may be crucial for charge trapping and separation of charge carriers on the surface; (iii) interaction of molecules with surface defects; and (iv) surface potential differences (such as work function differences measured in vacuum or flat band potentials in aqueous solution) [10].

These surface effects have prominence in photocatalytic and solar energy applications [11,12]. However, a systematic correlation of these surface properties and TiO_2 phase to tribological properties is hitherto missing. This chapter aims to briefly enumerate the properties with a view to correlate them to tribological properties.

19.2.1 STRUCTURE OF TiO_2

TiO_2 exists as anatase, rutile, brookite, and srilankite phases. Srilankite is a high-pressure, metastable phase [13]. In general, TiO_2 is made up of TiO_6 octahedrons, anatase joined by vertices, rutile joined by edges, and brookite joined by vertices and edges (Figure 19.1a, b, and c). Anatase is made up of corner (vertex)-sharing octahedrons which form (001) planes, resulting in a tetragonal structure. In rutile, the octahedrons share edges at (001) planes to give a tetragonal structure, and in brookite, both edges and corners are shared to give an orthorhombic structure [14]. Anatase is more distorted than rutile phase and has Ti–O bonds, shorter than rutile.

19.2.2 PROPERTIES OF TiO_2

Anatase and rutile are commonly studied phases of TiO_2 as compared to brookite and TiO_2 II phases, which are metastable. The properties of anatase and rutile are elaborated in various reviews [4,13,15,16]. The important properties are enumerated here. For solar applications, the anatase

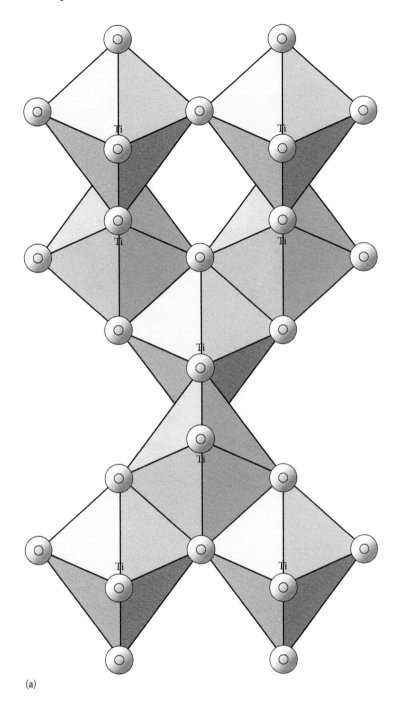

(a)

FIGURE 19.1   (a) Structure of anatase phase of TiO$_2$.   (*Continued*)

structure is preferable instead of other polymorphs because of its higher electron mobility, low dielectric constant, and lower density. Its increased photoreactivity is because of slightly higher Fermi level, lower capacity to adsorb oxygen, and higher degree of hydroxylation in the anatase phase [17]. Anatase has lower hardness than rutile (anatase, 5.5–6; rutile, 6–6.5 on Moh's scale). Anatase undergoes transition to rutile above 600°C, as enthalpy for conversion of anatase to rutile is low, −1.3 to −6.0 KJ/mol. The anatase-to-rutile transition, sometimes referred to as the ART, is

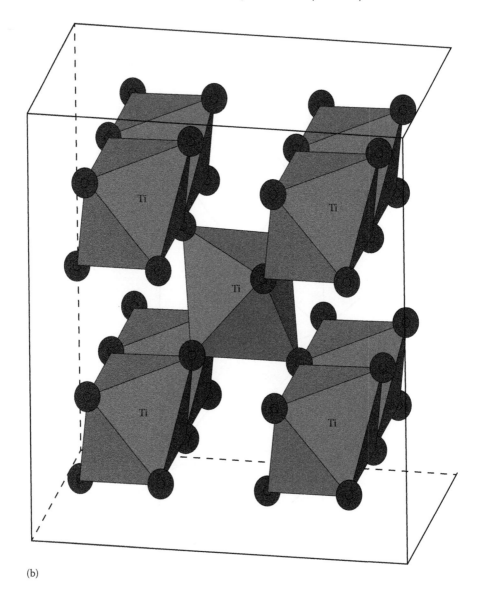

(b)

FIGURE 19.1 (CONTINUED)   (b) Structure of rutile phase of TiO$_2$.   (*Continued*)

a nucleation and growth process. The kinetics of this transition are dependent on variables such as impurities, morphology, sample preparation method, and heat flow conditions. Hanaor and Sorrell [18] in their review reported the various parameters and the various effect of dopents on anatase-to-rutile transformation. An important parameter is oxygen defect levels, where oxygen vacancies enhance the transformation of anatase to rutile phase. The oxygen defect levels are influenced by atmospheric conditions, reduction or oxidation reactions, unintentional impurities, intentional dopants and synthetic reaction conditions.

19.2.3   PHASE TRANSFORMATION IN TiO$_2$

The stability of anatase and rutile phases at varying temperatures and pressures [18] is shown in Figures 19.2 and 19.3. The stability of anatase to pressure-induced transformation is size dependent.

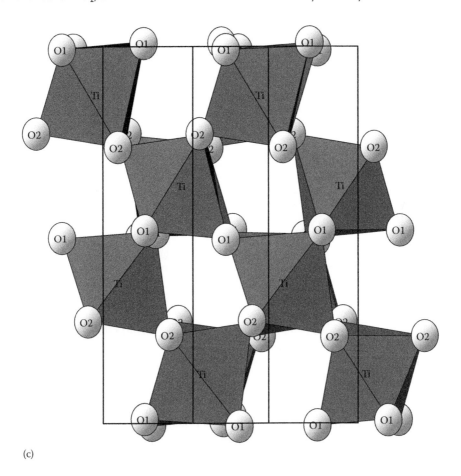

(c)

FIGURE 19.1 (CONTINUED)   (c) Structure of brookite phase of TiO$_2$. (Reprinted from *Applied Catalysis B: Environmental*, 125, Pelaez, M. et al., A review on the visible light active titanium dioxide photocatalysts for environmental applications, 331–349, Copyright (2012), with permission from Elsevier.)

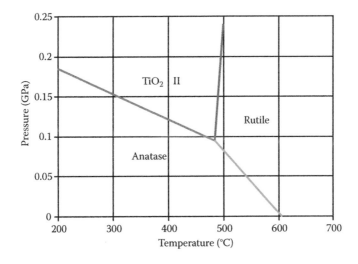

FIGURE 19.2   Reaction boundary conditions for various phases. (With kind permission from Springer Science+Business Media: *Journal of Materials Science*, Review of the anatase to rutile phase transformation, 46, 2011, 855–874, Hanaor, D. A., and C. C. Sorrell.)

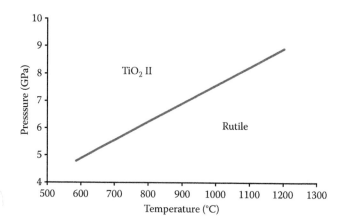

FIGURE 19.3 Transformation of rutile to TiO$_2$ II. (With kind permission from Springer Science+Business Media: *Journal of Materials Science*, Review of the anatase to rutile phase transformation, 46, 2011, 855–874, Hanaor, D. A., and C. C. Sorrell.)

Nanocrystals of various sizes were subjected to a pressure of 30–40 GPa. Three crystallite size regimes exist with regard to pressure-induced phase transitions of anatase at room temperature: the first regime is characterized by the pressure-induced amorphization of nanocrystals typically less than 12 nm; in the second regime size range, 12–50 nm anatase transforms directly to the baddeleyite structure; and in the third regime crystal sizes 40–50 nm, anatase transforms directly to the thermodynamically stable orthorhombic structure like α-PbO$_2$ [19].

TiO$_2$ is intrinsically insulating in nature. But, TiO$_2$ having vacancies or defects or dopants favoring defects/oxygen vacancies (like nitrogen) are conducting due to electrons serving as carriers. Hence, such nitrogen-doped TiO$_2$ exhibits thermoelectric effect, under specific conditions [20].

19.2.4 Synthesis Methods of TiO$_2$

TiO$_2$ synthesis can be classified as bottom-up and top-down approaches. The bottom-up approach involves synthesis of nanosized material from basic building blocks like atoms or molecules. In this approach, conditions/parameters can be tuned to obtain narrower and homogeneous particle distribution, desired morphology, phase, porosity, surface area, particle size, and hydroxyl content. The second approach, the top-down method, involves restructuring a bulk material in order to create a nanomaterial. The tailoring of particle size of the final product is not feasible/reproducible [21].

The synthetic techniques can also be classified as chemical and physical methods. TiO$_2$ can be synthesized by various techniques by chemical methods like coprecipitation/polyol, sol-gel, combustion, hydrothermal, solvothermal, microemulsion, sonochemical, microwave method, and physical methods like laser ablation, vapor deposition, electrodeposition, thermal oxidation, and ball milling [4,15,17,21]. Various authors have enumerated the effects of synthetic conditions on properties of TiO$_2$ (Table 19.1).

Chemical methods generally involve hydrolysis of titanium precursors like TiCl$_4$, or titanium isopropoxide and titanium butoxide to form titanous hydroxide. Sol-gel involves the hydrolysis, polycondensation, drying, and sintering of the precursor to obtain TiO$_2$ nanomaterials. The sol obtained by hydrolysis upon evaporation of solvent gets transformed into gel. The effect of hydrolyzing agent, reactant molar ratio, aging temperature, aging time, and calcination temperature can be altered to tune the properties of TiO$_2$ [4,21,22].

TABLE 19.1

Effects of Method on Properties of TiO$_2$

Method	Effect
1 Thermal decomposition of ammonium titanyl sulfate	Properties of TiO$_2$ tuned by varying the gas atmosphere and reaction temperature; mesoporous texture with mean pore diameter = 15 nm; particle size = 20–30 nm; surface area ca. 64 m^2/g
2 Precipitation of a mixture of titanium (IV) isopropoxide, calculated amount of stearic acid and 1-propanol, followed by calcination at different temperatures	Pore size tuned by adjusting the molar composition of stearic acid; pore diameter = 5–15 nm; surface area = 92–130 m^2/g
3 Four different synthesis routes involving the hydrolysis of titanium (IV) isopropoxide or TiCl$_4$ followed by calcination at different temperatures	Samples made from TiCl$_4$ exhibited the highest photoactivity; 100% anatase phase TiO$_2$ was obtained with crystallite size = 7–30 nm; surface area = 100 m^2/g; pore size = 7–14 nm
4 Combustion of aqueous titanyl nitrate with stoichiometric amounts of glycine at 350°C; precursor—Titanium (IV) isopropoxide	100% anatase phase TiO$_2$ was obtained; particle size = 8 ± 2 nm; bandgap = 2.21 and 2.85 eV; surface area = 246 m^2/g; thermogravimetric analysis (TGA) wt; loss = 15.5%; high surface acidity
5 Combustion of aqueous titanyl nitrate with stoichiometric amounts of glycine at 350°C; precursor—Tetrabutyl titanate	100% anatase phase TiO$_2$ was obtained; crystallite size = 4–6 nm; surface area = 4–6 nm; surface area = 257 m^2/g; bandgap = 2.62 eV
6 Hydrothermal synthesis using TiCl$_4$ using cationic surfactants like CTAB and CPB	Anatase phase TiO$_2$ nanopillar arrays were grown on Ti substrate; ca. 250 nm in width and ca. 700 nm in length with a tetrahedral bipyramidal tip; optimum conditions—1 M tetramethyl ammonium hydroxide, 200°C, 8 h

Source: Vinu, R., and G. Madras, *Journal of the Indian Institute of Science*, 90, 189–230, 2012. With permission.

Polyol process involves heating a titanium precursor in high-boiling solvent like glycol to give titanous hydroxide, which on sintering yields TiO$_2$ nanomaterials. Here, ethylene glycol acts as reducing as well as capping agent to produce TiO$_2$ nanoparticles with narrow size distribution [23].

TiO$_2$ nanostructures with fascinating morphologies like cubes, spheres, and rods were synthesized by a simple microwave irradiation technique. Tuning of different morphologies was achieved by changing the pH and the nature of the medium or the precipitating agent. Anatase nanocubes had the highest surface area and pore size as compared to rutile nanospheres and rutile nanorods [24]. A recent review compiles the work done by various groups and explores a green, versatile, microwave technique for synthesis of various nanomaterials [25].

Microemulsion method actually involves two immiscible phases and a surfactant with amphiphilic properties to be added to a titanium precursor. The surfactant will form micelles under specific conditions, which restricts the growth of particles to nanoregime. The water/surfactant, water/precursor, and reaction temperatures are significant parameters for tailoring the size of TiO$_2$ [4,21].

Hydrothermal method is used for obtaining one-dimensional nanomaterials like nanorods, nanowires, or hierarchical structures like nanowhiskers, nanospheres, and nanopeanuts [26,27]. One-dimensional nanostructures are defined as nanostructures at least one dimension of which lies between 1 and 100 nm, more often the lateral dimension, and clearly include nanowires and nanorods as suggested by Xia et al. [28]. This method involves basically heating a titanium precursor under controlled temperature and/or pressure in an aqueous reaction medium (hydrothermal) and in a nonaqueous reaction medium (solvothermal). TiO$_2$ precursor initially gets converted into lamellar nanosheet structure, which on bending and rolling, changes to nanorods and nanowires. The hydrothermal method can be used in combination with other techniques like microwave-hydrothermal and microwave-solvothermal. The temperature and the amount of precursor solution determine the internal pressure produced and they can be used to tailor the morphology/phase properties of TiO_2. In some cases, templates/surfactants are used to obtain one-dimensional material. A template may be defined as a central structure within which a network forms in such a way that removal of the template creates a filled cavity with morphological and/or stereochemical features related to those of the template [29]. Surfactants like cetyl trimethylammonium bromide (CTAB), cetyl pyridinium chloride (CPB), and sodium dodecyl sulfonate or polymers like Pluronic F-127, polyvinyl pyrrolidone, and polyethylene glycol are added to a precursor to obtain porous nanomaterials [4,21]. These not only restrict the particle size but also inhibit the aggregation of particles through stabilization of their surface energy. On sintering, these surfactants/templates can be removed, resulting in porous nanomaterials. A recent review gives an elaborate account of mechanism, methodology, and parameters for the synthesis of various one-dimensional materials [30].

Solution combustion method is an integrated approach (breaking down and building up) for the synthesis of nanocrystalline oxide materials, as the desired oxide products nucleate and grow from the combustion residue. This is actually an exothermic, self-propagating reaction requiring the use of oxidizer like water-soluble metal nitrates and fuels like urea, glycine, metal acetates, and hydrazides. Synthesis of metal oxides of desired composition, phase, and structure is achieved by rapidly heating aqueous precursor solution with stoichiometric quantities of redox mixture. Thus, properties can be tailored by varying the oxidizer/fuel, precursor/fuel ratios, and reaction temperature and altering different fuels [13].

All solution methods basically involve nucleation, growth, and aging. The main parameters governing these processes are the solution properties, including the solvent viscosity, the dielectric constant, and the presence of adsorbing anions, the solubility of the metal oxide, and the metal oxide surface energy. The interplay of all these parameters will determine the final particle size and size distribution [31].

Chemical vapor deposition and physical vapor deposition (PVD) are common methods for synthesis of films and coating. Here, thermal energy heats up gases in the vacuum chamber, which are condensed to form solid-phase nanomaterials on deposition. When a chemical reaction occurs, it is called chemical vapor deposition; otherwise, it is called PVD. PVD encompasses a host of other techniques like sputtering, laser ablation, thermal deposition, and ion implantation. The reaction temperature, substrate/deposition temperature, and gaseous composition can be used to tune the properties of nanomaterials. For example, after deposition on silicon or bare fused silica substrates at 630°C and 560°C under 5 Torr pressure, single crystalline rutile and anatase nanorods were formed, respectively; while at 533°C under 3.6 Torr, anatase nanowalls composed of well-aligned nanorods were formed [4,32].

Sonochemical method utilizes acoustic cavitation for synthesis. This involves generation, growth, and collapse of bubbles in the liquid precursor solution due to the action of ultrasound. This results in intense localized heating with a rise of temperature (~5000 K) and high pressure (~1000 atm). Thus, enormous heating and cooling rates are responsible for the formation of nanomaterials [21].

Ball milling is a convenient top-down approach for obtaining nanosized powders. By vigorously shaking or high-speed rotation, a high mechanical energy is applied on the powders because of collision with heavy balls. The milling process embraces a complex mixture of fracturing, grinding, high-speed plastic deformation, cold welding, thermal shock, and intimate mixing. Thus, TiO_2 thus prepared has narrow particle size, high specific surface area, and disordered phases like amorphous TiO_2 and srilankite phase [33].

Direct anodization of Ti metal in the presence of ethanol [34] or acetone also gives TiO_2 nanostructures. Also, electrochemical etching is a metal-etching process which relies on surface reduction/oxidation (redox) reactions to selectively remove certain portions of the bulk material and leave one-dimensional nanostructures [30]. There is no particular generalization that a specific synthetic process will give a specific phase of TiO_2 [18].

19.3 TiO$_2$ LUBRICATION

19.3.1 TiO$_2$ as Solid Lubricant

Liquid lubrication provides the means to minimize the tangential resistance when heavily loaded counterparts roll or slide over each other. When this function is performed by a solid substance [35] in the absence of a liquid lubricant or in dry condition, it is called solid lubrication and the material used is called solid lubricant. Conventional solid lubricants have layered structure and primarily hexagonal lattices such as in graphite and molybdenum disulfide.

Song et al. [36] studied friction and wear behavior of TiO$_2$ nanotube reinforced polyurethane (PU) composite coatings. Two PU composite coatings, i.e., unmodified TiO$_2$ nanotube reinforced (TiNTs/PU) and hexamethylene diisocynate (HDI)-modified TiO$_2$ nanotube reinforced (TiNTs-HDI/PU), were synthesized. These coatings were applied on steel substrates with thickness ranging between 30 and 50 µm. In this study, friction and wear tests were conducted using MHK-500 wear tester (Jian Testing Machine Factory, China) in ring-on-block configuration using speeds between 1.28 and 3.84 m/s and applied load between 320 and 1620 N for 60 minutes for each test. A steel ring was rotated against a composite coating block. A ratio of sliding distance in meters (m) and corresponding coating thickness in microns (µm) is reported as specific wear life (m/µm). The COF consistently reduced with increasing load for both types of composite coatings (Figure 19.4a). TiNTs-HDI/PU showed lower COF compared to TiNTs/PU. The specific wear life of both composite coatings was reduced as the applied load increased. TiNTs-HDI showed longer wear life (Figure 19.4b) compared to TiNTs/PU. Authors suggested that the factor contributing to the reduced COF and increased wear life was the dispersion of TiO$_2$ nanotubes due to HDI modification. The interfacial adhesion between modified TiNTs and PU might have increased.

The synergic effect of lubricant additives, i.e., nano-MoS$_2$ and anatase nano-TiO$_2$ on lubrication behavior in liquid paraffin (LP), was studied by Hu et al. [37]. The nano-TiO$_2$ used in this study was commercial grade available from Zixilai Company, China. Synthesis method of nano-MoS$_2$ and TiO$_2$/MoS$_2$ nanoclusters is reported in the work of Hu et al. [37] and references cited therein. The chosen additive of 1.0 wt% was added in LP. The solution was then mixed using ultrasonic bath for 10 minutes. Tribological tests were conducted using MQ-800 four-ball tribometer (Jinan Shijin Group Co., China) at room temperature. The applied load was 300 N for 30 min with sliding velocity

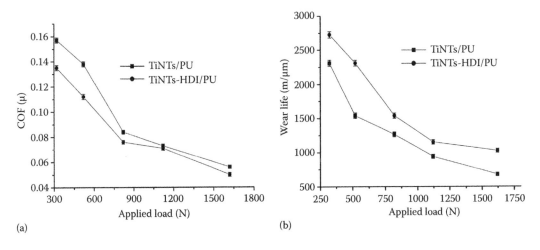

(a)

(b)

FIGURE 19.4 Effects of load on (a) COF and (b) wear life. TiNTs/PU: polyurethane composite coating with unmodified TiO$_2$ nanotube reinforcement; TiNTs-HDI/PU: polyurethane composite coating with TiO$_2$ modified with hexamethelene diisocynate. (Reprinted from *European Polymer Journal*, 44, Song, H.-J., Z.-Z. Zhang, and X.-H. Men, Tribological behavior of polyurethane-based composite coating reinforced with TiO$_2$ nanotubes, 1012–1022, Copyright (2008), with permission from Elsevier.)

of 0.556 m/s. SAE52100 bearing balls with 12.7 mm diameter were used. Wear scar diameter (WSD) was measured using optical microscope and used to estimate the wear loss (wear resistance/antiwear property). It was reported that a nanocluster with 2:1 weight ratio of MoS_2/TiO_2 provided the lowest COF (Figure 19.5a) when compared with other ratios in MoS_2/TiO_2 nanoclusters, nano-MoS_2, and nano-TiO_2. It was reported that there was no correlation between the average WSD and COF. The nanocluster with the lowest COF produced average WSD close to that of nano-MoS_2. It can be inferred that the nanocluster reduced the COF with similar wear characteristic to that of nano-MoS_2. An increase in nano-TiO_2 content in the nanocluster reduced the antiwear property (Figure 19.5b).

Another application of solid lubricants is as low-friction materials for surface coatings or reaction agents for the synthesis of composite materials. There are several studies on TiO_2 chemistry to develop coatings [38–42] and/or composite reinforcement [43] for improved tribological performance. TiO_2 microporous coating formed on titanium alloy (Ti6Al4V) was studied for tribological performance [44]. The coatings were developed using microarc oxidation process, which is a combination of electrochemical oxidation and a high-voltage treatment. The surfaces were treated for 30 min, which produced approximately 3 μm sized microspores (Figure 19.6a). The coating thickness obtained was 10 μm with a mixture of rutile and anatase structures. In this study, dry friction

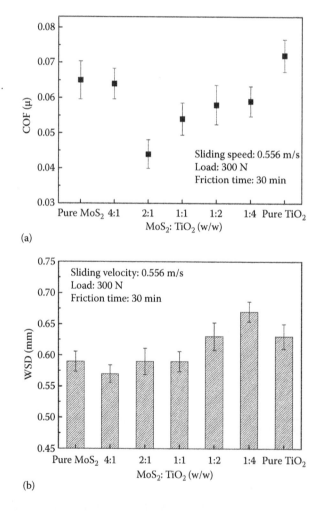

(a)

(b)

FIGURE 19.5 Effects of MoS_2-to-TiO_2 ratio (MoS_2/TiO_2) on (a) COF and (b) average WSD. (With kind permission from Springer Science+Business Media: *Tribology Letters*, Synergistic effect of nano-MoS_2 and anatase nano-TiO_2 on the lubrication properties of MoS_2/TiO_2 nano-clusters, 43, 2011, 77–87, Hu, K. H. et al.)

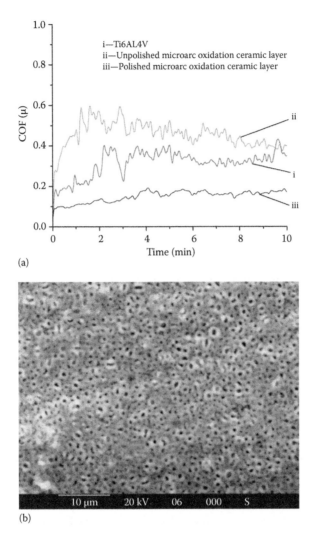

(a)

(b)

FIGURE 19.6 (a) COFs of (i) Ti6Al4V substrate and (ii) unpolished and (iii) polished surfaces; (b) surface morphology of TiO$_2$ coating. (Reprinted from *Progress in Organic Coatings*, 64, Fei, C. et al., Study on the tribological performance of ceramic coatings on titanium alloy surfaces obtained through microarc oxidation, 264–267, Copyright (2009), with permission from Elsevier.)

tests were conducted using ball-on-disk test configuration on Ti6Al4V alloy surface and coated surface in unpolished and polished conditions. The counterface was a SAE52100 bearing ball with 4.75 mm diameter. The normal load, the rotation speed, and the sliding time used during these tests were 100 N, 1000 rpm, and 10 min, respectively. Abrasive papers of up to 800 grit size were used to polish the coating surface. The COF of the specimen tested is shown in Figure 19.6a. It is evident that TiO$_2$ coating with a mixture of rutile and anatase phase structures reduced the COF. Polished coating showed further reduction in COF.

Similar coatings were tested by Wang et al. [45] under lubricated conditions using fretting test configuration. During the fretting test (Plint fretting fatigue machine), a SAE52100 bearing ball was slid against the lubricated coating with vibrating amplitude of 60 μm and 100 N normal load. For lubrication, the coated specimens were dipped in the Shell Tellus T46 oil for 2 min and removed for fretting tests. Each test was conducted for 10,000 cycles at room temperature. COF was found to be 0.15. Relatively smooth wear tracks were observed during the lubricated test conditions without significant material transfer. It was suggested that microcavities that were formed on the surface might have served as the lubricant reservoir.

Aluminum alloys are successfully used in industrial applications because of their light weight, superior corrosion resistance, and high electrical and thermal conductivities. Strength comparable to steels can be achieved through age hardening of aluminum alloys. However, aluminum alloys have poor wear properties due to scoring, adhesive wear, and plastic deformation [46,47].

Sun studied the tribological properties of rutile TiO_2 coating synthesized on selected aluminum alloy [47]. Commercially available 6061 aluminum alloy was initially coated with pure titanium using direct current (DC) magnetron sputtering technique. Pure titanium coating was converted to TiO_2 using thermal oxidation at 550°C for 5- and 11-hour duration in an air furnace. This type of coating showed four structural zones. These zones consist of the first layer (top layer) of rutile TiO_2, which is followed by the second layer, i.e., α-Ti with dissolved oxygen and nitrogen. The intermetallic phases of Ti and Al were located in the third layer. The fourth layer was a Ti diffusion zone above the aluminum alloy substrate. The author conducted dry friction tests using coal supply and transportation model (CSTM) pin-on-disk tribometer at room temperature for aluminum alloy substrate, pure titanium coating, and rutile TiO_2 coatings. The specimens were rotated at 100 rpm (0.05 m/s) against stationary 6 mm diameter alumina ball counterface with 1 and 2 N of applied load. The wear volume (mm^3) was calculated using surface profile of wear track using surface profiler. The wear volume normalized with sliding distance (m) and applied load (N) was used to calculate the wear rates ($mm^3/N\ m$). Figure 19.7a shows the COF of aluminum alloy substrate and pure titanium coating (applied load of 1 N). It is evident that higher COF for both the materials (i.e., aluminum alloy substrate and pure titanium coating) is fluctuating with a large COF value. Both rutile TiO_2 coatings (5 and 11 h) showed lower COF when compared to aluminum alloy substrate and pure titanium coating. The fluctuation of COF values was significantly reduced (Figure 19.7b). The thermal oxidation time showed influence on the COF of the rutile TiO_2 coating. The longer time of oxidation provided lubricious coating (Figure 19.7b).

The aluminum alloy substrate and pure titanium coating demonstrated severe wear behavior as compared to both rutile TiO_2 coatings (Figure 19.8).

Surface protection of aluminum alloys with rutile TiO_2 can extend their applications where superior tribological properties, energy efficiency, and weight reduction are some of the design criteria. This will consequently lead to reduced usage of lubrication.

Also, thermal oxidation improves the interfacial strength by metallurgical bond formation between the substrate and the intermetallic and solid solutions of Ti and Al [47].

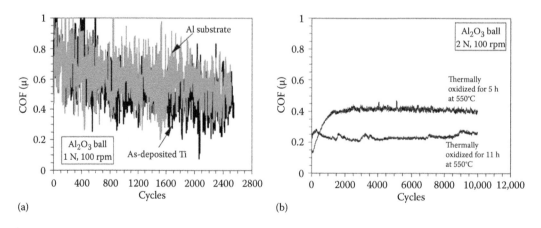

(a)   (b)

FIGURE 19.7 COFs of (a) pure aluminum and as-deposited Ti and (b) thermally oxidized TiO_2 layer tested against 6 mm diameter alumina ball counterface. (Reprinted from *Applied Surface Science*, 233, Sun, Y., Tribological rutile-TiO_2 coating on aluminium alloy, 328–335, Copyright (2004), with permission from Elsevier.)

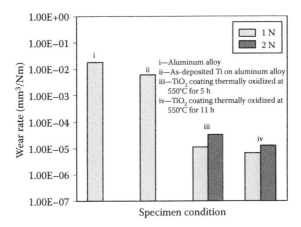

FIGURE 19.8 Wear rates of aluminum alloy, pure titanium coating, and thermally oxidized rutile TiO₂ coatings. (Reprinted from *Applied Surface Science*, 233, Sun, Y., Tribological rutile-TiO₂ coating on aluminium alloy, 328–335, Copyright (2004), with permission from Elsevier.)

Krishna and Sun [48] studied the tribological properties of TiO₂ coatings. These coatings are developed with similar method that Sun [47] used and as described previously. Krishna and Sun used surgical grade AISI 316L austenitic stainless steel as substrate materials. Titanium coating that was 1 μm thick was developed on substrate using DC magnetron sputtering at 300 W DC power and 300°C. Thermal oxidation of titanium coating for 5 hours in air at 550°C gave rutile TiO₂. The coating showed three zones; i.e., the first zone (top layer) was rutile TiO₂, the second zone underneath the rutile TiO₂ consisted of oxygen and nitrogen dissolved in α-Ti, and the third zone was Ti-diffused interface with steel substrate. Unlubricated friction tests were conducted using CSTM pin-on-disk tribometer for substrate, titanium coating, and rutile TiO₂ coating. The specimen was rotating with a linear speed of 5 cm/s at room temperature and ambient atmosphere against 6 mm diameter alumina ball counterface. Tests were conducted for 200 m of sliding distance using 1 N applied load. The COFs of steel substrate, as-deposited coating, and rutile TiO₂ coating are shown in Figure 19.9. Higher COF and fluctuation of COF values are evident in substrate and as-deposited coating. The rutile TiO₂ coating drastically reduced the COF.

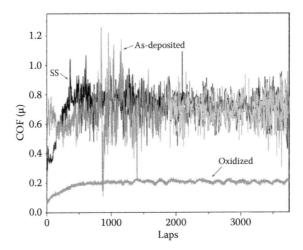

FIGURE 19.9 COFs of AISI 316L austenitic stainless steel (SS), as-deposited Ti coating, and thermally oxidized coating (rutile-TiO₂). (Reprinted from *Applied Surface Science*, 252, Krishna, D. S. R., and Y. Sun, Thermally oxidised rutile-TiO₂ coating on stainless steel for tribological properties and corrosion resistance enhancement, 1107–1116, Copyright (2005), with permission from Elsevier.)

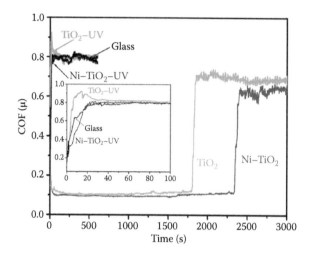

FIGURE 19.10 COFs of glass substrate, TiO_2 film, Ni-TiO_2 film, UV-irradiated TiO_2 film, and Ni-TiO_2 film against SAE52100 bearing ball. (Reprinted from *Applied Surface Science*, 258, Wan, Y. et al., Effect of UV irradiation on wear protection of TiO_2 and Ni-doped TiO_2 coatings, 4347–4350, Copyright (2012), with permission from Elsevier.)

The effect of ultraviolet (UV) irradiation on wear protection of TiO_2 coatings was studied by Wan et al. [49]. TiO_2 and Ni-doped TiO_2 thin film were prepared on glass plates with sol-gel and spin-coating method at 2000 cycles/s speed for 30 s duration. As the coated films initially dried for 15 min at 80°C and then sintered at 480°C for 1 h, nickel-doped TiO_2 exhibited anatase phase with no impurity peaks for rutile or NiO. UV irradiation at room temperature was performed by using Xe excimer lamp (172 nm wavelength; Ushio Inc., Tokyo, Japan) for 20 min. The intensity of irradiated light was 7 mW/cm². Friction and wear tests were conducted at room temperature and ambient atmosphere using Center for Tribology Inc. UMT-3 tribometer in reciprocating configuration. The applied load of 0.01 N with 10 mm/s sliding speed and 5 mm amplitude of reciprocation were used. The countersurface was a 4 mm diameter SAE52100 bearing ball for all tested materials in this study. Authors reported the wear performance (wear life) based on the time for constant COF values. COF was obtained for glass substrate, Ni-TiO_2 film, TiO_2 film, and UV-irradiated Ni-TiO_2 and TiO_2 films. The average COF of glass substrate was 0.8. The protective film of TiO_2 reduced the COF to 0.1 until 1650 s and it was abruptly increased to 0.7. Ni-TiO_2 film also reduced COF to 0.1 until 2300 s and then it was drastically increased to 0.6. The wear properties deteriorate on exposure to UV light for both the films, with wear life being only 20 s (Figure 19.10). The authors suggested the deteriorated wear life was due to UV-induced hydrophilicity of TiO_2 films [49].

19.3.2 TiO_2 NANOPARTICLES AS LUBRICANT ADDITIVES

Solid lubricant in particle form as additives can also be used in greases and liquid lubricant. Reducing the size of TiO_2 particles from micro- to nanometer can have a significant effect on lubrication. The size of TiO_2 showed the influence on wear resistance when used as reinforcement in composites [43]. Study showed that severe adhesion and abrasion resulting in surface failure when micro-TiO_2 particles were used for reinforcement. The nanosized TiO_2 reinforcement resulted in slight abrasion of composites. As the size of the particles reduces, the surface area increases and the particles show better dispersion. Microsized particles showed higher hardness than nanosized TiO_2, which explains the abrasive wear result from microsized particles. Nanosized TiO_2 tends to deposit in the concave of rubbing surface, which can result in reduced COF when added to lubricating oil [50].

Some studies are discussed here that used the four-ball test to evaluate the application of TiO$_2$ nanoparticles as lubricant additives [50–54]. The particles were prepared with different synthesis methods and dispersion media.

Gao et al. studied the tribological properties of oleic acid-modified nano-TiO$_2$ (OA-TiO$_2$) (30 nm) using a four-ball test in double distilled water [51]. The tests were conducted at room temperature for a 30 min duration with 1450 rpm. The four balls used in this investigation were SAE52100 bearing balls with 12.7 mm diameter. The tribological performance was compared between double distilled water, water with dispersant, and different concentrations of OA-TiO$_2$ additives dispersed in double distilled water. The authors did not list the specific dispersant used during this investigation; however, it is reported that OA-TiO$_2$ dispersed in nonpolar and weakly polar solvents such as toluene, acetone, and chloroform, and also dispersed in water with small addition of dispersant (0.5%). The applied load varied between 100 and 600 N. The maximum nonseizure load which is the representation of the load-bearing capacity was reported along with antiwear properties and COF. It was reported that the maximum nonseizure load was improved (700%) with the addition of a small amount (0.1%) of OA-TiO$_2$ additive. The load-bearing capacity further improved; i.e., for 1% OA-TiO$_2$ additive addition, maximum nonseizure load reached 1000 N (900%), and for 2% OA-TiO$_2$ additive addition, it achieved a value of 1046 N. The antiwear property was studied using WSD. WSD reduced with the addition of OA-TiO$_2$ of up to about 0.5%, showing improved antiwear property; thereafter, WSD gradually increased (Figure 19.11a). These reported results were for 300 N applied load. WSD increased when applied load was increased (Figure 19.11b). WSDs obtained with water and water with dispersant were always larger than WSDs obtained with 0.5% of OA-TiO$_2$ in water for corresponding loads of up to 400 N. It showed that 0.5% addition of OA-TiO$_2$ in water improved the antiwear performance. Figure 19.11c shows the graph between COF and varying amounts of OA-TiO$_2$ addition to water. It is clear that the addition of OA-TiO$_2$ to water reduced COF.

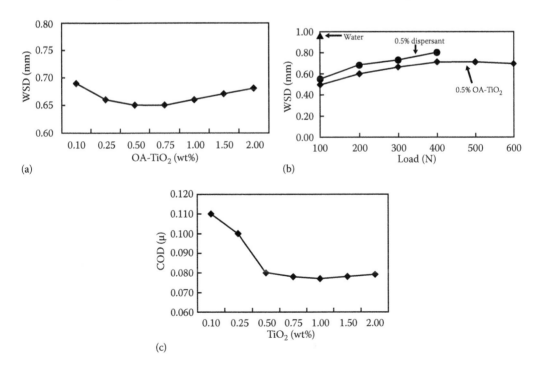

FIGURE 19.11 Antiwear property: (a) WSD variation with addition of weight percentage of TiO$_2$, (b) WSD as a function of applied load, and (c) COF of OA-TiO$_2$ additive in double distilled water. (Reprinted from *Materials Science and Engineering: A*, 286, Gao, Y. et al., Tribological properties of oleic acid-modified TiO$_2$ nanoparticle in water, 149–151, Copyright (2000), with permission from Elsevier.)

Tribological properties of TiO$_2$ nanoparticles modified with tetrafluorobenzoic acid were studied by Ye et al. [53]. The average particle size of TiO$_2$ particles used in this investigation was 40 nm. The tetrafluorobenzoic acid-modified TiO$_2$ (FA-TiO$_2$) particles were dispersed in LP. Four-ball tests were carried out with 1450 rpm speed for a 30 min test duration at room temperature. The balls used in these tests were SAE52100 with 12.7 mm diameter. WSD, as an indicative of antiwear property, drastically reduced with 0.2% addition of FA-TiO$_2$ to LP (Figure 19.12a). For pure LP, the WSD was 0.79 mm and it was reduced to 0.36 mm when 0.25% FA-TiO$_2$ was added to LP. Further addition increased WSD. Addition of FA-TiO$_2$; 0.75%, 1%, and 1.5%, WSD remained relatively constant (approximately 0.475 mm) but was lower than WSD obtained for pure LP. The authors suggested that antiwear properties enhanced due to the formation of thin film of FA-TiO$_2$ on the sliding surfaces (for 0.25% addition). As the TiO$_2$ content increased, abrasive wear dominated surface damage. The WSD variation as a function of applied load (Figure 19.12b) for LP with the addition of 0.25% FA-TiO$_2$ showed that the WSD slightly changed for loads between 200 and 400 N. The WSD was 0.75 mm when 500 N load was applied and suggested that the system was effectively lubricated for this high load. From Figure 19.12c, it can be seen that addition of FA-TiO$_2$ to LP increased the maximum nonseizure load. With the addition of 1% FA-TiO$_2$, the maximum nonseizure load obtained was 700 N. This observation demonstrates the better antiwear property of LP when FA-TiO$_2$ additives were added.

Hu and Dong studied antiwear and frictional behavior of nano-TiO$_2$ additive in 500 SN base oil [50]. They used ethanol supercritical fluid drying preparation technique to synthesize the nanoparticles of TiO$_2$. Nano-TiO$_2$ was dispersed in a base oil using sorbitol monostearate as the dispersing agent. MQ-800 four-ball tribotester was used to measure the maximum nonseizure load. The COF was measured for base oil and base oil with additive (0.95 wt% nano-TiO$_2$ and 1 wt% dispersant) with 300 N applied load using HQ-I block-on-ring tribometer. CrWMn steel ring (49 mm diameter, 13 mm height) was used against a steel block at 1500 rpm. The maximum nonseizure load for base oil with additive addition was increased with additive contents (wt%). It was 594 N for base oil and it achieved the highest value with the 1.5 wt% addition of additive (Figure 19.13a). The COF of the base oil when compared to the base oil with additive showed lower value in the beginning of the test but gradually increased and stayed high. The COF of the base oil with additive was constant and lower compared to that of the base oil (Figure 19.13b). The authors suggested that the enhanced antiwear property and the reduced COF of the base oil with additive content are due to the deposition of nano-TiO$_2$ particles on the rubbing surfaces, which reduced the shearing stress at the sliding interface [50].

Qian et al. [54] compared the tribological performances of LP with and without the addition of stearic acid-modified microspheres of TiO$_2$ (SA-TiO$_2$). In this investigation, the anatase phase TiO$_2$ microspheres were synthesized using facile solvothermal method. The microspheres were added to an anhydrous ethanol solution of 0.02 M stearic acid to modify their surfaces. The mixture was filtered, washed, and vacuum dried at 50°C after heating for 3 h under reflux. Tribological tests were carried out using a four-ball tester (Jian instrument manufacturer, China) using 12.7 mm SAE52100 bearing balls at 1450 rpm and ambient temperature. The applied load was 200 N for a 30 min test duration. Figure 19.14 shows the graph of COF and WSD as a function of the weight percentage of SA-TiO$_2$ added to LP and of LP alone. It can be concluded from this figure that addition of SA-TiO$_2$ to LP of up to 2 wt% effectively reduced COF. The COF increased when the SA-TiO$_2$ content increased beyond 2 wt%. The WSD reduced with the addition of SA-TiO$_2$ of up to 1.5 wt%. It increased when the SA-TiO$_2$ content was higher than 1.5 wt%. The authors explained the reduced COF values and the increased antiwear property (i.e., reduced WSD) based on adsorption of long organic chain segments through polar carbonyl group and TiO$_2$ microspheres on sliding surfaces. They suggested that tribochemical reactions on the sliding surfaces also contributed to improve the tribological properties.

Zhang et al. studied the tribological properties of stearic acid-modified anatase TiO$_2$ nanoparticles (SA-nTiO$_2$) in LP [55]. The nanoparticles were synthesized using solvothermal method.

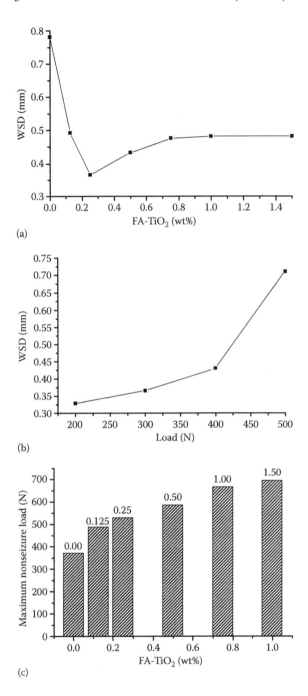

FIGURE 19.12 Antiwear property: (a) WSD variation with addition on FA-TiO$_2$ to LP, (b) WSD variation with applied load, and (c) maximum nonseizure load variation with addition FA-TiO$_2$ in LP. (Reprinted from *Materials Science and Engineering: A*, 359, Ye, W. et al., Preparation and tribological properties of tetrafluorobenzoic acid-modified TiO$_2$ nanoparticles as lubricant additives, 82–85, Copyright (2003), with permission from Elsevier.)

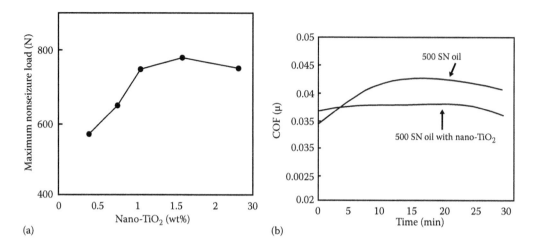

(a)

(b)

FIGURE 19.13 Effects of nano-TiO$_2$ weight percentage on (a) maximum nonseizure load and (b) COF. (Reprinted from *Wear*, 216, Hu, Z., and J. Dong, Study on antiwear and reducing friction additive of nanometer titanium oxide, 92–96, Copyright (1998), with permission from Elsevier.)

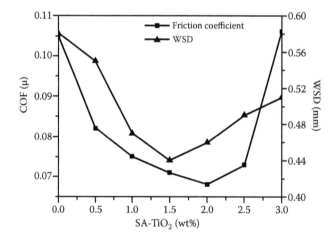

FIGURE 19.14 COF and WSD as a function of weight percentage of SA-TiO$_2$ in LP and LP alone. (Reprinted from *Applied Surface Science*, 258, Qian, J. et al., Preparation and tribological properties of stearic acid-modified hierarchical anatase TiO$_2$ microcrystals, 2778–2782, Copyright (2012), with permission from Elsevier.)

Nanoparticles were dispersed in 0.056 M stearic acid solution in 250 mL ethanol for surface modification for 4 h at 75°C followed by filtration, washing, and drying in vacuum desiccators. A four-ball tester was used with 12.7 mm diameter SAE52100 bearing balls, and tests were conducted at ambient temperature. SA-nTiO$_2$ particles were dispersed in LP using ultrasonic mixing. The test speed was 1450 rpm for a 30 min duration. The effect of SA-nTiO$_2$ added to LP on COF and WSD was studied at the applied load of 200 N. The COF and the WSD of LP were also studied at the applied load of 200 N. It can be found from Figure 19.15a that 0.25 wt% addition of SA-nTiO$_2$ to LP reduced the COF to minimum among all weight percentage additions. The COF increased when SA-nTiO$_2$ addition was increased to 0.5 wt% and to 1.0 wt%. This COF was comparable to the COF of LP.

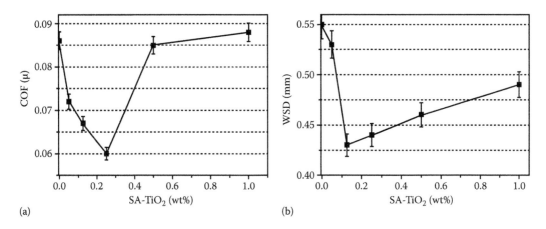

FIGURE 19.15 (a) COF and (b) WSD, i.e., antiwear property as a function of stearic acid-modified anatase TiO$_2$ nanoparticles (SA-nTiO$_2$) content in LP). (With kind permission from Springer Science+Business Media: *Tribology Letters*, Synthesis and tribological properties of stearic acid-modified anatase (TiO$_2$) nanoparticles, 41, 2011, 409–416, Zhang, L. et al.)

WSD significantly reduced when 0.15 wt% SA-nTiO$_2$ was added to LP, and gradually increased with further addition of SA-nTiO$_2$ to LP (Figure 19.15b).

The better performance was due to TiO$_2$ nanoparticles depositing on the surface and resulting in third body sliding and reduced shear stress. The debris generated was unstable compared to the stable deposited nanoparticles at the sliding interfaces [50].

TiO$_2$, due to its high surface area and surface defect states, enables efficient dissipation of heat generated due to friction. Excellent load-bearing capacity and effective heat dissipation enhanced the antiwear properties. Thus, nano-TiO$_2$ with its versatility, nontoxicity, low cost of synthesis, and biocompatibility [6–8] has a good potential to become an environmentally friendly lubricant additive.

Ingole et al. [56] performed a comparative study of the tribological properties of two types of nano-TiO$_2$ (i.e., anatase and mixed anatase–rutile) particle addition to base oil. They used TR-281M-M6 reciprocating friction and wear monitors (Ducom Instruments Pvt. Ltd.). Both sliding surfaces were SAE52100 steel. A specimen that was 20 mm in diameter and 5 mm thick was reciprocating against a 5 mm diameter bearing ball with applied load of 14.715 N, sliding velocity of 0.05 m/s, and sliding amplitude of 5 mm in ambient temperature for 30 minutes. The base oil used in this investigation was mineral oil which was recycled (refined) using total vacuum distillation system. Two types of nano-TiO$_2$ particles used were Degussa® P25 (P25) (mixture of rutile and anatase phases) and anatase TiO$_2$ (TiO$_2$). P25 was acquired from Evonik Degussa India Pvt. Ltd. Anatase TiO$_2$ was synthesized using polyol-mediated coprecipitation method. The COF of the base oil obtained was 0.1. The addition of 0.25 and 1 wt% of TiO$_2$ slightly reduced COF compared to that of the base oil (Figure 19.16a). The COF of the base oil was fluctuating with time during most of the test duration. This fluctuation was minimized when TiO$_2$ was added to the base oil. The COF (when P25 was added to the base oil) was higher for all contents (i.e., 0.25, 1, and 2 wt%) when compared to the COF of the base oil and the COF of the base oil with all TiO$_2$ weight percentage additions (Figure 19.16a). The authors explained the reduced fluctuation in COF within the test period based on worn surface observation. Figure 19.16b shows that the adsorption of TiO$_2$ film on the surface of the sliding component that provided the lower resistance and COF fluctuation was reduced. The worn surface showed abrasive wear when P25 was used (Figure 19.16c).

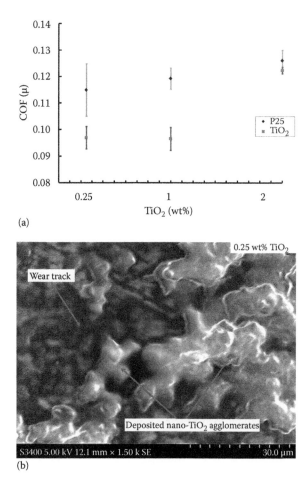

(a)

(b)

FIGURE 19.16 (a) Comparison of COFs of base oil with and without addition of TiO$_2$ (P25 and anatase TiO$_2$); (b) scanning electron micrograph of anatase TiO$_2$ film formed on the worn surface. (Reprinted from *Wear*, 301, Ingole, S. et al., Tribological behavior of nano-TiO$_2$ as an additive in base oil, 776–785, Copyright (2013), with permission from Elsevier.)

19.4 CONCLUSIONS

Micro- and nanosized TiO$_2$ particles as solid lubricant and a lubricant additive are reviewed. Synthesis methods, properties of TiO$_2$, and studies conducted on the tribological properties such as COF and antiwear properties are described. TiO$_2$ as a solid lubricant in the form of coatings or composite reinforcement showed reduced COF. The results of four-ball tests of lubricant with TiO$_2$ as an additive are also reviewed. The results included load-bearing capacity, maximum antiseizure load, and antiwear properties based on the WSD shown by TiO$_2$. The enhanced load-bearing capacity, friction reduction, improved antiwear property, and other properties such as nontoxicity, low cost of synthesis, and biocompatibility also add benefit toward its applications as lubricant additive. The above-mentioned tribological property improvements were from the dispersion of TiO$_2$ in the lubricant and the adsorption of the nano-TiO$_2$ on sliding interfaces. However, there is still a need for careful considerations and additional scientific enquiries on the tribological properties of nano-TiO$_2$. These tribological properties include the definitive effect of the phase of triboperformance, in relation to its photocatalytic activities to friction, and wear behavior.

REFERENCES

1. Rudnick, L. R., *Lubricant Additives: Chemistry and Applications*. CRC Press, Boca Raton, FL, 2003.
2. Bakunin, V. N. et al., Synthesis and application of inorganic nanoparticles as lubricant components—A review. *Journal of Nanoparticle Research*, 6(2): pp. 273–284, 2004.
3. Bakunin, V. N. et al., Tribological behavior and tribofilm composition in lubricated systems containing surface-capped molybdenum sulfide nanoparticles. *Tribology Letters*, 22(3): pp. 289–296, 2006.
4. Chen, X., and S. S. Mao, Titanium dioxide nanomaterials: Synthesis, properties, modifications, and applications. *Chemical Reviews*, 107(7): pp. 2891–2959, 2007.
5. Carp, O., C. L. Huisman, and A. Reller, Photoinduced reactivity of titanium dioxide. *Progress in Solid State Chemistry*, 32(1): pp. 33–177, 2004.
6. Sun, T., and M. Wang, Mechanical performance of apatite TiO₂ composite coatings formed on Ti and NiTi shape memory alloy. *Applied Surface Science*, 255(2): pp. 404–408, 2008.
7. Siu, H., and H. Man, Fabrication of bioactive titania coating on nitinol by plasma electrolytic oxidation. *Applied Surface Science*, 274: pp. 181–187, 2013.
8. Chen, H.-T. et al., Microscopic observations of osteoblast growth on micro-arc oxidized β titanium. *Applied Surface Science*, 266: pp. 73–80, 2013.
9. Liqiang, J. et al., Review of photoluminescence performance of nano-sized semiconductor materials and its relationships with photocatalytic activity. *Solar Energy Materials and Solar Cells*, 90(12): pp. 1773–1787, 2006.
10. Cao, G., *Synthesis, Properties, and Applications*. World Scientific, Singapore, 2004.
11. Luttrell, T. et al., Why is anatase a better photocatalyst than rutile?—Model studies on epitaxial TiO₂ films. *Science Reports*, 4, 2014.
12. Yan, J. et al., Understanding the effect of surface/bulk defects on the photocatalytic activity of TiO₂: Anatase versus rutile. *Physical Chemistry Chemical Physics*, 15(26): pp. 10978–10988, 2013.
13. Patil, K. et al., *Chemistry of Nanocrystalline Oxide Materials: Combustion Synthesis, Properties and Applications*. World Scientific, Singapore, 2008.
14. Pelaez, M. et al., A review on the visible light active titanium dioxide photocatalysts for environmental applications. *Applied Catalysis B: Environmental*, 125: pp. 331–349, 2012.
15. Farr, J., Molybdenum disulphide in lubrication: A review. *Wear*, 35(1): pp. 1–22, 1975.
16. Diebold, U., The surface science of titanium dioxide. *Surface Science Reports*, 48(5): pp. 53–229, 2003.
17. Gupta, S. M., and M. Tripathi, A review of TiO₂ nanoparticles. *Chinese Science Bulletin*, 56(16): pp. 1639–1657, 2011.
18. Hanaor, D. A., and C. C. Sorrell, Review of the anatase to rutile phase transformation. *Journal of Materials Science*, 46(4): pp. 855–874, 2011.
19. Swamy, V. et al., Finite-size and pressure effects on the Raman spectrum of nanocrystalline anatase TiO₂. *Physical Review B*, 71(18): Article 184302, 2005.
20. Mikami, M., and K. Ozaki, Thermoelectric properties of nitrogen-doped TiO$_{2-x}$ compounds. *Journal of Physics: Conference Series*, 379: Article 012006, 2012.
21. Tavakoli, A., M. Sohrabi, and A. Kargari, A review of methods for synthesis of nanostructured metals with emphasis on iron compounds. *Chemical Papers*, 61(3): pp. 151–170, 2007.
22. Vinu, R., and G. Madras, Environmental remediation by photocatalysis. *Journal of the Indian Institute of Science*, 90(2): pp. 189–230, 2012.
23. Charanpahari, A. et al., Enhanced photocatalytic activity of multi-doped TiO₂ for the degradation of methyl orange. *Applied Catalysis A: General*, 443: pp. 96–102, 2012.
24. Suprabha, T. et al., Microwave-assisted synthesis of titania nanocubes, nanospheres, and nanorods for photocatalytic dye degradation. *Nanoscale Research Letters*, 4(2): pp. 144–152, 2009.
25. Zhu, Y.-J., and F. Chen, Microwave-assisted preparation of inorganic nanostructures in liquid phase. *Chemical Reviews*, 114(12): pp. 6462–6555, 2014.
26. Zhang, Q., and L. Gao, Preparation of oxide nanocrystals with tunable morphologies by the moderate hydrothermal method: Insights from rutile TiO₂. *Langmuir*, 19(3): pp. 967–971, 2003.
27. Chatterjee, S. et al., Photocatalytic properties of one-dimensional nanostructured titanates. *The Journal of Physical Chemistry C*, 114(20): pp. 9424–9430, 2010.
28. Xia, Y. et al. One-dimensional nanostructures: Synthesis, characterization, and applications. *Advanced Materials*, 15(5): pp. 353–389, 2003.
29. Huczko, A., Template-based synthesis of nanomaterials. *Applied Physics A*, 70(4): pp. 365–376, 2000.

30. Wang, X. et al., One-dimensional titanium dioxide nanomaterials: Nanowires, nanorods, and nanobelts. *Chemical Reviews*, 114(19): pp. 9346–9384, 2014.

31. Oskam, G., Metal oxide nanoparticles: Synthesis, characterization and application. *Journal of Sol-Gel Science and Technology*, 37(3): pp. 161–164, 2006.

32. Wu, J.-J., and C.-C. Yu, Aligned TiO$_2$ nanorods and nanowalls. *The Journal of Physical Chemistry B*, 108(11): pp. 3377–3379, 2004.

33. Saitow, K.-i., and T. Wakamiya, 130-fold enhancement of TiO$_2$ photocatalytic activities by ball milling. *Applied Physics Letters*, 103(3): Article 031916, 2013.

34. Daothong, S. et al., Size-controlled growth of TiO$_2$ nanowires by oxidation of titanium substrates in the presence of ethanol vapor. *Scripta Materialia*, 57(7): pp. 567–570, 2007.

35. Singer, I. L., Solid lubrication processes, in *Fundamentals of Friction: Macroscopic and Microscopic Processes*, I. L. Singer and H. M. Pollock (Eds.). pp. 237–261 Springer, Dordrecht, 1992.

36. Song, H.-J., Z.-Z. Zhang, and X.-H. Men, Tribological behavior of polyurethane-based composite coating reinforced with TiO$_2$ nanotubes. *European Polymer Journal*, 44(4): pp. 1012–1022, 2008.

37. Hu, K. H. et al., Synergistic effect of nano-MoS$_2$ and anatase nano-TiO$_2$ on the lubrication properties of MoS$_2$/TiO$_2$ nano-clusters. *Tribology Letters*, 43(1): pp. 77–87, 2011.

38. Kusoglu, I. M. et al., Wear behavior of flame-sprayed Al$_2$O$_3$–TiO$_2$ coatings on plain carbon steel substrates. *Surface and Coatings Technology*, 200(1–4): pp. 1173–1177, 2005.

39. Habib, K. A. et al., Comparison of flame sprayed Al$_2$O$_3$/TiO$_2$ coatings: Their microstructure, mechanical properties, and tribology behavior. *Surface and Coatings Technology*, 201(3–4): pp. 1436–1443, 2006.

40. Lin, X. et al., Effects of temperature on tribological properties of nanostructured and conventional Al$_2$O$_3$–3 wt.% TiO$_2$ coatings. *Wear*, 256(11–12): pp. 1018–1025, 2004.

41. Krishna, D. S. R., Y. Sun, and Z. Chen, Magnetron sputtered TiO$_2$ films on a stainless steel substrate: Selective rutile phase formation and its tribological and anti-corrosion performance. *Thin Solid Films*, 519(15): pp. 4860–4864, 2011.

42. Berger, L. M. et al., Dry sliding up to 7.5m/s and 800°C of thermally sprayed coatings of the TiO$_2$–Cr$_2$O$_3$ system and (Ti,Mo)(C,N)–Ni(Co). *Wear*, 267(5–8): pp. 954–964, 2009.

43. Shao, X., W. Liu, and Q. Xue, The tribological behavior of micrometer and nanometer TiO$_2$ particle-filled poly(phthalazine ether sulfone ketone) composites. *Journal of Applied Polymer Science*, 92(2): pp. 906–914, 2004.

44. Fei, C. et al., Study on the tribological performance of ceramic coatings on titanium alloy surfaces obtained through microarc oxidation. *Progress in Organic Coatings*, 64(2–3): pp. 264–267, 2009.

45. Wang, Y. et al., Fretting wear behaviour of microarc oxidation coatings formed on titanium alloy against steel in unlubrication and oil lubrication. *Applied Surface Science*, 252(23): pp. 8113–8120, 2006.

46. Zhang, J., and A. Alpas, Transition between mild and severe wear in aluminium alloys. *Acta Materialia*, 45(2): pp. 513–528, 1997.

47. Sun, Y., Tribological rutile-TiO$_2$ coating on aluminium alloy. *Applied Surface Science*, 233(1–4): pp. 328–335, 2004.

48. Krishna, D. S. R., and Y. Sun, Thermally oxidised rutile-TiO$_2$ coating on stainless steel for tribological properties and corrosion resistance enhancement. *Applied Surface Science*, 252(4): pp. 1107–1116, 2005.

49. Wan, Y. et al., Effect of UV irradiation on wear protection of TiO$_2$ and Ni-doped TiO$_2$ coatings. *Applied Surface Science*, 258(10): pp. 4347–4350, 2012.

50. Hu, Z., and J. Dong, Study on antiwear and reducing friction additive of nanometer titanium oxide. *Wear*, 216(1): pp. 92–96, 1998.

51. Gao, Y. et al., Tribological properties of oleic acid-modified TiO$_2$ nanoparticle in water. *Materials Science and Engineering: A*, 286(1): pp. 149–151, 2000.

52. Chang, H. et al., Tribological property of TiO$_2$ nanolubricant on piston and cylinder surfaces. *Journal of Alloys and Compounds*, 495(2): pp. 481–484, 2010.

53. Ye, W. et al., Preparation and tribological properties of tetrafluorobenzoic acid-modified TiO$_2$ nanoparticles as lubricant additives. *Materials Science and Engineering: A*, 359(1): pp. 82–85, 2003.

54. Qian, J. et al., Preparation and tribological properties of stearic acid-modified hierarchical anatase TiO$_2$ microcrystals. *Applied Surface Science*, 258(7): pp. 2778–2782, 2012.

55. Zhang, L. et al., Synthesis and tribological properties of stearic acid-modified anatase (TiO$_2$) nanoparticles. *Tribology Letters*, 41(2): pp. 409–416, 2011.

56. Ingole, S. et al., Tribological behavior of nano TiO$_2$ as an additive in base oil. *Wear*, 301(1–2): pp. 776–785, 2013.

20 Biodiesel
A Fuel, a Lubricant, and a Solvent

Gerhard Knothe

CONTENTS

ABSTRACT

Biodiesel is well known as a biogenic alternative to conventional diesel fuel derived from petroleum. It is produced from feedstocks such as plant oils consisting largely of triacylglycerols through transesterification with an alcohol such as methanol. The properties of biodiesel are largely competitive with those of petrodiesel. These properties, however, make biodiesel also suitable for other applications such as lubricants and solvents. This chapter presents an overview of biodiesel and its properties in light of these various applications.

20.1 INTRODUCTION

Renewable resources, including those derived from vegetable/plant oils or other triacylglycerol-containing materials, are of interest for many common commercial industrial products ranging from fuels and lubricants to paints, coatings, solvents, polymers and many other applications. For most of these applications, suitable derivatives need to be synthesized and/or specific formulations need to be developed.

Some materials, however, may be useful for several applications without significant changes in product formulation, as the major components possess properties that make these materials suitable for such multiple uses. Prime examples of such a material are the fatty acid methyl (or other alkyl) esters of triacylglycerol-containing materials such as plant oils and animal fats. As a result of their most common use, the methyl esters are also often referred to as biodiesel [1,2]. Indeed, biodiesel in a standard such as ASTM D6751 [3] is defined as "comprised of mono-alkyl esters of long chain fatty acids derived from vegetable oils or animal fats." This definition thus includes esters other than methyl, although methyl esters are the most common form of biodiesel because methanol is the least expensive alcohol for biodiesel production in most countries around the world. Biodiesel has become generally recognized as a renewable fuel alternative to petrodiesel, the conventional diesel fuel derived from petroleum.

Historical developments in the area of diesel fuels can serve to illustrate the multifaceted use of biodiesel. The first test of a vegetable oil, peanut oil, in a diesel engine dates from 1900 with extensive literature existing on this subject from the 1920s to the 1940s [4]. The suitability of biodiesel for use in diesel engines was established by research in the 1930s, at which time Belgian Patent 422,877 [5] was probably the first document to describe the synthesis of what is now known as biodiesel in the form of ethyl esters. A later report [6] describes the first likely use of this fuel in the form of ethyl esters of palm oil in a passenger bus in Belgium in 1938. During what may be termed as the *age of petroleum* after World War II, interest in and research on renewable feedstocks waned, only to rekindle with ever-increasing intensity since the late 1970s sparked by the energy crises of that time.

The 1970s and the years prior also marked the onset of concerns regarding the environmental effects of fuel use, particularly the contribution of exhaust emissions to urban smog and lung disease as well as later to global warming. These concerns resulted in increasingly stringent regulation of exhaust emissions, particularly from vehicular fuel use. One aspect of renewables in this connection is that they are regarded as carbon neutral, thus not contributing to the increase of carbon dioxide in the atmosphere, an effect held largely responsible for the increase of global surface temperatures. On the other hand, these increasingly stringent regulations led to changes in the composition of the conventional petroleum-derived fuels. A salient issue in this area is the reduction of sulfur levels in petrodiesel to address the issue of acid rain as well as to reduce the poisoning of catalysts necessary for reducing other regulated exhaust emissions. The reduction of sulfur levels in petrodiesel to give what is now the ultralow sulfur diesel (ULSD) fuel had the unintended consequence of reducing the lubricity of petrodiesel fuels by also removing oxygen- and nitrogen-containing species which were the actual lubricity-imparting materials [7,8]. Historically, diesel engines had relied on the self-lubricating effect of diesel fuels to provide the necessary engine lubrication. Consequently, the introduction of ULSD fuels in some cases led to injector and fuel pump failures [9,10].

At about the same time that (ultra)low sulfur diesel fuels were arising, interest in biodiesel increased. It was soon noticed that neat biodiesel possesses excellent lubricity and can act as lubricity-imparting additive to low-lubricity petrodiesel [11,12]. In order to provide a market for biodiesel in light of its inherently higher cost than petrodiesel and the more limited production and supply, the use of low-level blends of biodiesel with petrodiesel was promoted, for example, 2% biodiesel blended with petrodiesel. Blending biodiesel at such a level into low-lubricity ULSD fuel significantly improved the lubricity of the ULSD fuel. Biodiesel is of additional interest for such blends as it also exhibits fuel properties that are largely competitive with those of petrodiesel/ULSD and which lubricity additives do not necessarily possess. This development demonstrates the multifunctional nature of the alkyl esters of vegetable oils (biodiesel) as both a fuel and a lubricant.

With the expanding use of fatty acid methyl esters (FAMEs) as biodiesel, some attention was also devoted to other applications. Among the first uses other than fuel to be investigated were solvent-related uses [13], which was specifically related to applying FAMEs in the cleanup and the remediation of shorelines contaminated by petroleum or even vegetable oil spills [14,15]. A later study listed the numerous solvent-related applications of soybean oil methyl esters [16], but these results hold for other FAMEs, too. Thus, biodiesel has considerable application potential beyond fuel and lubricant uses.

20.2 BIODIESEL PRODUCTION

As mentioned previously, biodiesel is defined as the monoalkyl esters of vegetable oils or other triacylglycerol-containing materials. Biodiesel, with glycerol as coproduct, is produced through a transesterification reaction, as depicted in Figure 20.1. The most common process for producing biodiesel is base catalysis using alkoxide (methoxide in case of methanol) or hydroxide as catalyst.

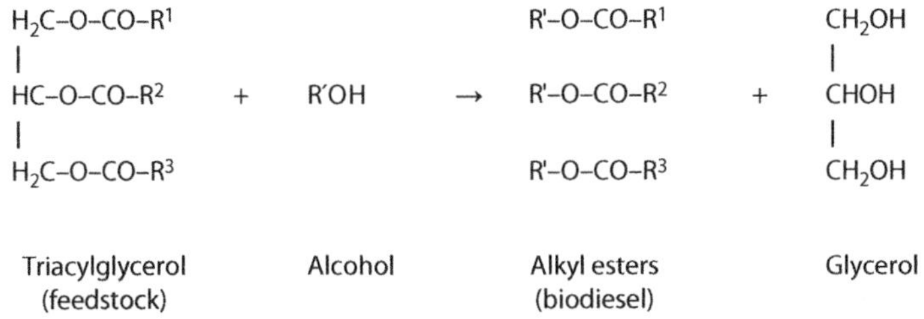

$$
\begin{array}{llll}
\text{H}_2\text{C–O–CO–R}^1 & & \text{R'–O–CO–R}^1 & \text{CH}_2\text{OH} \\
| & & & | \\
\text{HC–O–CO–R}^2 \quad + \quad \text{R'OH} \quad \rightarrow & \text{R'–O–CO–R}^2 \quad + \quad \text{CHOH} \\
| & & & | \\
\text{H}_2\text{C–O–CO–R}^3 & & \text{R'–O–CO–R}^3 & \text{CH}_2\text{OH}
\end{array}
$$

| Triacylglycerol (feedstock) | Alcohol | Alkyl esters (biodiesel) | Glycerol |

FIGURE 20.1 The transesterification reaction. In the feedstock, R^1, R^2, and R^3 represent different fatty acid chains; thus, R^1, R^2, and R^3 may or may not be identical. Methanol ($R' = CH_3$) is the most commonly used alcohol with sodium methoxide being the preferable catalyst.

As the reaction system and the product should be as moisture free as possible [17], alkoxides are advantageous catalysts because the water-forming reaction

$$XOH + ROH \rightarrow ROX + H_2O \quad (X = \text{Na or K}; R = \text{methyl or other alkyl})$$

when using hydroxide does not occur. Typical reaction conditions are a molar ratio of alcohol to vegetable oil of 6:1 at 60–65°C (when using methanol; generally, slightly below the boiling point of the alcohol) for 1 h [17]. As indicated above, the reaction system should be as moisture free as possible and the FFAs should also be kept to a minimum [17] as otherwise undesirable side reactions such as soap formation can occur. In case of feedstocks with high FFA content, an acid pretreatment [18] is necessary in which the FFAs are reacted in the presence of an acid, sulfuric acid, as catalyst with methanol to give methyl esters and then, as second step, the remaining triacylglycerols are transesterified in the conventional procedure to the methyl esters.

The transesterification reaction system is biphasic, alcohol and vegetable oil, at the beginning of the reaction and also biphasic, methyl esters (in case of methanol) and glycerol, at the end [17]. The reaction proceeds stepwise from the triacylglycerol-containing starting material via the di- and monoacylglycerols to glycerol, with fatty acid alkyl ester formed at each step. Besides alkoxide and hydroxide, many other catalysts or catalytic systems or variations of the transesterification process have been described in the literature, but apparently none has replaced the conventional transesterification reaction using alkoxide (or hydroxide) catalysts.

20.3 FEEDSTOCKS

Virtually every oil or fat that is mainly composed of triacylglycerols is a potential biodiesel feedstock and thus a potential source for the other uses, too. Apparently, the first triacylglycerol feedstock to be investigated as neat fuel was peanut oil as mentioned previously. In the "historical" times, numerous feedstocks ranging from fish oils to various vegetable oils were tested neat as fuels in diesel engines [4]. Palm oil was used as feedstock for the first fuel meeting the current definition of biodiesel. Upon the "rediscovery" of biodiesel around 1980, sunflower oil was used to produce the corresponding methyl esters [4]. With increasing interest in Europe and the United States in the early 1980s, oils of local production such as rapeseed (as low-erucic rapeseed) and soybean oils were of primary interest. In the United States, the excess production of soybean oil was a major driving force. With expanding interest in and production of biodiesel around the world, numerous other feedstocks of local provenience have been explored with a notable example being palm oil in countries with more tropical climate.

Several issues have beset biodiesel feedstocks. One is that all vegetable oils combined cannot potentially afford enough biodiesel to fully replace the petroleum-derived diesel fuel. Another issue is economics, namely, the observation that biodiesel from vegetable oils is more expensive than petrodiesel. This issue has been addressed by a plethora of legislative and regulatory measures in many countries around the world mandating or creating incentives for the use of fuels of biological origin. In recent years, some concerns have been raised about using commodity vegetable oils which often have nutritional applications as feedstocks for fuel. This aspect together with the other two issues has been major driving force in the search for additional feedstocks for biodiesel which are not or less affected by these issues. Feedstocks that have received attention under these aspects include a variety of plant oils of more or less sustainable nature, animal fats, used cooking oils, and algae-derived oils, with the latter currently finding special interest.

Plant oils of alternative nature that have been studied include camelina [19], castor [20,21], coriander [22], cuphea [23], desert date [24], jatropha [25,26], kapok [27], kenaf [28], mahua (*Madhuca indica*) [29], moringa [30], mustard species [31], okra [32], pennycress [33], pongamia [34], rubber seed [35], and thespesia [36]. Probably the most discussed of these plant oils is jatropha and it may be noted that, besides the promise associated with it, problems have been identified when growing it on a commercial scale [37,38]. Thus, such issues may be latent for other alternative feedstocks too. In the last 10 years, oils from various algae have prominently figured as potential biodiesel feedstocks with proponents touting advantages such as potential high production and sustainability, while skeptics have been pointing to production problems, economics, and, photosynthetic efficiency [39–46].

Many alternative plant oils have fatty acid profiles resembling those of common commodity oils, thus leading to biodiesel fuels with similar properties. The most common fatty acids in these oils are palmitic (hexadecanoic; C16:0), stearic (octadecanoic; C18:0), oleic (9(Z)-octadecenoic; C18:1 $\Delta 9c$), linoleic (9(Z),12(Z)-octadecadienoic; C18:2 $\Delta 9c,\Delta 12c$), and linolenic (9(Z),12(Z),15(Z)-octadecatrienoic; C18:3 $\Delta 9c,\Delta 12c,\Delta 15c$) acids (see Figure 20.2 for a structural depiction of the methyl esters of these acids).

On the other hand, there are also many potential biodiesel feedstocks with varying fatty acid profiles. Castor oil containing a hydroxy fatty acid (ricinoleic acid, 12-hydroxy-9(Z)-octadecenoic acid) and many algal oils containing highly polyunsaturated fatty acids such as eicosapentaenoic and docosahexaenoic acids are examples of potential feedstocks with fatty acid profiles deviating from those of most commodity oils and many alternative oils. This variation leads to fuels with varying

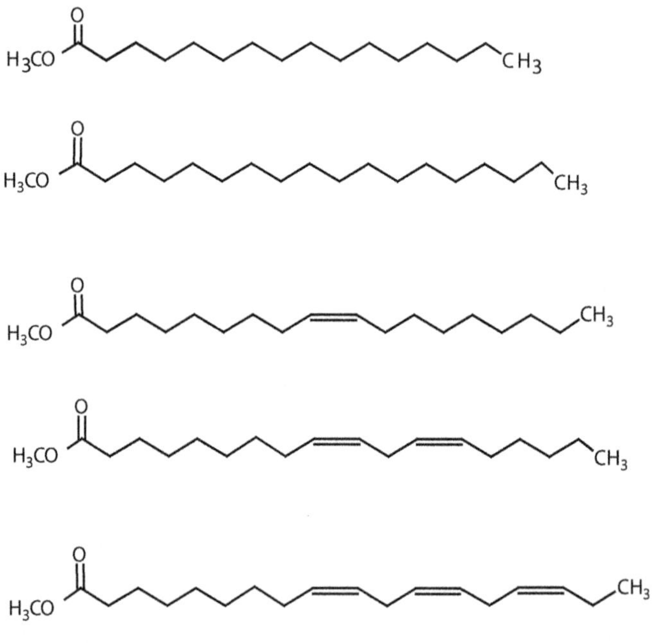

FIGURE 20.2 *Top to bottom:* The structures of the five most common FAMEs (methyl palmitate, methyl stearate, methyl oleate, methyl linoleate, methyl linolenate).

properties as the individual fatty esters have different properties and occur in varying amounts in the various biodiesel fuels. Besides this aspect, the original quality of the various oil feedstocks can vary, for example, used cooking oils usually contain high amounts of FFAs.

20.4 FUEL-RELATED PROPERTIES

Table 20.1 lists the fuel properties of individual FAMEs discussed in the following.

TABLE 20.1
Fuel-Related Properties of Some Common Fatty Acid Alkyl Esters[a]

FAME	CN	Heat of Combustion (kJ/g)	Kinematic Viscosity (40°C) (mm²/s)	Melting Point (°C)	Oxidation Stability (h)	Density (15°C; 40°C) (kg/m³)
C10:0	51.6	36.59	1.72	−13.5	>24	876.4; 855.9
C12:0	66.7	37.91	2.43	4.3	>24	873.0; 853.4
C14:0			3.30	18.1	>24	−; 851.8
C16:0	85.9	39.63	4.38	28.5	>24	−; 849.8
C16:1 Δ9c	56.6/51.0		3.67	−34.1	2.11	881.0; 862.5
C18:0	101	40.06	5.85	37.7	24	
C18:1 Δ9c	57	39.93	4.51	−20.2	2.79	877.5; 859.4
C18:2 Δ9c, Δ12c	38.2	39.65	3.65	−43.1	0.94	890.2; 872.0
C18:3 d9,12,15	22.7		3.14		0.0	901.7; 883.2
C20:0	112–115[b]		(7.40)[b]	46.4		
C20:1 Δ11c	73.2		5.77	−7.8		876.6; 858.9
C22:0	125–130[b]		(9.31)[b]	53.2		
C22:1 Δ13c	74.2		7.33	−3.0	74.2	874.4; 856.9
C22:6 Δ4,7,10,13,16,19	24.4		2.97		0.07	923.6; 905.2
C24:0			(11.53)	58.6		
C18:1 Δ9c, 12-OH	37.4		15.44		0.67	929.5; 911.4

Source: Reprinted with permission from B. R. Moser et al., *Energy Fuels*, 23, 4149–4155. Copyright 2009 American Chemical Society; With kind permission from Springer Science+Business Media: *J. Am. Oil Chem. Soc.*, Heats of combustion of fatty acids and fatty acid esters, 91, 2014, 235–249, Levine, F., R. V. Kayea III, R. Wexler, D. J. Sadvary, C. Melick, and J. La Scala; Reprinted from *Fuel*, 84, Knothe, G., and K. R. Steidley, Kinematic viscosity of biodiesel fuel components and related compounds: Influence of compound structure and comparison to petrodiesel fuel components, 1059–1065, Copyright (2005), with permission from Elsevier; Reprinted from *Fuel*, 90, Knothe, G., and K. R. Steidley, Kinematic viscosity of fatty acid methyl esters: Prediction, calculated viscosity contribution of esters with unavailable data, and carbon-oxygen equivalents, 3217–3224, Copyright (2011), with permission from Elsevier; With kind permission from Springer Science+Business Media: *J. Am. Oil Chem. Soc.*, A comprehensive evaluation of the melting points of fatty acids and esters determined by differential scanning calorimetry, 86, 2009, 843–856, Knothe, F. G., and R. O. Dunn; With kind permission from Springer Science+Business Media: *J. Am. Oil Chem. Soc.*, A comprehensive evaluation of the density of neat fatty acids and esters, 91, 2014, 1711–1722, Knothe, G., and K. R. Steidley; Reprinted with permission from G. Knothe, *Energy Fuels*, 22, 1358–1364. Copyright 2008 American Chemical Society.

[a] See references for information on the methods used for data determination.

[b] Extrapolated or calculated data. See studies by Knothe, *Fuel*, 119, 6–13, 2014, for CN and corresponding publications and Knothe and Steidley, *Fuel*, 90, 3217–3224, 2011, for kinematic viscosity. The extrapolated data for kinematic viscosity are given in parentheses because these compounds are solids at room temperature and the values reflect the contribution to a mixture that is liquid at 40°C.

20.4.1 CETANE NUMBER (CN) AND OTHER COMBUSTION-RELATED PROPERTIES

The CN is a dimensionless descriptor of the ignition quality of a diesel fuel related to the ignition delay time a fuel experiences when it is injected into the combustion chamber of a diesel engine. Its name is derived from the trivial name of the long-chain hydrocarbon hexadecane ($C_{16}H_{34}$), which has a high assigned CN of 100 on the cetane scale for which it is the high-quality reference compound. Highly branched 2,2,4,4,6,8,8-heptamethylnonane (HMN; also $C_{16}H_{34}$) is the low-quality reference compound on the cetane scale [47]. The structural difference between hexadecane and HMN illustrates that the branching in a hydrocarbon chain reduces the CN. Similarly, reducing the chain length of a hydrocarbon reduces the CN, as does increasing unsaturation, i.e., an increasing number of double bonds [48]. Aromatic compounds also tend to possess low CNs [49]. These structural effects can also be observed in fatty acid alkyl esters that are the major components of biodiesel [50–52]. Saturated fatty acid alkyl esters exhibit high CN; for example, the CN of methyl stearate is approximately 100, with the CN decreasing with decreasing chain length. Thus, the CN of methyl palmitate is approximately 85 and that of methyl laurate about 65. The effect of increasing unsaturation as mentioned above is shown by the CN of methyl oleate being about 58, that of methyl linoleate about 40, and that of methyl linolenate approximately 25. The minimum CNs prescribed in biodiesel standards are 47 in ASTM D6751 [47] and 51 in EN 14214 [53]. The minimum CN in the ASTM standards is especially higher for biodiesel than for petrodiesel, for which the minimum CN is 40 in ASTM D975 [54]. The CN is also a property that shows the suitability of fatty acid alkyl esters as diesel fuel caused by the long hydrocarbon chain found, of course, also in the ideal long-chain hydrocarbons such as hexadecane. More CNs are given in Table 20.1.

The experimental CNs of biodiesel fuels from vegetable oils are in the range from the upper 40s to the mid-60s depending on the fatty acid profile. Thus, biodiesel from soybean oil has a CN in the upper 40s to the lower 50s as it contains slightly more than 50% methyl linoleate as well as approximately 8% methyl linolenate, while the CN of biodiesel from palm oil is well above 60 due to its content of more than 40% methyl palmitate. The highly polyunsaturated fatty acid esters found in many methyl esters derived from algal oils have low CNs; for example, methyl eicosatetraenoate (methyl arachidonate) showed a CN of 29.6 and methyl docosahexaenoate showed a CN of 24.4 [51], leading to low overall CN for many algae-derived FAMEs containing highly polyunsaturated FAMEs. The overall CN of a biodiesel fuel can be straightforwardly calculated [52] by proportional contribution of the components using the following equation:

$$CN_{mix} = \sum A_C \times CN_C, \tag{20.1}$$

where CN_{mix} = overall CN of the mixture, A_C = the relative amount (vol%) of the individual components, and CN_C = CN of the individual neat esters.

The final product of combustion in an engine is the exhaust emissions vented from the engine through the exhaust system and the tailpipe. While carbon dioxide and water are the ideal products of combustion, nonideal (nonstoichiometric) combustion causes the formation of other species that are also emitted through the exhaust system. Several of these species resulting from nonideal combustion are subject to regulations limiting their concentration in the exhaust emissions. These species are nitrogen oxides (NO_x), particulate matter (PM), carbon monoxide (CO), and hydrocarbons. The existence of such specimens and subsequent regulations have caused a significant amount of research to reduce emissions as far as possible through improved combustion in the engine itself and development of a variety of exhaust emission control systems.

The various species of exhaust emissions are also observed when using biodiesel instead of petrodiesel as fuel, although their relative amounts differ. This observation shows that the compound structure has a significant effect on the nature of the exhaust emissions formed during the combustion process. Thus, numerous studies since the beginning of the interest in biodiesel as alternative to petrodiesel have shown that biodiesel reduces most regulated exhaust emission species with exception of NO_x, which are slightly elevated when using biodiesel. Results of these tests have been compiled and summarized [55]. PM species are especially considerably reduced when using biodiesel compared to when using petrodiesel [56,57] with apparently little to no effect to the structure of the FAME. The reduction of CO and hydrocarbons, however, depends on the chain length of the methyl esters with the emission-reducing effect versus petrodiesel diminishing at shorter chain lengths. The level of NO_x, on the other hand, depends on the level of unsaturation of the fatty compounds with higher unsaturation levels increasing the NO_x compared to petrodiesel. Thus, the CN can be correlated with NO_x exhaust emissions to some extent with fuels with higher CN giving lower NO_x [58]. Neat saturated fatty compounds reduced NO_x versus a petrodiesel reference fuel but the effect of saturated hydrocarbons was even greater.

The fact that esters of fatty acids are suitable as fuels is also shown by the heat of combustion besides the CN. The heat of combustion of biodiesel fuels is, however, slightly reduced compared to that of petrodiesel due to the oxygenated nature of biodiesel. The heat of combustion of fatty esters has been given in prior literature [59,60].

20.4.2 Kinematic Viscosity

The major reason why biodiesel is used as a fuel in diesel engines is its lower kinematic viscosity compared to that of the parent vegetable oil or other feedstock. The high viscosity of vegetable oils or other triacylglycerol-containing materials can result in operational problems such as the formation of engine deposits. The viscosity of biodiesel is approximately an order of magnitude less than that of its parent oil and is thus close to that of conventional petrodiesel. This is reflected in the kinematic viscosity specifications of biodiesel and petrodiesel standards. Thus, the kinematic viscosity ranges specified in biodiesel standards are 1.9–6.0 and 3.5–5.0 mm^2/s in ASTM D6751 and EN 14214, respectively, while the kinematic viscosity ranges of petrodiesel are 1.9–4.1 and 2.0–4.5 mm^2/s in the standards ASTM D975 and EN 590, respectively. The kinematic viscosity of an individual fatty acid alkyl ester depends on its structure, with increasing saturation and increasing chain length leading to increased viscosity [61]. Compounds with *trans* double bonds, however, exhibit kinematic viscosity close to that of the saturated compounds of the same chain length [61]. Thus, the kinematic viscosity of saturated FAMEs decreases from 5.85 mm^2/s for methyl stearate to 4.38 mm^2/s for methyl palmitate, to 3.30 mm^2/s for methyl myristate, to 2.43 mm^2/s for methyl laurate, and to 1.72 mm^2/s for methyl decanoate (see Table 20.1 for more data). In comparison, the kinematic viscosity of unsaturated FAMEs with C18 chain length is 4.51 mm^2/s for methyl oleate, 3.65 mm^2/s for methyl linoleate, and 3.14 mm^2/s for methyl linolenate. Differences in the fatty acid profiles of biodiesel fuels will also entail differences in their kinematic viscosities. The kinematic viscosity of most biodiesel fuels is in the range of about 4.0–4.5 mm^2/s, with that of soybean oil-derived biodiesel at approximately 4.0–4.1 mm^2/s and that of rapeseed oil (canola oil type as commonly used for biodiesel in Europe) at approximately 4.4 mm^2/s. Similar to CN, the overall kinematic viscosity of FAMEs can be calculated from an equation using proportional contribution of the individual components [62].

20.4.3 Cold Flow

The behavior of biodiesel at low temperatures is one of the two major technical issues affecting its use. The problem is documented by the melting points of FAMEs, especially the saturated compounds. The melting points of fatty compounds (Table 20.1) increase with chain length. Thus, the melting point of methyl decanoate is −13.5°C; methyl laurate, 4.3°C; methyl myristate, 18.1°C; methyl palmitate, 28.5°C; and methyl stearate, 37.7°C [63]. Unsaturation significantly reduces the melting point; thus, the melting point of methyl oleate is −20.2°C and that of methyl linoleate is −43.1°C. When cooling a biodiesel fuel, the components with higher melting points tend to crystallize/precipitate first; thus, solids appear in the fuel, which can clog fuel filters and fuel lines. Thus, the fatty acid profile again plays a significant role in determining the cold-flow properties with the small amounts of saturated fatty compounds having a stronger influence than the low amounts may indicate [64].

Several tests exist for assessing the cold-flow properties of diesel fuels and biodiesel [65]. These include filtration-type tests such as the low-temperature flow test, the cold filter plugging point, and the cold soak filtration test, as well as flow-type values such as the cloud point and the pour point. The cloud point is likely the most stringent value as it defines the temperature at which the first solids or crystals in cooling biodiesel become visible. The pour point is the temperature below which the fuel does not readily flow [65]. Overall, the cloud point is likely the most used value for indicating the cold-flow properties of biodiesel and is the value on which efforts for improving cold flow should focus [66]. The cloud point of biodiesel from soybean oil is approximately 0°C, while that of biodiesel from palm oil is about 16°C [67], a difference reflective of the different fatty acid profiles.

Several approaches have been devised for improving the low-temperature properties of biodiesel. These include development of additives; winterization, i.e., removal of the higher-melting components by means of several cooling cycles; blending with petrodiesel; and also the use of branched esters. The latter approach [68,69] inherently reduces the cloud point as the branched esters, such as isopropyl esters, possess lower melting points than the methyl esters.

Other constituents of biodiesel can also cause cold-flow problems. These are also materials of high-melting nature such as remaining monoacylglycerols formed as intermediates in the transesterification reaction [70] and steryl glucosides that are not only naturally present but also in esterified form resulting from the transesterification reaction [71]. For example, the melting point of monostearin is >70°C and that of steryl glucosides is >240°C [71].

20.4.4 OXIDATION STABILITY

Besides cold flow, the stability of biodiesel during storage and toward air (oxygen) is the other major technical issue affecting its use. This is a result of the content of unsaturated, especially polyunsaturated, FAMEs, which are prone to oxidation reactions upon contact with air. The oxidation reaction commences at the allylic CH_2 positions in the unsaturated fatty acid chains; thus, *bis*-allylic CH_2 positions as they occur in linoleic and linolenic acids are even more susceptible. If the relative oxidation rate of methyl oleate is set to 1, the relative oxidation rate of methyl linoleate is 41 and that of methyl linolenate is 98 [72]. For FAMEs with even higher degrees of polyunsaturation, the relative rate is even higher, thus for methyl stearidonate (C18:4), the relative rate has been given as 198 [72]. This is of significance for algal oils as they often contain elevated amounts of more highly polyunsaturated fatty acids. Similar to cold flow, minor amounts of the more oxidation-prone polyunsaturated fatty acid chains may have a greater influence on oxidation stability than their minor amounts would indicate.

Table 20.1 contains oxidation stability data obtained with the so-called Rancimat method, in which the sample is subjected at 110°C to an airflow, with the effluent being swept into a water bath, the conductivity of which is measured. The maximum rate change in the conductivity of the water is given as the induction time for oxidation. The limits for this time are 3 h in ASTM D6751 and 8 h in EN 14214. No neat unsaturated FAME, including methyl oleate, meets the requirements of these standards.

Oxidation-promoting factors include presence of oxygen (air), light, elevated temperature, and presence of extraneous materials, such as certain metals. This may include the material of which the storage tank is constructed. Copper is especially effective in promoting biodiesel oxidation [73]. Storage conditions therefore play a major role for biodiesel. Optimized conditions would be storage at low temperatures under exclusion of light and air (storage under nitrogen), exclusion of extraneous oxidation-promoting materials, and minimum headspace. On the other hand, antioxidants such as butylated hydroxytoluene (BHT; 2,6-*tert*-butyl-4-hydroxytoluene) can delay oxidation, although oxidation will commence once the antioxidant has been consumed.

20.4.5 DENSITY

The density of biodiesel has not been a significant issue as most biodiesel fuels have densities within a tight range. This range is illustrated by the density specification in the European biodiesel standard EN 14214 of 860–900 kg/m^3 at 15°C. The density values given in Table 20.1 for most common FAMEs are in this range [74]. Only some biodiesel fuels containing significant amounts of highly polyunsaturated FAMEs may approach the upper limit of this specification. It therefore serves to additionally distinguish the methyl esters from the triacylglycerol feedstocks which have density values of >900 kg/m^3. The density of a FAME mixture can be calculated by proportional contribution of the components similar to CN and kinematic viscosity [74]. Table 20.1 contains density values at 15°C and 40°C because of density being the factor between kinematic viscosity and dynamic viscosity. Values are often given for dynamic viscosity in the literature instead of kinematic viscosity and data at 40°C aid in the conversion as kinematic viscosity is specified at this temperature in standards.

20.4.6 IMPROVING BIODIESEL FUEL PROPERTIES

Overall, there are five approaches to potentially improving the fuel properties of biodiesel [75], which are use of additives, production of biodiesel from an alcohol other than methyl, change of the fatty acid profile by physical means such as winterization, or use of feedstocks with inherently different fatty acid profiles in the form of alternative feedstocks or through genetic modification. An issue of interest resulting from such studies and data on the properties of fatty acid alkyl esters as fuel is that of what may be called as a designer fuel [76] and if feedstocks approaching an optimized fatty acid composition are available. Thus, it was discussed that methyl decanoate (C10:0) or methyl palmitoleate (C16:1 $\Delta 9c$) may be ideal biodiesel components. Studies on some feedstocks containing these materials showed that biodiesel from cuphea oil, which contains high amounts of decanoic acid, indeed showed overall improved properties [77]. Unfortunately, the cuphea plant has several agronomic problems which impede the large-scale utilization of its oil for this purpose.

20.5 LUBRICANT-RELATED PROPERTIES

As discussed in Section 20.1, lubricity is an important fuel property, but to emphasize the potential use of biodiesel as lubricant, this issue is separately treated here. Also as mentioned above, the issue of petrodiesel lubricity arose with the introduction of (ultra)low-sulfur diesel fuels.

The most commonly used equipment for assessing the lubricity of diesel fuels in general is the high-frequency reciprocating rig (HFRR) tester as defined in standards such as ASTM D6079 [78] or ISO 12156 [79]. This test is also applied to biodiesel. In this test, a steel ball reciprocates at a defined frequency over a steel disk immersed in 2 mL of the sample with the diameter of the wear scar generated on the ball used to assess the lubricity of the sample. Typically, the test is conducted at 60°C for 75 min at a frequency of 50 Hz and a stroke length of 1 mm. The ball WSD thus generated is limited in the European petrodiesel standard EN 590 [80] to 460 µm and in the American standard D975 [54] to 520 µm. Neat biodiesel generally meets these requirements, regardless of the feedstock, with typical values being in the range of 140–200 µm. Common ULSD fuels usually exceed these prescribed maximum values in diesel fuel standards.

Table 20.2 contains some selected data from a publication discussing the lubricity of biodiesel and its origin [81]. Here it was shown that neat FAMEs generate larger wear scars than that of biodiesel. On the other hand, some minor constituents of biodiesel, FFAs, and monoacylglycerols in particular generate smaller wear scars than that of biodiesel itself. Their effect is largely responsible for the wear scar of biodiesel being smaller than that of the methyl esters. In an extension of this observation, these minor constituents are largely responsible for the lubricity-imparting properties of biodiesel to petrodiesel at low blend levels. At slightly elevated blend levels, the lubricity-enhancing

TABLE 20.2
WSDs (60°C) Generated from the HFRR Test of FAMEs and Other Fatty Compounds[a]

Compound/Material	WSD (μm)
Petrodiesel (ULSD)	651, 636
Soy methyl esters	129, 134
Hexadecane	572, 571
Addition to ULSD	
1% biodiesel	292, 292
2% biodiesel	281, 258
1% methyl oleate	597, 515
2% methyl oleate	384, 368
1% methyl oleate with 1% oleic acid[b]	356, 344
1% methyl oleate with 1% monoolein[b]	335, 303
Methyl palmitate	357, 362
Methyl oleate	290, 342
Methyl linoleate	236, 219
Methyl linolenate	183, 185
Oleic acid	0, 0
Triolein	143, 154
Diolein	186 [81], 163
Monoolein	139, 123
Methyl ricinoleate	191, 174

Source: Reprinted with permission from G. Knothe and K. R. Steidley, *Energy Fuels*, 19, 1192–1200. Copyright 2005 American Chemical Society.

[a] Some values for diesel fuel and hydrocarbons are included.
[b] For these samples, the second-named material was added at 1% level to the first-named material and then this mixture was added to ULSD.

effect of the methyl esters suffices to impart sufficient lubricity. The effect of the minor constituents was demonstrated by experiments in which a small amount of the minor constituent was added to methyl oleate resulting in a smaller wear scar [79]. These results coincide with similar observations by other authors [82,83]. It may also be noted that fatty esters with hydroxyl groups, such as methyl ricinoleate contained in castor oil methyl esters, are also good lubricity enhancers [80,84]. Overall, the excellent lubricity exhibited by biodiesel and its components therefore implies that biodiesel may serve as a lubricity enhancer for other purposes. For example, FAME has been discussed as lubricants in metalworking [85].

20.6 SOLVENT-RELATED PROPERTIES

As mentioned previously, biodiesel has found solvent-related use, documented in the technical literature as an agent for remediating shorelines after petroleum spills. Biodiesel is, however, also sold as a solvent as it has been employed for various cleaning and degreasing purposes [13,16]. Solvent use is likely the most common use of biodiesel/fatty acid alkyl esters beyond application as fuel. Such applications include paint strippers which replace methylene chloride, printing ink cleaner which replaces toluene, adhesive remover which replaces acetone, and graffiti remover which replaces mineral solvents. Besides inherent solvency, the application of fatty acid alkyl esters as a solvent also takes advantage of the high flash point of biodiesel and its low levels of volatile organic compounds. Besides the aforementioned applications, biodiesel has also been used as a polymerization solvent [86,87] and as a replacement for organic solvents in liquid–liquid extraction [88,89].

One descriptor for solvent strength is the kauri-butanol (KB) value, with a higher KB value indicating stronger solvency. A standard for determining the KB value is ASTM D1133 [90]. The method is based on titrating the sample with KB solution until turbidity sets in. While the KB value of straight-chain hydrocarbons is in the range from approximately 25 to low 30s decreasing with chain length, the KB value of neat FAMEs increases from C1 to C4/C5 and then decreases with increasing chain length [91]. The KB values of FFAs, octanoic acid, and oleic acid were found to be significantly higher [91]. Accordingly, the KB value of soybean methyl esters was determined as 59.6 [91], agreeing well with a previous statement that it is around 58 [13]. On the other hand, neat soybean oil had a low KB value of 19.1 [91]. Some KB values as determined in the literature are given in Table 20.3.

A more elaborate and reliable predictor of solubility are the Hansen solubility parameters (HSPs). The HSPs are used to predict if one material will dissolve another and form a solution. In this approach, each material possesses three Hansen parameters. In simple terms, the closer these parameters for two given materials are, the more likely they will form a solution. The HSP parameters are δ_d (energy from dispersion forces between molecules), δ_p (energy from dipolar intermolecular force between molecules), and δ_h (energy from hydrogen bonds between molecules). Individual compounds not only possess HSP but they can also be applied to innumerable materials, including of biological or other nature. A detailed discussion of these parameters and the approach is beyond

TABLE 20.3
Properties of Neat Methyl Esters Related to Solvent Use

Compound/Material	Kauri-Butanol Value[a]	HSP[b]		
		δ_D	δ_P	δ_H
C8:0	122.6	15.4	2.7	5.9
C10:0	96.1	15.9	2.4	5.7
C12:0	77.0	16.0	2.1	5.2
C14:0	63.5	16.0	1.9	4.2
C16:0	nd	16.0	1.6	3.6
C18:0	nd	15.9	1.4	3.2
C18:1	52.1	16.1	1.5	3.5
C18:2	58.2	16.2	1.6	3.9
C18:3	nd	16.2	1.7	4.2
C18:1 12-OH	nd	16.4	3.5	9.3
Soybean	59.3	15.03	3.69	8.92
Coconut		15.12	3.99	9.25
Palm		15.43	5.28	6.61
Castor		16.10	6.72	9.11
Soybean oil	19.1			

Source: Reprinted with permission from G. Knothe and K. R. Steidley, *Ind. Eng. Chem. Res.,* 50, 4177–4182, and M. M. Batista et al., *Energy Fuels,* 27, 7497–7509. Copyright 2011 and 2013 American Chemical Society.

Note: nd: not determined.

[a] From G. Knothe and K. R. Steidley, *Ind. Eng. Chem. Res.,* 50.

[b] HSPs are the mean of two approaches; Reprinted with permission from M. M. Batista et al., *Energy Fuels,* 27, 7497–7509. Copyright 2013 American Chemical Society.

the scope of this chapter. Hansen parameters of methyl soyate [92,93] and some other vegetable oil esters as well as individual FAMEs [93] have been studied. These parameters as given in recent literature [93] for some vegetable oil esters and individual FAMEs are also listed in Table 20.3.

20.7   OTHER USES

Besides the three uses (fuel, lubricant, and solvent) discussed previously, the methyl (or other alkyl) esters of vegetable oils and related feedstocks have been suggested or studied for a variety of other uses which include as a plasticizer [94], a combustion enhancer in candles [95], and an absorbent for gas purification [96].

REFERENCES

1. M. Mittelbach and C. Remschmidt, *Biodiesel—The Comprehensive Handbook*, M. Mittelbach (Ed.), Graz, Austria, 2004.
2. G. Knothe, J. Krahl, and J. Van Gerpen (Eds.), *The Biodiesel Handbook*, Second edition, AOCS Press, Urbana, IL, 2010.
3. American Society for Testing and Materials (ASTM), ASTM D6751: Standard specification for bio-diesel fuel blend stock (B100) for middle distillate fuels. ASTM, West Conshohocken, PA.
4. G. Knothe, Historical perspectives on vegetable oil-based diesel fuels, *INFORM*, 12, 1103–1107, 2001.
5. C. G. Chavanne, Procédé de transformation d'huiles végétales en vue de leur utilisation comme carburants (Procedure for the transformation of vegetable oils for their uses as fuels), Belgian Patent 422,877, August 31, 1937.
6. M. van den Abeele, L'huile de palme: Matière première pour la préparation d'un carburant lourd utilisable dans les moteurs à combustion interne (Palm oil as raw material for the production of a heavy motor fuel), *Bull. Agr. Congo Belge*, 33, 3–90, 1942.
7. D. Wei and H. A. Spikes, The lubricity of diesel fuels, *Wear*, 111, 217–235, 1986.
8. R. H. Barbour, D. J. Rickeard, and N. G. Elliott, Understanding diesel lubricity, SAE Tech. Pap. Ser. 2000-01-1918, 2000.
9. P. I. Lacey and S. J. Lestz, Effect of low-lubricity fuels on diesel injection pumps—Part I: Field performance, SAE Tech. Pap. Ser. 920823, 1992.
10. M. Nikanjam and P. T. Henderson, Lubricity of low sulfur diesel fuels, SAE Tech. Pap. Ser. 932740, 1993.
11. P. I. Lacey and S. R. Westbrook, Diesel fuel lubricity, SAE Tech. Pap. Ser. 950248, 1995.
12. H. H. Masjuki and S. M. Sapuan, Palm oil methyl esters as lubricant additive in a small diesel engine, *J. Am. Oil Chem. Soc.*, 72, 609–612, 1995.
13. S. Wildes, Methyl soyate: A new green alternative solvent, *Chem. Health Safety*, 9, 24–26, 2002.
14. S. M. Mudge and G. Pereira, Stimulating the biodegradation of crude oil with biodiesel preliminary results, *Spill Sci. Technol. Bull.*, 5, 353–355, 1999.
15. S. M. Mudge, Shoreline treatment of spilled vegetable oils, *Spill Sci. Technol. Bull.*, 5, 303–304, 1999.
16. S. G. Wildes, *Solvents: A Market Opportunity Study*, United Soybean Board/Omni Tech International, Midland, MI, 2007. http://soynewuses.org/wp-content/uploads/pdf/final_SolventsMarketStudy.pdf.
17. B. Freedman, E. H. Pryde, and T. L. Mounts, Variables affecting the yields of fatty esters from trans-esterified vegetable oils, *J. Am. Oil Chem. Soc.*, 61, 1638–1643, 1984.
18. M. Canakci and J. Van Gerpen, Biodiesel production from oils and fats with high free fatty acids, *Trans. ASAE*, 44, 1429–1436, 2001.
19. B. R. Moser and S. F. Vaughn, Evaluation of alkyl esters from *Camelina sativa* oil as biodiesel and as blend components in ultra low-sulfur diesel fuel, *Bioresour. Tech.*, 101, 646–653, 2010.
20. V. Scholz and J. N. da Silva, Prospects and risks of the use of castor oil as a fuel, *Biomass Bioenergy*, 32, 95–100, 2008.
21. G. Knothe, S. C. Cermak, and R. L. Evangelista, Methyl esters from vegetable oils with hydroxy fatty acids: Comparison of lesquerella and castor methyl esters, *Fuel*, 96, 535–540, 2012.
22. B. R. Moser and S. F. Vaughn, Coriander seed oil methyl esters as biodiesel fuel: Unique fatty acid composition and excellent oxidative stability, *Biomass Bioenergy*, 34, 550–558, 2010.
23. G. Knothe, S. C. Cermak, and R. L. Evangelista, Cuphea oil as source of biodiesel with improved fuel properties caused by high content of methyl decanoate, *Energy Fuels*, 23, 1743–1747, 2009.

24. B. P. Chapagain, Y. Yehoshua, and Z. Wiesman, Desert date (*Balanites aegyptiaca*) as an arid lands sustainable bioresource for biodiesel, *Bioresour. Technol.*, 100, 1221–1226, 2009.

25. N. Foidl, G. Foidl, M. Sanchez, M. Mittelbach, and S. Hackel, *Jatropha curcas* L. as a source for the production of biofuel in Nicaragua, *Bioresour. Technol.*, 58, 77–82, 1996.

26. S. Shah, S. Sharma, and M. N. Gupta, Biodiesel preparation by lipase-catalyzed transesterification of *Jatropha* oil, *Energy Fuels*, 18, 154–159, 2004.

27. U. Rashid, G. Knothe, R. Yunus, and R. L. Evangelista, Kapok oil methyl esters, *Biomass Bioenergy*, 66, 419–425, 2014.

28. G. Knothe, L. F. Razon, and F. T. Bacani, Kenaf oil methyl esters, *Ind. Crops Prod.*, 49, 568–571, 2013.

29. S. Puhan, N. Vedaraman, G. Sankaranarayanan, and B. V. B. Ram, Mahua oil (Madhuca Indica seed oil) methyl ester as biodiesel—Preparation and emission characteristics, *Biomass Bioenergy*, 28, 87–93, 2005.

30. U. Rashid, F. Anwar, B. R. Moser, and G. Knothe, *Moringa oleifera* oil: A possible source of biodiesel, *Bioresour. Technol.*, 99, 8175–8179, 2008.

31. G. N. Jham, B. R. Moser, S. N. Shah, R. A. Holser, O. D. Dhingra, S. F. Vaughn, M. A. Berhow et al., Wild Brazilian mustard (*Brassica juncea* L.) seed oil methyl esters as biodiesel fuel, *J. Am. Oil Chem. Soc.*, 86, 917–926, 2009.

32. F. Anwar, U. Rashid, M. Ashraf, and M. Nadeem, Okra (*Hibiscus esculentus*) seed oil for biodiesel production, *Appl. Energy*, 87, 779–785, 2010.

33. B. R. Moser, G. Knothe, S. F. Vaughn, and T. A. Isbell, Production and evaluation of biodiesel from field pennycress (*Thlaspi arvense* L.) oil, *Energy Fuels*, 23, 4149–4155, 2009.

34. P. T. Scott, L. Preglj, N. Chen, J. S. Hadler, M. A. Djordjevic, and P. M. Grasshoff, *Pongamia pinnata*: An untapped resource for the biofuels industry of the future, *Bioenergy Res.*, 1, 2–11, 2008.

35. A. S. Ramadhas, S. Jayaraj, and C. Muraleedharan, Biodiesel production from high FFA rubber seed oil, *Fuel*, 4, 335–340, 2005.

36. U. Rashid, F. Anwar, and G. Knothe, Biodiesel from milo (*Thespesia populnea* L.) seed oil, *Biomass Bioenergy*, 35, 4034–4039, 2011.

37. W. M. J. Achten, L. Verchot, Y. J. Franken, E. Mathijs, V. P. Singh, R. Aerts, and B. Muys, Jatropha biodiesel production and use, *Biomass Bioenergy*, 32, 1063–1084, 2008.

38. P. Kant and S. Wu, The extraordinary collapse of jatropha as a global biofuel, *Environ. Sci. Technol.*, 45, 7114–7115, 2012.

39. Y. Chisti, Biodiesel from microalgae, *Biotechnol. Adv.*, 25, 294–306, 2007.

40. Q. Hu, M. Sommerfeld, E. Jarvis, M. Ghirardi, M. Posewitz, M. Seibert, and A. Darzins, Microalgal triacylglycerols as feedstocks for biofuel production: Perspectives and advances, *Plant J.*, 54, 621–639, 2008.

41. S. A. Scott, M. P. Davey, J. S. Dennis, I. Horst, C. J. Howe, D. J. Lea-Smith, and A. G. Smith, Biodiesel from algae: Challenges and prospects, *Curr. Opin. Biotechnol.*, 21, 277–286, 2010.

42. J. B. van Beilen, Why microalgal biofuels won't save the internal combustion machine, *Biofuels Bioprod. Bioref.*, 4, 41–52, 2010.

43. P. J. le B. Williams and L. M. L. Laurens, Microalgae as biodiesel & biomass feedstocks: Review & analysis of the biochemistry, energetics & economics, *Energy Environ. Sci.*, 3, 554–590, 2010.

44. G. Knothe, A technical evaluation of biodiesel from vegetable oils vs. algae: Will algae-derived biodiesel perform? *Green Chem.*, 13, 3048–3065, 2011.

45. G. Petkov, A. Ivanova, I. Iliev, and I. Vaseva, A critical look at the microalgae biodiesel, *Eur. J. Lipid Sci. Technol.*, 114, 103–111, 2012.

46. G. B. Leite, A. E. M. Abdelaziz, and P. C. Hallenbeck, Algal biofuels: Challenges and opportunities, *Biores. Technol.*, 145, 134–141, 2013.

47. American Society for Testing and Materials (ASTM), ASTM D613: Standard test method for cetane number of diesel fuel oil. ASTM, West Conshohocken, PA.

48. K. J. Harrington, Chemical and physical properties of vegetable oil esters and their effect on diesel fuel performance, *Biomass*, 9, 1–17, 1986.

49. A. D. Puckett and B. H. Caudle, Ignition qualities of hydrocarbons in the diesel-fuel boiling range, *U.S. Bureau of Mines, Information Circular*, 7474, 1948.

50. G. Knothe, A. C. Matheaus, and T. W. Ryan III, Cetane numbers of branched and straight-chain fatty esters determined in an ignition quality tester, *Fuel*, 82, 971–975, 2003.

51. G. Knothe, Fuel properties of highly polyunsaturated fatty acid methyl esters: Prediction of fuel properties of algal biodiesel, *Energy Fuels*, 26, 5265–5273, 2012.

52. G. Knothe, A comprehensive evaluation of the cetane numbers of fatty acid methyl esters, *Fuel*, 119, 6–13, 2014.

53. CEN (European Committee for Standardization), EN 14214: Automotive fuels diesel—Fatty acid methyl esters (FAME)—Requirements and test methods. CEN, Brussels, Belgium.

54. American Society for Testing and Materials (ASTM), ASTM D975: Standard specification for diesel fuel oils. ASTM, West Conshohocken, PA.

55. U.S. Environmental Protection Agency (EPA), A comprehensive analysis of biodiesel impacts on exhaust emissions, Draft Technical Report EPA420-P-02-001, 2002.

56. R. L. McCormick, M. S. Graboski, T. L. Alleman, and A. M. Herring, Impact of biodiesel source material and chemical structure on emissions of criteria pollutants from a heavy-duty engine, *Environ. Sci. Technol.*, 35, 1742–1747, 2001.

57. G. Knothe, C. A. Sharp, and T. W. Ryan III, Exhaust emissions of biodiesel, petrodiesel, neat methyl esters, and alkanes in a new technology engine, *Energy Fuels*, 20, 403–408, 2006.

58. N. Ladommatos, M. Parsi, and A. Knowles, The effect of fuel cetane improver on diesel pollutant emissions, *Fuel*, 75, 8–14, 1996.

59. B. Freedman and M. O. Bagby, Heats of combustion of fatty esters and triglycerides, *J. Am. Oil Chem. Soc.*, 66, 1601–1605, 1989.

60. F. Levine, R. V. Kayea III, R. Wexler, D. J. Sadvary, C. Melick, and J. La Scala, Heats of combustion of fatty acids and fatty acid esters, *J. Am. Oil Chem. Soc.*, 91, 235–249, 2014.

61. G. Knothe and K. R. Steidley, Kinematic viscosity of biodiesel fuel components and related compounds: Influence of compound structure and comparison to petrodiesel fuel components, *Fuel*, 84, 1059–1065, 2005.

62. G. Knothe and K. R. Steidley, Kinematic viscosity of fatty acid methyl esters: Prediction, calculated viscosity contribution of esters with unavailable data, and carbon–oxygen equivalents, *Fuel*, 90, 3217–3224, 2011.

63. G. Knothe and R. O. Dunn, A comprehensive evaluation of the melting points of fatty acids and esters determined by differential scanning calorimetry, *J. Am. Oil Chem. Soc.*, 86, 843–856, 2009.

64. H. Imahara, E. Minami, and S. Saka, Thermodynamic study on cloud point of biodiesel with its fatty acid composition, *Fuel*, 85, 1666–1670, 2006.

65. R. E. Manning and M. R. Hoover, Cold flow properties. In *Fuels and Lubricants Handbook*, G. E. Totten, S. R. Westbrook, and R. J. Shah (Eds.), pp. 879–883. ASTM International, West Conshohocken, PA, 2003.

66. R. O. Dunn, M. W. Shockley, and M. O. Bagby, Improving the low-temperature properties of alternative diesel fuels: Vegetable oil-derived methyl esters, *J. Am. Oil Chem. Soc.*, 73, 1719–1728, 1996.

67. P. Benjumea, J. Agudelo, and A. Agudelo, Basic properties of palm oil biodiesel-diesel blends, *Fuel*, 87, 2069–2075, 2008.

68. I. Lee, L. A. Johnson, and E. G. Hammond, Use of branched-chain esters to reduce the crystallization temperature of biodiesel, *J. Am. Oil Chem. Soc.*, 72, 1155–1160, 1995.

69. T. A. Foglia, L. A. Nelson, R. O. Dunn, and W. N. Marmer, Low-temperature properties of alkyl esters of tallow and grease, *J. Am. Oil Chem. Soc.*, 74, 951–955, 1997.

70. L. Yu, I. Lee, E. G. Hammond, L. A. Johnson, and J. H. Van Gerpen, The influence of trace components on the melting point of methyl soyate, *J. Am. Oil Chem. Soc.*, 75, 1821–1824, 1998.

71. F. Lacoste, F. Dejean, H. Griffon, and C. Rouquette, Quantification of free and esterified steryl glucosides in vegetable oils and biodiesel, *Eur. J. Lipid Sci. Technol.*, 111, 822–828, 2009.

72. E. N. Frankel, *Lipid Oxidation*, The Oily Press, Bridgwater, UK, 2005.

73. G. Knothe and R. O. Dunn, Dependence of oil stability index of fatty compounds on their structure and concentration and presence of metals, *J. Am. Oil Chem. Soc.*, 80, 1021–1026, 2003.

74. G. Knothe and K. R. Steidley, A comprehensive evaluation of the density of neat fatty acids and esters, *J. Am. Oil Chem. Soc.*, 91, 1711–1722, 2014.

75. G. Knothe, Improving biodiesel fuel properties by modifying fatty ester composition, *Energy Environ. Sci.*, 2, 759–766, 2009.

76. G. Knothe, "Designer" biodiesel: Optimizing fatty ester composition to improve fuel properties, *Energy Fuels*, 22, 1358–1364, 2008.

77. G. Knothe, S. C. Cermak, and R. L. Evangelista, Cuphea oil as source of biodiesel with improved fuel properties caused by high content of methyl decanoate, *Energy Fuels*, 23, 1743–1747, 2009.

78. American Society for Testing and Materials (ASTM), ASTM D613: Standard test method for cetane number of diesel fuel oil. ASTM, West Conshohocken, PA.

79. International Organization for Standardization (ISO) 12156, Diesel fuel—Assessment of lubricity using the high-frequency reciprocating rig (HFRR).

80. CEN (European Committee for Standardization), EN 590, Automotive fuels—Diesel—Requirements and test methods. CEN, Brussels, Belgium.

81. G. Knothe and K. R. Steidley, Lubricity of components of biodiesel and petrodiesel: The origin of biodiesel lubricity, *Energy Fuels*, 19, 1192–1200, 2005.

82. G. Hillion, X. Montagne, and P. Marchand, Methyl esters of plant oils used as additives or organic fuel, (In Fr.) *Ol. Corps Gras Lipides*, 6, 435–438, 1999.

83. J. Hu, Z. Du, C. Li, and E. Min, Study on the lubrication properties of biodiesel as fuel lubricity enhancers, *Fuel*, 84, 1601–1606, 2005.

84. J. W. Goodrum and D. P. Geller, Influence of fatty acid methyl esters from hydroxylated vegetable oils on diesel fuel lubricity, *Bioresour. Technol.*, 96, 851–855, 2005.

85. H. Wichmann and M. Bahadir, Bio-based ester oils for use as lubricants in metal working, *Clean*, 35, 49–51, 2007.

86. S. Salehpour and M. A. Dubé, Biodiesel: A green polymerization solvent, *Green Chem.*, 10, 321–326, 2008.

87. S. Salehpour, M. A. Dubé, and M. Murphy, Solution polymerization of styrene using biodiesel as a solvent: Effect of biodiesel feedstock, *Can. J. Chem. Eng.*, 87, 129–135, 2009.

88. S. K. Spear, S. T. Griffin, K. S. Granger, J. G. Huddleston, and R. D. Rogers, Renewable plant-based soybean oil methyl esters as alternatives to organic solvents, *Green Chem.*, 9, 1008–1015, 2007.

89. L. Adhami, B. Griggs, P. Himebrook, and K. Taconi, Liquid–liquid extraction of butanol from dilute aqueous solutions using soybean-derived biodiesel, *J. Am. Oil Chem. Soc.*, 86, 1123–1128, 2009.

90. American Society for Testing and Materials (ASTM), ASTM D1133: Standard test method for kauri-butanol value of hydrocarbon solvents. ASTM, West Conshohocken, PA.

91. G. Knothe and K. R. Steidley, Fatty acid alkyl esters as solvents: Evaluation of the kauri-butanol Value: Comparison to hydrocarbons, dimethyl diesters, and other oxygenates, *Ind. Eng. Chem. Res.*, 50, 4177–4182, 2011.

92. K. Srinivas, T. M. Potts, and J. W. King, Characterization of solvent properties of methyl soyate by inverse gas chromatography and solubility parameters, *Green Chem.*, 11, 1581–1588, 2009.

93. M. M. Batista, R. Guirardello, and M. A. Krähenbühl, Determination of the solubility parameters of biodiesel from vegetable oils, *Energy Fuels*, 27, 7497–7509, 2013.

94. J. Wehlmann, Use of esterified rapeseed oil as plasticizer in plastics processing, *Fett/Lipid*, 101, 249–256, 1999.

95. Stepan Co., J. Schroeder, I. Shapiro, and J. Nelson, Candle mixtures comprising naturally derived alkyl esters, WO/2004/046286, June 3, 2004.

96. K. Bay, H. Wanko, and J. Ulrich, Biodiesel—Hoch siedendes Absorbens für die Gasreinigung (Biodiesel—High-boiling absorbent for gas purification), *Chem. Ing. Technik.*, 76, 328–333, 2004.

21 Corrosion Protection of Steel by Thin Coatings of Starch–Oil Dry Lubricants*

Victoria L. Finkenstadt, James A. Kenar, and George F. Fanta

CONTENTS

ABSTRACT

Corrosion of materials is one of the most serious and challenging problems faced worldwide by the industry. The research in this chapter investigated the inhibition of corrosive behavior by a dry lubricant formulation consisting of jet-cooked cornstarch and soybean oil on SAE 1010 steel. Electrochemical impedance spectroscopy was used to evaluate the corrosion inhibition of starch–soybean oil coatings containing 0.0%, 11.6%, 26.8%, and 53.4% w/w of soybean oil loading at 0.6 and 2.0 mg/cm^2 surface coverage. At high soybean oil applications, coatings with higher surface coverage exhibited oil separation from the dried starch film. Control starch samples (0% soybean oil) showed minor protection up to approximately twice that observed on bare steel. The coatings were highly variable in coating integrity and some phase separation occurred at higher oil loadings; thus, no linear correlation for oil loading or surface coverage was noted. The coating comprising 26.8 wt% soybean oil applied at 0.5 mg/cm^2 showed maximum protection from corrosion, which was approximately 10 times that of bare steel.

21.1 INTRODUCTION

The corrosion of metal is a global issue with huge economic and safety consequences. Corrosion is an electrochemical process in which metal is oxidized. For steel (iron) substrates, corrosion manifests

---

* Names are necessary to report factually on available data. However, the USDA neither guarantees nor warrants the standard of the product, and the use of the name by the USDA implies no approval to the exclusion of others that may also be suitable.

itself as rust. The rate of corrosion of steel is mainly determined by the environment in which it is exposed. Corrosion causes enormous direct and indirect economic losses, which are estimated at over $900 billion annually in the United States by the National Association of Corrosion Engineers. Corrosive environments include exposure to acidic, neutral, or basic aqueous solutions, especially in the presence of oxygen. The metal surface has an electrochemical potential where, in the presence of ionic species, oxidation–reduction reactions occur, leading to oxidation at the metal surface which manifests itself as rust. Solutions containing dissolved salts are much more corrosive than pure water since the electrically charged ionic species contained within the water provide an easy pathway for the flow of electrons, thus accelerating the rusting of metal. Typically, the control of the corrosion process involves interference against the electrochemical mechanism with a physical barrier or coating [1]. The coating may be another metal (galvanizing), a liquid such as oil, or a dry film such as paint. This barrier acts to prevent the diffusion of ionic species, moisture, and oxygen toward the metal surface. Many anticorrosion coatings on the market today are expensive and toxic to the environment [2]. This chapter evaluates the anticorrosion potential of biobased dry film lubricants.

Biobased lubricants are environmentally friendly, easy to clean, and potentially biodegradable [3,4]. Unlike petroleum-based lubricants, vegetable oil-based lubricants adsorb on metal surfaces [5]. Lubricants reduce friction behavior on metal surfaces, reducing wear and extending tool or part life. Traditional lubricants are made from petroleum and they usually include an additive to enhance the oxidative stability of the lubricant and prevent corrosion of the metal.

This study evaluates the anticorrosive properties of a dry film lubricant composed of starch and soybean oil.

Starch is a glucose polymer that is abundant in nature and can be harvested from many agricultural crops. The cornstarch used in this study contains both linear (amylose) and branched (amylopectin) polysaccharide chains with a 20:80 ratio. Soybean oil, a mixture of saturated and unsaturated fats, contains fatty acids that can complex with starch [6]. Chemically modified starch containing canola oil was shown to reduce friction when applied as a dry film lubricant [7]. When blended with canola oil, the COF dramatically reduced. Recently, effective dry film lubricants were developed based on renewable starch–soybean oil composites applied as a coating on metals [8–11]. Dry film lubricants are applied to metal during metalworking applications [8]. Aqueous dispersions of jet-cooked starch and soybean oil were sprayed onto metal surfaces and the resulting dried films were evaluated for tribological properties [9]. Surface coverage of the coating on the metal surface was varied between 0.15 and 0.50 mg/cm². Starch–oil composites containing 4–5 wt% soybean oil relative to the starch were found to effectively decrease the COF to a minimum and the constant value that was not further reduced by increasing the oil loading of the composite [10]. The mechanism postulated for these dry film lubricants of starch–oil composites was determined to be the on-demand release of embedded oil contained within the dry starch matrix at the point of contact between the friction surfaces [10].

Jet cooking with steam is an industrial process in which aqueous starch dispersions can be prepared in a rapid and continuous manner [12] as first described in 1940 [13]. Work in our laboratory has shown that when a mixture of starch and oil are co-jet cooked, the high temperature and the intense mechanical shear of the steam jet cooking process not only substantially solubilize the starch components, but also convert the hydrophobic lipid oil component to micrometer-sized droplets [14]. The resulting oil droplets, which can be incorporated in amounts of approximately 40–50% relative to starch, do not coalesce even when aqueous dispersions are dried and redispersed into the water [15–17]. The individual oil droplets do not coalesce because a thin film of adsorbed starch is present at the oil–water interface despite the fact that the native starch is not surface active and surface-active materials are not used during the preparation of these starch–lipid composites [18]. This spontaneous formation of interfacial starch is best explained by a process referred to as *prewetting*. Prewetting occurs when the starch is dissolved in a thermodynamically poor solvent (i.e., water), and when the accumulation of starch at the oil/water interface leads to a reduction in interfacial tension [17,18].

Biobased dry film lubricants with the additional functionality of preventing or inhibiting corrosion would be advantageous. Many natural polymers have been shown to offer some type of electroactivity, specifically corrosion resistance [19]. Potato starch coatings, both oxidized and organosiloxane grafted, were shown to protect aluminum from corrosion [20,21]. The adsorption of tapioca starch onto aluminum from seawater has also been shown to inhibit corrosion [22]. Modified cassava starch adsorbed onto carbon steel from salt solutions also resulted in corrosion inhibition [23]. These studies suggest that the observed corrosion inhibition resulted from the starch molecules competing more successfully against the corrosive species for adsorption onto the metal substrate or its oxide.

Starch–soybean oil composites were formulated for dry lubricant applications and evaluated for their ability to inhibit corrosion on steel substrates. Starch and soybean oil are abundant, relatively inexpensive, and renewable agricultural commodities in the United States.

21.2 EXPERIMENTAL

21.2.1 MATERIALS

Normal cornstarch (Dent) containing 23% amylose was obtained from Cargill (Minneapolis, Minnesota). Percentage of moisture in starch samples was gravimetrically calculated from weight loss after drying at 100°C under vacuum, and all starch weights are given on a dry weight basis. Soybean oil was purchased from Fisher Scientific (Pittsburgh, Pennsylvania) and used without modification. Low-carbon, cold-rolled steel panels (SAE 1010) were obtained from Q-Lab (Cleveland, Ohio) and cleaned with hexane.

21.2.2 PREPARATION OF SOYBEAN OIL COMPOSITES

A slurry of normal dent cornstarch (600.0 g; moisture content, 10.8%; dry wt basis, 535.2 g) and deionized water (3000 mL) was stirred in a 4 L stainless steel Waring blender (model 37BL84; Dynamics Corporation of America, New Hartford, Connecticut). The resulting slurry was delivered to the jet cooker utilizing a Moyno progressive cavity pump (Robbins Meyers, Springfield, Ohio) at a flow rate of about 1.0 L/min. The starch slurry and the steam were combined in a Penick and Ford hydroheater (Penford Corp., Cedar Rapids, Iowa). The cooking temperature was 140°C using steam supplied at 550 kPa (65 psig), and the hydroheater backpressure set at 377 kPa (40 psig). The cooked starch solution was collected along with additional steam-heated water to obtain a total volume of 3900 mL (solid content: 13.6% ± 1.4%, as determined by freeze-drying accurately weighed amounts of the dispersion in duplicate). The hot starch paste was divided into three portions (1000 mL, containing 136.0 g starch). To each starch paste, the following amounts of soybean oil was added: composite 1 = 17.0 g; composite 2 = 50.9 g; and composite 3 = 150.5 g. Each mixture was separately blended with a high-speed Waring blender for approximately 2 min. Each oil–starch mixture was then fed through the jet cooker under the conditions previously described. A center cut (approximately 800 mL) from each run was collected from the jet cooker, cooled, and then freeze-dried to give three white starch–oil composite powders. These composites were subsequently milled to a fine powder using a rotary beater mill (Model ZM 200; Retsch Inc., Newtown, Pennsylvania) operating at 300 Hz (18,000 rpm). The amount of soybean oil contained in each starch–oil composite was determined by pulsed NMR [24], and found to contain the following amounts of soybean oil: composite 1 = 11.6%; composite 2 = 26.8%; and composite 3 = 53.4%. All percentages are determined against the dry weight of starch.

21.2.3 SPRAY APPLICATION OF STARCH–SOYBEAN OIL COMPOSITES TO METAL SURFACES

Each of the dried starch–oil composites were each reconstituted in room-temperature distilled water (200 mL) utilizing a high-speed blender to give solutions containing 8.5 wt% starch. The redispersed composites were immediately used to coat the metal specimens by spraying.

A Badger Model 400 detail/touch-up siphon feed spray gun (Badger Air-Brush Co., Franklin Park, Illinois) with a medium spray needle valve tip (0.85 mm) and a vertical fan pattern spray nozzle was operated at 207 kPa (30 psi). The fan spray height and width was approximately 5–6 cm × 2 cm. The nozzle was held at about 10 cm from the metal substrate. The steel panels were placed vertically at a slight angle and the aqueous dry film lubricant formulation was applied by spraying parallel to the metal surface. The spray gun was moved from left to right and back again while progressively moving down the plate. A cool stream of air from a dryer was blown onto the panels after each pass to quickly dry the panels. This process was repeated until the desired surface coverage of the coating was obtained. Surface coverage and film thickness were gravimetrically determined and expressed in milligrams per square centimeter. The spraying process can be highly variable and care was taken to assure smooth, even coatings. Average coating thicknesses of 0.6 mg/cm² (range: 0.43–0.80 mg/cm²) and 2.0 mg/cm² (range: 1.54–2.5 mg/cm²) were obtained. Two levels of coverage (referred to as *thin* or *thick* in the results) were selected based on earlier work that showed that the starch–oil composites could be used as dry lubricants.[11] A starch control containing 0% soybean oil was also constructed.

21.2.4 Electrochemical Analysis

The dried films of the starch–soybean oil composites were evaluated for corrosion inhibition on the metal substrates under an electrolyte solution using a potentiostat. A potentiostat is an instrument that controls the voltage difference between a working electrode and a reference electrode. Electrochemical measurements were performed using a PARSTAT 2273 Advanced Potentiostat (Princeton Applied Research, Oak Ridge, Tennessee). Electrochemical impedance was recorded at an alternating current voltage of 10 mV for a frequency range of 0.1 to 10 MHz. The electrochemical cell was equipped with a conventional three-electrode system with a working electrode, a graphite counterelectrode, and a reference electrode (PTC1 Gamry Instruments, Warminster, Pennsylvania). In the corrosion experiment, the metal substrate functions as the working (i.e., corroding) electrode. The test area was approximately 14 cm². Figure 21.1 shows the experimental setup for the electrochemical cell and potentiostat.

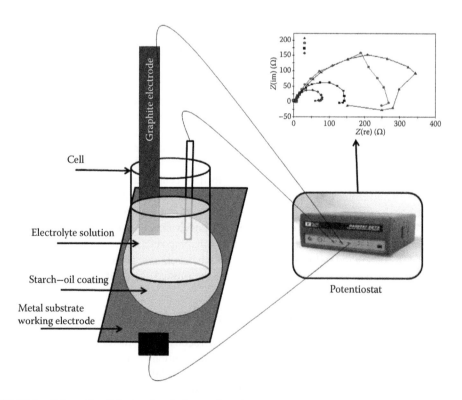

FIGURE 21.1 Schematic of electrochemical experiment.

Sodium chloride (NaCl; 5% w/v) was used as the electrolyte solution. Analysis was performed using electrochemistry software (PowerCorr, Princeton Applied Research, Oak Ridge, Tennessee) that includes modules specifically designed for corrosion measurements and analysis.

Corrosion inhibition is measured using electrochemical impedance spectroscopy (EIS). EIS measures the impedance and the capacitance of the coating as a corrosive species, in this case sodium chloride, diffuses through the coating, and interacts with the surface of the metal. The impedance (Z) is the barrier in the circuit that the electrical current experiences when voltage is applied and has both magnitude and phase. In direct current measurements, the impedance only has magnitude and is known as the resistance (R). With alternating current, there are additional impeding mechanisms such as inductance and capacitance. Inductance deals with the self-induced magnetic field current in the circuit. The capacitance is the electrostatic storage of the circuit. Both phase and magnitude of the impedance (known as Bode plots) are recorded and, for convenience, are converted into complex impedance (Tafel) plots. Plotted on the y axis, the imaginary part $Z(im)$ represents both inductance and capacitance in the circuit. Plotted on the x axis, the real part of complex impedance $Z(re)$ is resistance. It is this value that is used to compare the relative inhibition of corrosion by coatings. Tafel plots, therefore, indicate the probable mechanism of corrosion through the shape of the curve and the protection against corrosion imparted by the coating on the metal substrate through the diameter of the curve along the x axis.

21.3 RESULTS

21.3.1 THIN COATINGS

At 0.6 mg/cm^2 coverage, composites containing 26.8 wt% soybean oil loading (composite 2) had better corrosion protection than the starch control (0% soybean oil), the composite containing 11.6 wt% soybean oil loading (composite 1), or the composite loaded with 53.4 wt% soybean oil (composite 3) (see Figure 21.2). The corrosion protection of composite 2 was 10 times better than that of the bare steel and 3 times better than that of the starch control. Figure 21.3 shows starch–oil

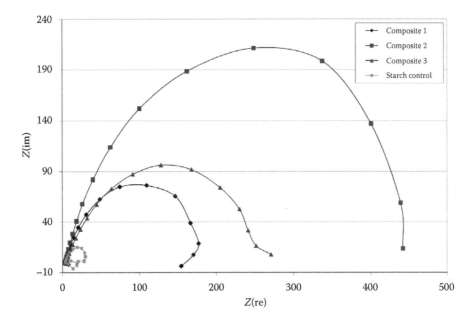

FIGURE 21.2 Electrochemical impedance measurements of thin (0.5 mg/cm^2) starch–soybean oil film coatings on steel.

composites coatings at 0.6 mg/cm^2 coverage with different oil loadings. Overall, the spray-coated substrates have uniform coverage and show that the soybean oil is incorporated within the bulk of the starch film. However, at 53.4 wt% soybean oil loading, it appears that the dried coating has undergone some phase separation where the oil has separated causing loss of adhesion at the metal–coating interface (Figure 21.3d). This probably contributed to decreasing the inhibition with higher soybean oil.

21.3.2 THICK COATINGS

At higher coverage (2.0 mg/cm^2) and thicker coatings, composite 1 (11.6% soybean oil) had better corrosion protection than the starch control or the other two composites as indicated by the EIS (Figure 21.4). Examining the micrographs of the coatings in Figure 21.5 shows that the coating becomes disturbed at higher levels of soybean oil loading and in thicker coatings.

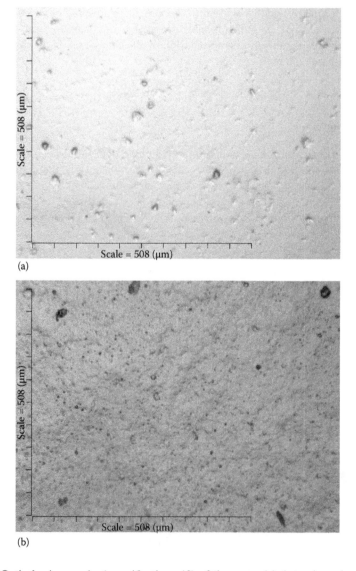

(a)

(b)

FIGURE 21.3 Optical micrographs (magnification ×10) of the spray-dried starch–soybean oil films with thin coatings containing (a) 0% soybean oil and (b) 11.6 wt% soybean oil. *(Continued)*

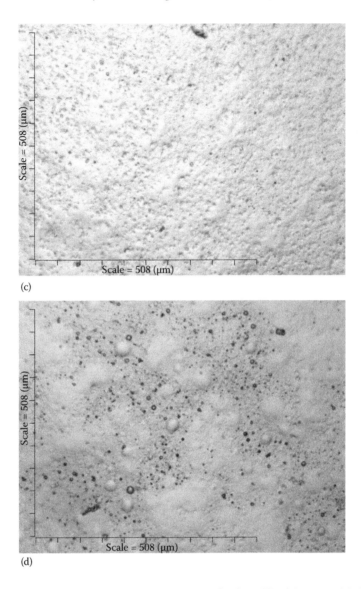

(c)

(d)

FIGURE 21.3 (CONTINUED)   Optical micrographs (magnification ×10) of the spray-dried starch–soybean oil films with thin coatings containing (c) 26.8 wt% soybean oil and (d) 53.4 wt% soybean oil.

Unexpectedly, the total protection for the 11.6 wt% soybean oil composite coatings regardless of coverage was very similar, approximately $Z(re) = R = 150$. At 26.8 and 53.4 wt% soybean oil, thinner composite coatings offered more protection (three to four times) than thicker coatings. The metal/starch–soybean oil coating/electrolyte solution circuit is a very complex system consisting of a multicomponent coating that changes over time from internal and external phenomena. This will be discussed in Section 21.4.

21.3.3 Starch Control

EIS of the control starch sample (0% soybean oil) indicated minor protection around twice that of bare steel. In starch control samples (green line in Figures 21.2 and 21.4), the thicker coating offered

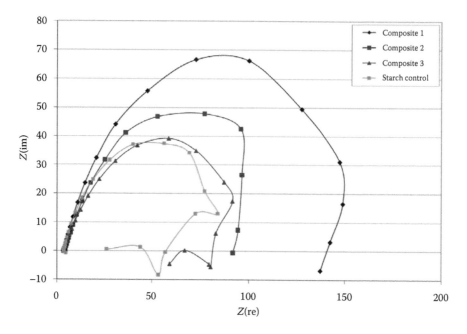

FIGURE 21.4 Electrochemical impedance measurements of thick (2.0 mg/cm²) starch–soybean oil film coatings on steel.

more protection, which is consistent with other studies. For example, corrosion inhibition increased as the concentration of tapioca starch increased and the adsorbed layer became thicker using aluminum substrates [22].

21.4 DISCUSSION

The adsorption of starch–soybean oil composites onto a steel substrate after spray coating is strong because the soybean oil is distributed both in the bulk of the starch matrix and onto the outer surface of the starch coating. This allows the vicinal hydroxyl groups of starch to chelate with iron and iron oxide at the metal interface while providing a progressively hydrophobic surface to the environment. In an EIS experiment reported in the literature using cassava starch solutions containing electrolytes, the modified starch was adsorbed onto the steel electrode and provided some protection from corrosion through an iron oxide layer chelated with starch [23]. In other experiments with dextrans, iron oxide-chelated glucans were suggested as the mechanism for corrosion protection of low carbon steel [25]. The starch molecules are able to chelate the iron oxides, allowing the soybean oil to be closely associated with the steel surface. In fact, the larger surface energy differential between the starch and the metal favors the adsorption of starch to the metal [9]. In studies using starch and canola oil, the incorporation of the canola oil into the starch matrix was found to have an upper concentration limit determined by the number of free hydroxyl groups available for the triglycerides to bind [7]. There is a similar limit here as the increasing amount of oil overcomes the preferred interaction of the starch and the metal surface. Phase separation of the oil onto the outer surface of the starch coating, however, should not necessarily be considered a defect in a dry lubricant because the oil will impart greater hydrophobicity and lubrication. In addition, examining the shape of the complex impedance curves (semicircle) in Figures 21.2 and 21.4 indicate that the corrosion process is mainly controlled through mass-transfer reactions as the electrolytes diffuse through the coating to the steel surface.

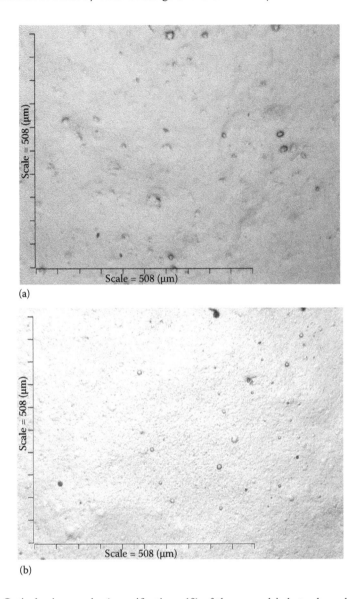

(a)

(b)

FIGURE 21.5 Optical micrographs (magnification ×10) of the spray-dried starch–soybean oil films with thick coatings containing (a) 0% soybean oil and (b) 11.6 wt% soybean oil.   (*Continued*)

21.5 CONCLUSION

Starch–soybean oil composites can be sprayed onto metals in thin layers and strongly adhere to the surface of low carbon steel through polar bonds and chelation. Starch–soybean oil composites offer multiple mechanisms to protect metal from corrosion: strong binding of the coating to the metal through adsorption of both soybean oil and starch molecules, chelation of the oxide layer by starch, decreased diffusion of corrosive species through the starch–soybean oil coating, and increased overall hydrophobicity of the coating. The electrochemical experiments were performed on coatings which had not been subjected to friction tests and so the results should serve only as a guideline for further investigation into starch–soybean oil composites as anticorrosive dry lubricants.

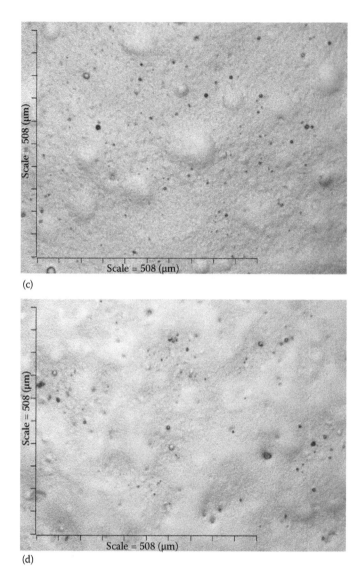

(c)

(d)

FIGURE 21.5 (CONTINUED) Optical micrographs (magnification ×10) of the spray-dried starch–soybean oil films with thick coatings containing (c) 26.8 wt% soybean oil and (d) 53.4 wt% soybean oil.

ACKNOWLEDGMENTS

The authors wish to thank Jeanette Little and Rick Haig for technical support. VLF wishes to thank Dr. Gordon Selling for helpful discussions on the study.

REFERENCES

1. Hare, C. H., Corrosion and its control by coatings. In *Protective Coatings*, C. H. Hare, Ed., Technology Pub. Co., Pittsburgh, PA: 1994; pp. 331–359.
2. Raja, P. B., and Sethuraman, M. G., Natural products as corrosion inhibitor for metals in corrosive media—A review. *Materials Letters* 2008, *62* (1), 113–116.
3. Alam, M., Akram, D., Sharmin, E., Zafar, F., and Ahmad, S., Vegetable oil based eco-friendly coating materials: A review article. *Arabian Journal of Chemistry* 2014, *7* (4), 469–479.

4. Kesavan, D., Gopiraman, M., and Sulochana, N., Green inhibitors for corrosion of metals: A review. *Chemical Science Review and Letters* 2012, *1* (1), 1–8.

5. Biresaw, G., Surfactants in lubrication. In *Lubricant Additives: Chemistry and Applications*, L. S. Rudnick, Ed., CRC Press, Taylor & Francis, New York: 2008; p 399.

6. Fanta, G. F., Felker, F. C., Byars, J. A., Kenar, J. A., and Shogren, R. L., Starch–soybean oil composites with high oil: Starch ratios prepared by steam jet cooking. *Starch* 2009, *61* (10), 590–600.

7. Biresaw, G., and Shogren, R. L., Friction properties of chemically modified starch. *Journal of Synthetic Lubrication* 2008, *25* (1), 17–30.

8. Biresaw, G., and Erhan, S. M., Solid lubricant formulations containing starch–soybean oil composites. *JAOCS* 2002, *79* (3), 291–296.

9. Biresaw, G., Kenar, J. A., Kurth, T. L., and Felker, F. C., Erhan, S. M., Investigation of the mechanism of lubrication in starch–oil composite dry film lubricants. *Lubrication Science* 2007, *19*, 41–55.

10. Kenar, J. A., Felker, F. C., Biresaw, G., and Kurth, T. L., Properties of dry film lubricants prepared by spray application of aqueous starch–oil composites. *Industrial Crops and Products* 2008, *29* (1), 45–52.

11. Biresaw, G., Biobased dry-film metalworking lubricants. *Journal of Synthetic Lubrication* 2004, *21*, 43–57.

12. Klem, R. E., and Brogley, D. A., Methods for selecting the optimum starch binder preparation system. *Pulp & Paper* 1981, *55*, 98–103.

13. Coppock, P. D., *Manufacture of Starch Paste*. Edinburgh, Scotland. US Patent 2202573, 1940.

14. Byars, J. A., Fanta, G. F., and Felker, F. C., Rheological properties of starch–oil composites with high oil-to-starch ratios. *Cereal Chemistry* 2011, *88* (3), 260–263.

15. Eskins, K., Fanta, G. F., Felker, F. C., and Baker, F. L., Ultrastructural studies on microencapsulated oil droplets in aqueous gels and dried films of a new starch–oil composite. *Carbohydrate Polymers* 1996, *29* (3), 233–239.

16. Fanta, G. F., and Eskins, K., Stable starch–lipid compositions prepared by steam jet cooking. *Carbohydrate Polymers* 1995, *28*, 171–175.

17. Fanta, G. F., Felker, F. C., Eskins, K., and Baker, F. L., Aqueous starch–oil dispersions prepared by steam jet cooking: Starch films at the oil–water interface. *Carbohydrate Polymer* 1999, *39*, 25–35.

18. Fanta, G. F., Felker, F. C., Shogren, R. L., and Knutson, C. A., Starch–paraffin wax compositions prepared by steam jet cooking: Examination of starch adsorbed at the paraffin–water interface. *Carbohydrate Polymers* 2001, *46* (1), 29–38.

19. Finkenstadt, V. L., Natural polysaccharides as electroactive polymers. *Applied Microbiology and Biotechnology* 2005, *67* (6), 735–745.

20. Sugama, T., and DuVall, J. E., Polyorganosiloxane-grafted potato starch coatings for protecting aluminum from corrosion. *Thin Solid Films* 1996, *289* (1–2), 39–48.

21. Sugama, T., Oxidized potato-starch films as primer coatings of aluminum. *Journal of Material Science* 1997, *32*, 3995–4003.

22. Rosliza, R., and Wan Nik, W. B. Improvement of corrosion resistance of AA6061 alloy by tapioca starch in seawater. *Current Applied Physics* 2010, *10* (1), 221–229.

23. Bello, M., Ochoa, N., Balsamo, V., Lopez-Carrasquero, F., Coll, S., Monsalve, A., and Gonzalez, G., Modified cassava starches as corrosion inhibitors of carbon steel: An electrochemical and morphological approach. *Carbohydrate Polymers* 2010, *82* (3), 561–568.

24. Kenar, J. A., Direct determination of the lipid content in starch–lipid composites by time-domain NMR. *Industrial Crops & Products* 2007, *26*, 77–84.

25. Finkenstadt, V. L., Cote, G. L., and Willett, J. L., Corrosion protection of low-carbon steel using exopolysaccharide coatings from *Leuconostoc mesenteroides*. *Biotechnology Letters* 2011, *33*, 1093–1100.

Index

Page numbers followed by f and t indicate figures and tables, respectively.

Printed and bound by CPI Group (UK) Ltd, Croydon, CR0 4YY

01/11/2024

01782601-0009